Discovering the Solar System

Second Edition

523.2

J

Discovering the Solar System

Second Edition

Barrie W. Jones
The Open University,
Milton Keynes, UK

John Wiley & Sons, Ltd

Copyright © 2007   John Wiley & Sons Ltd, The Atrium, Southern Gate, Chichester,
West Sussex PO19 8SQ, England

Telephone   (+44) 1243 779777

Email (for orders and customer service enquiries): cs-books@wiley.co.uk
Visit our Home Page on www.wiley.com

Other Wiley Editorial Offices

John Wiley & Sons Inc., 111 River Street, Hoboken, NJ 07030, USA

Jossey-Bass, 989 Market Street, San Francisco, CA 94103-1741, USA

Wiley-VCH Verlag GmbH, Boschstr. 12, D-69469 Weinheim, Germany

John Wiley & Sons Australia Ltd, 33 Park Road, Milton, Queensland 4064, Australia

John Wiley & Sons (Asia) Pte Ltd, 2 Clementi Loop #02-01, Jin Xing Distripark, Singapore 129809

John Wiley & Sons Canada Ltd, 6045 Freemont Blvd, Mississauga, Ontario, L5R 4J3, Canada

Wiley also publishes its books in a variety of electronic formats. Some content that appears in print may not be available in electronic books.

Anniversary Logo Design: Richard J. Pacifico

Library of Congress Cataloging in Publication Data

Jones, Barrie William.
    Discovering the solar system / Barrie W. Jones. — 2nd ed.
       p.   cm.
    ISBN 978-0-470-01830-9
    1. Solar system.   I. Title.
    QB501.J65 2007
    523.2—dc22

                                        2007008860

British Library Cataloguing in Publication Data

A catalogue record for this book is available from the British Library

ISBN 978-0-470-01830-9 (HB)
ISBN 978-0-470-01831-6 (PB)

Typeset in 10/12pt Times by Integra Software Services Pvt. Ltd, Pondicherry, India
Printed and bound in Great Britain by Antony Rowe Ltd, Chippenham, Wiltshire
This book is printed on acid-free paper responsibly manufactured from sustainable forestry
in which at least two trees are planted for each one used for paper production.

*To my wife Anne, the rest of my family,
and to my friend and colleague Nick Sleep*

Contents

List of Tables

Preface and Study Guide to the First Edition

In *Discovering the Solar System* you will meet the Sun, the planets, their satellites, and the host of smaller bodies that orbit the Sun. On a cosmic scale the Solar System is on our doorstep, but it is far from fully explored, and there continues to be a flood of new data and new ideas. The science of the Solar System is thus a fast-moving subject, posing a major challenge for authors of textbooks.

A major challenge for the student is the huge range of background science that needs to be brought to bear—geology, physics, chemistry, and biology. I have tried to minimise the amount of assumed background, but as this book is aimed at students of university-level science courses I do assume that you have met Newton's laws of motion and law of gravity, that you know about the structure of the atom, and that you have met chemical formulae and chemical equations. Further background science is developed as required, as is the science of the Solar System itself, and it is therefore important that you study the book in the order in which the material is presented. There is some mathematics—simple algebraic equations are used, and there is a small amount of algebraic manipulation. It is assumed that you are familiar with graphs and tables. There is no calculus.

To facilitate your study, there are 'stop and think' questions embedded in the text, denoted by '☐'. The answer follows immediately as part of the development of the material, but it will help you learn if you *do* stop and think, rather than read straight on. There are also numbered questions (Question 1.1, etc.). These are at the end of major sections, and it is important that you attempt them before proceeding—they are intended to test and consolidate your understanding of some of the earlier material. Full answers plus comments are given at the end of the book. Another study aid is the Glossary, which includes the major terms introduced in the book. These terms are emboldened in the text at their first appearance. Each chapter ends with a summary.

The approach is predominantly thematic, with sequences of chapters on the interiors, surfaces, and atmospheres of the major bodies (including the Earth). The first three chapters depart from this scheme, with Chapter 2 on the origin of the Solar System, and Chapter 3 on the small bodies—asteroids, comets, and meteorites. Chapter 1 is an overview of the Solar System, and this is also where most of the material on the Sun is located. Though the Sun is a major body indeed, it is very singular, and it is therefore treated separately. It also gets only very brief coverage, biased towards topics that relate to the Solar System as a whole. There is a significant amount of material on how the Solar System is investigated. The 'discovering' in the title thus has a double meaning—not only can *you* discover the Solar System by studying this book, you will also learn something about how it has been discovered by the scientific community in general.

A large number of people deserve thanks for their assistance with this book. Nick Sleep and Graeme Nash each commented on a whole draft, and Nick Sleep also made a major contribution to generating the figures. Coryn Bailer-Jones, George Cole, Mark Marley, Carl Murray, Peter Read, and Lionel Wilson commented on groups of chapters. Information and comments on specific matters have been received from Mark Bailey, Bruce Bills, Andrew Collier Cameron, Apostolos Christou, Ashley Davies, David Des Marais, Douglas Gough, Tom

Haine, Andy Hollis, David Hughes, Don Hunten, Pat Irwin, Rosemary Killen, Jack Lissauer, Mark Littmann, Elaine Moore, Chris Owen, Roger Phillips, Eric Priest, Dave Rothery, Gerald Schubert, Alan Stern, George Wetherill, John Wood, and Ian Wright. Jay Pasachoff supplied data for the Electronic Media list. Material for some of the figures was made available by Richard McCracken, Dave Richens, and Mark Kesby. John Holbrook loaned me some meteorite samples to photograph.

Good luck with your studies.

Preface to the Second Edition

Much has been added to, or changed, in our knowledge and understanding of the Solar System since the first edition of this book was completed in 1998 (and published in early 1999). The book has been thoroughly revised accordingly, though the overall organisation into chapters and sections is much the same.

In the preparation of this second edition, *particular* thanks are due to Nick Sleep, who read and commented on a draft of the whole book. Many people have provided information and comments on specific matters. They include (in alphabetical order) Steve Blake, Alan Boss, John Chambers, Michele Dougherty, Michael Drake, Bruce Fegley, Martyn Fogg, Bernard Foing, Tristan Guillot, James Head, Robert Hutchison, Andrew Ingersoll, Patrick Irwin, Noel James, Joe Kirschvink, Chris Kitchin, Ulrich Kolb, Robert Kopp, Stephen Lewis, Ralph Lorenz, Neil McBride, Adam Morris, John Murray, Richard Nelson, Carolyn Porco, Eric Priest, Janna Rodionova, Dave Rothery, Sean Ryan, Chuck See, Peter Skelton, Sean Solomon, Anne Sprague, Fred Taylor, Nick Teanby, Ashwin Vasavada, Iwan Williams, and Ian Wright.

1 The Sun and its Family

Imagine that you have travelled far into the depths of space. From your distant vantage point the Sun has become just another star amongst the multitude, and the Earth, the other planets, and the host of smaller bodies that orbit the Sun are not visible at all to the unaided eye. The Sun is by far the largest and most massive body in the Solar System, and is the only one hot enough to be obviously luminous. This chapter starts with a description of the Sun. We shall then visit the other bodies in the Solar System, but only briefly, the purpose here being to establish their main characteristics – each of these bodies will be explored in much more detail in subsequent chapters. Chapter 1 then continues with an exploration of the orbits of the various bodies. Each of them also rotates around an axis through its centre, and we shall look at this too. The chapter concludes with aspects of our view of the Solar System as we see it from the Earth.

1.1 The Sun

This is only a very brief account of the Sun, and it is biased towards topics of importance for the Solar System as a whole. Fuller accounts of the Sun are in books listed in Further Reading.

1.1.1 The Solar Photosphere

The bright surface of the Sun is called the **photosphere** (Plate 1). Its radius is 6.96×10^5 km, about 100 times the radius of the Earth. It is rather like the 'surface' of a bank of cloud, in that the light reaching us from the photosphere comes from a range of depths, though the range covers only about one-thousandth of the solar radius, and so we are not seeing very deep into the Sun. It is important to realise that whereas a bank of cloud scatters light from another source, the photosphere is *emitting* light. It is also emitting electromagnetic radiation at other wavelengths, as the solar spectrum in Figure 1.1 demonstrates. The total power radiated is the area under the solar spectrum, and is 3.85×10^{26} watts (W). This is the solar luminosity. The photosphere, for all its brilliance, is a tenuous gas, with a density of order 10^{-3} kg m^{-3}, about 1000 times less than that of the air at the Earth's surface.

The spectrum in Figure 1.1 enables us to estimate the mean photospheric temperature. This is done by comparing the spectrum with that of an **ideal thermal source**, sometimes called a black body. The exact nature of such a source need not concern us. The important point is that its spectrum is uniquely determined by its temperature. Turning this around, if we can fit an ideal thermal source spectrum reasonably well to the spectrum of any other body, then we can estimate the other body's temperature. Figure 1.1 shows a good match between the solar spectrum and the spectrum of an ideal thermal source at a temperature of 5770 K. Also shown is the poor match with an ideal thermal source at 4000 K, where the peak of the spectrum is

Discovering the Solar System, Second Edition   Barrie W. Jones
© 2007 John Wiley & Sons, Ltd

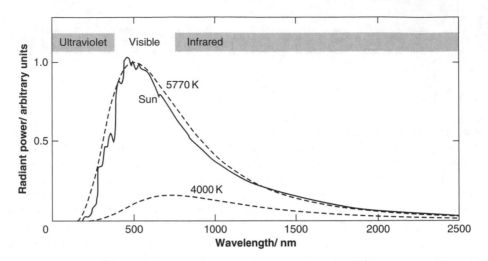

Figure 1.1 The solar spectrum, and the spectra of ideal thermal sources at 5770 K and 4000 K (1 nm = 10^{-9} m).

at longer wavelengths. Also, the power emitted by this source is a lot less. The power shown corresponds to the assumption that the 4000 K source has the same area as the source at 5770 K, and thus brings out the point that the temperature of an ideal thermal source determines not only the wavelength range of the emission, but the power too. Note that 5770 K is a *representative* temperature of the Sun's photosphere; the local temperature varies from place to place.

At a finer wavelength resolution than in Figure 1.1 the solar spectrum displays numerous narrow dips, called spectral absorption lines. These are the result of the absorption of upwelling solar radiation by various atoms and ions, mainly in the photosphere, and therefore the lines provide information about chemical composition. Further information about the Sun's composition is provided by small rocky bodies that continually fall to Earth. They are typically 1–100 cm across, and constitute the **meteorites** (Section 3.3). At 5770 K significant fractions of the atoms of some elements are ionised, and so it is best to define the composition at the photosphere in terms of atomic nuclei, rather than neutral atoms. In the photosphere, hydrogen and helium dominate, with hydrogen the most abundant – all the other chemical elements account for only about 0.2% of the nuclei. Outside the Sun's fusion core (Section 1.1.3) about 91% of the nuclei are hydrogen and about 9% are helium.

Plate 1 shows that the most obvious feature of the photosphere is dark spots. These are called (unsurprisingly) **sunspots**. They range in size from less than 300 km across to around 100 000 km, and their lifetimes range from less than an hour to 6 months or so. They have central temperatures of typically 4200 K, which is why they look darker than the surrounding photosphere. Sunspots are shallow depressions in the photosphere, where strong magnetic fields suppress the convection of heat from the solar interior, hence the lower sunspot temperatures. Their number varies, defining a sunspot cycle. The time between successive maxima ranges from about 8 years to about 15 years with a mean value of 11.1 years. From one cycle to the next the magnetic field of the Sun reverses. Therefore, the magnetic cycle is about 22 years.

Sunspots provide a ready means of studying the Sun's rotation, and reveal that the rotation period at the equator is 25.4 days, increasing with latitude to about 36 days at the poles. This differential rotation is common in fluid bodies in the Solar System.

1.1.2 The Solar Atmosphere

Above the photosphere there is a thin gas that can be regarded as the solar atmosphere. Because of its very low density, at most wavelengths it emits far less power than the underlying photosphere, and so the atmosphere is not normally visible. During total solar eclipses, the Moon just obscures the photosphere, and the weaker light from the atmosphere then becomes visible. In Plate 2 the atmosphere just above the photosphere is not visible, whereas in Plate 3 the short exposure time has emphasised the inner atmosphere. The atmosphere can be studied at other times, either by means of an optical device called a coronagraph that attenuates the radiation from the photosphere, or by making observations at wavelengths where the atmosphere is brighter than the photosphere.

Figure 1.2 shows how the temperature and density in the solar atmosphere vary with altitude above the base of the photosphere. A division of the atmosphere into two main layers is apparent, the chromosphere and the corona, separated by a thin transition region.

The chromosphere

The chromosphere lies immediately above the photosphere. It has much the same composition as the photosphere, so hydrogen dominates. The density declines rapidly with altitude, but the temperature *rises*. The red colour that gives the chromosphere its name ('coloured sphere') is a result of the emission by hydrogen atoms of light at 656.3 nm. This wavelength is called Hα ('aitch-alpha').

The data in Figure 1.2 are for 'quiet' parts of the chromosphere. Its properties are different where magnetic forces hold aloft filamentary clouds of cool gas, extending into the lower corona. The filaments are the red prominences above the limb of the photosphere in Plate 3. Prominences are transitory phenomena, lasting for periods from minutes to a couple of months.

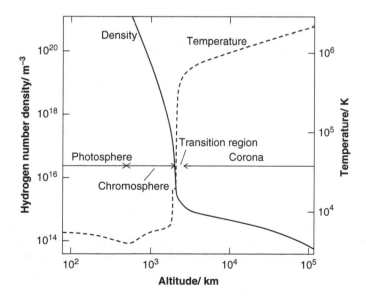

Figure 1.2 The variation of temperature and density in the Sun's atmosphere with altitude above the base of the photosphere.

The chromosphere is also greatly disturbed in regions where a flare occurs. This is a rapid brightening of a small area of the Sun's upper chromosphere or lower corona, usually in regions of the Sun where there are sunspots. The increase in brightness occurs in a few minutes, followed by a decrease taking up to an hour, and the energy release is spread over a very wide range of wavelengths. Flares, like certain prominences, are associated with bursts of ionised gas that escape from the Sun. Magnetic fields are an essential part of the flare process, and it seems probable that the electromagnetic radiation is from electrons that are accelerated close to the speed of light by changes in the magnetic field configuration. As with so many solar phenomena, the details are unclear.

The corona

Above the chromosphere the density continues to fall steeply across a thin transition region that separates the chromosphere from the corona (Figure 1.2).

☐ What distinctive feature of the transition region is apparent in Figure 1.2?

A distinctive feature is the enormous temperature gradient. This leads into the corona, where the gradient is not so steep. The corona extends for several solar radii (Plate 2), and within it the density continues to fall with altitude, but the temperature continues to rise, reaching $3\text{–}4 \times 10^6$ K, sometimes higher. Conduction, convection, and radiation from the photosphere cannot explain such temperatures – these mechanisms would not transfer net energy from a body at *lower* temperature (the photosphere) to a body at *higher* temperature (the corona). The main heating mechanism seems to be magnetic – magnetic fields become reconfigured throughout the corona, and induce local electric currents that then heat the corona. Waves involving magnetic fields (magnetohydrodynamic waves) also play a role in certain regions.

The corona is highly variable. At times of maximum sunspot number it is irregular, with long streamers in no preferred directions. At times of sunspot minimum, the visible boundary is more symmetrical, with a concentration of streamers extending from the Sun's equator, and short, narrow streamers from the poles. Coronal 'architecture' owes much to solar magnetic field lines. The white colour of the corona is photospheric light scattered by its constituents. Out to two or three solar radii the scattering is mainly from free electrons, ionisation being nearly total at the high temperatures of the corona. Further out, the scattering is dominated by the trace of fine dust in the interplanetary medium.

The solar wind

The solar atmosphere does not really stop at the corona, but extends into interplanetary space in a flow of gas called the **solar wind**, which deprives the Sun of about one part in 2.5×10^{-14} of its mass per year. Because of the highly ionised state of the corona, and its predominantly hydrogen composition, the wind consists largely of protons and electrons. The temperature of the corona is so high that if the Sun's gravity were the only force it would not be able to contain the corona, and the wind would blow steadily and uniformly in all directions. But the strong magnetic fields in the corona act on the moving charged particles in a manner that reduces the escape rate. Escape is preferential in directions where the confining effect is least strong, and an important type of location of this sort is called a coronal hole. This is a region of exceptionally low density and temperature, where the solar magnetic field lines reach huge distances into interplanetary space. Charged particles travel in helical paths around magnetic field lines, so the outward-directed lines facilitate escape. The escaping particles constitute the fast wind. Elsewhere, where the field lines are confined near the Sun, there is an additional outward flow, though at lower speeds, called the slow wind.

Solar wind particles (somehow) gain speed as they travel outwards, and at the Earth the speeds range from 200 to $900\,\mathrm{km\,s^{-1}}$. The density is extremely low – typically about 4 protons and 4 electrons per $\mathrm{cm^3}$, though with large variations. Particularly large enhancements result from what are called coronal mass ejections, often associated with flares and prominences, and perhaps resulting from the opening of magnetic field lines. If the Earth is in the way of a concentrated jet of solar wind, then various effects are produced, such as the aurorae (the northern and southern lights – Plate 26). The solar wind is the main source of the extremely tenuous gas that pervades interplanetary space.

Solar activity

Solar activity is the collective term for those solar phenomena that vary with a periodicity of about 11 years.

☐ What two aspects of solar activity were outlined earlier?

You have already met the sunspot cycle, and it was mentioned that the form of the corona is correlated with it. Prominences (filaments) and flares are further aspects of solar activity, both phenomena being more common at sunspot maximum. The solar luminosity also varies with the sunspot cycle, and on average is about 0.15% *higher* at sunspot maximum than at sunspot minimum. This might seem curious, with sunspots being cooler and therefore less luminous than the rest of the photosphere. However, when there are more sunspots, a greater area of the photosphere is covered in bright luminous patches called faculae.

All the various forms of solar activity are related to solar magnetic fields that ultimately originate deep in the Sun. The origin of these fields will be considered briefly in the following description of the solar interior.

1.1.3 The Solar Interior

To investigate the solar interior, we would really like to burrow through to the centre of the Sun, observing and measuring things as we go. Alas! This approach is entirely impractical. Therefore, the approach adopted, in its broad features, is the same as that used for all inaccessible interiors. A model is constructed and varied until it matches the major properties that we either *can* observe, or can obtain fairly directly and reliably from observations. Usually, a *range* of models can be made to fit, so a model is rarely unique. Many features are, however, common to all models, and such features are believed to be correct. This modelling process will be described in detail in Chapter 4, in relation to planetary interiors. Here, we shall present the *outcome* of the process as applied to the Sun.

A model of the solar interior

Figure 1.3 shows a typical model of the Sun as it is thought to be today. Hydrogen and helium predominate throughout, as observed in the photosphere. Note the enormous increase of pressure with depth, to 10^{16} pascals (Pa) at the Sun's centre – about 10^{11} times atmospheric pressure at sea level on the Earth! The central density is less extreme, 'only' about 14 times that of solid lead as it occurs on the Earth, though the temperatures are so high that the solar interior is everywhere fluid – there are no solids. Another consequence of the high temperatures is that at all but the shallowest depths the atoms are kept fully ionised by the energetic atomic collisions

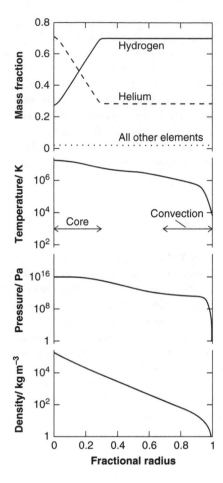

Figure 1.3 A model of the solar interior.

that occur. A highly ionised medium is called a **plasma**. The central temperatures in the Sun are about 1.4×10^7 K, sufficiently high that nuclear reactions can sustain these temperatures and the solar luminosity, and can have done so for the 4600 million years (Ma) since the Sun formed (an age based on various data to be outlined in Chapter 3, notably data from radiometrically dated meteorites). This copious source of internal energy also sustains the pressure gradient that prevents the Sun from contracting.

Though nuclear reactions sustain the central temperatures today, there must have been some other means by which such temperatures were initially attained in order that the nuclear reactions were triggered. This must have been through the gravitational energy released when the Sun contracted from some more dispersed state. With energy being radiated to space only from its outer regions, it would have become hotter in the centre than at the surface. Nuclear reaction rates rise so rapidly with increasing temperature that when the central regions of the young Sun became hot enough for nuclear reaction rates to be significant, there was a fairly sharp boundary between a central core where reaction rates were high, and the rest of the Sun where reactions rates were negligible. This has remained the case ever since. At present the central core extends to about 0.3 of the solar radius (Figure 1.3). This is a fraction $(0.3)^3$ of the Sun's volume, which is only 2.7%. However, the density increases so rapidly with depth that a far greater fraction of the Sun's mass is contained within its central core.

The Sun was initially of uniform composition, many models giving proportions by mass close to 70.9% hydrogen, 27.5% helium, and 1.6% for the total of all the other elements. In such a mixture, at the core temperatures that the Sun has had since its birth, there is only one group of nuclear reactions that is significant – the pp chains. The name arises because the sequence of reactions starts with the interaction of two protons (symbol p) to form a heavier nucleus (deuterium), a proton being the nucleus of the most abundant isotope of hydrogen (^{1}H). When a heavier nucleus results from the joining of two lighter nuclei, this is called **nuclear fusion**. The details of the pp chains will not concern us, but their net effect is the conversion of four protons into the nucleus of the most abundant isotope of helium (^{4}He), which consists of two protons and two neutrons.

The onset of hydrogen fusion in the Sun's core marks the start of its **main sequence** lifetime. A main sequence star is one sustained by core hydrogen fusion, and ends when the core hydrogen has been used up. The main sequence phase occupies most of a star's active lifetime. In the case of the Sun it will be another 6000 Ma or so until it ends, with consequences outlined in Section 11.5.

Various other subatomic particles are involved in the pp cycles, but of central importance are the gamma rays produced – electromagnetic radiation with very short wavelengths. These carry nearly all of the energy liberated by the pp chains' reactions. The gamma rays do not get very far before they interact with the plasma of electrons and nuclei that constitutes the solar core. To understand the interaction, it is necessary to recall that although electromagnetic radiation can be regarded as a wave, it can also be regarded as a stream of particles called **photons**. The wave picture is useful for understanding how radiation gets from one place to another; the photon picture is useful for understanding the interaction of radiation with matter. The energy e of a photon is related to the frequency f of the wave via

$$e = h\,f \tag{1.1}$$

where h is Planck's constant. The frequency of a wave is related to its wavelength via

$$f = c/\lambda \tag{1.2}$$

where c is the wave speed. For electromagnetic radiation in space c is the speed of light, $3.00 \times 10^5\,\mathrm{km\,s^{-1}}$. Table 1.6 lists values of c, h, and other physical constants of relevance to this book. (For ease of reference, the Chapter 1 tables are located at the end of the chapter.)

On average, after only a centimetre or so, a gamma ray in the core either bounces off an electron or nucleus, in a process called scattering, or is absorbed and re-emitted. This maintains the level of random motion of the plasma: in other words, it maintains its high temperature. The gamma ray photons are not all of the same energy. They have a spectrum shaped like that of an ideal thermal source at the temperature of the local plasma. This is true throughout the Sun, so as the photons move outwards their spectrum moves to longer wavelengths, corresponding to the lower temperatures, until at the photosphere the spectrum is that shown in Figure 1.1 (Section 1.1.1). The number of photons is greater than in the core, but they are of much lower average energy. From the moment a gamma ray is emitted in the core to the moment its descendants emerge from the photosphere, a time of several million years will have elapsed.

☐   What is the direct travel time?

The direct travel time at the speed of light c across the solar radius of 6.96×10^5 km is $6.96 \times 10^5 \, \text{km}/3.00 \times 10^5 \, \text{km s}^{-1}$, i.e. 2.23 seconds!

The transport of energy by radiation is, unsurprisingly, called **radiative transfer**. This occurs *throughout* the Sun. Another mechanism of importance in the Sun is convection, the phenomenon familiar in a warmed pan of liquid, where energy is transported by currents of fluid. When the calculations are done for the Sun, then the outcome is as in Figure 1.3. **Convection** is confined to the outer 29% or so of the solar radius, where it *supplements* radiative transfer as a means of conveying energy outwards. The tops of the convective cells are seen in the photosphere as transient patterns called granules. These are about 1500 km across, and exist for 5–10 minutes. There are also supergranules, about 10 000 km across and extending about as deep.

Because convection does not extend to the core in which the nuclear reactions are occurring, the core is not being replenished, and so it becomes more and more depleted in hydrogen and correspondingly enriched in helium. The core itself is unmixed, and so with temperature increasing with depth, the nuclear reaction rates increase with depth, and therefore so does the enrichment. This feature is apparent in the solar model in Figure 1.3.

The solar magnetic field

The source of any magnetic field is an electric current. If a body contains an electrically conducting fluid, then the motions of the fluid can become organised in a way that constitute a net circulation of electric current, and a magnetic field results. This is just what we have in the solar interior – the solar plasma is highly conducting, and the convection currents sustain its motion. We shall look more closely at this sort of process in Section 4.2. Detailed studies show that the source of the solar field is concentrated towards the base of the convective zone. The differential rotation of the Sun contorts the field in a manner that goes some way to explaining sunspots and other magnetic phenomena.

The increase of solar luminosity

Evolutionary models of the Sun indicate that the solar luminosity was only about 70% of its present value 4600 Ma ago, that it has gradually increased since, and will continue to increase in the future. This increase is of great importance to planetary atmospheres and surfaces, as you will see in later chapters.

1.1.4   The Solar Neutrino Problem

There is one observed feature of the Sun that solar models had difficulty in explaining. This is the rate at which solar neutrinos are detected on the Earth. Solar neutrinos are so unreactive that most of them escape from the Sun and so provide one of the few direct indicators of conditions deep in the solar interior. A neutrino is an elusive particle that comes in three kinds, called flavours. The electron neutrino is produced in the pp chains of nuclear reactions that occur in the solar interior. The rates at which electron neutrinos from the Sun are detected by various installations on the Earth are significantly below the calculated rate. Are the calculated pp reaction rates in the Sun too low?

No, they are not. It is now known that neutrinos oscillate between the three flavours. If, in their 8 minute journey at the speed of light from the solar core to the terrestrial detectors, they settle into this oscillation, then at any instant only some of the neutrinos arriving here are of the electron type. The earlier neutrino detectors could only detect the electron type. Now, all three can and have been detected coming from the Sun, giving a greater flux. This accounts for most of the discrepancy. The rest of it has been accounted for by improvements in solar models that have modified the predictions of the solar neutrino flux.

Question 1.1

The Sun's photospheric temperature, as well as its luminosity, has also increased since its birth. What is the combined effect on the solar spectrum in Figure 1.1?

1.2 The Sun's Family – A Brief Introduction

Within the Solar System we find bodies with a great range of size, as Figure 1.4 shows. The Sun is by far the largest body. Next in size are the four **giant planets**: Jupiter, Saturn, Uranus, and Neptune. We then come to a group of bodies of intermediate size. Prominent are the Earth, Venus, Mars, and Mercury. These four bodies constitute the **terrestrial planets**, so called because they are comparable in size and composition, and are neighbours in space. This intermediate-sized group has an arbitrary lower diameter which we shall take to be that of the planet Pluto, the ninth planet. At least one body well beyond Pluto is slightly larger than Pluto – Eris, of which, more later. Seven planetary satellites are larger than Pluto. As their name suggests, planetary satellites are companions of a planet, bound in orbit around it and with a smaller mass. In spite of their size, this binding means that they are classified as planetary bodies, rather than as planets.

There are plenty of bodies smaller than Pluto: the remaining satellites, of which one of Uranus's satellites Titania is the largest; a swarm of **asteroids**, of which Ceres ('series') is easily the largest; a huge number of comets, or bodies that become comets; and a continuous range of even smaller bodies, right down to tiny particles of dust.

Tables 1.1–1.3 display the radius, and several other properties, of Solar System bodies and of their orbits. Table 1.1 covers the nine planets and Ceres. Table 1.2 covers the planetary satellites,

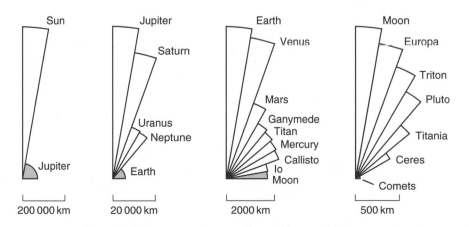

Figure 1.4 Sizes of bodies in the Solar System.

excluding the many satellites of Jupiter and Saturn less that 5 km mean radius, plus a few others of Uranus and Neptune. Table 1.3 covers the 15 largest asteroids.

Figure 1.5 shows the orbits of the planets. These orbits are roughly circular, and lie more or less in the same plane. The plane of the Earth's orbit is called the **ecliptic plane**. The planets move around their orbits at different rates, but in the same direction, anticlockwise as viewed from above the Earth's North Pole – this is called the **prograde direction**. The asteroids are concentrated in the space between Mars and Jupiter, in the asteroid belt. The distances in Figure 1.5 are huge compared even to the solar radius of 6.96×10^5 km. A convenient unit of distance in the Solar System is the average distance of the Earth from the Sun, 1.50×10^8 km, which is given a special name, the **astronomical unit** (AU). Between them, Figures 1.4 and 1.5 provide a map of the Solar System's planetary domain.

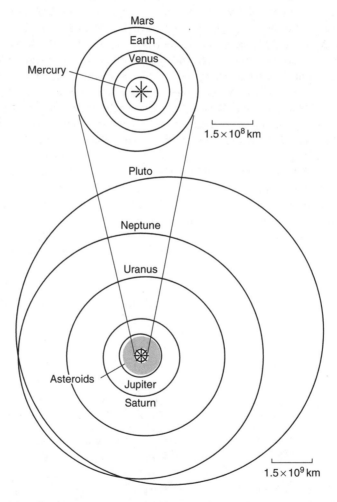

Figure 1.5 The orbits of the planets as they would appear from a distant viewpoint perpendicular to the plane of the Earth's orbit.

1.2.1 The Terrestrial Planets and the Asteroids

The terrestrial planets occupy the inner Solar System (Figure 1.5). They consist largely of rocky materials, with iron-rich cores. Most of the Earth's core is liquid, and this is probably the case for Venus too. Each core is overlain by a mantle of rocky materials (silicates), overlain in turn by a silicate crust. Mercury's surface is heavily cratered by the accumulated effects of impacts from space (Plate 4), indicating little geological resurfacing since the planet was formed. It has a negligible atmosphere. Venus is the Earth's twin in size and mass, and like the Earth it is geologically active, with volcanic features common (Plate 5), but it differs from the Earth in that it has no oceans. The surface of Venus, at a mean temperature of 740 K, is far too hot for liquid water, a consequence of its proximity to the Sun, and its massive, carbon dioxide (CO_2) atmosphere. The Earth is further from the Sun and has an atmosphere about 100 times less massive, mainly nitrogen (N_2) and oxygen (O_2). It is thus cool enough to have oceans, but not so cold that they are frozen (Plate 6). Unlike Mercury and Venus, the Earth has a satellite – the Moon. Figure 1.4 shows that it is a considerable world, larger than Pluto. It is devoid of an appreciable atmosphere and has a heavily cratered surface (Plate 7).

Beyond the Earth we come to Mars, smaller than the Earth but larger than Mercury. It has a thin CO_2 atmosphere through which its cool surface is readily visible (Plate 8). About half of the surface is heavily cratered. The other half is less cratered, and shows evidence of the corresponding past geological activity. Plate 9 is a view at the surface. Mars has two tiny satellites, Phobos and Deimos (Table 1.2). These orbit very close to the planet, and might be captured asteroids.

It is the domain of the asteroids – the asteroid belt – that we cross in the large gulf of space that separates Mars from Jupiter. Asteroids are rocky bodies of which Ceres is by far the largest (Table 1.3), although it is still a good deal smaller than Pluto (Figure 1.4). It is thought that there are about 10^9 asteroids larger than 1 km, and Plate 10 shows just one with a typically irregular shape at this small size. At a size of 1 metre there is a switch in terminology, with smaller bodies being called meteoroids, and these are even more numerous. Those that fall to Earth constitute the meteorites, which have provided much information about the origin, evolution, and composition of the Solar System. Below about 0.01 mm there is another switch in terminology – smaller particles are called dust. This is widely distributed within and beyond the asteroid belt, and is predominantly submicrometre in size (less that 10^{-6} m across). The asteroids are sometimes called minor planets.

1.2.2 The Giant Planets

The giant planets are very different from the terrestrial planets, not just in size (Figure 1.4) but also in composition. Whereas the terrestrial planets are dominated by rocky materials, including iron, Jupiter and Saturn are dominated by hydrogen and helium. There are also materials, notably water (H_2O). The icy materials tend to concentrate towards the centres, where it is so hot, typically 10^4 K, that the icy materials are liquids not solids. Rocky materials make up only a small fraction of the mass of Jupiter and Saturn, and they also tend to concentrate towards the centres. Uranus and Neptune are less dominated by hydrogen and helium, and the central concentration of icy and rocky materials is more marked. All four giant planets are fluid throughout their interiors.

☐ What other body in the Solar System is dominated by hydrogen and helium, and is fluid throughout?

The Sun is also a fluid body, dominated by hydrogen and helium (Section 1.2).

Jupiter is the largest and most massive of the planets. Plate 11 shows the richly structured uppermost layer of cloud, which consists mainly of ammonia (NH_3) particles, coloured by traces of a wide variety of substances, and patterned by atmospheric motions. The prominent banding is parallel to the equator.

Jupiter has a large and richly varied family of satellites. Figure 1.6 is a plan view, drawn to scale, of the orbits of the four largest by far of Jupiter's satellites – Io, Europa, Ganymede, Callisto. They are called the **Galilean satellites**, after the Italian astronomer Galileo Galilei (1564–1642) who discovered them in 1610 when he made some of the very first observations of the heavens with the newly invented telescope. They orbit the planet close to its equatorial plane. These remarkable bodies are shown in Plates 12–15. They range in size from Ganymede, which is somewhat larger than Mercury and is the largest of all planetary satellites, to Europa, which is somewhat smaller than the Moon. Io is a rocky body. The other three contain increasing amounts of water (mainly as ice) with increasing distance from Jupiter. Table 1.2 includes all but the smallest satellites of Jupiter.

We move on to Saturn, which is somewhat smaller than Jupiter, but is otherwise not so very different (Plate 16). We shall say no more about the planet in this chapter, but turn to its family of satellites, and in particular to its largest satellite Titan, an icy–rocky body larger than Mercury, and second only to Ganymede among the satellites. A remarkable thing about Titan is that it has a massive atmosphere. Indeed, per unit area of surface, it has about 10 times more mass of atmosphere than the Earth. The atmosphere is well over 90% N_2 with a few per cent of methane (CH_4), but contains so much hydrocarbon cloud and haze that the surface is almost invisible from outside it (Plate 17).

Saturn is most famous for its rings (Plate 18). These lie in the planet's equatorial plane, and consist of small solid particles. The rings are extremely thin, probably no more than a few hundred metres. They are, however, so extensive that they were observed by Galileo in 1610, though it was the Dutch physicist Christiaan Huygens (1629–1693) who, in 1655, was first to

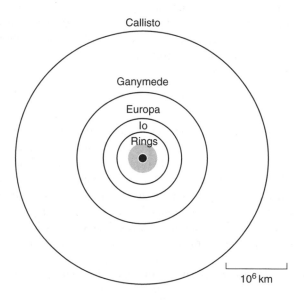

Figure 1.6   The orbits of the Galilean satellites of Jupiter.

realise that they are rings encircling the planet. Plate 18 shows that each main ring is broken up into many ringlets, to form a structure of exquisite complexity. The other three giant planets also have ring systems, but they are far less substantial.

Beyond Saturn we head off across another of the increasingly large gulfs of space that separate the planets as we move out from the Sun. We come to Uranus, a good deal smaller than Saturn, and with a smaller proportion of hydrogen and helium and a large icy–rocky core. In spite of its size it was unknown until 1781 when it was discovered accidentally by the Germano-British astronomer William Herschel (1738–1822) during a systematic survey of the stars. This was the first planet to be discovered in recorded history. It had escaped earlier detection because it is at the very threshold of unaided eye visibility, owing to its great distance from us. Its bands are generally not as strong as those of Jupiter and Saturn (Plate 19).

Neptune, like Uranus, was discovered in recorded history, but the circumstances were very different. Whereas Uranus was discovered accidentally, Neptune was discovered as a result of predictions made by two astronomers in order to explain slight departures of Uranus from its expected orbit. The British astronomer John Couch Adams (1819–1892) and the French astronomer Urbain Jean Joseph Le Verrier (1811–1877) independently predicted that the cause was a previously unknown planet orbiting beyond Uranus, and in 1846 Neptune was discovered by the German astronomer Johann Gottfried Galle (1812–1910) close to its predicted positions. Neptune, the last of the giants, is not so very different from Uranus (Plate 20), and so in the spirit of this quick tour we shall say no more here about the planet itself.

Uranus and Neptune have many satellites. The largest among them by far, Neptune's satellite Triton, is a rocky–icy body slightly larger than Pluto, and it is the only satellite other than Titan that has a significant atmosphere, though it is fairly tenuous, and allows the icy surface of Triton to be seen (Plate 21). Among Neptune's other satellites, Nereid has a huge and extraordinarily eccentric orbit (Table 1.2). The orbit of Triton is curious in a different way – though it is nearly circular it is **retrograde**, which is the opposite direction to the prograde orbital motion of the planets and all other large satellites.

1.2.3 Pluto and Beyond

Beyond Neptune lies Pluto, in an orbit where sunlight is 1600 times weaker than at the Earth. Pluto was discovered in 1930 by the American astronomer Clyde William Tombaugh (1906–1997) during a systematic search of a band of sky straddling the orbital planes of the known planets. It is a small world (Figure 1.4) and has not yet been visited by a spacecraft. Consequently we know rather little about Pluto and its comparatively large satellite Charon. Pluto is an icy world, with about half of its volume consisting of frozen water and other icy substances, and the remainder consisting of rock. Charon probably has a broadly similar composition. Pluto also has two tiny satellites, Nix and Hydra, of unknown composition.

Beyond Pluto space is not empty, and we have certainly not come to the edge of the Solar System. One type of body abundant beyond Pluto is the **comets**. These are small icy–rocky bodies that, through the effect of the Sun, develop huge fuzzy heads and spectacular tails when their orbits carry them into the inner Solar System (Plate 22). In the outer Solar System they have no heads and tails, and are not called comets there. There are two main populations. One of these has bodies in prograde orbits concentrated towards the ecliptic plane, and occupying orbits ranging from around the size of Pluto's orbit (39.8 AU from the Sun, on average) to far larger. This is the **Edgeworth–Kuiper belt**, and its occupants are called E–K objects (EKOs). Over 1000 have been seen, the largest at present being Eris, which Hubble Space Telescope

(HST) images have shown to have a radius about 20% larger than Pluto. It is currently (2006) 97 AU from the Sun, and when closest to the Sun lies at a distance of 38 AU. It is estimated that more than 10^5 EKOs are larger than 100 km across, and lie in orbits out to about 50 AU. There are more EKOs further away, Eris among them, and there are certainly many more that are smaller than 100 km.

The Edgeworth–Kuiper belt might blend into the second population of icy–rocky bodies, a swarm of 10^{12}–10^{13} in a thick spherical shell surrounding the Solar System, extending from about 10^3 to 10^5 AU. This is the **Oort cloud** (also called the Öpik–Oort cloud). Its outer boundary is at the extremities of the Solar System, where passing stars can exert a gravitational force comparable with that of the Sun. The Oort cloud has not been observed directly, but its existence is inferred from the comets that we see in the inner Solar System. These are a small sample of the Oort cloud and also of the Edgeworth–Kuiper belt, but in orbits that have been greatly modified. Table 1.4 lists some properties of selected comets.

Definition of a planet

That Eris, and several other EKOs, are larger or comparable in size with Pluto, has raised the issue of whether there are several more planets in the Solar System, or whether large EKOs, including Pluto, should not be regarded as planets.

At its triennial meeting in Prague in 2006, the International Astronomical Union faced this issue, and passed resolutions defining what, in the Solar System, determines whether a body is a planet. You might be surprised that previously there was no formal definition. The least controversial parts of the definition are that a planet is in its own orbit around the Sun and is large enough for its own gravity to overcome the strength of its materials, which, for a non-rotating, isolated body, would make it spherical. On this basis, Pluto, Eris, and Ceres would be planets. But the IAU added a further criterion, that to be a planet a body has to have cleared material in the neighbourhood of its orbit. This is a tricky concept. The important point is that Pluto, Eris, and Ceres do not meet it, and are therefore to be regarded as **dwarf planets**.

However, the debate is not over. Many astronomers are unhappy with the IAU resolutions, and therefore the definition of what is a planet might well be revised in the near future. Consequently, in this book, Pluto will continue to be regarded as a planet and also as a large EKO. Eris, and other large EKOs, will not, for now, be labelled as (dwarf) planets, and Ceres will continue to be regarded as the largest asteroid.

Question 1.2

In about 100 words, discuss whether there is any correlation between the size of a planet and its distance from the Sun.

1.3 Chemical Elements in the Solar System

With most of the mass in the Solar System in the Sun, and the Sun composed almost entirely of hydrogen and helium, the chemical composition of the Solar System is dominated by these two elements. Hydrogen is the lightest element. Its most common isotope (by far) has a nucleus consisting of a single proton. You saw in Section 1.1.3 that this isotope is represented as ^{1}H.

Helium is the next lightest element, with the nucleus of its most common isotope (again by far) consisting of two protons and two neutrons. Recall that an element is defined by the number of protons in its nucleus – this is the atomic number – and that the isotopes are distinguished by different numbers of neutrons. To denote a particular isotope the number of neutrons plus protons is included with the chemical symbol, as you have seen for helium's common isotope, ^{4}He (Section 1.1.3).

The Solar System contains all 92 naturally occurring chemical elements with atomic numbers from 1 (hydrogen) to 92 (uranium). The relative abundances of these elements have been determined through observations of the Sun and through analyses of primitive meteorites (Section 3.3.2).

Most of the mass outside the Sun is in Jupiter and Saturn, and these are also composed largely of hydrogen and helium, though they contain larger proportions of the other elements – the so-called **heavy elements**. For the Solar System as a whole, Table 1.5 gives the relative abundances of the 15 most abundant of the chemical elements. Note that the value for helium is for the Sun outside its fusion core. This region has not been depleted in helium by its conversion into hydrogen by nuclear fusion, such as occurs in the core of the Sun.

Except in very high-temperature regions, most of the atoms of most elements are combined with one or more other atoms, either of the same element, or of other elements. The important exceptions are helium, neon, argon, krypton, and xenon, which are so chemically unreactive that they remain monatomic and have been given the name **inert gases** or noble gases. If an element is combined with itself, as in H_2, then we have the element in molecular form, whereas if it is combined with other elements, then we have it as a chemical compound.

Water (H_2O) is the most abundant chemical compound of hydrogen in the Solar System. Table 1.5 suggests the reason.

☐   What is the reason?

It is because oxygen has a high abundance. But hydrogen is so overwhelmingly abundant that there is plenty left over after the formation of hydrogen compounds. Most of the uncompounded hydrogen outside of the Sun is in the giant planets, as H_2, or as a fluid of hydrogen with metallic properties. Water is the main repository of hydrogen in most of the other bodies.

1.4 Orbits of Solar System Bodies

1.4.1 Kepler's Laws of Planetary Motion

Each planet orbits the Sun as shown in plan view in Figure 1.5. As a crude approximation, the planetary orbits can be represented as circles centred on the Sun, with all the circles in the same plane, and each planet moving around its orbit at a constant speed; the larger the orbit, the slower the speed. A far better approximation is encapsulated in three empirical rules called **Kepler's laws of planetary motion**. These were announced by the German astronomer Johannes Kepler (1571–1630), the first two in 1609, the third in 1619.

Kepler's first law   Each planet moves around the Sun in an ellipse, with the Sun at one focus of the ellipse.

Kepler's second law   As the planet moves around its orbit, the straight line from the Sun to the planet sweeps out equal areas in equal intervals of time.

We shall come to the third law shortly.

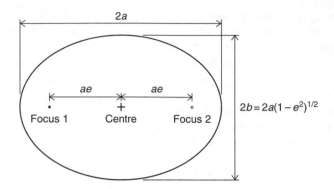

Figure 1.7 An ellipse, though far more eccentric than the orbit of any planet. This is the shape of the orbit of the comet 21P Giacobini–Zinner (Table 1.4).

Figure 1.7 shows an **ellipse**. The shape is that of a circle viewed obliquely: the more oblique the view, the greater the departure from circular form. The important features of an ellipse are marked in Figure 1.7, and are that

- it has a major axis of length $2a$ and a minor axis of length $2b$ – unsurprisingly, a and b are called, respectively, the **semimajor axis** and the semiminor axis;
- there are two foci that lie on the major axis, each a distance ae from the centre of the ellipse, where e is the **eccentricity** of the ellipse; note that the foci are in the plane of the ellipse, and that $e = \sqrt{1 - b^2/a^2}$.

The eccentricity is a measure of the departure from circular form. If e is zero, then the foci coalesce at the centre, a equals b, and the ellipse has become a circle of radius a. If e approaches one then the ellipse becomes extremely elongated.

Kepler's first law tells us that the shape of a planetary orbit is an ellipse, and that the Sun is at one focus. Figure 1.8 shows the orbit of Pluto, which among planetary orbits has the greatest eccentricity, $e = 0.254$. Note that whereas the shape is very close to a circle, the Sun, which

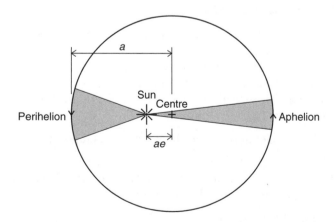

Figure 1.8 The orbit of Pluto.

is at one of the foci, is distinctly off centre. Note also that the semimajor axis is less than the maximum distance of a body from the Sun, but is greater than the minimum distance, and it is therefore some sort of average distance. At its greatest distance from the Sun the body is at a point in its orbit called **aphelion**; the closest point is called **perihelion**. These terms are derived from the Greek words *Helios* for the Sun, and *peri-* and *apo-* which in this context mean 'in the vicinity of' and 'away from' respectively. The length of the semimajor axis of the Earth's orbit is called the astronomical unit (AU), mentioned earlier.

Kepler's laws don't apply just to planets. Figure 1.7 is in fact the shape of the orbit of the comet 21P/Giacobini–Zinner (Table 1.4).

❑ Where should the Sun be marked in Figure 1.7?

The Sun should be shown at either one of the two foci. This is an orbit of fairly high eccentricity, $e = 0.7057$. The non-circular form is now very clear, and the foci are greatly displaced from the centre.

Kepler's *second* law tells us how a planet (or comet) moves around its orbit. For the case of Pluto the shaded areas within the orbit in Figure 1.8 are equal in area, and so by Kepler's second law these are swept out in equal intervals of time. Thus, around aphelion the body is moving slowest, and around perihelion it is moving fastest. The difference in these two speeds is larger, the greater the eccentricity.

❑ What are the speeds at different positions in a *circular* orbit?

In a circular orbit the equal areas correspond to equal length arcs around the circle, so the body moves at a constant speed around its orbit.

So far, Kepler's laws have described the orbital motion around the Sun of an *individual* body. The third and final law compares the motion of one body to another:

Kepler's third law If P is the time taken by a planet to orbit the Sun once, and a is the semimajor axis of the orbit, then

$$P = ka^{3/2} \tag{1.3}$$

where k has the same value for each planet.

P is called the orbital period or the period of revolution. It is the period as observed from a non-rotating viewpoint, which, for practical purposes, is any viewpoint fixed with respect to the distant stars. This leads to the term **sidereal** ('= star-related') **orbital period** for P. For the Earth this period is called the sidereal year. Therefore, with $P = 1$ (sidereal) year and $a = 1\,\mathrm{AU}$, $k = 1$ year $\mathrm{AU}^{-3/2}$. According to Kepler's third law, this is the value of k for all the planets.

Equation (1.3) tells us that the larger the orbit, the longer the orbital period. This is partly because the planet has to travel further, and partly because the planet moves more slowly. We can see that the planet moves more slowly from the simple case of a circular orbit of radius a. The circumference of the orbit is $2\pi a$, so if the orbital speed were independent of a then P would be proportional to a, not, as observed, to $a^{3/2}$. Therefore, the orbital speed must be proportional to $a^{-1/2}$. In an elliptical orbit the circumference still increases as a increases, and now it is the *average* speed that decreases.

Kepler's third law enables us to obtain relative distances in the Solar System. If we measure the orbital periods of bodies A and B, then the ratio of the semimajor axes of their orbits is obtained from equation (1.3):

$$\frac{a_A}{a_B} = \left(\frac{P_A}{P_B}\right)^{2/3}$$

If one of the two bodies has a in AU, then we can express the other semimajor axis in astronomical units. This can be repeated for all orbits. Moreover, from the shape and orientation of the orbits, we can draw a scale plan of the Solar System, and at any instant we can show where the various planets lie. At any instant we can thus express in astronomical units the distance between any two bodies. If at the same instant we can measure the distance between any two bodies in metres, we can then obtain the value of the astronomical unit in metres.

Today, the astronomical unit is best measured using radar reflections. Radar pulses travel at the speed of light c, which is known very accurately (Table 1.6). Time intervals can also be measured very accurately, so if we measure the time interval Δt between sending a radar pulse from the Earth to a planet and receiving its echo, then the distance from the Earth to the planet is $c\Delta t/2$. Accurate measurements of distances in the Solar System have revealed that the semimajor axis of the Earth's orbit is subject to very slight variations. As a consequence the AU is now defined as *exactly* equal to $1.495\,978\,706\,9 \times 10^8$ km. The Earth's semimajor axis is currently (2006) 0.999 985 AU.

Question 1.3

The asteroid Fortuna is in an orbit with a period of 3.81 years. Show that the semimajor axis of its orbit is 2.44 AU.

Question 1.4

Suppose that when the Earth is at perihelion Venus lies on the straight line between the Earth and the Sun. The time interval between sending a radar pulse from the Earth to Venus and receiving its echo is 264 s. Taking the speed of light in space as 3.00×10^5 km s^{-1}, calculate to two significant figures the astronomical unit in metres. Proceed as follows.

For the instant of measurement

- from the orbital details calculate the distance between the Earth and Venus in AU;
- from the radar data calculate the distance in km between the Earth and Venus.

Hence calculate the number of metres in 1 AU.

Note: *For two-significant-figure accuracy Venus is sufficiently close to perihelion when the Earth is at perihelion for you to use the perihelion distance of Venus.*

1.4.2 Orbital Elements

The quantities a and e are two of the five quantities – of the five **orbital elements** – that are needed to specify the elliptical orbit of a body. P is *not* normally among the three remaining elements.

☐ Why is P (normally) redundant?

The orbital period is redundant because it can usually be obtained with sufficient accuracy from a via Kepler's third law. The need for three further elements is illustrated in Figure 1.9, which shows the plane of the Earth's orbit plus the orbit of another body. Note that, for clarity, the orbit of the Earth is not shown, though the direction of the Earth's orbital motion is indicated by an arrow. The plane of the Earth's orbit acts as a reference plane for all other orbits and, as noted earlier, is called the ecliptic plane. The position of the Earth in its orbit at a certain

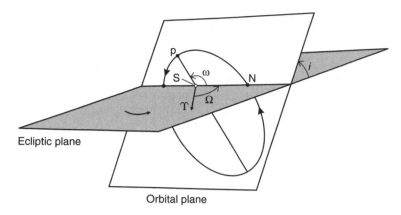

Figure 1.9 The three orbital elements i, ω, Ω, used to specify the orientation of an elliptical orbit with respect to the ecliptic plane.

moment in the year provides a reference direction. The direction chosen is that from the Earth to the Sun when the Earth is at the **vernal** (March) **equinox**. The direction points to the stars at a location called the **first point of Aries**. The direction (and the location) has the symbol ♈. The basis of these names will be given later.

For the other body in Figure 1.9, its orbital plane intersects the ecliptic plane to form a line. The Sun lies on this line at the point S – the Sun must lie in both orbital planes (Kepler's first law). Another point on the line is marked N, and this is where the body crosses the ecliptic plane in going from the south side to the north side, north and south referring to the sides of the ecliptic plane on which the Earth's North and South Poles lie. N is called the **ascending node** of the body's orbit. The angle Ω is measured in the direction of the Earth's motion, from ♈ to the line SN. This is the orbital element called the **longitude of the ascending node**. It can range from 0° to 360°. The orbital plane of the planet makes an angle i with respect to the ecliptic plane, and this is the element called the **orbital inclination**. It can range from 0° to 180° – values greater than 90° correspond to retrograde orbital motion.

☐ What is the inclination of the Earth's orbit, and why is the longitude of the ascending node an inapplicable notion?

The Earth's orbit lies in the ecliptic plane. With the ecliptic plane as the reference plane, the inclination of the Earth's orbit is therefore zero. An ascending node is one of the two points where an orbit intersects the ecliptic plane. The Earth's orbit lies in this plane and therefore the ascending node is undefined.

The last of the five elements that are needed to specify the elliptical orbit of a body is the angle ω, measured from SN to the line Sp, where p (Figure 1.9) is the perihelion position of the body. The angle ω is measured in the direction of motion of the body, and can range from 0° to 360°. It is called the **argument of perihelion**. However, it is somewhat more common to give as the fifth element the angle $(\Omega + \omega)$. This is called the **longitude of perihelion**. It is a curious angle, being the sum of two angles that are not in the same plane. Note that if the sum exceeds 360°, then 360° is subtracted.

To specify exactly where a body will be in its orbit at some instant we need to know when it was at some specified point at some earlier time. For example, we could specify one of the times at which the body was at perihelion. This sort of specification is a sixth orbital element.

Table 1.1 lists the values of the orbital elements for each planet and for the largest asteroid, Ceres. Note that

- the orbital inclinations are small: the planets' orbital planes are almost coincident, Pluto's inclination of 17.1° being by far the greatest;
- except for Pluto and Mercury, and to a lesser extent Mars, the orbital eccentricities are also small, and the exceptions are not dramatic.

Question 1.5

(a) Comet Kopff has the following orbital elements: inclination 4.7°, eccentricity 0.54, argument of perihelion 163°, longitude of the ascending node 121°. Sketch the orbit with respect to the ecliptic plane and the direction ♈. (An accurate drawing is *not* required.)

(b) The distance of Comet Kopff from the Sun at its perihelion on 2 July 1996 was 1.58 AU. Calculate the semimajor axis of the orbit, and hence calculate: its aphelion distance, its orbital period, and the month and year of the first perihelion in the twenty-first century (given that there are 365.24 days per year).

(c) The perihelion and aphelion distances of Mars are 1.38 AU and 1.67 AU, and yet the orbits of Mars and Comet Kopff do *not* intersect. In a few sentences, state why not. (A proof is *not* required.)

1.4.3 Asteroids and the Titius–Bode Rule

Nearly all of the asteroids are in a belt between Mars and Jupiter, and though their orbital inclinations and eccentricities are more diverse than for the planets (Table 1.3), the asteroids in the asteroid belt do, by and large, partake in the nearly circular swirl of prograde motion near to the ecliptic plane.

If we compare the semimajor axes of the planets, and include the asteroids, then something curious emerges. One way of making this comparison is shown in Figure 1.10. The planets have been numbered in order from the Sun: Mercury is numbered 1, Venus 2, Earth 3, Mars 4, the asteroids 5, Jupiter 6, and so on. The semimajor axes of the orbits have been plotted versus each planet's number. For the asteroids the dot is Ceres and the bar represents the range of semimajor axes in the main belt, a concentration within the broader asteroid belt. The curious thing is that, with a logarithmic scale on the 'vertical' axis, the data in Figure 1.10 lie close to a straight line. This means that the semimajor axes increase by about the same factor each time we go from one planet to the next one out. This is one of several ways of expressing the **Titius–Bode rule**, named after the German astronomers Johann Daniel Titius (1729–1796), who formulated a version of the rule in 1766, and Johann Elert Bode (1747–1826) who published it in 1772. Theories of the formation of the Solar System (Chapter 2) can give rise to an increase in spacing of planetary orbits as we go out from the Sun, so the Titius–Bode rule is an expression of this feature of the theories.

1.4.4 A Theory of Orbits

Kepler's laws are empirical rules that describe very well the motion of the planets around the Sun. One of the many achievements of the British scientist Isaac Newton (1642–1727) was that he was able to explain the rules in terms of two universal theories that he had developed. One

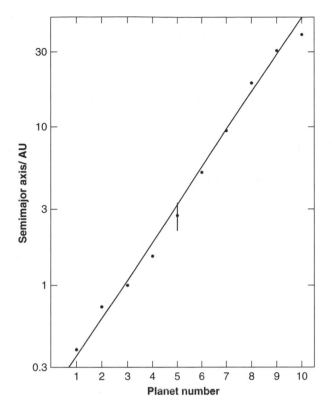

Figure 1.10 The semimajor axes of the planets versus the planets in order from the Sun: 1 = Mercury, 2 = Venus, etc., until 10 = Pluto. The vertical line at 5 is the asteroid belt.

theory is encapsulated in **Newton's laws of motion**, and the other in **Newton's law of gravity**. I state these laws here on the assumption that you have met them before, and will concentrate on using them to explore motion in the Solar System.

Newton's first law of motion   An object remains at rest or moves at constant speed in a straight line unless it is acted on by an unbalanced force. (In other words, an unbalanced force causes acceleration, i.e. either a change of speed or a change of direction, or a change of both speed and direction.)

Newton's second law of motion   If an unbalanced force of magnitude (size) F acts on a body of mass m, then the acceleration of the body has a magnitude given by

$$a = F/m \qquad (1.4)$$

and the direction of the acceleration is in the direction of the unbalanced force.

Newton's third law of motion   If body A exerts a force of size F on body B, then body B will exert a force of the same magnitude on body A but in the opposite direction.

Newton's law of gravity If two point masses M and m are separated by a distance r then there is a gravitational force of attraction between them with a magnitude given by

$$F = GMm/r^2 \tag{1.5}$$

where G is the universal gravitational constant (its value is given in Table 1.6).

A point mass has a spatial extent that is negligible compared with r. For extended bodies the net gravitational force is the sum of the gravitational forces between all the points in one body and all the points in the other.

To derive Kepler's laws from Newton's laws three conditions have to be met:

(1) The only force on a body is the gravitational force of the Sun.
(2) The Sun and the body are **spherically symmetrical**. This means that their densities vary only with radius from the centre to the (spherical) surface. In this case they interact gravitationally like point masses with all the mass of each body concentrated at its centre.
(3) The mass of the orbiting body is negligible compared with the Sun's mass.

The detailed derivation of Kepler's laws from Newton's laws can be found in books on celestial mechanics, and will not be repeated here, but we can illustrate some links between the two sets of laws.

Kepler's first and second laws

Take the first and second laws together and consider a body A in an elliptical orbit such as orbit 1 in Figure 1.11. Newton's law of gravity tells us that the Sun attracts A. Thus, from the second law of motion, A accelerates towards the Sun, its speed increasing as its distance from the Sun decreases. Because it has a component of motion other than towards the Sun, it does not fall directly towards the Sun. It therefore misses the Sun and swings through perihelion (p)

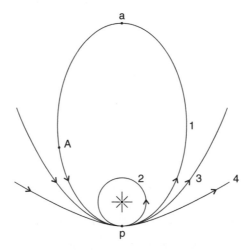

Figure 1.11 A body in a variety of orbits around the Sun.

at its maximum speed. It is then slowed down by the Sun's gravity as it climbs away from the Sun, and has its minimum speed as it passes through aphelion (a). The mathematical details show that under the three conditions the precise shape of the orbit is elliptical with the Sun at a focus (Kepler's first law) and that the increase in speed with decreasing solar distance gives the equal areas law (Kepler's second law).

Consider the body now in the circular orbit 2 in Figure 1.11. This orbit has the same perihelion distance as orbit 1, but the body is now moving more slowly at p than it was in orbit 2, and so it does not climb away from the Sun. It still accelerates towards the Sun in that its motion is always curving towards the Sun, but its overall motion is just right to keep it at the same distance from the Sun. Consequently its speed in its orbit is constant, and its acceleration is entirely in its change of direction. If a body had *no* sideways motion then it would accelerate straight into the Sun.

Parabolic and hyperbolic orbits

Now consider the body with a speed at the perihelion distance of p greater than that of the body in orbit 1 in Figure 1.11.

☐   What would be the orbit were the speed at p only slightly greater?

In this case the body would climb slightly further away at aphelion – the semimajor axis would be greater. If we increase the speed further then Newton's laws predict that we will reach a value at which the body climbs right away from the Sun, never to return. This threshold is met in orbit 3 in Figure 1.11. This is a parabolic orbit. It is not a closed curve – the two arms become parallel at infinity. Orbits with even greater perihelion speeds are even more opened out, and one example is orbit 4. These are **hyperbolic orbits**. At infinity, the two arms of a hyperbola become tangents to diverging straight lines; the greater the perihelion speed, the greater the angle between the lines. Parabolic and hyperbolic orbits are called unbound orbits, whereas an elliptical orbit is a bound orbit.

Are there any Solar System bodies in unbound orbits? Yes there are. Table 1.4 shows that the orbital eccentricities of two of the comets listed are indistinguishable from 1, a value that corresponds to a parabolic orbit. Two of those listed are in hyperbolic orbits. If a comet is in an unbound orbit then, unless its orbit is suitably modified to become bound, e.g. by a close encounter with a planet, it will leave the Solar System. Also, unless its orbit has been modified on its way inwards, it must have come from beyond the Solar System. Comets are a major topic in Chapter 3.

Kepler's third law

For Kepler's third law $(P = ka^{3/2})$ we have to consider bodies in orbits with different semimajor axes. You saw earlier that the $a^{3/2}$ dependence is the combined result of an increase in the distance around the larger orbit, and a lower orbital speed. This lower speed is explained by the decrease of gravitational force with distance (Newton's law of gravity, equation (1.5)) and the corresponding decrease in acceleration, a result derived in detail in standard texts. Such texts also show that, under the conditions 1 and 2 above, Newton's laws give

$$P = \left(\frac{4\pi^2}{G(M_\odot + m)} \right)^{1/2} a^{3/2} \tag{1.6}$$

where $M_\odot$ is the mass of the Sun and m is the mass of the other body. This is not quite Kepler's third law.

☐   What further condition is needed?

To get Kepler's third law $4\pi^2/G(M_\odot + m)$ must be a constant for the Solar System. With m being the property of the non-solar body, this condition is met if m is negligible compared with the Sun's mass. This is condition (3) above. In the Solar System Jupiter is by some way the most massive planet, but even so is only 0.1% the mass of the Sun. Therefore, condition (3) is met to a good approximation, and Kepler's third law is explained satisfactorily by Newton's laws.

Question 1.6

From the orbital data for the Earth in Table 1.1, calculate the mass of the Sun. Work in SI units, and note that 1 year $= 3.156 \times 10^7$ s. Repeat the calculation using the data for Jupiter's orbit. State any approximations you make, and whether your calculated masses seem to bear them out.

1.4.5   Orbital Complications

Conditions (1)–(3) in Section 1.4.4 are met only approximately in the Solar System, and because of this, complications arise, as follows.

The mass of the orbiting body is not *negligible compared with the Sun's mass*

Consider a single planet and the Sun, as in Figure 1.12(a). You can see that they each orbit a point on a line between them. This point is called the **centre of mass** of the system comprising the Sun and the planet. For any system of masses the centre of mass is the point that accelerates under the action of a force external to the system *as if* all the mass in the system were concentrated at that point. Thus if the external forces are negligible then the centre of mass is unaccelerated. By contrast both the Sun and the planet accelerate the whole time because of their orbital motions with respect to the centre of mass. In Figure 1.12(b) the same planet is shown in its orbit with respect to the Sun. This orbit is bigger than the two in Figure 1.12(a) but all three orbits have the same eccentricity and orbital period. Kepler's first two laws apply to the planetary orbit with respect to the Sun, as in Figure 1.12(b), and are *not* invalidated by the non-negligible planet's mass.

For two spherically symmetrical bodies, such as the Sun and planet in Figure 1.12, the centre of mass is at a position such that

$$r_\odot/r_{\rm p} = m_{\rm p}/M_\odot \tag{1.7}$$

where $r_\odot$ and $r_{\rm p}$ are the *simultaneous* distances of the Sun and planet from the centre of mass at any point in the orbits, and $m_{\rm p}$ and $M_\odot$ are the masses. Though we shall not prove this equation, it has reasonable features. For example, the greater the value of $m_{\rm p}/M_\odot$, the further the centre of mass is from the centre of the Sun. In Figure 1.12 $m_{\rm p}/M_\odot = 1/4$, corresponding to a planet far more massive than any in the Solar System.

☐   Where is the centre of mass if the mass of the planet is negligible compared with the solar mass?

It is then at the centre of the Sun.

Jupiter, the most massive planet, has a mass 0.0955% of that of the Sun. Jupiter is in an approximately circular orbit with a semimajor axis of 7.78×10^8 km, and so, from equation (1.7),

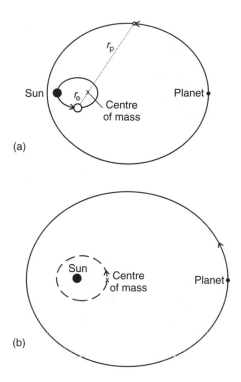

Figure 1.12 A planet in orbit around the Sun. (a) Motion with respect to the centre of mass. (b) Motion of the planet with respect to the Sun.

we can calculate that the centre of mass of the Jupiter–Sun system is 740 000 km from the Sun's centre. Thus, if Jupiter were the only planet in the Solar System the Sun's centre would move around a nearly circular orbit of radius 740 000 km – not much more than the solar radius. The effects of the other planets are to make the Sun's motion complicated, though the excursions of the Sun's centre are confined to within a radius of about 1.5×10^6 km.

The Sun and the body are not *spherically symmetrical*

Though the Sun and the planetary bodies are close to spherical symmetry, they are not perfectly so. One cause is the rotation of the body. No body is rigid and so the rotation causes the equatorial region to bulge, as in Figure 1.13(a), to give a tangerine shape. The rotational distortion of Saturn is clear in Plate 16. Another cause of departure from spherical symmetry is a gravitational force that varies in magnitude and/or direction across a body. From Newton's law of gravity (equation (1.5)) we can see that the parts of a planet closer to the Sun experience a slightly larger gravitational force than the parts further away, and so the planet stretches. An additional distortion arises from the change in direction to the Sun across the body perpendicular to the solar direction – this results in a 'squeeze'. The outcome (exaggerated) is shown in Figure 1.13(b) – a shape somewhat like a rugby ball, or an American football. The differential force (stretch and squeeze) is called a **tidal force**, and the distortion is called a tide. The Sun produces a tide in the body of the Earth, and a larger tide in the oceans. The Moon also produces tides in the Earth and actually raises greater tides than the Sun does, in spite of the Moon's far lower mass. This

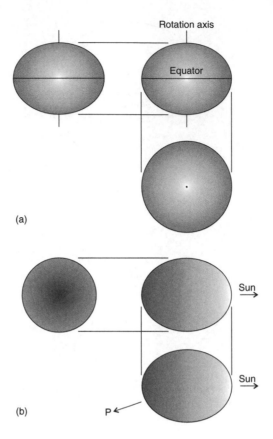

Figure 1.13 Departures from spherical symmetry in a planet due to (a) rotation and (b) the tidal force of the Sun.

is because it is so much closer than the Sun that the differential force it exerts across the Earth is greater than the differential force exerted by the Sun: the gravitational force of the Sun is almost uniform across the Earth, whereas that of the Moon is less so.

The importance of departures from spherical symmetry, however caused, is that they enable one body to exert a **torque** – a twisting – on another body. For example, a planet in Figure 1.13(b) in the direction P is slightly closer to the left end of the distorted planet than to the right end. It therefore exerts a greater overall gravitational force to the left than to the right, and so there is a torque. It can be shown that orbital changes result from such torques.

There are forces on a body additional *to the gravitational force of the Sun*

☐   List some gravitational forces on a planet other than the gravitational force of the Sun.

Most obviously there is the gravitational force exerted by the *other* planets. The planets have much smaller masses than the Sun, and are relatively well separated. Therefore, from Newton's law of gravity (equation (1.5)), it is clear that the combined gravitational force of the other planets is small, giving only slight effects on the planet's orbit. In contrast, a comet can approach a planet fairly closely, in which case the comet's orbit will be greatly modified. Planetary satellites also have an effect – it is the centre of mass of a planet–satellite system that follows an elliptical orbit around the Sun, in accord with Kepler's laws. The planet and each satellite thus follow a slightly wavy path.

As well as other *gravitational* forces there are non-gravitational forces. For example, when a comet approaches the Sun, icy materials are vaporised – it is these that give rise to the head and the tails. But they also exert forces on the comet, rather in the manner of rocket engines, and considerable orbital changes can result.

Because of additional forces and a lack of spherical symmetry the planetary orbits are therefore not quite as described by Kepler's three laws. However, the departures from the laws are usually sufficiently slight that we can regard the orbits as ellipses in which the orbital elements change, usually slowly, and often chaotically, i.e. without pattern, although the semimajor axes, eccentricities, and inclinations are usually confined to narrow ranges of values. The values given in Table 1.1 apply in 2006, but the values, almost to the precision given, will be unchanged for many decades. The values for a, e, and i in particular will not wander far from the values given, for millennia, except perhaps for the least massive planet Pluto.

The word 'usually' has been used several times in the preceding paragraph, which raises the question 'what about the exceptions?' In Section 1.4.6 we consider exceptions arising from the gravitational interaction between two bodies orbiting the Sun.

Question 1.7

Explain briefly why the orbital elements of Venus would be subject to greater variation than at present, if

(a) the Sun rotated more rapidly;
(b) the mass of Jupiter were doubled;
(c) the Sun entered a dense interstellar cloud of gas and dust.

1.4.6 Orbital Resonances

The gravitational interaction between two bodies orbiting the Sun gives rise to what are called orbital resonances. These can greatly affect the stability of an orbit. There are two types of resonance, mean motion resonances and secular resonances. Here we present a minimal account, sufficient to serve later needs.

A **mean motion resonance** (mmr) occurs when the ratio of the orbital periods P_J and P_A of bodies J and A is given by

$$\frac{P_J}{P_A} = \frac{p+q}{p} \tag{1.8}$$

where p and q are integers. Figure 1.14 illustrates the case of Jupiter J and an asteroid A when $P_J/P_A = 2$, i.e. for every one orbit of Jupiter the asteroid completes two orbits. This is called a 2:1 mmr. In Figure 1.14(a) the perihelion of the asteroid occurs when it is in line between the Sun and Jupiter (the eccentricity of Jupiter's orbit is small). Therefore, the asteroid is never very close to Jupiter, and its orbit is likely to be stable. In Figure 1.14(b) the asteroid's aphelion occurs when it is in line between the Sun and Jupiter. It therefore approaches Jupiter more closely and suffers a strong gravitational tug. Crucially, this is repeated in every Jovian orbit,

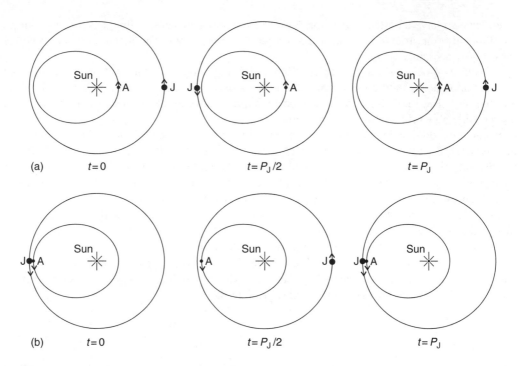

Figure 1.14 A 2:1 mean motion resonance (mmr) between Jupiter and an asteroid. (a) The perihelion of the asteroid occurs when it is in line between the Sun and Jupiter (probably stable). (b) The aphelion of the asteroid occurs when it is in line between the Sun and Jupiter (probably unstable).

so the effect of the tugs builds up, probably leading to ejection of the asteroid from its orbit. Many mmr effects are seen in the Solar System, as you will see in later chapters.

The other type of resonance is the **secular resonance**. 'Secular' in this context means a long-term interaction. Thus, rather than looking at the *instantaneous* interaction between two bodies in orbit around a star as in Figure 1.14, we consider the *averaged interaction* over a long period, In effect, it is as if each body has been smeared out along its orbit and the gravitational interaction is between these rings. There is a great variety of secular resonances.

Figure 1.15 illustrates just one type for the case of two bodies orbiting in the same plane. For the sake of clarity the orbit of each body has been replaced by its semimajor axis. Note that the interval between each configuration corresponds to many orbital periods. The gravitational interaction between the two bodies causes the semimajor axis of each of them to move around in the plane of the orbit (shown in grey). This means that the perihelion of each body also moves around – this is called **precession of the perihelion**. This is a general phenomenon when there are more than two bodies orbiting a star. But in this particular case you can see that the angle between the semimajor axes oscillates around zero, and that it never gets large. This confined difference is an example of a secular resonance. In this case it enhances the stability of the orbits. Other secular resonances lead to instability. Later chapters outline examples of secular resonances in the Solar System. Precession of the perihelion does not always correspond to a secular resonance, as you will see in the next section.

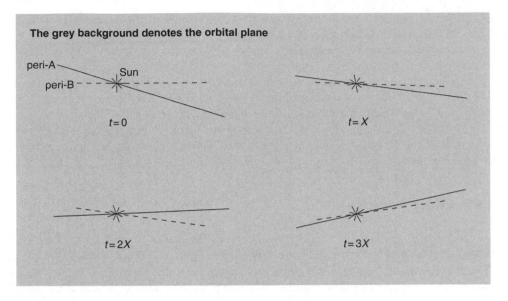

Figure 1.15 A secular resonance in which the angle between the semimajor axes never gets large.

Question 1.8

From the orbital periods of Neptune and Pluto in Table 1.1, deduce whether these two planets are in a resonance, and, if so, whether it is a secular resonance or a mean motion resonance.

1.4.7 The Orbit of Mercury

As for all the planetary orbits, the orbit of Mercury is not *quite* an ellipse fixed in space. An important departure is the precession of the perihelion that you encountered in Section 1.4.6. For Mercury it is illustrated in Figure 1.16. The actual precession (with respect to a coordinate

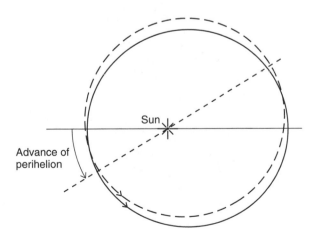

Figure 1.16 Precession of the perihelion of the orbit of Mercury. The two orbits are separated by 2000 years.

system fixed with respect to the distant stars) is through an angle of 574 arc seconds (arcsec) per century (3600 arcsec= 1°). The effect of all the other planets, and of the slight departure of the Sun from spherical symmetry, leaves a discrepancy of 43 arcsec per century. This discrepancy (at rather less precision) was a great puzzle when it was identified in the nineteenth century, and it was not accounted for until 1915 when the German–Swiss physicist Albert Einstein (1879–1955) applied his newly developed theory of **general relativity** to the problem. General relativity is *not* a modification of Newton's laws, but a very different sort of theory. Fortunately, for most purposes in the Solar System, the far simpler theory of Newton suffices. Einstein's theory accounts for the observed rate of precession of the perihelion of Mercury to within the measurement uncertainties.

1.5 Planetary Rotation

Each planet rotates around an axis that passes through its centre of mass. In the case of the Earth this rotation axis is shown in Figure 1.17. It intersects the Earth's surface at the North and South Poles, and the equator is the line half way between the Poles. You can see that the rotation axis is not perpendicular to the Earth's orbital plane (the ecliptic plane) but has an **axial inclination** of 23.4° from the perpendicular.

As the Earth moves around its orbit the rotation axis remains (very nearly) fixed with respect to the distant stars. This is shown (from an oblique viewpoint) in Figure 1.18. The axis is *not* fixed with respect to the Sun, and so the aspect varies around the orbit. At A the North Pole is maximally tilted towards the Sun. This is called the June **solstice**, and it occurs on or near 21 June each year. Six months later, at C, the North Pole is maximally tilted away from the Sun. This is the December solstice, which occurs around the 21st of the month. At B and D we have the only two moments in the year when the Earth's rotation axis is perpendicular to the line from the Earth to the Sun. Over the whole Earth, day and night are of equal length, which gives us the name for these two configurations – the **equinoxes**. The direction from the Earth

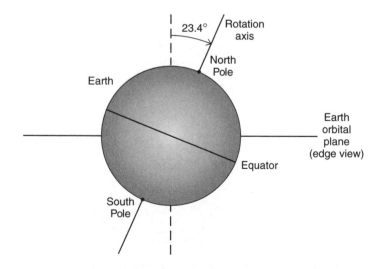

Figure 1.17   The axial inclination of the Earth.

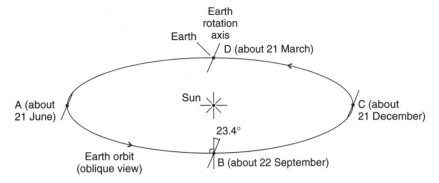

Figure 1.18 The Earth's rotation axis as the Earth orbits the Sun. This is an oblique view of the orbit, which is nearly circular.

to the Sun at the vernal (March) equinox is used as the reference direction Υ in the ecliptic plane that you met in Section 1.4.2.

We now turn to the *period* of rotation. Figure 1.19 shows the Earth moving around a segment of its orbit. As it does so it also rotates, and the arrow extending from a fixed point on the Earth's surface enables us to monitor this rotation. Between positions 1 and 2 the Earth has rotated just once with respect to a distant star. This is the **sidereal rotation period**. The distant stars, to sufficient accuracy, provide a non-rotating frame of reference (just as for the sidereal orbital period in Section 1.4.1). For the Earth, the sidereal rotation period is actually called the mean rotation period – astronomical terminology can be perverse. However, the Earth has not yet rotated once with respect to the Sun. The Earth has to rotate further to complete this rotation, and in the extra time taken it moves further around its orbit, to position 3. The period of rotation of the Earth with respect to the Sun is called the **solar day**. It is clearly longer than the mean rotation period, though only by a few minutes.

◻ State in what way the motions in Figure 1.19 are *not* shown to scale.

In Figure 1.19 the Earth's motion around its orbit between positions 1, 2, and 3 has been exaggerated for clarity. As there are just over 365 days in a year, the Earth should only proceed about 1° around its orbit in the time it takes the Earth to rotate once.

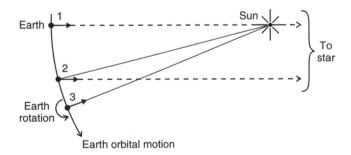

Figure 1.19 The rotation of the Earth with respect to the Sun and with respect to the distant stars (*not* to scale).

The mean rotation period does not vary significantly through the year, but the solar day does. This is a consequence of the eccentricity of the Earth's orbit and the inclination of its rotation axis (we shall not go into details). By contrast, the mean solar day is defined to be fixed in duration, and has the mean length of the solar days averaged over a year. If solar time and mean solar time coincide at some instant, they will coincide again a year later, but in between, differences develop, sometimes solar time being ahead of mean solar time and sometimes behind. The maximum differences are about 15 minutes ahead or behind. The day that we use in our everyday lives, as marked by our clocks, is the mean solar day. Even this varies in length, *very* slightly, and so for scientific purposes a standard day is defined, very nearly the same as the current length of the mean solar day. It is this standard **day** that appears in Tables 1.1–1.4 and elsewhere. It is exactly $24 \times 60 \times 60$ seconds in length, and thus consists of exactly 24 hours of 60 minutes, with each minute consisting of 60 seconds.

The mean rotation period is 23 h 56 min 4 s, i.e. 3 min 56 s shorter than the mean solar day. Over one sidereal year, this difference must add up to one extra rotation of the Earth with respect to the distant stars. You can convince yourself of this by considering a planet that is rotating as in Figure 1.20. In this case there are three rotations per orbit with respect to the Sun and four with respect to the stars. For the Earth, during the sidereal year there are 365.26 mean solar days and 366.26 mean rotation periods.

Table 1.1 gives the axial inclination and sidereal rotation period of each planet and also of the Sun. The inclination of each planet is with respect to the plane of its orbit, whereas in the case of the Sun it is with respect to the ecliptic plane. Note that, with three exceptions, the inclinations are fairly small. This means that the prograde swirl of motion of the orbits, almost in one plane, is shared by planetary and solar rotation. The exceptions are Venus, Uranus, and Pluto. The inclination of Venus is not far short of 180°.

☐   What is the difference between an axial inclination of 180° and 0°?

The difference is that 0° is prograde rotation whereas 180° is retrograde rotation, in each case with the rotation axis perpendicular to the orbital plane. Any inclination greater than 90° is retrograde, and so Pluto and Uranus are also in retrograde rotation, though Uranus's inclination

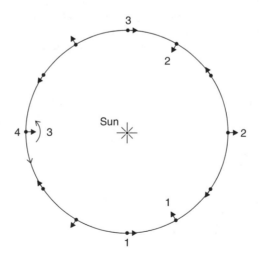

Figure 1.20   A fictitious planet rotating three times per orbit with respect to the Sun.

of 97.8° means that its rotation axis is almost in its orbital plane. We shall return to these oddities when we discuss the origin of the Solar System in Chapter 2.

As in the case of the orbital elements, the axial inclinations and rotation periods of a body are subject to changes, and for the same basic reason – the forces applied by the other bodies in the Solar System. For example, the sidereal rotation period of the Earth is currently increasing, somewhat erratically, by 1.4×10^{-3} seconds per century, largely because of the torque exerted by the Moon on the Earth's tidal distortion. The Earth has had a similar effect on the Moon, and has slowed down the Moon so that it is now locked into a rotation period that keeps it facing the Earth. When one body rotates so that it keeps one face to the body it orbits, it is said to be in **synchronous rotation**.

Seasons

Figure 1.21 is an edge view of the Earth's orbit with the positions A and C in Figure 1.18 marked, and the size of the Earth *greatly* exaggerated. When the North Pole of the Earth is maximally tilted towards the Sun, as at A, there is summer in the northern hemisphere because the surface there is receiving its greatest solar radiation. This is not only because the Sun reaches high in the sky, but also because of the long duration of daylight. By contrast, the southern hemisphere is maximally tilted *away* from the Sun.

☐ What season is this hemisphere experiencing?

It is winter in this hemisphere, because solar radiation is thinly spread over the surface and daylight is short. Six months later, at C, the December solstice, the seasons are reversed. It is thus the axial inclination that is responsible for seasonal changes. The eccentricity of the Earth's orbit has only a secondary effect. The Earth is at perihelion in early January, with the northern hemisphere in the depths of winter, and so, as a result of the orbital eccentricity, the seasonal contrasts are reduced in the northern hemisphere, and increased in the southern hemisphere.

Question 1.9

Discuss whether you would expect seasonal changes on Venus.

1.5.1 Precession of the Rotation Axis

So far, the direction of the Earth's rotation axis has been regarded as fixed with respect to the distant stars. This is not quite the case. In fact, it cones around in the manner of Figure 1.22, a

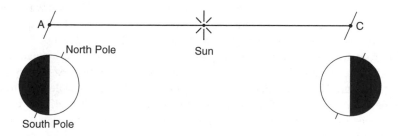

Figure 1.21 Seasonal changes in the solar radiation at the Earth's surface.

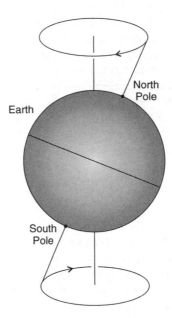

Figure 1.22 Precession of the Earth's rotation axis.

motion called the **precession of the rotation axis**. It is a result of the torques exerted by other bodies in the Solar System on the slightly non-spherical form of the Earth. The Moon and the Sun account for almost the whole effect. All planets are slightly non-spherical, so all of them are subject to precession. For the Earth, one complete coning takes 25 800 years, an interval called the precession period of the Earth.

One consequence of precession is that the positions of the equinoxes and solstices move around the orbit, giving rise to the term **precession of the equinoxes**. In the case of the Earth this motion is in a retrograde direction (taking 25 800 years to move around once). Figure 1.23 compares the present configuration (dashed lines) with the configuration 12 900 years from now (solid lines) – each equinox and solstice has moved half way around the orbit. Recall that the reference direction in the ecliptic plane is the line from the Earth to the Sun when the Earth is at the vernal equinox. Therefore, with respect to the distant stars, this reference direction has moved through 180° in Figure 1.23. At present, when the Earth is at the vernal equinox, the direction is to a point in the constellation Pisces, but about 2000 years ago, when precession became widely recognised, it was in the constellation Aries, when its location was called the first point of Aries. The name sticks, even though the point long ago moved into the constellation Pisces, and is now not far from the boundary with the constellation Aquarius.

The slow retrograde motion of the vernal equinox around the Earth's orbit means that the time taken for the Earth to traverse its orbit from one vernal equinox to the next is very slightly less than the sidereal year. The time interval between vernal equinoxes is called the tropical year, and it is the year on which our calendars are based. Its duration is 365.242 190 days, whereas the sidereal year is 365.256 363 days. From now on, the term **year** will mean the tropical year. It is this year that is the unit of time measurement in Tables 1.1, 1.3, and 1.4, and elsewhere. It is denoted by the symbol 'a', from the Latin word for year, *annus*.

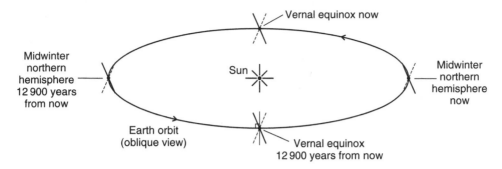

Figure 1.23 The effect of the precession of the Earth's rotation axis on the position of the equinoxes and solstices. The dashed line is the Earth's rotation axis now, and the solid line the axis 12 900 years from now.

Question 1.10

What would be the problem with basing our calendar on the *sidereal* year?

1.6 The View from the Earth

1.6.1 The Other Planets

The way that the planets appear in our skies depends on whether the orbit of the planet is larger or smaller than the orbit of the Earth. Figure 1.24 shows Venus representing the two planets (Venus and Mercury) with smaller orbits, and Mars representing those with larger orbits. The planets are shown at three instances. In position 1 all three planets are lined up with the Sun, a very rare occurrence but a useful one for describing the view from the Earth. The planets move at different rates around their orbits, so this alignment lasts only for an instant.

In position 1, Venus is between the Earth and the Sun. It is then at what is called inferior conjunction. The alignment is rarely exact, because of the inclination of the orbit of Venus. Exact or not, our view of Venus is drowned by the overwhelming light of the Sun. The greater angular speed of Venus in its orbit then causes it to draw ahead of the Earth and we start to see part of the hemisphere illuminated by the Sun as an ever-thickening crescent. At position 2, Venus has reached its greatest angle from the Sun and is at what is called its maximum western elongation. It is now relatively easy to see (before sunrise) and half of its illuminated hemisphere is visible. As it moves on we see even more of its sunlit hemisphere, but it is getting further away from the Earth, and closer in direction to the Sun, until at superior conjunction Venus is pretty well in the direction of the Sun again, but now on the far side of the Sun. Subsequently, it moves towards maximum eastern elongation, then again to inferior conjunction, and the whole cycle is repeated.

For planets beyond the Earth, such as Mars in Figure 1.24, the sequence of events is different. The line-up with Mars and Earth on the same side of the Sun does not result in an inferior conjunction, but in what is called an **opposition**, Mars being in the opposite direction in the sky from the Sun, as viewed from the Earth. Mars is then well seen, with the illuminated hemisphere facing us, and the separation between the planets being comparatively small – though this distance is different from opposition to opposition.

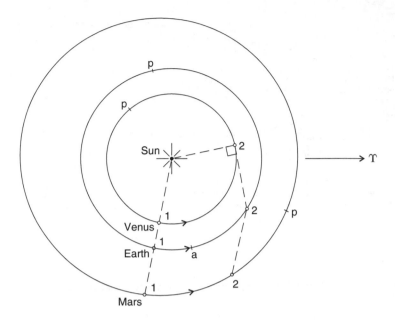

Figure 1.24 The motion of Venus and Mars with respect to the Earth. Perihelia are denoted by 'p', and Earth's aphelion by 'a'.

▢ When will the opposition distance between Mars and the Earth be a minimum?

It will be a minimum when opposition occurs with the Earth near aphelion, and Mars near perihelion. After opposition the greater angular orbital speed of the Earth causes it to overtake Mars 'on the inside track' as the configuration moves towards superior conjunction, with Mars on the far side of the Sun as seen from Earth.

The time interval between similar configurations of the Earth and another planet is called the **synodic period** of the planet. Opposition and inferior conjunction are important types of configuration. For any type of configuration the synodic period varies slightly, mainly because of the variations in the rate at which the Earth and the planet move around their respective orbits, as described by Kepler's second law. It is thus the *mean* value of the synodic period that is normally quoted, as in Table 1.1. For a particular planet the mean synodic period is the same for all types of configuration. These mean periods are not simple multiples or simple fractions of the sidereal year, and so successive configurations have the Earth at different points in its orbit.

Question 1.11

Discuss why the opposition distance to Mars is least when oppositions occur in mid August.

1.6.2 Solar and Lunar Eclipses

Figure 1.25 shows an oblique view of the nearly circular orbit of the Moon around the Earth, and part of the orbit of the Earth around the Sun (strictly, the orbit of the centre of mass of the Earth–Moon system around the Sun). The size and inclination of the lunar orbit, and the sizes of the Sun, Earth, and Moon, have all been exaggerated. When the Moon is at A (Figure 1.25(a)) its

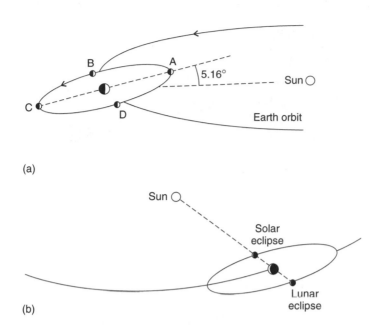

(a)

(b)

Figure 1.25 The Moon's motion around the Earth, as the Earth orbits the Sun (*not* to scale).

unilluminated hemisphere faces the Earth, and we have a new Moon. A quarter of an orbit later, at B, half of its illuminated hemisphere is facing us, and we see a half Moon, also called first quarter. At C the fully lit hemisphere faces the Earth, and the Moon is full. At D, three-quarters of the way around its orbit from A, we see another half Moon, the third quarter. We then get another new Moon at A, on average 29.53 days after the previous new Moon.

The plane of the Moon's orbit is inclined by 5.16° to the ecliptic plane, which it crosses at the two nodes labelled B and D in Figure 1.25(a). This orbital inclination means that for this configuration, the Moon, as seen from the Earth, cannot pass in front of the Sun, nor can the Moon be touched by the Earth's shadow. However, as the Earth moves around the Sun, the Moon's orbit stays (almost) fixed with respect to the distant stars, so that about a quarter of an Earth orbit later (3 months) the nodes lie on or near the line that joins the Earth and the Sun, as in Figure 1.25(b). If, at this time, the Moon is sufficiently near the node between the Earth and the Sun, then, as seen from the Earth, part or all of the Moon will pass in front of the Sun, and we get a **solar eclipse**. If the Moon is at or near the other node then the Earth's shadow will fall on part or all of the Moon, and we get a **lunar eclipse**. The nodes line up twice a year, and usually the Moon is sufficiently near a node for there to be an eclipse of some sort.

There are different types of solar eclipses. Figure 1.26(a) shows umbral and penumbral shadows of the Moon on the Earth. If we are at a point on the Earth's surface within the umbral shadow then the photosphere of the Sun is completely obscured and we see a **total solar eclipse**. With the photosphere obscured, we see the pearly white solar corona (Plate 2), the chromosphere, and prominences (Plate 3). It is worth making a considerable effort to see a total solar eclipse, which is a most magnificent spectacle. If we are in the penumbral shadow the Sun is only partly obscured and we see a partial solar eclipse. If the Moon is too far from the node, then the umbral shadow misses the Earth completely, and nowhere on Earth can we see a total solar eclipse.

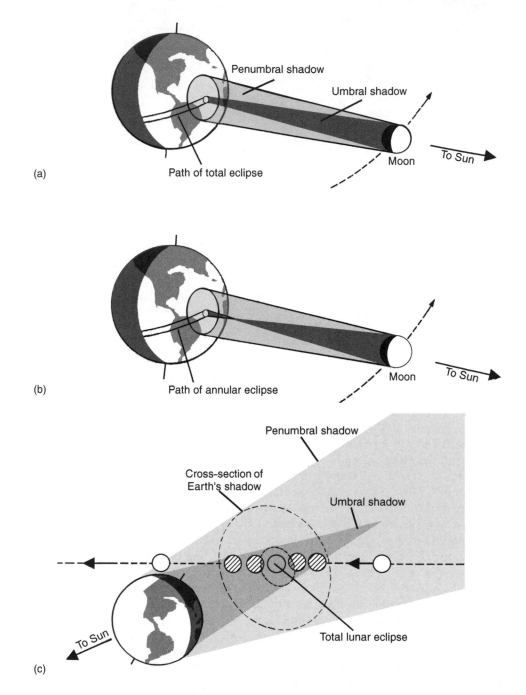

Figure 1.26 Different sorts of eclipses (*not* to scale). (a) Total and partial solar eclipses. (b) An annular solar eclipse. (c) A lunar eclipse. (From *Foundations of Astronomy* 3rd Edition, by Seeds, 1994. Reprinted with permission of Brooks/Cole, a division of Thomson Learning: www.thomsonrights.com, Fax 800 730–2215)

Even with the Moon at the node, not all solar eclipses are total. The Sun is about 400 times the diameter of the Moon, but it is also about 400 times further away, so the angular diameters of the two bodies are nearly the same, about 0.5°. Because of the eccentricity of the orbits of the Earth and the Moon, this coincidence means that sometimes an eclipse occurs with the Moon's angular diameter a bit smaller than that of the Sun, as in Figure 1.26(b). The umbral shadow does not reach the Earth, and from the Earth's surface, at the centre of the penumbral shadow, a thin ring of the photosphere is still exposed. This is called an annular solar eclipse. Largely because of tidal interactions with the Moon, the distance between the Moon and the Earth is currently increasing at a rate of about 25 mm per year, and so, from about 1000 Ma in the future, the Moon will never be close enough to the Earth to produce a total solar eclipse, and all solar eclipses will then be partial or annular.

Figure 1.27 shows the umbral shadow paths for the years 2001–2025. Within these narrow paths a total solar eclipse occurs. The paths are determined by the line-up of the Earth, Moon, and Sun, and by the combined effect of the orbital motion of the Moon and the rotation of the Earth, which together sweep the umbral shadow across the Earth. The duration of totality is longest at the centre of the path, and varies from eclipse to eclipse. The longest durations, approaching 7.5 minutes, occur when the Earth is at aphelion, and the Moon is closest to the Earth, at what is called perigee.

Figure 1.26(c) shows a total lunar eclipse, which occurs when the Moon is entirely within the umbral shadow of the Earth. Where the umbral shadow falls on the Moon, the lunar surface

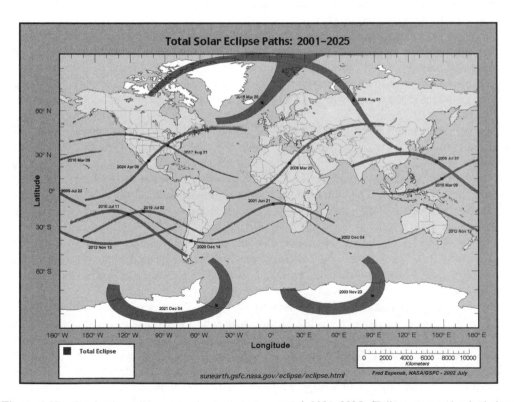

Figure 1.27 Total solar eclipse paths (umbral shadow paths) 2001–2025. (Eclipse map and calculations courtesy Fred Espenak, NASA-GSFC)

is not completely dark. Sunlight is refracted by the Earth's atmosphere, red light more than the other visible wavelengths, which can give the Moon a coppery tint. At any moment the eclipsed Moon can be seen from half the Earth's surface. The view *from* the Moon would be of the black, night side of the Earth, surrounded by a thin red ring of sunlight refracted by the Earth's atmosphere.

Question 1.12

The nodes of the lunar orbit are not quite fixed but move around the lunar orbit in the retrograde direction in 18.6 years. How does this explain why eclipses are not confined to particular months?

1.7 Summary of Chapter 1

The Solar System consists of the Sun, nine planets with their satellites and rings, many asteroids (about 10^9 greater than 1 km across), the Edgeworth–Kuiper belt (more than 10^5 objects larger than 100 km), 10^{12}–10^{13} small icy-rocky bodies in the Oort cloud, and an interplanetary medium of tenuous gas and small solid bodies ranging in size down to less than 10^{-6} m.

Meteoroids are small rocky bodies, and those that fall to Earth are called meteorites, These have provided much information about the origin, evolution, and composition of the Solar System and the ages of events within it.

The planets orbit the Sun in one direction – the prograde direction – in approximately circular, coplanar orbits with the Sun near the centre. The orbital planes of the asteroids have a wider range of inclinations and eccentricities. The rotation of the Sun is prograde, as is that of most of the planets. If the inclination of the rotation axis to the orbital plane is more than a few degrees then the surface of the planet will experience seasonal changes. Rotation axes are subject to precession.

Some comet orbits reach to within a few AU of the Sun, but the great majority spend most or all of their time at far greater distances, where they are dormant icy–rocky bodies. There are two reservoirs. The Edgeworth–Kuiper belt lies immediately beyond the planetary domain, and contains bodies (EKOs) in orbits that are predominantly prograde and that are concentrated towards the ecliptic plane. The Oort cloud is more far flung and consists of bodies in a spherical distribution around the Sun, reaching out to the edge of interstellar space, about 10^5 AU from the Sun.

The Sun is by far the largest, the most massive, and the most luminous body in the Solar System. It is fluid throughout, and consists largely of hydrogen and helium. Its luminosity is sustained by the nuclear fusion of hydrogen deep in its interior where temperatures reach 1.4×10^7 K.

The four planets closest to the Sun – Mercury, Venus, the Earth, and Mars – are the terrestrial planets. They are comparable with the Earth in size, and consist of iron-rich cores overlain by rocky materials. The Earth is the largest of these bodies. The asteroids are rocky bodies concentrated between Mars and Jupiter.

The giant planets – Jupiter, Saturn, Uranus, and Neptune – are considerably larger and more massive than the terrestrial planets, Jupiter by some margin being the most massive planet of all. The giants consist largely of hydrogen, helium, and icy–rocky materials, and (like the Sun) are fluid throughout. The giants have richly varied families of satellites, and all giants have rings, those of Saturn being by far the most substantial. Beyond Neptune we come to the outermost planet, Pluto, smaller in size than the terrestrial planets, and more icy in its composition. The comets are also icy–rocky bodies. Beyond Pluto there is at least one body somewhat larger than Pluto – Eris. Pluto and Eris are regarded as large members of the Edgeworth–Kuiper belt.

The orbits of the planets are described to a very good approximation by Kepler's laws of planetary motion.

First law Each planet moves around the Sun in an ellipse, with the Sun at one focus of the ellipse.

Second law As the planet moves around its orbit, the straight line from the Sun to the planet sweeps out equal areas in equal intervals of time.

Third law If P is the sidereal period of a planet, and a is the semimajor axis of the orbit, then

$$P = ka^{3/2} \tag{1.3}$$

where $k = 1$ year $AU^{-3/2}$.

Elliptical orbits are characterised by five orbital elements: the semimajor axis a, the eccentricity e, the inclination i, the longitude of the ascending node Ω, and the longitude of perihelion $(\Omega + \omega)$. To calculate the position of a body in its orbit we need a sixth element – a single position at any known time.

Newton's laws of motion and law of gravity account for Kepler's laws, and go further by accounting for the motion of comets and of other bodies, and for slight departures from Kepler's laws that have various causes. Major effects on orbits are caused by mean motion resonances and secular resonances.

The precession of the perihelion of Mercury shows that at the highest level of precision Einstein's theory of general relativity is superior to Newton's laws.

Our view from the Earth of the apparent motion of a planet depends on whether it is in a smaller or larger orbit than our own. Solar and lunar eclipses result when the Moon, Sun, and Earth line up. Figure 1.27 shows the umbral tracks of forthcoming total solar eclipses.

Tables 1.1–1.6 list basic data on the Solar System.

Table 1.1 Orbital elements in 2006 and some physical properties of the Sun, the planets, and Ceres

| Object | Mean orbital elements^{a,b} | | | | | | From the Earth | | | Physical properties | | | |
	Semimajor axis/ 10^6 km		Sidereal period/ years	Eccentricity	Inclination/ °	Long. of ascending node/ °	Long. of perihelion/ °	Mean syn. period/ days	Date, oppn, or inf. conj.	Axial inclin- ation/ $^{\circ c}$	Sid. rotn period/ days^d	Equat. radius/ km^e	Mass/ kg	Mean density/ kgm^{-3}
	AU													
Sun	—	—	—	—	—	—	—	—	—	7.2	25.4	696 000	1.9891×10^{30}	1410
Mercury	0.387	57.9	0.2409	0.206	7.00	48.3	77.5	116	08 Nov 06	0.0	58.646	2440	3.302×10^{23}	5430
Venus	0.723	108.2	0.6152	0.0068	3.39	76.7	131.2	584	20 Aug 07	177.4	243.019	6052	4.869×10^{24}	5240
Earth	1.000	149.6	1.0000	0.017	0	—	103.0	—	—	23.4	0.9973	6378	5.974×10^{24}	5520
Mars	1.524	227.9	1.8808	0.093	1.85	49.5	336.1	780	24 Dec 07	25.2	1.0260	3396	6.419×10^{23}	3940
Ceres	2.766	413.8	4.5993	0.080	10.59	80.4	153.6	466	12 Sept 06	54	0.378	479	9.4×10^{20}	2100
Jupiter	5.202	778.2	11.859	0.049	1.31	100.5	14.7	399	4 May 06	3.1	0.4135	71490	1.8988×10^{27}	1330
Saturn	9.552	1430.0	29.521	0.055	2.49	113.6	93.8	378	27 Jan 06	26.7	0.4440	60270	5.684×10^{26}	690
Uranus	19.173	2868.3	83.957	0.048	0.77	74.0	172.4	370	5 Sept 06	97.8	0.7183	25560	8.684×10^{25}	1270
Neptune	30.091	4501.6	165.07	0.0069	1.77	131.8	48.0	367	11 Aug 06	28.3	0.6712	24765	1.0245×10^{26}	1640
Pluto	39.777	5950.6	250.88	0.254	17.1	110.3	224.5	367	16 June 06	123	6.3872	1153	1.30×10^{22}	2030

^{a} For all bodies except Ceres, the orbital data are for August 2006. For Ceres, the data are for September 2006.

^{b} Another orbital element is the time of any perihelion passage. For the Earth in 2006 this was at 15:00 UT on 4 January.

^{c} The Sun's axial inclination is with respect to the ecliptic plane.

^{d} The rotation period of the Sun is at the equator: it increases with latitude, reaching about 36 days near the poles of the Sun. For Jupiter and Saturn it is for the deep interior.

^{e} For the giant planets the equatorial radius is to the atmospheric altitude where the pressure is 10^5 Pa (1 bar).

Table 1.2 Some properties of planetary satellites^{a,b}

Object^c	Some mean orbital elements				Size, mass, mean density		
	Semimajor axis/ 10^3 km	Sidereal period/days	Eccentricity	Inclination/ $^{\circ d}$	Radius/km^e	Mass/10^{18} kg	Mean density/kg m^{-3}
Earth							
Moon	384.4	27.322	0.0554	5.16	1738	73 490	3 340
Mars							
Phobos	9.38	0.319	0.0151	1.1	11	0.0107	1870
Deimos	23.46	1.262	0.0002	1.8 var	6	0.0022	2300
Jupiter							
Metis	128.0	0.295	0.0012	0.02	22	?	?
Adrastea	129.0	0.298	0.0018	0.39	8	?	?
Amalthea	181.4	0.498	0.0031	0.39	84	2.08	860
Thebe	221.9	0.675	0.0177	1.07	50	?	?
Io	421.8	1.769	0.0041	0.04	1822	89 330	3530
Europa	671.1	3.551	0.0094	0.47	1561	48 000	3010
Ganymede	1 070	7.155	0.0011	0.17	2631	148 200	1940
Callisto	1 883	16.689	0.0074	0.19	2410	107 600	1830
Leda	11 165	241	0.16	27	~10	?	?
Himalia	11 461	251	0.16	27	93	?	?
Lysithea	11 717	259	0.11	28	~18	?	?
Elara	11 741	260	0.22	26	43	?	?
Ananke	21 276	630	0.24	149	~15	?	?
Carme	23 404	734	0.25	165	~23	?	?
Pasiphae	23 624	744	0.41	151	~30	?	?
Sinope	23 939	759	0.25	158	~19	?	?
Saturn							
Pan	133.6	0.575	0.0002	0.007	13	?	?
Atlas	137.7	0.602	0.0012	0.01	10	?	?
Prometheus	139.4	0.613	0.002	0.006	50	?	?
Pandora	141.7	0.629	0.004	0.052	43	?	?
Epimetheus	151.4	0.694	0.010	0.35	60	0.5	610
Janus	151.5	0.695	0.007	0.17	90	1.9	660
Mimas	185.6	0.942	0.0206	1.57	199	38	1160
Enceladus	238.1	1.370	0.0001	0.01	253	84	1120
Tethys	294.7	1.888	0.0001	0.168	530	627	960
Telesto	294.7	1.888	~0	1.16	13	?	?
Calypso	294.7	1.888	~0	1.47	10	?	?
Dione	377.4	2.737	0.0002	0.002	560	1097	1480
Helene	377.4	2.737	~0	0.21	15	?	?
Rhea	527.1	4.518	0.0009	0.327	765	2 308	1230
Titan	1 221.9	15.945	0.0288	1.634	2575	134 570	1880
Hyperion	1 464.1	21.276	0.0175	0.57	142	?	?
Iapetus	3 560.8	79.331	0.028	14.7	718	1 590	1090
Kiviuq	11 365	449	0.33	46	7	?	?
Ijiraq	11 442	451	0.32	47	5	?	?
Phoebe	12 944	548	0.1644	175	~110	?	?
Paaliaq	15 198	687	0.36	45	10	?	?
Albiorix	16 394	783	0.48	34	13	?	?
Siarnaq	18 195	896	0.30	46	16	?	?
Tarvos	18 239	926	0.54	33	7	?	?

(continued)

Table 1.2 (Continued)

Object^c	Some mean orbital elements				Size, mass, mean density		
	Semimajor axis/ 10^3 km	Sidereal period/days	Eccentricity	Inclination/ ^{od}	Radius/km^e	Mass/10^{18} kg	Mean density/kg m^{-3}
Uranus							
Cordelia	49.8	0.335	~ 0	0.08	13	?	?
Ophelia	53.8	0.376	0.01	0.10	15	?	?
Bianca	59.2	0.435	0.001	0.19	23	?	?
Cressida	61.8	0.464	~ 0	0.01	33	?	?
Desdemona	62.7	0.474	~ 0	0.11	30	?	?
Juliet	64.4	0.493	0.001	0.06	43	?	?
Portia	66.1	0.513	~ 0	0.06	55	?	?
Rosalind	69.9	0.558	~ 0	0.28	30	?	?
2003 U2	74.8	0.618	~ 0	~ 0	6	?	?
Belinda	75.3	0.624	~ 0	0.03	34	?	?
1986 U10	76.4	0.638	~ 0	~ 0	~ 20		
Puck	86.0	0.762	~ 0	0.32	78	?	?
Mab	97.7	0.923	~ 0	~ 0	8	?	?
Miranda	129.9	1.413	0.0013	4.34	236	66	1200
Ariel	190.9	2.520	0.0012	0.04	579	1 350	1700
Umbriel	266.0	4.146	0.0035	0.0	585	1 170	1400
Titania	436.3	8.704	0.0024	0.0	789	3 520	1700
Oberon	583.4	13.463	0.0007	0.0	762	3 010	1600
Caliban	7 231	580	0.16	141	30	?	?
Sycorax	12 179	1288	0.32	159	60	?	?
Neptune							
Naiad	48.2	0.294	~ 0	4.74	30	?	?
Thalassa	50.1	0.311	~ 0	0.21	40	?	?
Despina	52.5	0.335	~ 0	0.07	75	?	?
Galatea	62.0	0.429	~ 0	0.05	80	?	?
Larissa	73.5	0.555	0.001	0.20	95	?	?
Proteus	117.6	1.122	~ 0	0.039	210	?	?
Triton	354.8	5.877	~ 0	157	1353	21 400	2060
Nereid	5 513	360	0.75	7.23	170	?	?
2002 N1	15 686	1875	0.57	134	25	?	?
Pluto							
Charon	19.57	6.387	0.000	96.15	603	1 518	1660
Nix	48.68	24.856	~ 0.002	96.18	tiny	?	?
Hydra	64.78	38.207	0.0052	96.36	tiny	?	?

^{a} These data are as published in 2006.

^{b} The inclinations of the rotation axes of the satellites and the rotation periods are not given, but in most cases the inclinations are small. Many of the satellites, like the Moon, are in synchronous rotation around their planet.

^{c} Very small satellites of the giant planets are not included. The excluded satellites are: Jupiter and Saturn, all are < 5 km mean radius; Uranus, all those beyond Oberon smaller than Caliban; Neptune, all those beyond Nereid smaller than 2002 N1.

^{d} Note that in most cases the orbital inclination is with respect to the *equatorial* plane of the planet. The exceptions are the Moon and the outer satellites of the giant planets: Jupiter, beyond Callisto; Saturn, beyond Iapetus; Uranus, beyond Oberon; Neptune, beyond Triton. In these cases the inclination is with respect to the *orbital* plane of the planet. This is because the inclination with respect to the equatorial plane changes periodically through a fairly large range of values. Inclinations greater than 90° indicate retrograde orbital motion, i.e. opposite to the direction of rotation of the planet.

^{e} Values less than a few hundred km are average radii of irregularly shaped bodies. For many of these satellites the size is based on an assumed albedo of 0.04.

Table 1.3 Some properties of the largest 15 asteroids

Object/number and name	Some orbital elements[a]					Rotation and size	
	Semimajor axis/ AU	10^6 km	Sidereal period/years	Eccentricity	Inclination/ °	Sid. rotn period/h	Radius/km[b]
1 Ceres	2.766	413.8	4.599	0.080	10.59	9.07	479
2 Pallas	2.772	414.7	4.615	0.231	34.84	7.81	262
4 Vesta	2.361	353.2	3.629	0.089	7.13	5.34	256
10 Hygiea	3.137	469.3	5.555	0.118	3.84	27.62	222
704 Interamnia	3.061	457.9	5.357	0.150	17.29	8.69	165
511 Davida	3.166	473.6	5.633	0.186	15.94	5.13	163
15 Eunomia	2.643	395.4	4.298	0.187	11.74	6.08	160
52 Europa	3.102	464.1	5.464	0.103	7.47	5.63	151
3 Juno	2.668	399.1	4.357	0.258	12.97	7.21	137
87 Sylvia	3.489	522.0	6.519	0.080	10.86	5.18	131
31 Euphrosyne	3.150	471.2	5.591	0.226	26.32	5.53	128
16 Psyche	2.920	436.8	4.989	0.139	3.10	4.20	120
88 Thisbe	2.768	414.1	4.605	0.165	5.22	6.04	116
65 Cybele	3.433	513.6	6.362	0.105	3.55	6.1	115
324 Bamberga	2.682	401.2	4.394	0.338	11.11	29.41	114

[a] In September 2006.
[b] The asteroids are in order of decreasing size. Values less than a few hundred km are average radii of irregularly shaped bodies.

Table 1.4 Some properties of selected comets

| Name | Some orbital properties | | | | | | Associated |
	Semimajor axis/ AU	Sidereal period/ years	Eccentricity	Inclination/ °	Perihelion distance/ AU	Date, last perihelion passage	meteor shower(s)
Short period							
21P/Giacobini–Zinner	3.53	6.62	0.7057	31.81	1.0337	July 2005	Giacobinids	Oct
3D/Biela[a]	3.53	6.62	0.756	12.55	0.861	1852 (lost)	Andromedids	Nov
1P/Halley	17.94	75.98	0.9673	162.24	0.5871	Feb 1986	Eta Aquarids	May
109P/Swift–Tuttle	26.32	135.01	0.9636	113.43	0.9582	Dec 1992	Perseids	July–Aug
2P/Encke	2.22	3.30	0.8473	11.77	0.3385	Dec 2003	Taurids	Oct–Nov
36P/Whipple	4.17	8.51	0.2590	9.93	3.0882	July 2003	—	
29P/Schwassmann–Wachmann 1	5.99	14.65	0.0442	9.39	5.7236	July 2004	—	
39P/Oterma	7.24	19.49	0.2446	1.94	5.4707	Dec 2002	—	
23P/Brorsen–Metcalfe	17.07	70.53	0.9720	19.33	0.4789	Sept 1989	—	
Long period (bright)							
C/1843 D1 Great Comet of 1843	640	16 000	0.99 99 914	114.35	0.005 527	1843	—	
C/1858 L1 Donati	~150	~1700	0.99 6	116.96	0.578	1858	—	
C/1956 R1 Arend–Roland	Large	Long	1.00 0	119.95	0.316	1957	—	
C/1957 P1 Mrkos	Large	Long	0.99 9	93.94	0.355	1957	—	
C/1962 C1 Seki–Lines	Large	Long	1.00 0	65.01	0.031 397	~1962/3	—	
C/1965 S1 Ikeya–Seki[b]	92	880	0.999 915	141.86	0.007 786	1965	—	
C/1969 Y1 Bennett	141.9	1690	0.996 210 6	90.04	0.537 606 3	1971	—	
C/1973 E1 Kohoutek	—	—	1.000 007 9	14.30	0.142	1973	—	
C/1975 V1 West[c]	—	—	> 1.000	43.07	0.197	1976	—	
C/1996 B2 Hyakutake[d]	923.8	2808	0.999 750 8	124.92	0.230 220 7	1996	—	
C/1995 O1 Hale–Bopp[d]	185.3	2521	0.995 065 4	89.43	0.914 141 0	1997	—	

[a] Split into two in 1846, and not seen since 1852.
[b] Split into three. Orbital elements are for one of the two larger fragments. The other two fragments had similar elements.
[c] Split into four.
[d] The orbital elements apply after the last perihelion passage.

Table 1.5 Relative abundances of the 15 most abundant chemical elements in the Solar System

Chemical element			Relative atomic mass $(^{12}C \equiv 12)$	Relative abundance[a]	
Atomic number	Name	Symbol		By number of atoms	By mass
1	Hydrogen	H	1.0080	1 000 000	1 000 000
2	Helium[b]	He	4.0026	97 700	388 000
6	Carbon	C	12.0111	331	3 950
7	Nitrogen	N	14.0067	83.2	1 160
8	Oxygen	O	15.9994	676	10 730
10	Neon	Ne	20.179	120	2 410
11	Sodium	Na	22.9898	2.09	48
12	Magnesium	Mg	24.305	38.0	917
13	Aluminium	Al	26.9815	3.09	83
14	Silicon	Si	28.086	36.3	1 010
16	Sulphur	S	32.06	15.9	504
18	Argon	Ar	39.948	2.51	100
20	Calcium	Ca	40.08	2.24	89
26	Iron	Fe	55.847	31.6	1 750
28	Nickel	Ni	58.71	1.78	104

[a] Abundances are given to 3–4 significant figures. Many are known to better than this.
[b] The helium values correspond to those before the conversion of some of the hydrogen in the Sun's core to helium, i.e. to the Sun at its formation.

Table 1.6 Some important constants

Name	Symbol	Value
Speed of light (in a vacuum)[a]	c	$2.99792458 \times 10^8 \, \mathrm{m\,s^{-1}}$
Gravitational constant[b]	G	$6.672 \times 10^{-11} \, \mathrm{N\,m^2\,kg^{-2}}$
Boltzmann's constant	k	$1.38065 \times 10^{-23} \, \mathrm{JK^{-1}}$
Planck's constant	h	$6.62607 \times 10^{-34} \, \mathrm{Js}$
Stefan's constant	σ	$5.6704 \times 10^{-8} \, \mathrm{Wm^{-2}\,K^{-4}}$
Astronomical unit	AU	$1.4959787069 \times 10^{11} \, \mathrm{m}$
Light year[c]	ly	$9.460536 \times 10^{15} \, \mathrm{m}$
Parsec	pc	$3.085678 \times 10^{16} \, \mathrm{m}$
Solar luminosity	$L_{\odot}$	$3.85 \times 10^{26} \, \mathrm{W}$
Day[d]	d	$86\,400 \, \mathrm{s}$ exactly
Tropical year	a	$365.242190 \, \mathrm{d}$
Pi	π	$3.14159\ldots$

[a] This is an exact value. The second (s) is now defined in terms of atomic vibrations, and the metre (m) as the distance travelled by light in a vacuum in $1/(2.99792458 \times 10^8)$ s.
[b] The kilogram (kg) is still defined as the mass of a metal cylinder at the International Bureau of Weights and Measures, Sèvres, France.
[c] This is the distance travelled by light in a vacuum in 1 year of 365.2425 days.
[d] The mean solar day is presently (2006) 86 400.0004 s.

2 The Origin of the Solar System

In Chapter 1 you met many of the broad features of the Solar System. It is these broad features that any theory of the origin must explain, and this chapter presents the type of theory that is very widely accepted. This is the solar nebular theory, in which the planets form from a disc of gas and dust around the Sun. Such a type of theory also accounts for many of the *details* of the Solar System, as you will see in subsequent chapters.

You might think that we could *deduce* the origin of the Solar System by working back from the state in which we observe the Solar System to be today. This cannot be done, for several reasons. First, our knowledge of the present state of the Solar System is incomplete. Second, there are areas of ignorance about the way the Solar System has interacted with its interstellar environment. Third, our understanding of the fundamental physical and chemical processes that operate on all matter, though extensive and deep, is incomplete. Fourth, and most profoundly, even if these three areas of ignorance were eliminated, it would still not be possible to 'reverse time' and deduce the origin. This is because an infinitesimal adjustment in the present state of the Solar System would lead to a very different journey into the past: it is not possible to have sufficiently accurate knowledge to deduce the origin. This is an example of the scientific phenomenon of **chaos**, and it is a barrier in principle, not just a barrier in practice.

Astronomers must therefore construct theories as best they can, guided by the broad features of the Solar System and by our knowledge of the rapidly growing number of other planetary systems – the **exoplanetary systems**. Observations of star formation and of young stars are also important, because these increase our understanding of the formation of the Sun, an event that was surely intimately involved in the formation of the rest of the Solar System.

2.1 The Observational Basis

2.1.1 The Solar System

Table 2.1 lists some of the broad features of the Solar System, most of which you met in Chapter 1. Any theory worthy of serious consideration really has to be able to account for most of these features, and for some others too. But it does not necessarily have to be able to account for them all. If there *are* any features that a theory cannot account for, this would not necessarily rule the theory out. For example, it might be that the theory has not yet been worked out in sufficient detail, perhaps because a physical process is insufficiently well understood, or because we do not know enough about the state of the substances from which the Solar System formed. It is however, fatal for a theory if it unavoidably produces features that are clearly unlike those

Discovering the Solar System, Second Edition Barrie W. Jones
© 2007 John Wiley & Sons, Ltd

Table 2.1 Some broad features of the Solar System today

1	The Sun consists almost entirely of hydrogen and helium
2	The orbits of the planets lie in almost the same plane, and the Sun lies near the centre of this plane
3	The planets all move around the Sun in the same direction that the Sun rotates (called the prograde direction)
4	The rotation axis of the Sun has a small but significant inclination, 7.2° with respect to the ecliptic plane (the Earth's orbital plane)
5	Whereas the Sun has 99.8% of the mass of the Solar System, it has only about 0.5% of its total angular momentum
6	The axial rotations of six of the nine major planets are prograde with small or modest axial inclinations. The rotations of Venus, Uranus, and Pluto are retrograde
7	The inner planets are of low mass and consist of rocky materials, including iron or iron-rich compounds; the closer to the Sun, the more refractory the composition
8	The giant planets lie beyond the inner planets, are of high mass, and are dominated by hydrogen, helium, and icy materials, with a decreasing mass and hydrogen–helium content from Jupiter to Saturn to Uranus/Neptune
9	The asteroids are numerous small rocky bodies concentrated between Mars and Jupiter
10	There are even more numerous small icy–rocky bodies concentrated beyond Neptune in two populations, the Edgeworth–Kuiper belt and the Oort cloud. These give rise to the comets
11	The giant planets have large families of satellites that are rocky or icy–rocky bodies

observed. For example, if a theory predicts that roughly half the planets should be in retrograde orbits then we can rule the theory out.

☐ What about a theory that predicts that there are no giant planets?

We can rule this out too!

2.1.2 Exoplanetary Systems

The number of exoplanetary systems presently known (13 January 2007) is 177, 20 with two or more planets, giving 205 exoplanets in total. Already, they have supplied valuable insights into the origin and evolution of the Solar System. Direct detection of exoplanets is at the limit of present instrumental capabilities, because a planet is a very faint object with a very small angular separation from a far brighter object – its star – and the planet's light therefore cannot be seen. Therefore, up to now detection has been almost entirely indirect.

Indirect detection techniques

Most of the exoplanets have been discovered through the motion they induce in the star they orbit. In our Solar System the planets cause the Sun to follow a small (complicated) orbit around the centre of mass of the system (Section 1.4.5). Therefore, if small orbital motion of other stars can be detected we can infer the presence of one or more planets even if they are too faint to be seen. One way is to measure repeatedly the position of the star with respect to much more distant stars. This is called the **astrometric technique**. An outcome is shown in Figure 2.1(a), where, for simplicity, it has been assumed that the centre of mass is fixed against the more distant stellar background. In reality, the motion of the centre of mass would add to that in Figure 2.1(a) to give a wavy stellar path. A second technique is possible if the angle i in Figure 2.1(b) is greater than zero. In this case the orbit is *not* presented face on to us, and therefore as the star moves around its orbit its speed along the direction to the Earth varies,

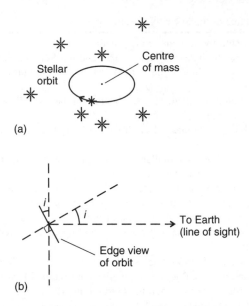

Figure 2.1 (a) The orbit of a star due to a single planet in orbit around it, in the simple case when the centre of mass of the system is stationary with respect to the distant stars. (b) The angle of inclination i of the normal to the plane of the stellar orbit with respect to our line of sight.

i.e. its line-of-sight speed varies. These speed variations cause variations in the wavelengths of the spectral lines of the star. This is due to the **Doppler effect**, whereby the observed wavelength depends on the speed of the radiation source with respect to the observer (Christian Johann Doppler, Austrian physicist, 1803–1853). The technique based on this effect is called the **radial velocity technique**. It has discovered the great majority of exoplanets to date.

From either technique we can obtain the mass of the planet(s) and some of the orbital elements. The details will not concern us except to note that whereas the astrometric technique gives the mass of the planet m_p, the radial velocity technique gives $m_p \times \sin(i)$. This is because we detect the component of the star's orbital velocity towards us and not the total orbital velocity. Thus, if i is unknown we obtain only a lower limit on the mass of the planet corresponding to $i = 90°$, as if we had an edge-on view of the orbit.

There are some other techniques for indirect detection of exoplanets, and though so far they have delivered a very small yield, this will rise, particularly in the case of the **transit technique**. This relies on the slight diminution of the light we receive from a star, if one of its planets passes between us and the star.

☐   If, from a distant observer's vantage point, Jupiter were to transit the face of the Sun, what decrease would it cause in the light received?

Jupiter's radius is about a tenth that of the Sun's, and so Jupiter would appear as a disc with an area about one-hundredth that of the solar disc. Therefore, the decrease would be 1%. For a transit to occur the orbit of an exoplanet must be presented edge on to us, or nearly so.

A much fuller account of the techniques for finding exoplanets can be found in books in Further Reading. We now consider what the exoplanets teach us about the origin and evolution of the Solar System.

Some characteristics of the known exoplanetary systems

The first exoplanets were discovered in 1992 in orbit around a pulsar. A pulsar is the remnant of a star that has suffered a catastrophic explosion – a supernova explosion. Such an explosion would surely have destroyed any planetary system, and so the planets are presumed to have formed subsequently. But interesting though these pulsar planets are, pulsars are rare objects, quite unlike the Sun. It was in 1995 that the first exoplanet was detected in orbit around a star other than a pulsar – the star was 51 Pegasi. Table 2.2 summarises the characteristics of the known exoplanetary systems that are of particular relevance to this chapter – the small number of pulsar planets is excluded.

Table 2.2 shows that, at present, the lowest known exoplanetary mass is 0.017 times the mass of Jupiter, or 5.4 times the mass of the Earth (this happens to be its actual mass, not the minimum mass). However, most exoplanets have (minimum) masses between a tenth and 10 times that of Jupiter. Table 2.2 also shows that the minimum semimajor axis is only $a = 0.0177$ AU, much less than Mercury's 0.387 AU. Indeed, nearly half of the exoplanets have $a < 0.387$ AU. Many of these have masses between 0.5 and 1.5 Jupiter masses, and are called 'hot Jupiters'.

The distance range in Table 2.2 needs to be put into perspective. Our Galaxy is about 100 000 light years across its disc, and contains roughly 2×10^{11} stars (**light year**, ly – see Table 1.6). The 300 ly in Table 2.2, compared with the size of our Galaxy, thus puts most exoplanetary systems in our cosmic backyard. This is because the closer a star, the brighter it appears and the easier it is to make observations. This is an example of an observational selection effect.

The stars in the great majority of the exoplanetary systems are main sequence stars not very different in mass from the Sun. Such stars have been the star of choice for observers, mainly because they have many narrow spectral lines suitable for the radial velocity technique, and because they are much brighter than low-mass main sequence stars. Higher mass main sequence stars are even brighter, but are rare and have short lives.

Only a few of the known exoplanetary systems are like the Solar System, with the giant planets several AU from the star. Is therefore the Solar System a rare type of planetary system? Not necessarily. This is because the easiest planets to detect with the radial velocity technique are those that induce the greatest orbital speed of the star, and these are massive planets close to the star – another example of an observational selection effect. Moreover, the orbital period increases with semimajor axis, and therefore data have to be accumulated for longer times to discover planets further out. In the case of Jupiter, with an orbital period of 11.86 years, an

Table 2.2 Some characteristics of the known exoplanetary systems^a

Characteristic	Data	Comment
Stellar mass	$0.34–1.5 M_\odot$	A substantial majority are main sequence stars
Stellar distance	10.5 ly and up	Very few are beyond 300 light years
Planet mass^b	$0.017–13^c m_J$	Most are in the range $0.1–10\, m_J$
Planet semimajor axis	0.0177–7.73 AU	The second largest value is 5.257 AU
Planet orbital eccentricity	0–0.92	Most hot Jupiters have values less than 0.1

^{a} At 13 January 2007: 205 planets in 177 exoplanetary systems, 20 systems with 2 or more planets; planets around pulsars are excluded.
^{b} These are minimum masses for those discovered by the radial velocity technique, in terms of Jupiter's mass m_J.
^{c} Above about $13 m_J$ the object is a brown dwarf, a 'failed star' not massive enough to attain central temperatures sufficiently high for hydrogen (^{1}H) fusion, but only a brief phase of fusion of the rare isotope ^{2}H (deuterium).

alien astronomer would have to observe the Sun for at least this time to discern the motion that Jupiter induces in it. Also, the probability of an edge-on view to give a transit decreases with the size of the orbit. It is therefore quite possible that planetary systems like ours are more common than in the presently known population of exoplanetary systems. Rather more than 10% of the stars investigated have planetary systems, so there is plenty of scope for this proportion to rise as the precision of observations increases, and as data are accumulated for longer times.

But already we reach the important conclusion that planetary systems are fairly common, at least around solar-type stars. Before 1995 this was *believed* to be the case. Now there is growing observational evidence that it is so.

Migration of planets in exoplanetary systems

Another important conclusion emerges from the exoplanetary systems, in particular from the presence of hot Jupiters. You will see in the remainder of this chapter that though it is beyond reasonable doubt that the giant planets formed within their systems, it is extremely unlikely that they could have formed so close in.

☐   In this case, what is the only logical alternative to formation where there are today?
The hot Jupiters must have formed further out, and then moved inwards.

Mathematical models show that the most common cause of inward movement is the gravitational effect of the (growing) giant planet on the circumstellar disc of gas and dust in which it is embedded and from which it has formed (Section 2.2). At first, this disc is symmetrical about an axis perpendicular to it and running through the growing star (protostar) at its centre. But as the mass of the embryonic giant grows, its gravity produces spiral density enhancements in the disc that destroy its symmetry. These spiral density waves have a net gravitational effect on the growing giant planet that causes it to migrate inwards. Figure 2.2 shows an advanced stage of migration.

Migration has to stop if a giant planet is to become a hot Jupiter rather that meet a fiery death. There are several plausible stopping mechanisms, such as tidal forces between the protostar and giant. Details are beyond our scope, but can be found in Further Reading. Ultimately, the disc is dispersed by the protostar as it becomes a main sequence star, as outlined in Section 2.1.3.

The question arises, why do some exoplanetary systems, including the Solar System, *not* have hot Jupiters? The answer is two-fold. First, the extent of migration depends on various properties of the circumstellar disc (density, thickness, temperature, and so on). Certain values give very low migration rates, with not a lot of inward movement before the disc is dispersed. Second, there will usually be more than one giant planet. Interaction between the gravitational effects they have on the disc can slow migration and even reverse it for some of the giants.

We thus reach the important conclusion: the giant planets in the Solar System might not have formed where we find them now. They could have formed elsewhere and migrated, with effects on the smaller bodies in the Solar System,

Question 2.1

Discuss why, in the astrometric technique,

(a) planets with large mass will be easier to detect than planets with small mass;
(b) it will be easier to detect planets around nearby stars than around distant ones.

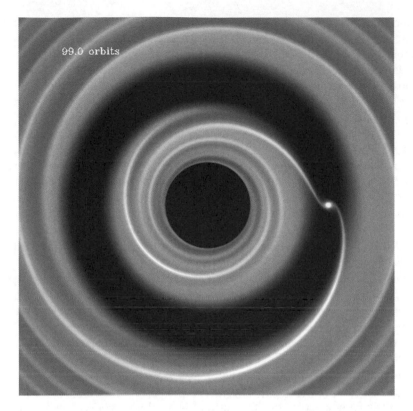

99.0 orbits

Figure 2.2 A computer simulation of an advanced stage of migration of a growing giant planet through the circumstellar disc of gas and dust from which it has formed. (Reproduced by permission from F Masset, CEA/CNRS/Université Paris, 2004)

Question 2.2

Discuss whether you would expect hot Jupiters to have orbits with eccentricities far larger than those of Jupiter and Saturn.

2.1.3 Star Formation

Observations of star formation provide further insight into the origin of the Solar System. Star formation is a relatively rapid process by astronomical standards, but it still takes many millions of years, and therefore the process has been pieced together by observing it at different stages in different locations, linking the observations together by physical theory. This is rather like observing a large number of people at a particular moment – they are seen at all stages of their lives, and it is therefore possible to use general biological principles to construct a theory of the complete human life cycle from an observation that occupied only a small fraction of the human lifespan.

From dense clouds to cloud fragments

Stars form from the interstellar medium (ISM) – the thin gas with a trace of dust that pervades interstellar space. Its chemical composition everywhere is dominated by hydrogen and helium.

In the region from which the Solar System formed, hydrogen typically accounts for about 71% of the mass, helium for about 27%, and all the other chemical elements (the 'heavy' elements) for only about 2%. Elsewhere in the ISM the proportion of helium is not very different, whereas the proportion of heavy elements can be as low as about 1% and as high as 5%, sometimes more. Almost all of the hydrogen and helium is in the form of gases, but a significant fraction of most of the heavy elements is condensed in the dust in a variety of compounds. Dust accounts for roughly 1% of the mass of the ISM.

The density and temperature of the ISM vary considerably from place to place. Star formation occurs in the cooler, denser parts of the ISM, because low temperatures and high densities each favour the gravitational contraction that must occur to produce a star from diffuse material. Low temperatures favour contraction because the random thermal motions of the gas that promotes spreading are then comparatively weak. High densities favour contraction because the gravitational attraction between the particles is then relatively strong. The cooler, denser parts of the ISM are called, unsurprisingly, dense clouds. They are often components of giant molecular clouds, 'molecular' because the predominant form of hydrogen throughout them is the molecular form, H_2.

Dense cloud temperatures are of order 10 K. They must not, however, be thought of as chunky things – a typical density at the high end of a wide range is only of order $10^{-14}\,\mathrm{kg\,m^{-3}}$, rather less than the density in a typical laboratory vacuum! A typical size is, however, a few light years across, and therefore most dense clouds are massive enough to form many hundreds of stars. They are also large enough for the dust content to make them opaque at visible wavelengths.

Though the conditions for gravitational contraction are best met in dense clouds, it is likely that in most cases they will contract only if they are subject to some external compression, particularly because magnetic fields and gas flows within the cloud hinder contraction. Compression can occur in one or more of a variety of ways, such as in a collision between two clouds, or by the impact of a shock wave from an exploding star, or by the action of a so-called spiral density wave that sweeps through the whole Galaxy (thereby sustaining its spiral arms). One way or another, a dense cloud, or a good part of it, becomes dense enough to become gravitationally unstable, and it starts to contract. As it contracts it becomes denser, to the point where the denser parts of the cloud, called dense cores, each contract independently, and are destined to become stars. This leads us to expect stars to form in clusters, and indeed the great majority of young stars *are* found in clusters (though a few form in isolation, from *small* dense clouds). Typically, a cluster contains a few hundred stars, and Plate 23 shows an example. Star clusters gradually disperse, and therefore older stars, like the Sun, are no longer in clusters.

From a cloud fragment to a star

Let us now follow the fate of a typical cloud fragment as it contracts. The gas molecules and dust particles gain speed as they fall inwards, and when they collide there is an increase in the random element of their motion. Temperature is a measure of the random motion of an assemblage of microscopic particles, and therefore the temperature rises. However, the rise is initially small because when the gas molecules collide they are raised into higher energy states of vibration or rotation. When the molecules return to lower energy states they get rid of their excess energy by emitting photons, usually at infrared (IR) wavelengths. Initially the density of the cloud fragment is so low that most of these photons escape. This loss of energy by the fragment retards the temperature rise.

The fragment continues to contract, and its density rises further. Detailed calculations show that the central regions of the fragment contract the most rapidly. It is therefore these central regions that become opaque to the photons emitted by the molecules. The temperature rise is then rapid and the central object is regarded as a protostar. Contraction continues, now more slowly, and a few million years after the fragment separated from the dense cloud the temperature in the core of the protostar has become high enough for nuclear fusion to occur – about 10^7 K. This fusion releases energy and creates a pressure gradient that halts the contraction of the protostar. At this point the protostar has become a star – a compact body sustained by nuclear fusion.

The fusion that dominates the nuclear reactions in the core of the star depends on its composition.

☐ What element accounts for most of the mass of the star?

As in the dense cloud, typically about 71% of the mass is hydrogen. It is also the case that nuclear fusion involving hydrogen occurs at a lower temperature than fusion involving helium and the other elements. Therefore it is the fusion of hydrogen nuclei that is by far the dominant source of energy. This fusion results in the creation of nuclei of helium, by the pp chains (Section 1.1.3).

You saw in Section 1.1.3 that the onset of core hydrogen fusion marks the start of the main sequence phase of a star's lifetime. It lasts longer the less massive the star, and for a star of solar mass it lasts about 10^{11} years. The Sun itself is 4600 Ma through its main sequence phase. In all stars it is a period of relative stability, but it is immediately preceded by a well-observed period of instability that is of considerable importance to the formation of any planetary system. This is the **T Tauri phase**, named after the protostar that was the first to be observed in this phase. For a protostar of solar mass it is thought to last for a few million years. It is marked by a considerable outflow of gas, called a T Tauri wind, a protostar of solar mass losing the order of 10% of its mass in this way, and by a high level of ultraviolet (UV) radiation from the protostar. The root causes of T Tauri activity are the final stages of infall of matter to the protostar, plus its strong interior convection and rapid rotation.

After the onset of hydrogen fusion the T Tauri activity quickly subsides. The UV radiation falls to a much lower level, and the wind declines to a much smaller rate of mass loss, called a stellar wind in general, and the solar wind in the case of the Sun (Section 1.1.2).

2.1.4 Circumstellar Discs

Meanwhile, that part of the fragment that has remained outside the protostar has also been evolving. As it contracts, a dense core starts to form, with more tenuous material outside it. But the fragment is rotating, and so it is to be expected that only the material on or near the rotation axis falls fairly freely towards the core – the infall of the remainder is moderated by its rotation around the core. A circumstellar disc should thus form in the plane perpendicular to the axis of rotation. Planetary systems are thought to form from such discs.

In recent decades circumstellar discs have been detected around many protostars. The discs have masses ranging from a few times the mass of Jupiter to hundreds of times Jupiter's mass, and diameters typically a few hundred AU. The gas component in the discs is readily imaged through its emission at radio and millimetre wavelengths. This gas component is largely removed during the T Tauri phase of the star (Section 2.1.3), in the case of solar mass stars in the 10 Ma or so that leads up to the main sequence phase.

Discs have also been detected around several hundred young main sequence stars through the IR emission from the dust in the discs. By this stage, the disc masses are considerably less than those around protostars. Dusty discs are observed around stars up to ages of about 10 Ma. There is a growing number of images of these dust discs, some utilising the dust emission at

IR and submillimetre wavelengths, others utilising IR and visible wavelengths with the disc in silhouette against a bright background. In Plate 24 the dust component in the disc around the young main sequence star Beta Pictoris is imaged through the light from the star that the dust scatters. There is good evidence that the dust in this disc is replenished by collisions of cometary bodies. This was one of the first discs to be imaged. A disc of dust around the star Rho[1] 55 Cancri, that has at least four planets, is thought to be sustained in the same way. The Beta Pictoris disc does not extend inwards of about 20 AU from the star – could the hole have been hollowed out by the formation of planets? This possibility is supported by a warping of the inner disc that could be caused by a giant planet just within the hole, in an orbit inclined at about 3° to the plane of the disc. Other discs have similar holes.

Thus, around protostars we have discs of material that could form a planetary system, and around young main sequence stars we have discs that seem to indicate that planetary formation has actually occurred. These observations lend strong support to solar nebular theories, to which we now turn.

Question 2.3

Identify the feature of the Solar System in Table 2.1 that is already present in circumstellar discs around protostars.

2.2 Solar Nebular Theories

Over the centuries there have been several different types of theory on the origin of the Solar System, but in recent decades one type, with antecedents in the eighteenth century, has emerged as the firm favourite. This is the solar nebular theory. Theories of this type are characterised by the formation of the planetary system from a disc of gas and dust encircling the young Sun – the **solar nebula**. This is clearly in accord with the relatively recent observations of circumstellar discs around protostars and young stars. Overall, such theories fit the observational data better than any other type of theory, and there are certainly no observations that rule them out. Within the general type there have been many variants, though there has been some degree of convergence so that most variants now differ only in relatively minor details. We shall concentrate on the *typical* features of solar nebular theories, pointing out where the variants differ significantly.

We pick up the story at the point where the proto-Sun is surrounded by a disc of gas and dust of order 100 AU across – the solar nebula. This is shown edge on in Figure 2.3(a). The disc would not have ended as abruptly as shown; it is the extent of the main bulk of the disc that is indicated. The plane of the disc coincides with the equatorial plane of the proto-Sun, and the disc and proto-Sun, being derived from a single dense cloud fragment, will rotate in the same direction. The disc rotates differentially, the orbital period increasing with distance from the centre, in accord with Kepler's third law. The elemental composition of the disc is much the same as the present Sun outside its core – we have adopted, by mass, 70.9% hydrogen, 27.5% helium, 1.6% the rest. The gas in the disc is predominantly hydrogen and helium, and a significant fraction of the other elements is in the various compounds that constitute the dust. The formation of planets in this disc will lead to their orbiting in the same plane, all in the same direction, the direction of solar rotation. These are features of the Solar System that any acceptable theory of its origin has to explain (Table 2.1).

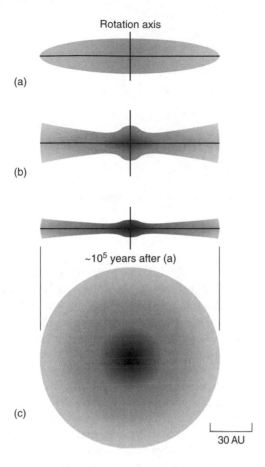

Figure 2.3 The solar nebula surrounding the proto-Sun. The proto-Sun is too small to show on this scale.

The starting point is, however, ill defined in one important respect: we do not know the initial mass of the disc. In some solar nebula theories the mass of the disc is about 1% of the present solar mass $M_\odot$. At the other extreme are versions in which the initial mass of the disc is comparable with $M_\odot$. A disc mass of about 1% $M_\odot$ is called the **minimum mass solar nebula**, MMSN. This is calculated from the estimate that there are about 65 Earth masses of heavy elements in the Solar System today, mainly in the interiors of the giant planets. To this is added the hydrogen and helium necessary to achieve solar composition. Much of the hydrogen and helium has been lost, mainly through the T Tauri wind. Some indication of an appropriate choice of mass is obtained by considering the angular momentum in the Solar System.

2.2.1 Angular Momentum in the Solar System

The magnitude l of the **angular momentum** of a body with mass m moving at a speed v with respect to an axis, as in Figure 2.4, is given by

$$l = mvr \tag{2.1}$$

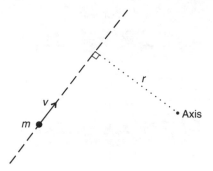

Figure 2.4 A body of mass m moving at a speed v at a perpendicular distance r from an axis perpendicular to the page.

where r is the perpendicular distance from the path of the body to the axis. The angular momentum of m is with respect to this axis. In the solar nebula a natural choice of axis is the rotation axis in Figure 2.3 – through the centre of the proto-Sun and perpendicular to the plane of the disc. In the Solar System today the natural choice is for the axis to go through the centre of mass of the Solar System and to be perpendicular to the ecliptic plane. It is the angular momenta with respect to these natural choices of axes that are of concern here.

Equation (2.1) applies to the mass m when its dimensions are small compared with r so that the whole of m can be regarded as being the same distance from the axis. This is closely approximated by a planet in orbit around the centre of mass of the Solar System, and the quantity is called the orbital angular momentum. If this condition is not met then the body is notionally subdivided into many small masses δm and the magnitude of its angular momentum is then a combination of the quantities $\delta m v r$. A simple case is when the angular momentum of a rotating planet or the Sun is calculated, as in Figure 2.5. The natural choice of axis is again the rotation axis, and because the paths of the δm around this axis are all circular and in the same set of parallel planes, the combination is simply the sum of $\delta m v r$ over the whole body. The quantity in this case is called the rotational (or spin) angular momentum.

In the Solar System today about 85% of the angular momentum is in the orbital motion of Jupiter and Saturn, and only about 0.5% is in the rotation of the Sun. Less than 0.5% is in the orbital motion of the Sun around the centre of mass of the Solar System. This is in sharp contrast to the Sun having about 99.8% of the mass of the Solar System. Thus today, 'where the mass is, the angular momentum is not'. The Sun's rotational angular momentum is small because it rotates slowly, about once every 26 days. Its orbital angular momentum is small for two reasons. First, the centre of mass of the Solar System is just outside the Sun, so r in equation (2.1) is small (call it $r_\odot$). Second, the orbital period $P_\odot$ for its small orbit is about 12 years, so its speed $v_\odot (= 2\pi r_\odot / P_\odot)$ is very small. For a planet, the average orbital angular momentum is well approximated by mva, where m is the mass of the planet, v is its average orbital speed, and a is the semimajor axis of the orbit (strictly, the distance from the centre of mass of the system should be used). By combining Kepler's third law

$$P = ka^{3/2} \tag{1.3}$$

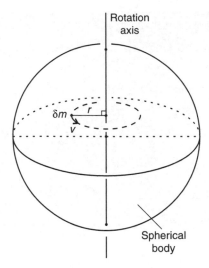

Figure 2.5 The rotation of an element δm of a spherical body.

with equation (2.1), and using $2\pi a/P$ for the average speed, where P is the planet's orbital period, we get

$$l_{orb} = \frac{2\pi}{k} m a^{1/2} \qquad (2.2)$$

where l_{orb} is the average orbital angular momentum of the planet (see Question 2.4).

☐   So, why are the orbital angular momenta of Jupiter and Saturn large?
They have large orbits, and they are by far the most massive of the planets.

This distribution of angular momentum today is in sharp contrast with that calculated for a contracting cloud fragment. The proto-Sun rotates rapidly, and has a correspondingly large fraction of the total angular momentum. Therefore we need to explain how most of the angular momentum of the proto-Sun could have been lost. One of the more convincing explanations involves turbulence in the disc at the time it still blended with the outer proto-Sun. Turbulence is the random motion of parcels of gas and dust and is expected to have been a feature of the contracting nebula. (Note that though the parcels can be small, they are much larger than atomic scale – this is not the random thermal motion that occurs at the atomic level.) Turbulent motions are superimposed on the orderly swirl of circular orbital motion around the proto-Sun. Turbulence transfers parcels radially, and it can be shown that the *net* transfer of disc mass is outwards in the outer part of the disc and inwards in the inner part of the disc. The associated net transfer of angular momentum is from the proto-Sun to the disc, and it is carried by a small fraction of the disc mass.

Further transfer arises from the solar wind. The ions that constitute the wind get snared by the Sun's magnetic field. Therefore, as they stream outwards they are forced to rotate with the Sun, and slow its rotation. There is thus a transfer of angular momentum from the Sun to the wind.

Note that the transfer of disc mass leads to the loss of mass from the disc. In the outer disc this loss is to interstellar space. In the inner disc it is to the proto-Sun, and also to interstellar space via an outflow along the rotation axis of the disc, perhaps enhanced by mass loss from the polar regions of the proto-Sun. Such bipolar outflow is observed from protostars, as in Figure 2.6,

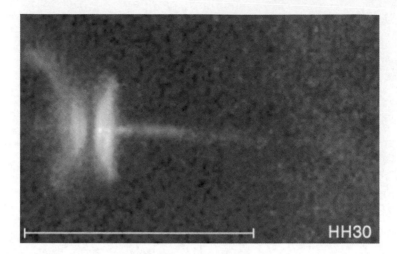

Figure 2.6 Bipolar outflow from the protostar in an object called HH-30. A disc (edge on) is also apparent. The scale bar is 1000 AU long. (Reproduced by permission of C Burrow, AURA/STScI ID Team, ESA)

where the disc is also apparent edge on. Why these outflows are so tightly collimated is not well understood, but the magnetic field of the protostar or of the disc itself, acting on electrically charged particles in the flow, might be important. Bipolar outflow would have carried off only a small proportion of the rotational angular momentum of the proto-Sun. But this is not the whole outflow, particularly in the strong T Tauri phase (Section 2.1.3), when a significant proportion of the proto-Sun's angular momentum could have been carried off.

How does the distribution of angular momentum cast light on the initial mass of the disc? If the initial disc mass were only about 1% of the solar mass $M_\odot$ (the MMSN) the angular momentum transfer would be weak, to the extent that it would be difficult to explain the necessary loss of angular momentum by the proto-Sun. At the other extreme, an initial disc mass comparable with $M_\odot$ is considerably more than is necessary and requires a huge proportion of disc mass to be lost to space. Therefore, many astronomers favour an intermediate value, a few times the MMSN. This will be adopted implicitly in Section 2.2.

Question 2.4

(a) Derive equation (2.2).
(b) Use equation (2.1) to calculate the magnitude of the average orbital angular momentum of the Earth, and then use equation (2.2) to calculate the magnitude of the average orbital angular momenta of Jupiter and Neptune.

2.2.2 The Evaporation and Condensation of Dust in the Solar Nebula

In nebular theories the formation of planets and other bodies occurs in a number of stages, the first of which is the evaporation of some of the initial complement of the dust in the solar nebula.

Evaporation of dust

You have seen (Section 2.1.3) that the proto-Sun only begins to increase greatly in temperature when it becomes dense enough to be opaque to its own radiation. The disc never becomes as dense as the proto-Sun and therefore the tendency for its temperature to rise as it contracts is more strongly moderated. Nevertheless the disc does rise in temperature, particularly in its inner regions where it is denser and thus more opaque, and where infall to the proto-Sun has caused greater frictional heating.

☐ What additional source of energy heats the inner disc more than the outer disc?

This is the proto-Sun when it becomes luminous.

In the inner disc, out to perhaps 1 AU, the calculated temperatures exceed about 2000 K, high enough to evaporate practically all of the dust in the disc – only those substances with very high sublimation temperatures escape evaporation. In **sublimation**, a substance goes directly from solid to gas, as does carbon dioxide ice (dry ice) at the Earth's surface. It occurs when the pressure is too low to sustain a liquid, and in the disc the pressures are well below his threshold. Above the sublimation temperature the substance is a gas, and below it, a solid. For any particular substance the sublimation temperature depends on the pressure: the higher the pressure, the higher the sublimation temperature. The value of the sublimation temperature for a particular substance at a given pressure is one measure of the **volatility** of the substance. The most volatile substances, such as hydrogen (H_2) and helium (He), have *extremely low* sublimation temperatures, whereas substances such as corundum (Al_2O_3) have *extremely high* sublimation temperatures, and are said to be **refractory**. With increasing distance from the proto-Sun the disc temperature decreases, and therefore increasingly more volatile substances avoid sublimation. Beyond the order of 10 AU even quite volatile substances escape sublimation, such as water ice.

Condensation of dust

So far the disc has evolved in completely the wrong direction to make planets – it has gained gas at the expense of solid material! However, at some point the contraction of the disc slows. Moreover, the luminosity of the proto-Sun declines as it contracts, its surface area decreasing greatly whilst its surface temperature increases only slightly. (In contrast, the protosolar core temperature is increasing enormously, because of the lower rate of energy transfer across the outer layers of the proto-Sun.) Heat generation within the disc also declines, and so the disc temperatures begin to fall as it continues to emit IR radiation. At some point fresh dust begins to condense, its composition depending on the composition of the gas and on the local temperature. Because of the low pressures, solids rather than liquids appeared.

Table 2.3 gives the condensation temperatures of representative substances (these are also the sublimation temperatures). The pressure for the data is 100 Pa, 0.1% of the atmospheric pressure at the Earth's surface. This is a theoretical value for the total gas pressure in the disc. The temperature at which a substance condenses will depend not only on this total pressure but also on the proportion of the disc accounted for by the substance, which determines its contribution to the total pressure, i.e. the **partial pressure**. It is the partial pressure that determines, albeit rather weakly, the condensation temperature of a substance. Also, the pressure might have been rather lower than 100 Pa, though the condensation temperatures are only slightly lower even at 10 Pa.

Table 2.3 A condensation sequence of some substances at 100 Pa nebular pressure

Temperature/K	Substance	Chemical formula
1758	Corundum	Al_2O_3
1471	Iron–nickel	Fe plus $\sim 6\%$ Ni by mass
1450	Diopside[a]	$CaMgSi_2O_6$
1444	Forsterite[b]	Mg_2SiO_4
< 1000	Alkali feldspars	$(Na, K)AlSi_3O_8$
700	Troilite	FeS
550–330	Hydrated minerals[c]	$X(H_2O)_n$ or $X(OH)_n$
190	Water	H_2O
135	Hydrated ammonia	$NH_3.H_2O$
77	Hydrated methane	$CH_4.7H_2O$
70	Hydrated nitrogen	$N_2.6H_2O$
37	Methane	CH_4
~ 8	Hydrogen	H_2
~ 1	Helium	He

[a] A particular form of pyroxene, $(Ca, Fe, Mg)_2Si_2O_6$.
[b] A particular form of olivine, $(Mg, Fe)_2SiO_4$.
[c] X can be a molecule of a variety of minerals, and n is greater than or equal to 1.

The disc temperatures are generally lower the further we are from the Sun. Therefore as the disc cools a substance condenses rather in the manner of a wave spreading inwards to some minimum distance within which the temperature is always too high.

In the innermost part of the disc the temperatures are probably always too high at the dust condensation stage for anything much less refractory than iron–nickel to condense. At greater distances less refractory dust components appear, including an important range of substances exemplified in Table 2.3 by diopside, forsterite, and alkali feldspars. These are examples of silicates. A **silicate** is a chemical compound that has a basic unit consisting of atoms of one or more metallic elements and atoms of the abundant elements silicon and oxygen. For example, olivine has the chemical formula $(Fe, Mg)_2SiO_4$. Therefore, in the basic unit there is one atom of silicon (Si), four of oxygen (O), and two atoms of iron or magnesium – either two iron atoms or two magnesium atoms, or one of each. A particle of dust consists of very many units, and so the proportion of iron to magnesium in the particle as a whole can be anywhere in the range 0–100%. The particular version of olivine in Table 2.3 (forsterite) has no iron at all. Particular versions are called **minerals**, naturally occurring substances with a basic unit that has a particular chemical composition and structure. Olivine is thus the name for a range of closely related minerals. The whole family of silicates cover a wide range of compositions.

Silicates are by far the most common refractory substances in the Solar System after iron–nickel, and they are common in rocks, a **rock** being an assemblage of one or more minerals in solid form. Together with iron–nickel, silicates account for most of the common **rocky materials**. Note that this is the name of a group of refractory substances and not an implication that they are solid.

In the disc, an extremely important boundary is the distance beyond which water condenses. This is important because there is a considerable mass of water in the disc, and where it condenses it becomes the dominant constituent of the dust grains. Water must have been abundant because oxygen, among the heavy elements, is particularly abundant (Table 1.5), and in a hydrogen-rich gas, at all but very high temperatures, most of the oxygen combines with hydrogen to form water molecules, H_2O. Figure 2.7 shows one model of the column mass of the disc versus distance from the proto-Sun at a time well into the dust condensation stage, when the disc probably resembled Figure 2.3(c). The **column mass** is the total mass in a cylinder of unit cross-sectional area with its axis running perpendicular to the plane of the disc. The increase in column mass at about 5 AU from the proto-Sun is due to the condensation of water beyond this distance. This distance is sometimes called the **ice line**. Note that the values in Figure 2.7 are illustrative, and not definitive. This applies to the column masses and also to the location of the ice line – in recent models this is around 4 AU.

Though water as H_2O condenses beyond 4–5 AU, the dust closer in is not devoid of water: **hydrated minerals** (Table 2.3) have higher condensation temperatures than water. These are substances that have one or more water molecules attached to their basic unit, or one or more hydroxyl molecules (OH) which are a fragment of the water molecule. Water is one of a group of substances called **icy materials**. As in the case of rocky materials this is the name of a group of volatile substances with no implication that they are present as solids. The solid form is called an ice. Other important icy materials include ammonia (NH_3) and nitrogen (N_2), which are shown in Table 2.3 in their hydrated forms, and methane (CH_4), shown in hydrated and non-hydrated forms. Carbon monoxide (CO) and carbon dioxide (CO_2) are also important icy

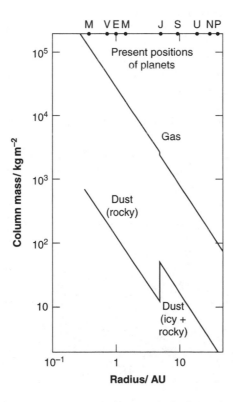

Figure 2.7 The column masses of gas and dust in the solar nebula disc well into the stage of dust condensation.

materials – CO_2 more volatile than water, CO more volatile than CO_2. Thus, all of these icy materials are more volatile than water and so have minimum condensation distances that are greater than that of water.

Some evidence for extensive though incomplete evaporation of dust, followed by recondensation, is provided by meteorites, where a proportion of their refractory and not-so-refractory substances have isotope ratios that differ markedly from the Solar System average. This would be the outcome if these proportions had survived evaporation – material that has not been recondensed from a nebular gas retains an imprint of its origin beyond the Solar System.

Question 2.5

What indications are there already that we will get two zones in the Solar System, with terrestrial planets in the inner zone and giant planets in the outer zone?

2.2.3 From Dust to Planetesimals

At the stage we have reached, the dust grains throughout the nebula are tiny, with sizes in the range $1–30\,\mu m(1\,\mu m = 10^{-6}\,m)$, as observed in circumstellar discs (Section 2.1.4). The grains now grow slightly by acquiring atoms and molecules from the gas, rather as raindrops grow by acquisition of water vapour. This slow-growth phase is accompanied by an increasing tendency for grains to settle to the mid plane of the disc, a result of the net gravitational field and gas drag. This tendency increases as the turbulence in the disc dies away and as the grains grow in size. There is thus an increasing concentration of dust around the mid plane of the nebula, forming a very thin sheet of order $10^4\,km$ thick. The rest of the nebula is much thicker and much more massive, and consists of gas, mainly H_2 and He, and remnant dust. This is shown in Figure 2.8 for the inner part of the dust sheet, where the thickness of the dust sheet has been greatly exaggerated. Note that the sheet gets thicker with increasing distance from the Sun, i.e. with increasing heliocentric distance.

☐ So how (except for the step up at the ice line) can the column mass of the dust decrease as in Figure 2.7?

As the heliocentric distance increases, the mass of dust per unit volume of the sheet decreases faster than the dust sheet thickness increases. This is mainly because the dust grains are further apart, and not because they are smaller.

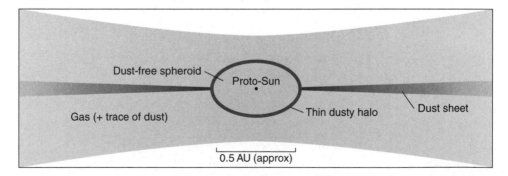

Figure 2.8 Edge-on view of the gas and dust around the proto-Sun after the dust has settled towards the mid plane. The thickness of the dust sheet has been greatly exaggerated.

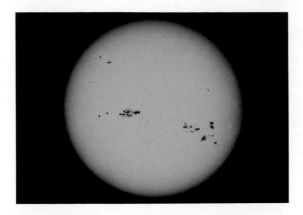

Plate 1 The Sun's photosphere, showing sunspots. (Reproduced by permission of Akira Fujii)

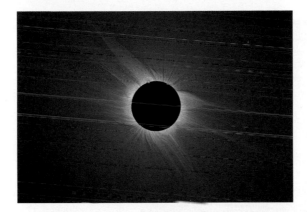

Plate 2 The corona of the Sun, imaged during the total solar eclipse of 29 March 2006. (Reproduced by permission of Hannah Drückmullerová and Miloslav Druckmüller)

Plate 3 The inner corona and chromosphere of the Sun, imaged during the total solar eclipse of 11 June 1983. Also visible are prominences (red), huge, transient clouds of ionised gas (plasma) arching above the chromosphere. (Reproduced by permission of Akira Fujii)

Plate 4 Mercury – part of the surface imaged by the flyby spacecraft Mariner 10 in 1974. The more heavily cratered areas are examples of heavily cratered terrain. (NASA/ NSSDC P14679)

Plate 5 The Venusian volcano Maat Mons (on the horizon), altitude 9 km, though the vertical scale is exaggerated 10 times. The sky should not be black but totally overcast by a high cloud layer. This is a radar image by the Magellan Orbiter. (NASA/NSSDC P40175)

Plate 6 Earth, showing oceans, continents, polar cap, and clouds. This view was obtained by the Apollo 17 astronauts in 1972. (NASA/ NSSDC AS17-148-22727)

Plate 7 The Moon – part of the surface that faces the Earth. (Reproduced by permission of Akira Fujii)

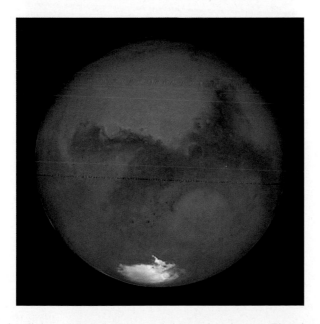

Plate 8 Mars, showing light and dark features, and the south polar cap. The triangular dark feature to the right is Syrtis Major, and the circular feature to the south of this is the Hellas basin. This image was obtained by the Hubble Space Telescope at Mars's closest approach, 27 August 2003, during the 2003 opposition. (NASA/STScI, J Bell (Cornell University), and M Wolff (SSI))

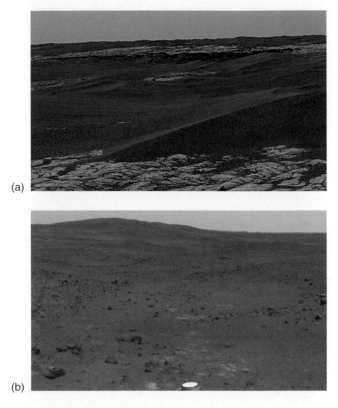

(a)

(b)

Plate 9 Mars: views from the surface, obtained by the Mars Exploration Rovers. (a) From *Opportunity*. Part of the rim of the crater Erebus, about 300 m in diameter, is visible in the upper part. (b) From *Spirit*. The high point on the horizon is McCool Hill, several hundred metres away, and is one of the Columbia Hills. (NASA/JPL-Caltech/Cornell PIA03622 and PIA03623)

Plate 10 The asteroid Gaspra, imaged by the Galileo space-craft en route to Jupiter. It is about 10 km across. (NASA/ NSSDC P40449)

Plate 11 Jupiter – a general view showing the richly structured clouds, including The Great Red Spot, one of the largest and most persistent atmospheric features. At about the same latitude the dark disc is the shadow cast by Jupiter's satellite Europa. This is a mosaic of four images obtained by the Cassini spacecraft in December 2000, on its way to Saturn. (NASA/JPL, University of Arizona PIΛ02873)

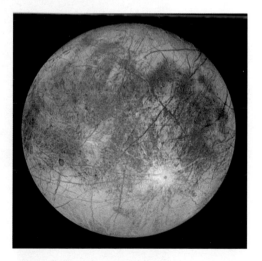

Plate 12 Io, imaged by the Galileo Orbiter. Note the volcanic plume lower left. (NASA/JPL MRPS85377)

Plate 13 Europa, imaged by the Galileo Orbiter. The icy surface is smooth, with only one prominent impact crater, Pwyll ('Poo-eel'), lower right. (NASA/JPL P48040)

Plate 14 Ganymede, imaged by the Galileo Orbiter. It has an ice-rich, variously cratered surface. (NASA/ JPL PIA00716)

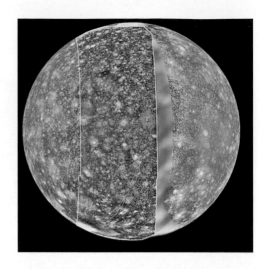

Plate 15 Callisto, imaged by Voyager 1 (left), Galileo Orbiter (centre), Voyager 2 (right). This is a heavily cratered icy surface. Note the different resolutions achieved by the three spacecraft. (NASA/JPL MRPS77654)

(a)

(b)

Plate 16 Saturn from Cassini in 2004. (a) A general view showing faintly structured clouds and the rings (PIA06193). (b) A false colour close-up, showing the banding and a storm (PIA06197). (NASA/JPL/Space Science Institute)

Plate 17 Titan, Saturn's largest satellite, imaged by Cassini in 2005, showing haze layers above the clouds. (NASA/JPL/Space Science Institute PIA06236)

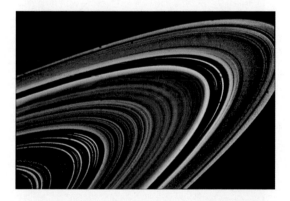

Plate 18 The rings of Saturn, imaged by Voyager 1. This is a false colour image. Note the many ringlets that make up each ring. (NASA/NSSDC P23953)

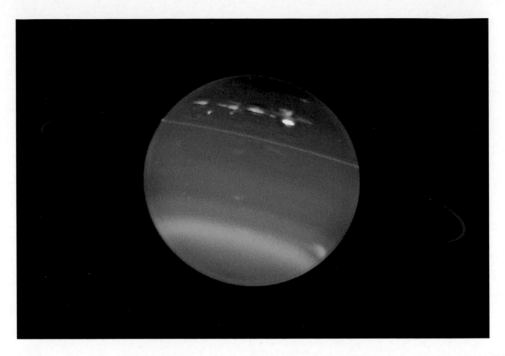

Plate 19 Uranus, imaged in 2006 by the Keck II telescopes at near-infrared wavelengths. At visible wavelengths it is a bland blue-green, due to methane in the atmosphere, with bands and clouds less prominent. One of the thin rings is clearly visible, in false colour – it lies in the equatorial plane of Uranus. (Reproduced by permission of L A Sromovsky, Space Science and Engineering Center, University of Wisconsin–Madison)

Plate 20 Neptune, imaged by Voyager 2 in August 1989. (NASA/JPL PIA 00048)

Plate 21 Triton's icy surface, imaged by Voyager 2 in 1989. (NASA/NSSDC PIA00317)

Plate 22 Comet Hale–Bopp, which passed through the inner Solar System in 1997. Note the dust tail (yellow) and the ion tail (blue). (Reproduced by permission of Akira Fujii)

Plate 23 The star cluster NGC 2024, about 5 ly across, and so young that the dense cloud that gave it birth is still evident. This is a near-infrared image, in which the longer wavelengths are red, the shorter, blue. Dense clouds are less opaque at infrared than at visible wavelengths. (Reproduced by permission of I S McLean)

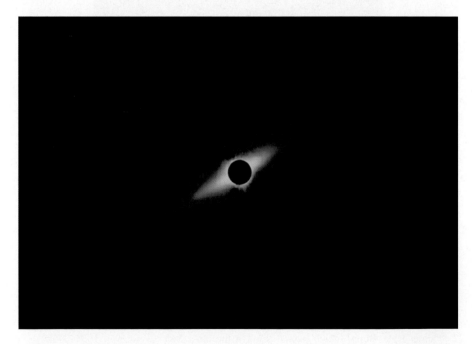

Plate 24 The dust component in a disc around the star Beta Pictoris. The disc is presented nearly edge on, and is nearly 2000 AU across in this red light image. The star image is blocked in the telescope. (Reproduced by permission of P Kalas and D Jewitt)

(a)

(b)

(c)

(d)

Plate 25 Various meteorites. The sizes in (a)–(c) are the widths. (a) A chondrite showing its fusion crust, and the interior revealed through a sample having been taken for analysis (115 mm). (Reproduced by permission of Dieter Heinlein, Augsburg, Germany) (b) An iron, showing the Widmanstätten pattern (115 mm). (The author) (c) A slice from a stony-iron (123 mm). (The author) (d) A thin section of a carbonaceous chondrite. The largest circular features (chondrules) are about 2 mm in diameter. (Reproduced by permission of Arabelle Sexton and OU-PSRI)

Plate 26 An aurora borealis over northern Sweden (Reproduced by permission of D H Rooks)

The concentration of the dust into a sheet leads to a greatly increased chance of a collision between two grains. Neighbouring grains tend to be in similar orbits and therefore a significant fraction of the collisions is at sufficiently low relative speed for the grains to stick together in a process called coagulation. Coagulation is more likely when one or both grains have a fluffy structure, and it is aided when the two grains have opposite electric charges, or when they contain magnetised particles. Gravitational instabilities in the dust sheet might also aid coagulation.

The outcome of coagulation is the gradual build-up of bodies of order 10 mm across. The time required for this to happen depends on the relative speeds of grains in slightly different orbits: the lower the relative speeds, the lower the collision rate and the slower the coagulation. These relative speeds are lower, the smaller the orbital speeds. Therefore, the coagulation time generally increases with increasing heliocentric distance. This tendency is reinforced because the coagulation time also depends on the average spacing of the grains: the greater the spacing, the slower the coagulation. This spacing increases as the column mass of the disc decreases, and so the coagulation time is further increased with increasing heliocentric distance. An exception is at the ice line, where the step up in column mass causes a significant reduction in the coagulation time in the Jupiter region. Broadly speaking, the coagulation times are in the approximate range of 1000 years to a few tens of thousands of years within about 4–5 AU of the proto-Sun, increasing to many hundreds of thousands of years at 30 AU. The earlier times for dust condensation and settling are shorter.

By the time bodies 10 mm or so in size are appearing at 30 AU, the bodies out to 4–5 AU have grown to 0.1–10 km across. These are called **planetesimals**, 'little planets', rocky within the ice line, and when they subsequently form, icy–rocky beyond it. There are at least two means of producing planetesimals, both of which might have been significant. The first is a continuation of coagulation, promoted by the continuing thinning of the dust sheet with the corresponding increase in its density. The second is a different consequence of this density increase. At a sheet thickness of order 100 km it is possible that the gravitational attraction between the bodies constituting the sheet leads to gravitational instability, the sheet breaking up into numerous fragments, each fragment forming a planetesimal.

Though the formation of planetesimals is a considerable step towards bodies of planetary size, there is clearly some way still to go. The theory of the remaining stages of planetary formation indicates that the process was rather different in the inner Solar System – within the ice line – than in the outer Solar System.

Question 2.6

(a) Use Figure 2.7 to estimate the total mass in the planetesimals between 0.8 AU and 1.2 AU from the proto-Sun. Compare your result with the mass of the Earth, and comment on the significance of the comparison.

(b) If the mean density of a planetesimal around 1 AU from the proto-Sun is $2500 \, kg \, m^{-3}$, calculate the number of planetesimals corresponding to the mass you calculated in part (a).

In both parts, state any assumptions that you make.

2.2.4 From Planetesimals to Planets in the Inner Solar System

A planetesimal about 10 km across has sufficient mass for it to exert a significant gravitational attraction on neighbouring planetesimals. This increases the collision rate between planetesimals, and models show that the net effect is growth of the larger planetesimals at the expense of the smaller ones. An essential condition for net growth is that the collisions are at low speed, thus requiring neighbouring planetesimals to be in low-eccentricity, low-inclination orbits. Such orbits could have been wrought by nebular gas drag on planetesimals in more eccentric, more inclined orbits. The acquisition by a larger body of smaller bodies is called **accretion**.

As a planetesimal gets more massive its accretional power increases, and consequently there is a strong tendency for a dominant planetesimal to emerge that ultimately accretes most of the mass in its neighbourhood. (This is an example of the Matthew effect: 'For unto every one that hath shall be given, and he shall have abundance: but from him that hath not, shall be taken away even that which he hath.' *The gospel according to St Matthew XXV, 29*. The Matthew effect is also familiar to players of Monopoly.) The outcome is runaway growth, in which the population of planetesimals in a neighbourhood evolves to yield a single massive planetesimal called an **embryo**, that accounts for over 90% of the original planetesimal mass in the neighbourhood, plus a swarm of far less massive planetesimals, the largest being perhaps a million times less massive than the embryo. The neighbourhood of an embryo is an annular strip covering a small range of heliocentric distances, and so we get a set of embryos each at a different heliocentric distance.

Figure 2.9 shows the embryo mass versus distance calculated from one model, the results from which must be regarded as illustrative and not definitive. In this model the orbits of the embryos

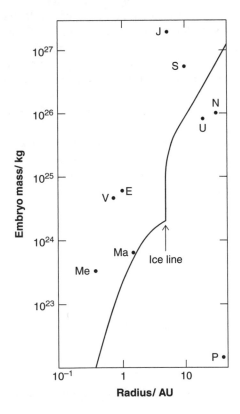

Figure 2.9 Embryo mass versus heliocentric distance as calculated in one model. The masses of the planets are also shown, in their present positions.

within the ice line are spaced by about 0.02 AU, and the time taken for the full development of a single embryo from a swarm of planetesimals is of order 0.5 Ma at 1 AU, and increases with heliocentric distance. These times are *very* uncertain, though a general increase in time with increasing heliocentric distance emerges in all models, largely because of the decrease in column mass. Another common feature is an increase in embryo spacing with increasing solar distance.

☐ In this model, how many embryos are there between 0.3 AU and 5 AU?

There are about $(5 - 0.3)/0.02$ embryos in this region, i.e. 200 or so. This is the region presently occupied by the terrestrial planets and the asteroids. From Figure 2.9 and the 0.02 AU spacing, their total mass can be estimated to be of order 10 times the mass of the Earth.

Figure 2.9 shows that (except perhaps for Mars) we have to put embryos together to form the terrestrial planets. However, assembly is a slow process because of the small number of embryos and their consequent large spacings. We have to rely on modest orbital eccentricities to produce collisions. Such collisions would produce fragmentation, but the gravitational field would be sufficient to assemble most of the fragments into a body with nearly the combined mass of the two colliding embryos. At some intermediate stage there could have been a few dozen Mars-sized embryos, and a host of less massive bodies. Collisions would usually have been off centre, and so even if many of the embryos initially had small inclinations of their rotation axes, larger inclinations could readily be imparted to some planets through the arrival of large embryos. This is in accord with item 6 in Table 2.1.

The time occupied by the transition from embryos to a terrestrial planet increases with increasing heliocentric distance. It is estimated that for the Earth the time was of order 100 Ma. This is by far the slowest stage in the formation of the terrestrial planets, though it is short compared with the 4500 Ma or so that have elapsed since. Figure 2.10 is a time line summary of the formation of a terrestrial planet. The relative durations of each stage are more reliable than the absolute durations, which vary considerably from model to model. Note the logarithmic scale.

After the last embryo collision we are left with planetesimals bombarding an essentially complete planet. There is widespread evidence that the terrestrial planets and the Moon suffered such a **heavy bombardment**, and that it tailed off about 3900 Ma ago, to be followed by a light bombardment by small bodies that persists to the present day.

If the terrestrial planets did indeed form as proposed here, then they should consist largely of substances more refractory than the hydrated minerals in Table 2.3, and the closer to the Sun the more refractory the composition should be. Note, however, that the compositional differences between the planets will have been moderated by embryos and planetesimals arriving from different regions. This is in accord with what we know about terrestrial planet composition.

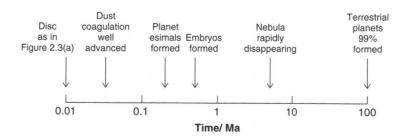

Figure 2.10 Time line of the formation of a terrestrial planet. The relative durations of each stage are more reliable than the absolute durations, which vary considerably from model to model.

Nebular gas would also have been captured, and though only to the extent of a tiny fraction of the planetary mass, it would have provided the planets with atmospheres rich in hydrogen and helium. In the theory, such atmospheres are removed by the T Tauri activity of the proto-Sun.

Throughout the T Tauri phase the nebular gas has been depleted by its accretion onto the proto-Sun. The T Tauri activity also removes nebular gas to space, driven out by copious UV and particle emission. This is the mechanism by which the last remnants of gas are swept away, and it happens rather rapidly. In Figure 2.10 this terminal sweeping is denoted by 'nebula rapidly disappearing', which is also around the end of the T Tauri phase.

The effects of Jupiter in the inner Solar System

Figure 2.9 and the embryo spacing of roughly 0.02 AU, shows that from 2 AU outwards towards Jupiter, but sufficiently far (about 4 AU) to evade direct capture, there are expected to have been several tens of embryos each with masses of the order of 10^{24} kg. Today this region is occupied by the asteroids, with a total mass only of order 10^{22} kg, the most massive being Ceres at 9.4×10^{20} kg. The answer to this seeming contradiction is the effect of Jupiter, the most massive planet, which is in orbit just beyond this region. In the theory, as Jupiter grows, its gravitational field 'stirs' the orbits of the planetesimals and embryos, producing a range of eccentricities, inclinations, and semimajor axes, so that most collisions occur between two bodies occupying substantially different orbits, with huge relative speed, as in Figure 2.11. The result is fragmentation and dispersal of the fragments, rather than accretion. Some of the dispersed fragments are flung into huge orbits, some are lost from the Solar System, and some are captured by the Sun and planets. Only a small fraction remains at 2–4 AU. This population continued to evolve, and we see the

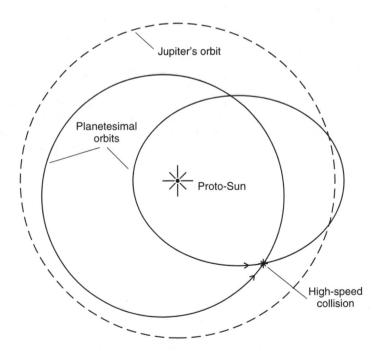

Figure 2.11   High relative speeds when two bodies in substantially different orbits collide.

survivors today as the asteroids, with perhaps only the order of 0.1% of the original mass in this region.

The growth of Mars at around 1.5 AU must also have been stunted by the stirring of the planetesimal and embryo orbits by Jupiter. Nearer the Sun, the effect of Jupiter might have been to speed up the final stages of growth of Mercury, Venus, and the Earth, partly through the provision of material from outside the terrestrial zone, and partly by increasing the eccentricity of the embryo orbits, thus increasing their collision rate without producing the huge relative speeds of the asteroid region.

Formation of the Moon (and of Mars's satellites)

Of the terrestrial planets only the Earth and Mars have satellites. The two tiny satellites of Mars (Table 1.2), Phobos and Deimos, are probably captured asteroids. Their densities are too small for them to be pieces of Mars, but the class C asteroids (Section 3.1.6) meet the requirements. Capture directly into orbits so near Mars is unlikely. Instead, a class C asteroid could have struck Mars, and the disc around Mars thus formed, consisting of a mixture of the asteroid material and the Martian crust, underwent accretion to form Phobos and Deimos.

The Moon, at 1.2% of the mass of the Earth, is far too massive for capture to be likely. In recent years widespread support has grown for the view that the Moon is the result of an embryo with a mass 10–15% that of the Earth, colliding with the nearly formed Earth at a grazing angle. All but the core of the embryo, and some of the outer part of the Earth, is scattered along an arc, predominantly as a gas produced by vaporization during the impact. Much of this material returns to Earth, some escapes, but a small fraction goes into orbit around the Earth, from where, in only a year or so, it forms the Moon. You will see in later chapters that the detailed composition and structure of the Earth and the Moon provide a good deal of support for this theory. One simulation of the process is illustrated in Figure 2.12.

The lunar orbit has a (nearly) fixed inclination of 5.16° with respect to the ecliptic plane (Figure 1.25), rather than to the equatorial plane of the Earth. This strong link to the ecliptic plane makes the Moon different from all the other large planetary satellites in the Solar System. Yet the models show that *initially* the inclination was (nearly) fixed with respect to the Earth's equatorial plane. This changed, because tidal forces between the Earth and the Moon caused the Moon to recede. As it did so, the Sun became more influential on the Moon, and the Earth less so, with the likely result that the lunar orbit acquired a (nearly) constant inclination with respect to the ecliptic plane.

Question 2.7

List the features of the Solar System in Table 2.1 that apply to the terrestrial planets. For each feature in your list state whether it can be explained by solar nebular theories.

2.2.5 From Planetesimals to Planets in the Outer Solar System

Figure 2.9 shows a huge increase in the embryo mass beyond the ice line. You might think this is simply the result of the huge mass of condensed water, but comparison with Figure 2.7 shows that this is not the only important factor. In Figure 2.7 there is certainly an important step up in the column mass of the solar nebula at 4–5 AU, i.e. at the ice line, but there is also an underlying decrease in column mass with heliocentric distance. As a result, the column

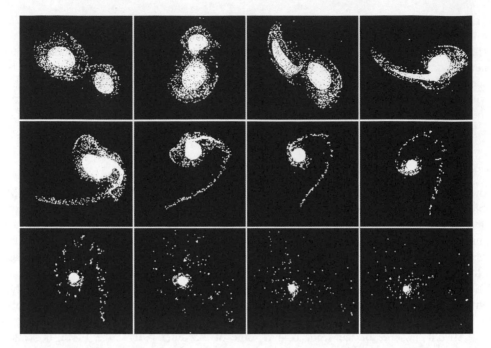

Figure 2.12 One way in which the Moon could have formed from a grazing embryo impact on the Earth. Note the decreasing scale from frame to frame – the Earth (the larger object) is about the same size in all frames. (Reproduced by permission of A G W Cameron)

mass in the neighbourhood of Jupiter is less than in the neighbourhood of the terrestrial planets. Figure 2.9 therefore indicates that, in the models, as the heliocentric distance increases, planetary embryos not only become more massive, but also become fewer in number. In other words, the 'feeding zone' of each embryo covers a wider annular strip of the disc. Consequently, embryo masses of order 10^{26} kg are typical in the models for the Jovian region, i.e. of order 10 times the Earth's final mass!

At greater heliocentric distances it also takes longer to form the embryos from planetesimals, about 0.5 Ma at 4–5 AU, and even longer further out. However, the time required is shorter for smaller planetesimals, and so if there is a trend whereby the greater the heliocentric distance, the smaller the planetesimals, then this would partly offset the increasing embryo formation times. In any case, many planetesimals are left over after the embryos have formed, enabling the embryos to grow.

The embryos are so few and far between beyond the ice line that embryo collisions are very rare, and so the slow embryo-to-final-planet phase that operates in the terrestrial region does not occur. Instead the embryos are massive enough to act as *kernels* that gravitationally capture large quantities of the considerable mass of gas that still dominates the solar nebula.

☐ Which two substances made up nearly all of the mass of this gas?

Hydrogen (as H_2) and helium (as He) together accounted for about 98% of the mass of the nebula, and for nearly all of the gas component. At first, the rate of capture of gas by the kernels is low, and it is estimated that it takes several times the kernel formation time for the capture of a mass of gas equal to the initial kernel mass. At this point, the capture rate is much higher and it is rising rapidly with further mass increase – there is a runaway.

As nebular gas is captured it undergoes self-compression, to yield an envelope with an average density that grows as its mass increases. As well as gas, the growing giants also capture a small but significant proportion of the surviving planetesimals, which still account for nearly 2% of the mass of the nebula. These icy–rocky bodies partially or wholly dissolve in the envelope, particularly in its later, denser stages. Icy materials dissolve more readily than rocky materials, so some preferential accretion of rocky materials onto the kernel might occur. On the other hand, convection in the envelope opposes core growth, so the further central concentration of icy–rocky materials might be slight. The (runaway) capture of gas is halted by the T Tauri phase of the proto-Sun, when the high radiation and particle fluxes sweep the remaining nebular gas into interstellar space.

We can thus account for the presence of giants in the outer Solar System. However, some critical timing is seen to be essential when we look at the *differences* between the giants, notably the decrease in mass with increasing heliocentric distance (item 8 in Table 2.1). In the models outlined so far, the key to understanding this trend is the increasing time it takes to reach the runaway stage with increasing heliocentric distance. If the T Tauri phase occurs *after* the onset of runaway at Jupiter and Saturn, but *before* it starts at Uranus and Neptune, then we can account qualitatively for the lower masses of Uranus and Neptune. This truncation of gas capture by Uranus and Neptune also explains their smaller proportion of hydrogen and helium; Chapter 5 presents incontrovertible evidence for this. Figure 2.13 is one possible time line for the formation of the giant planets. Again, this is illustrative, not definitive. As in Figure 2.10, the end of the T Tauri phase is around the time marked 'nebula rapidly disappearing'.

☐ What would have been the consequence of a much later T Tauri phase?

If the T Tauri phase had been much later, then all the giants would now be more massive than Jupiter.

After the T Tauri phase the giants must have captured further icy–rocky planetesimals. These will have added only very slightly to the total mass, but could have significantly enriched the envelopes in icy and rocky materials.

Non-zero axial inclinations of the giants could readily result from the off centre accumulation of mass – the same sort of explanation that can account for the axial inclinations of the terrestrial planets. But Saturn has a rather large inclination, 26.7°, and the planet is far too massive for off centre accumulation to be the cause. Instead, the explanation probably lies in the rate of Saturn's axial precession being equal to the rate of regression of the nodes of Neptune's orbit.

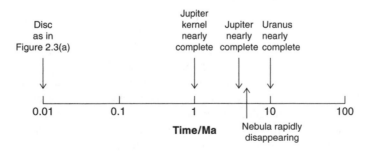

Figure 2.13 A possible time line for the formation of the giant planets. The times are far more uncertain than the sequence of events.

Quite why this resonance explains the large inclination of Saturn is beyond our scope. The inclination of Uranus is remarkable (97.8°), though for this less massive planet this can be explained by the accumulated effects of icy–rocky planetesimals that just happened to nudge the rotation axis predominantly in one direction. Alternatively, an impact with a large embryo could account for it.

This is the **core-accretion model** of giant planet formation.

The time line in Figure 2.13 does present us with some difficulties. The first is that Neptune's kernel might not have formed before the T Tauri phase had swept far too much of the gas away for Neptune to acquire anywhere near the amount of hydrogen and helium that it contains. The second is that the T Tauri phase could have been earlier, which gives us the same difficulty with Uranus as we have with Neptune.

Both of these difficulties can be overcome if we include giant planet migration in our models. These models also account neatly for other features of the Solar System.

Giant planet migration in the Solar System

If Uranus and Neptune formed closer to the Sun than we find them today (but not as close as Jupiter or Saturn), their kernels could have formed rapidly enough so that by the time the gas disc was removed by the T Tauri wind of the Sun they had acquired their modest hydrogen–helium envelopes. In this case they must have since migrated outwards.

The evidence for giant planet migration in exoplanetary systems (Section 2.1.2) lends credibility to the view that migration has indeed occurred in the Solar System. Note that the migration of terrestrial planets is slight – their masses are too small to excite appreciable spiral density waves in the disc. By contrast, the giant planets could well have migrated significant distances. Indeed, it is possible that one or more (growing) giants were consumed by the proto-Sun. But clearly four survived, possibly through gravitational interactions between them, via the spiral density waves they each induced in the disc.

But if Uranus and Neptune formed closer to the Sun than they are today, and remained there until the solar nebula had largely dissipated, how could they have moved outwards? The answer is that there is another way for migration to occur. This is via the scattering of planetesimals from one giant to another.

A recent computer simulation starts, after clearance of the gas disc, with Jupiter, Saturn, Uranus, and Neptune in circular orbits at the respective solar distances 5.5 AU, 8.2 AU, 14.2 AU, and 11.5 AU (yes, Neptune is placed closer to the Sun than Uranus).

☐ So, which way does each giant need to move?

Jupiter needs to move inwards to 5.2 AU, Saturn outwards to 9.6 AU, Uranus outwards to 19.2 AU, and Neptune outwards to 30.1 AU. The model crucially includes a large population of planetesimals in the range 15–35 AU, which constituted the Edgeworth–Kuiper (E–K) belt at this time, though more distant bodies are not excluded. From the details of the simulation it emerges that Uranus and Neptune scattered planetesimals predominantly inwards, and as a result these two giants gained angular momentum, and thus moved outwards. Saturn is also a net inward scatterer, so also moves outwards, though not by much owing to its large mass. Jupiter is a net outward scatterer, so moves inwards, again not by much. Many of the outward-scattered planetesimals escape into interstellar space; those that do not quite make it contribute to the Oort cloud. Uranus and Neptune move into the E–K belt, ejecting many objects, some of which make a major contribution to a late heavy bombardment in the inner Solar System, others of which become the major component of the Oort cloud. The migration of Jupiter and Saturn disturbs the asteroid belt, and this also contributes to the heavy bombardment.

The simulated migration of Jupiter and Saturn causes them to pass through their mutual 2:1 mean motion resonance. This causes the eccentricity of the orbits of Uranus and Neptune to increase to the extent that they interact, with the outcome that they exchange orbits. Their orbits are reduced in eccentricity by further interaction with planetesimals. After several million years, with the planetesimal population depleted, we end up with the four giant planets in their present orbits. This kind of simulation can also explain the high eccentricity of Pluto, via the outward migration of Neptune, and its capture into its 3:2 mmr with Neptune.

Formation of giant planets by gravitational instability in the disc

In the mid 1990s, before the problem of the slow formation of the Uranus and Neptune kernels had been solved by models that included migration, a radical alternative model was put forward, which solved the problem, and also tackled some other difficulties. In this alternative model, the gas in the outer nebula becomes gravitationally unstable, and fragments of higher density form. Each of these contracts to create, in at least some cases, what is called a protoplanet, rather in the manner that the Sun formed at the centre of the nebula. The fragment further contracts to form the giant. This one stage process is distinctly different from the core-accretion model elaborated above. Note that the gas in the inner nebula is too hot to become gravitationally unstable, so the model does not change the mode of formation of the terrestrial planets.

Initially, this **gravitational instability model** seemed promising, particularly because fragments appear in the models when the nebula is only a few hundred years old. But further modelling has revealed huge difficulties. First, with a solar nebula twice the minimum mass solar nebula, the gas becomes gravitationally unstable only beyond about 10 AU. Even at 14 times the minimum, instability extends inwards to only about 7 AU, still beyond the orbit of Jupiter. Worse, the fragments are themselves unstable, and usually do not form protoplanets. At best, the fragments might promote kernel formation. It is also difficult to see how Uranus and Neptune can be so different in composition from Jupiter and Saturn. Therefore, the core-accretion model is secure, for the time being.

Question 2.8

If the proto-Sun went through its T Tauri phase much *earlier* than in Figure 2.13, what might the planets in the outer Solar System be like today?

2.2.6 The Origin of the Oort Cloud, the E–K Belt, and Pluto

Regardless of the model used, the Oort cloud consists of icy–rocky planetesimals that were flung out by the giant planets, but not fast enough to escape from the Solar System. The remaining question is how they became confined to a thick shell rather than retrace the orbit of ejection. This is because their perihelion distances were increased, and the orbital eccentricities consequently reduced, by the overall gravitational force of the stars and interstellar matter that constitute our Galaxy. The force has this effect because it varies across a planetesimal orbit, i.e. it is a differential force.

☐   What is a suitable name for this force?

A differential force is a tidal force (Section 1.4.5), and so a suitable name is the Galactic tidal force, though it is usually called the **Galactic tide**. The planetesimal orbits were subsequently randomised in orientation by this tidal force and also by passing stars and giant molecular clouds, yielding a spherical shell of 10^{12}–10^{13} bodies greater than a kilometre across, 10^3–10^5 AU from the Sun. The Oort members were thus emplaced (Section 1.2.3), perhaps on a time scale as short as 10^3 Ma. In Section 3.2.6 you will see how the Oort cloud can account for some of the comets observed today in the inner Solar System.

In the migration model outlined in the previous section, Uranus and Neptune generated the majority of Oort cloud members, with Jupiter making a smaller contribution.

This model, and others like it, also show that few planetesimals within about 40 AU of the Sun survived the migration of Neptune to its final orbit, with a semimajor axis of 30.1 AU. Kepler's third law shows that an object with a semimajor axis of about 40 AU will orbit the Sun three times for every two orbits of Neptune – a 3:2 mmr. As Neptune migrated outwards it captured bodies into this resonance and swept them before it, thus clearing the space. The bodies beyond 40 AU are a mixture of planetesimals, even embryos, that formed from the solar nebula and have always resided there, and those scattered by the giant planets to modest distances. This mixture constitutes the E–K belt of icy–rocky bodies, with known sizes up to about 1500 km radius, and clustering around the mid plane of the erstwhile solar nebula.

That there are no giant planets beyond Neptune is readily explained by the low spatial density of objects in the E–K belt and their slow orbital motion, resulting in a very low collision rate, and the lack of sufficient nebular gas to reduce their eccentricities and hence their collision speeds – Figure 2.11 illustrates this in a different context. In Section 3.2.6 you will see that the E–K belt makes a further contribution to the observed comets.

In models that do not involve giant planet migration, the great majority of E–K objects (EKOs) are icy–rocky planetesimals that formed more or less where the E–K belt resides today.

The best-known member of the E–K belt is the planet Pluto, an icy–rocky body with a radius of 1153 km. With a semimajor axis of 39.8 AU it is in the 3:2 mmr with Neptune. The migration model shows that as Neptune migrated outwards it would have captured Pluto into this resonance when Neptune was at about 25 AU.

❒   What would have been the semimajor axis of Pluto's orbit at this time?

From Kepler's third law (equation (1.3)) this would have been $25 \times (3/2)^{2/3} = 33$ AU (to two significant figures). As Pluto was pushed outwards in this resonance a secular resonance would have increased the orbital inclination of Pluto to about that observed, 17.1°. Its orbital eccentricity would also have increased, again to about that observed, 0.25. This high eccentricity means that Pluto comes closer to the Sun than Neptune – see Figure 1.5. The orbits do not intersect because of their different orbital inclinations. Moreover, owing to the 3:2 resonance, Pluto is always near aphelion at the times the orbits are close, so Pluto and Neptune avoid the close approaches that would otherwise destabilise Pluto's orbit. As mentioned in Section 1.2.3, Pluto is not the largest EKO.

Interactions between large EKOs can account for Pluto's satellites, via capture, and the strange orbit of Triton (Section 2.3.1). Collisions of large EKOs with Neptune can explain its large axial inclination, 28.3°.

With Pluto and the E–K belt in place, there was ejection of some of the remaining objects in the giant region, and the collisional evolution of smaller objects everywhere, including the generation of dust. Thus, we have the Solar System as we see it today.

In Sections 3.2.6 and 3.2.7 we shall return to the Oort cloud and the E–K belt, and, in the case of the latter, explore its populations in more detail.

Question 2.9

In a few sentences, discuss whether the E–K belt could blend into the Oort cloud.

2.3 Formation of the Satellites and Rings of the Giant Planets

2.3.1 Formation of the Satellites of the Giant Planets

With the exception of Triton, all of the massive satellites of the giants, and many of their less massive satellites, orbit the planet in the same direction as the planet rotates, and in a plane tilted at only a small angle with respect to the equatorial plane of the planet (Table 1.2). This orderly arrangement is strong evidence against separate formation and capture, and strong evidence for formation in a disc of dust and gas around each planet, called a protosatellite disc. In some ways this mimics the formation of the planets from the disc of gas and dust around the proto-Sun, but it differs in one important respect – most of the angular momentum in the giant planet system is in the rotation of the planet and not in the orbits of the satellites. Therefore, there is no need to transfer angular momentum away from the planet, in contrast to the proto-Sun.

In the models, the protosatellite disc is composed of material attracted to the growing giant, but that fails to be incorporated into it. The material forms a cloud of gas, dust, and planetesimals. Interactions within the cloud and between the cloud and the planet cause the cloud to evolve into a thin disc in the equatorial plane of the planet, and orbiting in the same direction as the planet is rotating. Though much of the icy–rocky material in the disc is lost to interplanetary space, coagulation and accretion occur, building up the satellites. The time scale for satellite formation in this way is short, of order 1000 years, resulting in internal satellite temperatures up to about 1000 K, a consequence of the gravitational energy released during accretion. For satellites formed further out, the accretion is slower, so less heat is buried, and the accretion temperatures are correspondingly lower. Remnant gas in the system is lost during the T Tauri phase of the proto-Sun.

The surface temperature of Jupiter reaches about 1000 K – the result of infall of material from the nebula to Jupiter's outer envelope. The luminosity is then high enough for a significant rise in the temperature of the inner part of the protosatellite disc, supplementing the accretional heat. Therefore, the satellites that form close to Jupiter are expected to be more depleted in icy materials than those that form further away. In particular they should be depleted in the proportion of *water* that they contain, the temperatures in the solar nebula at Jupiter always being too high for appreciable condensation of the more volatile substances.

❐ Given that water is less dense than rocky materials, why are the densities of the Galilean satellites (Table 1.2) in accord with this prediction of a decreasing proportion of water with decreasing distance from Jupiter?

Table 1.2 shows that the densities of the Galilean satellites increase with decreasing distance from Jupiter, which is consistent with a decrease in water content. The lower internal temperatures further out have resulted in Callisto, the outermost Galilean, being undifferentiated, i.e. being a fairly uniform mixture of ice and rock, in contrast to Ganymede and Europa, where rocky materials form a core overlain by water (ice at the surface, liquid deeper down). Io has lost its water.

The known densities of the major satellites of Saturn, Uranus, and Neptune, i.e. with radii greater than a few hundred kilometres, show no trend with distance from the giant, and no density high enough to suggest a lack of icy materials. This could be the result of the lower masses of these giants and the correspondingly reduced heating of the inner protosatellite disc.

The smallest satellites are highly irregular in shape – rocky or icy–rocky bodies need to have radii greater than a few hundred kilometres for their own gravity to pull them close to spherical form. Small satellites are prone to collisional fragmentation, particularly the inner ones where the space is crowded with planetesimals gravitationally attracted to the giant and accelerated to high speeds. Therefore, any newly formed ice-poor satellite of modest mass in the inner region is readily disrupted. Satellites subsequently forming in this region are built of substances from across a wide range of distances from the giant, and compositional differences are consequently diluted. Among the smaller satellites of the giants there are some that seem to be collisional fragments. For example, Jupiter's Amalthea, irregularly shaped with a mean radius of 84 km, has a density so low that it must be a pile of rubble created by several collisions. Some of the medium-sized satellites display evidence that they were once disrupted but reformed from the fragments, e.g. Saturn's Enceladus (253 km radius).

Many small satellites far from their planet are in irregular orbits, i.e. with high inclinations and eccentricities, and a high proportion retrograde. For example, Saturn has at least 20 small irregular satellites, most of them in retrograde orbits (most of these are below the size threshold for inclusion in Table 1.2). These properties fit the capture model very well. Moreover, capture is easier far from the planet – large orbits require the captured body to lose a smaller fraction of its orbital energy than do small orbits, which is why the captured satellites are mainly far out. Capture requires the proximity of a third body, additional to the planet and the incoming object. This is typically a satellite already in the system. Disruption of the incomer and of the satellite is a likely outcome.

Just one *large* satellite might have been captured. This is Triton, one of the largest satellites, 1.6 times the mass of Pluto, and by far the largest satellite of Neptune. It is unique among the large satellites in that it orbits its planet in the retrograde direction (Table 1.2). This is strong evidence that Triton was indeed captured, presumably from the E–K belt. Immediately after capture its orbit would have been eccentric and perhaps inclined at a large angle with respect to Neptune's equatorial plane. Once captured, its gravitational interaction with Neptune would have reduced the eccentricity and the inclination of its orbit in about 500 Ma. Its eccentricity is now indistinguishable from zero. Tidal heating would have caused Triton to differentiate.

The capture of Triton could have been accomplished through its collision with one or more satellites each just a few per cent of Triton's mass. This event would have wreaked havoc with any emerging or fully formed satellite system. The orbit of another satellite of Neptune, Nereid, might bear witness to this. Its large eccentricity and large semimajor axis (Table 1.2) could be the result of the capture of Triton. Nereid's orbit would have remained peculiar because of its large average distance from Neptune. If Triton was captured, then its broad similarity in size and density to Pluto suggests that it might initially have been a large member of the E–K belt. This is also Triton's origin in an alternative explanation, in which Triton was originally one member of a binary E–K object. This system passed so close to Neptune that it was disrupted. As a result, Triton had its speed reduced to the extent that it was captured by Neptune. Its erstwhile companion had its speed correspondingly increased, with the likely outcome that further encounters with the giant planets resulted in its ejection from the Solar System.

Question 2.10

Examine Table 1.2 and list the satellites of the giant planets, additional to Triton, that are likely candidates for a capture origin. Justify your choices.

2.3.2 Formation and Evolution of the Rings of the Giant Planets

All four giants have rings (Figure 2.14), and these are particularly extensive in the case of Saturn (Plate 18). For all giants, the rings are close to the planet and lie in the equatorial plane. The

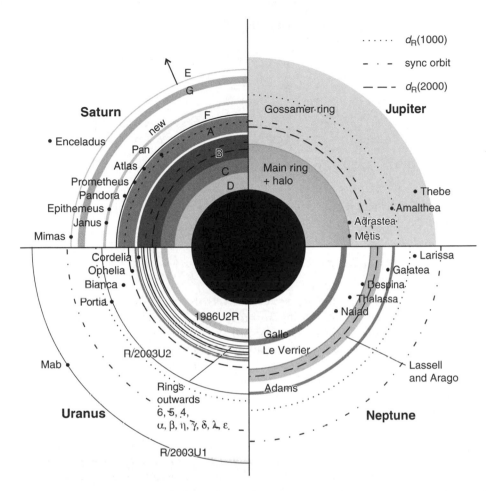

Figure 2.14 Simplified diagram of the ring systems of the giant planets, scaled so that all four giants are the same radius. $d_R(1000)$ and $d_R(2000)$ are, respectively, the Roche limits of bodies with densities $1000\,kg\,m^{-3}$ and $2000\,kg\,m^{-3}$. The synchronous orbits are also indicated.

rings consist of small bodies called ring particles. Very few are more than 1 m across, and the great majority are far smaller, down to less than $1\,\mu m$. The rings are very thin – even in the case of Saturn they are no more than about 100 m thick, and their total mass is only of order 10^{-5} times the mass of the Earth!

Rings and the Roche limit

An important concept relating to the origin of the rings is that of the Roche limit, named after the French scientist Edouard Albert Roche (1820–1883) who in 1847 derived the eponymous limit. It arises from the tidal force that one body exerts on another. Figure 2.15 shows the tidal distortion of a body of mass m a distance d from a body of mass M. If m is held together only

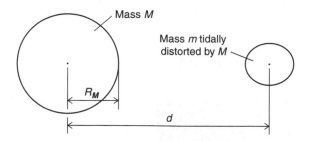

Figure 2.15　Tidal distortion, to illustrate the Roche limit.

by gravitational forces, and if both bodies are of uniform density, then it can be shown that m is torn apart if the distance d is less than d_R, where

$$d_R = 2.44 \times R_M \left(\frac{\rho_M}{\rho_m} \right)^{1/3} = 1.51 \times \left(\frac{M}{\rho_m} \right)^{1/3} \tag{2.3}$$

R_M is the radius of M, and ρ_m and ρ_M are respectively the densities of m and M. The second form is obtained from the first form using $M = \rho_M \times (4\pi/3)R_M^3$. The quantity d_R is the **Roche limit**. As one might expect, d_R increases as M increases and as ρ_m decreases.

Equation (2.3) applies to an initially uniform spherical body held together only by the gravitational attraction of one part on another. Bodies are also held together by non-gravitational forces. These operate at short range, binding molecule to molecule. By contrast, gravity, being proportional to $1/r^2$ (equation (1.5), Section 1.4.4), is a long-range force, and it is therefore the dominant cohesive force in large bodies. Therefore, the Roche limit applies only to bodies that are sufficiently large for gravitational cohesion to dominate – this explains why astronauts and satellites in Earth orbit are not torn apart by the tidal force of the Earth. For non-porous solid bodies made of icy or rocky materials, gravitational cohesion dominates only when they are more than a few hundred kilometres in radius. For poorly consolidated bodies, such as comets and loose aggregates, gravity dominates down to far smaller sizes.

Thus, any sufficiently large body that strays within the Roche limit will be torn apart. The resulting fragments will be numerous and in similar orbits, and therefore collisions among them will be frequent, resulting in further fragmentation. Any tendency to reassemble is counteracted by tidal disruption. After hundreds of millions of years the material evolves to a population of bodies that are predominantly smaller than a metre. This process is a plausible source of ring particles. But the Roche limit also provides a second source – the disruption of bodies already within this limit that are growing by accretion.

☐　How can this happen?

If such a body grows larger than the size at which gravitational cohesion predominates, it is then disrupted by tidal forces. Both processes continue today.

The importance of the Roche limit is illustrated by Figure 2.14, which shows that not much ring material exists outside the limits. The small quantity that does is readily explained by the inward spiralling of dust produced outside the limit.

Ring particles, ring lifetimes

Observations show that Jupiter's ring particles seem to be largely devoid of volatiles, and are probably composed mainly of silicates. Those of Saturn seem to be 'dirty snowballs' – mainly water ice mixed with a trace of less volatile substances, including rocky materials. This can be explained by the higher temperatures expected in the inner protosatellite disc of Jupiter than in Saturn's disc. This left Jupiter with ice-poor materials for its initial and subsequent populations of ring particles. The ring particles of Jupiter and Saturn also differ in size, with most of Saturn's particles being in the range 0.01–1 m, and most of Jupiter's being far smaller.

☐ How can the greater average size of Saturn's ring particles be explained?

This can be put down to the survival around Saturn of water ice, which is an abundant substance.

Little is known about the composition of the ring particles of Uranus and Neptune. They are very dark and for an unknown reason seem to be less icy than Saturn's particles. Their low reflectivity might be the result of solar wind action on **hydrocarbons** (compounds of carbon and hydrogen). Silicates are presumably also present.

Different-sized ring particles are affected differently by a variety of processes acting on them. One of several gravitational processes arises from the slight departure from spherical symmetry of the giant planet's gravitational field. The outcome depends on whether the orbital period of the particle is greater or less than the giant planet's rotation period. If these two periods are equal then the particle (or any other orbiting body) is said to be in a **synchronous orbit** (Figure 2.14). In such an orbit there is zero effect. In a closer orbit the outcome is a slow spiralling towards the giant, whereas in a larger orbit the outcome is a slow spiralling *outwards*. This effect tends to clear the rings of bodies of all sizes, but the replenishment rate is higher for small particles, and so the net effect is a downward trend in the size distribution.

Another gravitational effect occurs in close encounters between particles in nearly identical orbits. After the encounter is over the inner particle is in an even smaller orbit, and the outer one is in an even larger one. This effect is greater, the larger the mass of the particles, and thus it also causes a downward trend in the size distribution of the ring particles. The observed scarcity of ring bodies larger than a metre or so can be explained by these two gravitational effects.

Two other effects are greater, the *smaller* the body. As a body is swept by solar radiation it encounters the photons rather in the manner that you encounter raindrops when you are running – the front of you collects more raindrops than your back. The effect of the extra photon bombardment on the leading face of a body is to decelerate it. This is the **Poynting–Robertson effect**, named after the British physicist John Henry Poynting (1852–1914) and the US cosmologist Howard Percy Robertson (1903–1961). For a ring particle the effect is to cause it to spiral towards the giant. The effect is greater, the smaller the particle, because the magnitude F of the net force exerted by the bombardment is proportional to the surface area of the particle, whereas the magnitude of the deceleration (or acceleration) is given by F/m where m is the mass of the particle (equation (1.4), Section 1.4.4). The area, and hence F, are proportional to the square of the particle's mean radius r_m, and m is proportional to its cube, so F/m is proportional to r_m^{-1}. The Poynting–Robertson effect explains the sparseness of ring particles within the inner edge of the rings – particles of sizes that typify the rings traverse this inner region rapidly in their downward spiral.

The second effect is really a group of effects involving electromagnetic forces. A proportion of ring particles is electrically charged through the action on them of electrons and ions in the vicinity of the giant planet. These charged ring particles are then susceptible to electromagnetic forces exerted not only by the planet's magnetic field, but also by the electric and magnetic forces exerted by the ions and electrons that charged the ring particles. As for the Poynting–Robertson effect, small bodies suffer greater accelerations, and therefore electromagnetic forces are particularly important at micrometre sizes and below. Additional removal mechanisms that affect small particles are collisions, including collisions with micrometeorites sweeping in from interplanetary space. Collisions fragment or remove small particles. Bodies of all sizes are removed by the gravitational effects of satellites.

Ring particle lifetimes of 10–100 Ma have been estimated, which is much shorter than the 4600 Ma age of the Solar System. For Saturn, evidence of such a relatively short lifetime is the brightness of the rings, which darken under micrometeorite bombardment. The narrowness of Uranus's rings indicates youth, because rings tend to spread with age. In the case of Jupiter the particles are so small that they spiral into the planet in no more than about 1000 years.

Persistent sources of ring particles are therefore needed. Disruption within the Roche limit has already been described. The fragments from this disruption will collide and produce ring particles. A relatively recent disruption might explain why Saturn's rings are the most massive. Today, all the ring systems have small satellites interspersed among them, and perhaps100–1000 greater than 1 km in size await discovery. These small bodies are ground down, partly by existing ring particles, partly by micrometeorite bombardment, which is though to provide the major source of fresh ring particles. Micrometeorites themselves can become ring particles. In the case of Jupiter, a significant contribution comes from the volcanic emissions of Io, which consist mainly of silicates. The likely composition of the small satellites matches what we know about the composition of the rings.

Ring structures

The ring systems are structures of exquisite complexity (Plate 18, Figure 2.14). Electromagnetic forces and gravitational forces are responsible for this fine structure too. Of particular note are the gravitational effects of satellites, not only the large satellites well outside the rings, but also small satellites embedded within the rings. Their gravitational effects sustain much of the fine structure. The rings are a playground for modellers. Here, we merely list some of the types of structure seen. Further Reading contains publications where the rings are discussed in much greater detail.

- Narrow gaps between rings, either containing a small satellite or cleared by an mmr with a satellite.
- Narrow rings confined by small satellites.
- Dark radial rings where ring particles have been raised by electrostatic forces.
- Eccentric and inclined rings.
- Density variations around a ring.

And so on, including waves, kinked, and braided rings. What a feast!

Question 2.11

Discuss whether, at some time in the future, compared with today
(a) the ring system of Saturn could be much *less* extensive;
(b) the ring system of Jupiter could be much *more* extensive.

2.4 Successes and Shortcomings of Solar Nebular Theories

The solar nebular theory outlined in this chapter accounts for many of the features in Table 2.1. It accounts for other features too. Overall, it is a successful theory. Perhaps the most worrying aspect left unexplained is the 7.2° tilt of the solar rotation axis with respect to the ecliptic plane. Whereas it is not difficult to account for the axial tilts of many of the planets' axes by the effect of material added asymmetrically, it is less easy to understand how the addition of material to the proto-Sun could have been sufficiently asymmetrical. A possible explanation is that there was a close encounter between the proto-Sun and another young star in the cluster in which the Sun was born.

As with any scientific model, the nebular theory is not fully explored. Perhaps the most important area that needs further exploration is the timing of the T Tauri phase with respect to the evolution of the nebula. This is crucial to the final configuration of a planetary system.

❐ Why is this?

The T Tauri phase clears gas from the nebula, and thus halts the growth of the giant planets (Section 2.2.5).

Nevertheless, the great majority of astronomers believe that solar nebular theories are in fairly good shape, and offer by far the best type of theory that we have for the origin of the Solar System.

2.5 Summary of Chapter 2

The origin of the Solar System cannot be *deduced* from its present state, though this state is an essential guide for the construction of theories, as are observations of other planetary systems, of star formation, and of circumstellar discs.

Most astronomers believe that solar nebular theories offer the correct explanation of the origin of the Solar System. In these theories the Solar System, including the Sun, forms from a contracting fragment of a dense interstellar cloud. As the fragment contracts it becomes disc shaped, and at its centre the proto-Sun begins to form. Dust in the inner disc evaporates. As the temperatures in the disc decline, dust condenses and, along with pre-existing dust, settles towards the mid plane of the disc, where it coagulates into planetesimals, and these undergo accretion to form planetary embryos. In the inner Solar System embryos come together and accrete smaller bodies, ultimately to form the terrestrial planets, consisting of rocky materials. In current theories the Moon is the result of a collision between a massive embryo and the Earth late in the Earth's formation.

In the outer Solar System most embryos reach several Earth masses, the result of fewer embryos forming and the condensation of water beyond the ice line. These embryos – called kernels – thus have an icy–rocky composition. They are generally too far apart to come together but are massive enough to capture nebular gas, mainly hydrogen and helium, a process that stops when the proto-Sun goes through its T Tauri phase and blows the gas out of the Solar System. This is the core-accretion model. Icy–rocky planetesimals are also captured, and this capture continues at a low rate today. The rate of growth of the giants decreases with increasing solar distance, so the T Tauri phase, if correctly timed, can explain the decrease in mass from Jupiter, to Saturn, to Uranus and Neptune, and the associated decrease in mass of the hydrogen–helium envelopes. For Uranus and Neptune it is better for them to have formed closer to the Sun than we find them today, otherwise their kernels, particularly that of Neptune, could well have formed long after the nebular gas had disappeared. Such migration is easily attained in the models by outward migration of the fully formed Uranus and Neptune (after the nebular gas is cleared) through the scattering of planetesimals. This causes Saturn to migrate outwards slightly and

Jupiter inwards slightly. We can thus account not only for the existence of giants beyond the terrestrial planets, but also for the broad differences between them.

An alternative (less favoured) means of forming the giant planets (though still within the context of solar nebular theories) is by a one-stage process in which each giant forms from a fragment of the nebula that contracts to become a protoplanet, and then contracts further to become the giant. This is the gravitational instability model.

Estimates of the time it took for the Solar System to evolve from the formation of a nebular disc to the virtual completion of its formation are of the order of 100 Ma for the terrestrial planets, and about 10 Ma for the giant planets including migration.

The distribution of angular momentum in the Solar System is thought to be the result of the transfer of angular momentum by the proto-Sun to the disc through turbulence in the disc. The Sun also lost much of its angular momentum via its T Tauri wind, and through the trapping by the Sun's magnetic field of ions in the solar wind.

The disc of gas and dust that gave birth to the planets would have been rotating in the same direction as the solar rotation, giving rise to prograde planetary orbits roughly in the same plane. The axial inclinations are less well ordered partly because of off centre acquisition of material as the planets grew, and, at least in the case of Saturn, probably because of a resonance.

The asteroids are the result of failed accretion due to the gravitational influence of Jupiter. Jupiter also stunted the growth of Mars, though it speeded the final stages of growth of the other terrestrial planets. The comets are thought to be icy–rocky planetesimals that become active when they enter the inner Solar System. There are two distinct source populations. First, the far-flung Oort cloud, which is thought to be the result of icy planetesimals flung out by the giant planets during their migration. Second, the Edgeworth–Kuiper (E–K) belt extending from just beyond Neptune, which is thought to be a mixture of icy planetesimals and embryos, some having formed *in situ*, the others having been pushed out by any outward migration of Neptune. Pluto is probably a large member of the E–K belt, and so too might be Triton. There are a few EKOs known to be larger than Pluto.

The rings and most of the satellites of the giants are derived from discs of material in orbit around the giants. The Moon is thought to have originated from the impact on Earth of an embryo 10–15% of Earth's mass.

Solar nebular theories are successful in that they account for most of the broad features of the Solar System in Table 2.1.

3 Small Bodies in the Solar System

We turn now from the Solar System in general to look in more detail at the smallest bodies that orbit the Sun – asteroids and bodies that appear as comets in the inner Solar System. These are interesting in their own right, but they are also of importance in our attempts to understand the larger bodies, so it makes sense to consider the small bodies first. We shall start with the asteroids, then go on to comets, and conclude with meteorites – small interplanetary bodies that have reached the Earth's surface, where they can be collected and studied in much greater detail than we can study any body in space.

3.1 Asteroids

Until 1 January 1801, interplanetary space between Mars and Jupiter seemed empty, puzzlingly so, because the Titius–Bode rule (Section 1.4.3) had indicated the existence of a planet about 2.8 AU from the Sun. Therefore, a systematic search for the 'missing' planet was started in 1800 by 12 German astronomers. On 1 January 1801 the missing planet was discovered, not by a member of the German team, but by the Italian astronomer Giuseppe Piazzi (1746–1826) during routine stellar observations at Palermo. The new body was called Ceres, and though it was close to 2.8 AU from the Sun (2.766 AU today), it was regarded as disappointingly small. The modern value of its radius is 479 km – about a quarter that of the Moon. The German search therefore continued, and by 1807 had revealed three further asteroids: Pallas (at 2.772 AU), Juno (at 2.668 AU), and Vesta (at 2.361 AU). Each of these bodies is much smaller than Ceres, which is by far the largest asteroid (Table 1.3).

Even the larger asteroids are so small that in nineteenth-century telescopes they looked like points, as did the stars – 'asteroid' means 'resembling a star'. Alternative names are 'minor planet' and 'planetoid'. Now, we can see the largest asteroids as extended objects, and spacecraft have visited several. We can also detect very small ones, down to less than 1 km across. At a size of the order of 1 m there is a somewhat arbitrary change in terminology, smaller bodies being called **meteoroids**. At sizes below a few millimetres we have **micrometeoroids**, and below about 0.01 mm we have **dust**.

Modern catalogues list over 20 000 asteroids that have had their orbits accurately determined. These tend to be the larger bodies. Table 1.3 lists the orbital elements of the largest 15. Over 100 000 more asteroids have been *seen*, but have not had accurate orbits established. Overall, we have probably seen all of the asteroids greater than 100 km across (238), but only a tiny fraction of the small ones – it is estimated that there are about 10^9 with sizes greater than 1 km.

Discovering the Solar System, Second Edition   Barrie W. Jones
© 2007 John Wiley & Sons, Ltd

The smaller the size, the greater the number, but the total mass of all asteroids is dominated by the largest few. If an estimate of the order of 10^{22} kg for the total mass of the present-day asteroids between Mars and Jupiter is correct, then Ceres accounts for about 10% of this total. It is estimated that so much mass has been lost since the birth of the Solar System that a few times 10^{25} kg must have been present initially between Mars and Jupiter. The Earth's mass is 6×10^{24} kg (to one significant figure), so if there had been substantially less mass loss, Mars would have been more massive and there would have been one or two more terrestrial planets beyond Mars.

In Chapter 2 you saw that the asteroids are thought to be derived from the planetesimals and embryos that were in the space between Mars and Jupiter. Jupiter prevented the build-up of a major planet in this region, and scattered much of the material to other regions. Interactions between the asteroids and Jupiter have continued, and also between the asteroids themselves. This has resulted in considerable fragmentation and reduction of the asteroid mass over the 4600 Ma history of the Solar System.

Asteroid collisions continue, and the smaller fragments are subject to the **Yarkovsky effect** (Osipovich Yarkovsky, Russian civil engineer, 1844–1902). It results in the removal of bodies with sizes in the approximate range 0.1–100 m. The effect arises from the afternoon side of a rotating body being the hottest. It therefore radiates to space more photons than elsewhere on the surface. Photons carry momentum, and so these (infrared) photons act like a weak rocket. If the body orbits in the same direction as it rotates, i.e. both directions are prograde or retrograde, then the body is pushed in the direction of its orbital motion, and it gradually spirals outwards. Conversely, if the directions are opposite to each other, the body gradually spirals inwards. At a few AU from the Sun, in the asteroid region, the migration rate is about 0.1 AU in 10–50 Ma. In no great time, this causes the body to encounter a mean motion resonance with Jupiter, which usually results in ejection. On bodies outside the approximate 0.1–100 m size range, the Yarkovsky effect is not important: larger bodies are too massive, and smaller ones are more affected by the Poynting–Robertson effect (Section 2.3.2) and radiation pressure (Section 3.2.2)

3.1.1 Asteroid Orbits in the Asteroid Belt

Figure 3.1 shows the distribution of the semimajor axes of the orbits of the asteroids. You can see that the great majority of values lie in the range 1.7–4.0 AU, with a particular concentration in the range 2.2–3.3 AU. The asteroids with semimajor axes in these two ranges constitute respectively the **asteroid belt** and the **main belt**. The orbital inclinations in these belts are fairly small, with few values above 20°, so the asteroids are part of the prograde swirl of motion in the Solar System, though on the whole the inclinations are larger than those of the major planets' orbits. The orbital eccentricities are also somewhat larger, with values of 0.1–0.2 being typical.

☐   If the semimajor axis a of an asteroid's orbit is 3.0 AU, and the eccentricity e is 0.30, what are its perihelion and aphelion distances?

From Figure 1.7 (Section 1.4.1), the perihelion distance is $(a - ae)$, which is 2.10 AU, and the aphelion distance is $(a + ae)$, which is 3.90 AU. This shows that a main belt asteroid even with an atypically large orbital eccentricity does not stray from the space between Mars and Jupiter.

Figure 3.1 also shows that the semimajor axes are not distributed smoothly. A prominent feature is the **Kirkwood gaps**, named after the American astronomer Daniel Kirkwood (1814–1895) who first detected them. These gaps are semimajor axis values around which there are few asteroids (because of orbital eccentricity it does *not* follow that these are depleted zones *in space*). They correspond to mean motion resonances (mmrs) between the asteroid and Jupiter.

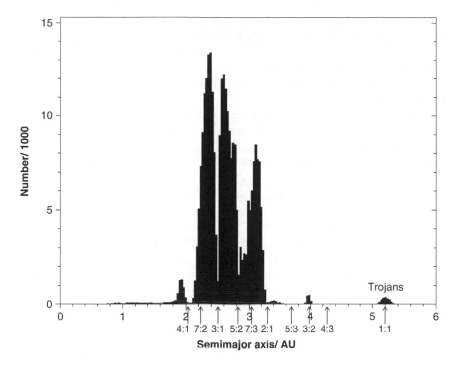

Figure 3.1 The distribution of the semimajor axes of the orbits of the asteroids in October 2006. (Adapted from data available at the Minor Planet Center)

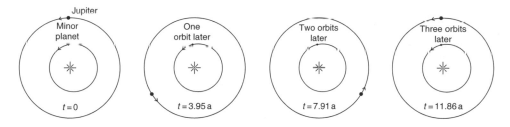

Figure 3.2 The 3:1 mmr of an asteroid with Jupiter. Figure 1.14 illustrates the 2:1 resonance.

Figure 3.2 illustrates the 3:1 mmr, in which an asteroid would orbit the Sun three times whilst Jupiter orbited the Sun once, i.e. the orbital periods are in the ratio 3:1. This means that if Jupiter and the asteroid are lined up, then three orbits later the line-up is exactly repeated. The periodically repeated alignments make many resonant orbits unstable, leading to an increase in eccentricity that might be abrupt and large, characteristic of chaotic behaviour. This results in the asteroid crossing the orbits of one or more of Mars, the Earth, and Jupiter. There is then a high ejection probability within a time of the order of only 0.1 Ma.

In Figure 3.1 the Kirkwood gaps are particularly noticeable at the mmrs 4:1, 3:1, 5:2, 7:3, and 2:1. Simulations show that at the 2:1 mmr the removal process is inefficient, and so the depletion of asteroids here might owe something to the original distribution of matter in the asteroid belt. The dearth of asteroids beyond 3.3 AU, which defines the outer edge of the main belt, can be explained by present-day resonances, plus an inward migration of Jupiter by a few tenths of an AU early in Solar System history (Section 2.2.5), which would have swept mmrs through this region. Close to Jupiter asteroids have been removed by one-shot processes, capture, or scattering, as can be performed by any planet. The location of the inner edge of the main belt near 2.2 AU seems also to be the result of Jupiter's gravity, though in this case orbital resonances are not involved.

In a few cases, resonant orbits have an *excess* of asteroids. In Figure 3.1 the 3:2 resonance with Jupiter shows just this effect, the corresponding asteroids constituting the Hilda group. A factor that helps explain why they have not been removed by Jupiter is that when the Hilda group have Jupiter near opposition they are near perihelion, and so close approaches are avoided (cf. Figure 1.14).

Another feature of asteroid orbits is grouping into families. The members of a family have similar semimajor axes, orbital inclinations, and eccentricities. In the early years of the twentieth century the Japanese astronomer Kiyotsugu Hirayama (1874–1943) discovered several such families, now called **Hirayama families**. Each family typically has several hundred known members. The orbital similarities within each family indicate that the members are the collisional fragments of a larger asteroid. This view is supported by the similar reflectance spectra (Section 3.1.6) observed across the members of most families, spectra that are distinct from those of other families. It is estimated that more than 90% of the asteroids in the asteroid belt are in families.

Major collisions, including the sort that produce families, are thought to be occurring once every few tens of million years on average. One outcome is that at least one of the two bodies involved in the collision reassembles as a gravitationally bound rubble pile, with a low density. Another outcome is that small fragments provide some asteroids with small satellites, e.g. Ida has the tiny Dactyl.

3.1.2 Asteroid Orbits Outside the Asteroid Belt

A few thousand asteroids are known to have orbits with semimajor axes outside the range 1.7–4.0 AU. They thus lie outside the asteroid belt, and several interesting groups have been identified.

Near-Earth asteroids

As their name suggests, **near-Earth asteroids** (NEAs) are asteroids that get close to the Earth. Some even share the Earth's orbit, each one with a distance from the Earth along the orbit that oscillates over a large range. Collisions between NEAs and the Earth cannot be ruled out. Over 600 NEAs are known, and various estimates based on this population put the total number up to several thousand greater than 1 km across. Of those discovered, 1950DA has the greatest chance of hitting the Earth. This would occur on 16 March 2880, but only with a 0.3% probability. This NEA's direction of rotation is unknown, but if the sense is opposite to that of the orbit, then the Yarkovsky effect will reduce the strike probability to zero.

The closest approach by an asteroid in recent decades was on 10 August 1972. A fireball was observed, and filmed, grazing the Earth's upper atmosphere over North America. The asteroid is estimated to have been 3–6 m across. Since then, the closest approach was on 18 March 2004. The Lincoln Near Earth Asteroid Research (LINEAR) survey saw an object about 30 m across that passed the Earth at a range of 42 700 km. In 1908 the most recent substantial encounter with Earth occurred, in the Tunguska region of Siberia, causing devastation in an area about

80 km across, thankfully lightly populated. The body is estimated to have been 50–75 m across and broke up in the atmosphere in a huge explosion. The average time between such encounters is estimated as roughly 1000 years. To reach the ground intact an asteroid would have to be considerably larger, depending on its composition. An intact body around 200 m across is a 1-in-100 000 year event, on average, perhaps less often, but would cause a global climate catastrophe, threatening human civilization.

A famous big hit was 65 Ma ago, when an asteroid 10–14 km across fell in Yucatan, and probably put enough debris into the atmosphere to cool the Earth to the extent that contributed to many species dying out, including all the dinosaurs.

Each NEA belongs to one of three well-known groups, each taking its name from a prominent group member. The Amors have semimajor axes greater than 1 AU, but perihelion distances between 1.017 AU and 1.3 AU.

☐ What is the significance of 1.017 AU for the Earth's orbit?

This is the aphelion distance of the Earth. The Apollos have perihelia less than 1.017 AU, so, if their orbital inclinations were zero, orbits with aphelia larger than Earth's perihelion distance of 0.983 would intersect the Earth's orbit. Even with non-zero inclination, intersection occurs if the ascending or descending node intersects the Earth's orbit (Section 1.4.2). The Atens have semimajor axes less than 1 AU. With non-zero eccentricity intersection with the Earth's orbit can occur.

Through the influence of the terrestrial planets, the orbital elements of the NEAs vary, and therefore most of them will collide with the Sun sooner or later, and others will collide with the Earth or with another terrestrial planet. It is estimated that the average orbital lifetime is only a few million year. This is very short compared with the 4600 Ma age of the Solar System, so replenishment must occur. The asteroid belt is undoubtedly a major source, the 3:1 orbital resonance being particularly copious. Comets are another source too, as you will see (Section 3.2.1).

Trojan asteroids

The **Trojan asteroids** share Jupiter's orbit – they are in a 1:1 mmr with Jupiter. Figure 3.3 shows where the Trojan asteroids are concentrated. Over 1000 are known, the largest being Hektor, 330 km by 150 km. There are probably more than 1000 very small ones yet to be discovered. The total mass is estimated to be of the order of 10^{21} kg, about 0.01% of the Earth's mass. The Trojans cluster around two of what are called the **Lagrangian points** of Jupiter and the Sun. Figure 3.3 shows all five of these points, labelled L_1–L_5. They are named after the Franco-Italian mathematician Joseph Louis Lagrange (1736–1813), who predicted their existence. They arise in a system of two bodies in low-eccentricity orbits around their centre of mass plus a third body with a much smaller mass. The five points are where the third body can be located and remain close to that position relative to the other two. Thus, the whole configuration can be thought of as rotating like a rigid body about the centre of mass. Regardless of the ratio of the masses of the two main bodies, the points L_4 and L_5 are located as shown in Figure 3.3. By contrast, the locations of L_1 and L_2 do depend on the mass ratio, and lie closer to the less massive body. For the remaining point, if, as in the case of the Sun and Jupiter, the mass of one of the two bodies is much greater than that of the other, then L_3 lies very close to the orbit of the less massive body. The positional stability of a small mass placed at L_1, L_2, or L_3 is poor, but at L_4 and L_5 it is much better. Objects at L_4 and L_5 need not remain exactly at the point, but can follow orbits around it. It is at L_4 and L_5 that the Trojans cluster.

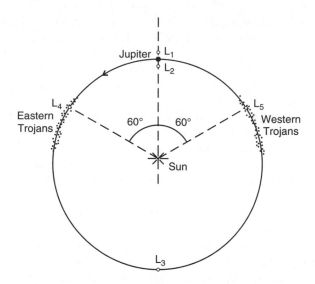

Figure 3.3　The five Lagrangian points with respect to Jupiter and the Sun, and the Trojan asteroids.

In the case of the Sun and Jupiter, the stability of L_4 and L_5 is disturbed by the presence of the other planets. Nevertheless, the Trojans can reside at and even orbit some way from these points for the present age of the Solar System. The origin of the Trojans is unknown. Each planet has Lagrangian points with respect to the Sun. By 2006, four Trojans of Neptune had been detected, but no other planet has yet had any Trojans confirmed. (The NEAs that share Earth's orbit wander very far from L_4 and L_5.)

The Centaurs

There is a handful of small bodies known with orbits that lie among the giant planets. The first of these to be discovered, Hidalgo in 1920, is a body about 15 km across, in an orbit with an inclination of 42.5°, and so eccentric that it extends from 2.01 AU to 9.68 AU from the Sun.

☐　What is Hidalgo's semimajor axis?

This is $(9.68\,\text{AU} + 2.01\,\text{AU})/2 = 5.85\,\text{AU}$ (Figure 1.7). Its highly eccentric, highly inclined orbit suggests that it might be the remains of a comet nucleus (Section 3.2.4). Small bodies with perihelia greater than that of Jupiter's semimajor axis (5.2 AU) and semimajor axes smaller than that of Neptune (30.1 AU) are called **Centaurs**. The first Centaur to be named as such was Chiron, discovered in 1977. It is about 180 km across, and ranges from 8.46 AU to 18.82 AU from the Sun. However, Chiron has low surface activity suggestive of a comet, so it is classified both as an asteroid and as a comet. In 1991 Pholus became the third asteroid to be discovered beyond Jupiter, in an orbit that takes it from 9 AU to 32 AU. It is also about 180 km across. A few tens have since been added to the list, ranging in size down to a few tens of kilometres. At such large distances from us, bodies smaller than this are difficult to discover – it has been estimated that there are a few thousand Centaurs greater than 75 km across. Several Centaurs are known to have (weak) surface activity.

Calculations indicate that their lifetime as Centaurs is about 1–10 Ma (perhaps 100 Ma in some cases) before they suffer huge orbital changes. Computer simulations show that their source is the E–K belt, from which they are drawn by the gravitational influence of Neptune. In this case they are icy–rocky in composition. Over about 1–10 Ma (perhaps 100 Ma) a Centaur will suffer a huge orbital change, resulting in its ejection from the Solar System or a reduction of its perihelion to the point where the Sun evaporates its ices and it becomes a short-period comet (Section 3.2.1).

Clearly, the Centaurs blur the distinction between asteroids and comets. They are probably best regarded as a population transient between the E–K belt and the short-period comets.

Question 3.1

For the 15 asteroids in Table 1.3
(a) identify any unusual orbital elements, and state why each is unusual;
(b) discuss whether any of the 15 could be at the Lagrangian points L_4 or L_5 of the Sun plus any planet.

3.1.3 Asteroid Sizes

With few exceptions, the asteroids are too small and too far away from the Earth to be seen as anything other than points of light in the sky. The exceptions include the largest few asteroids, a handful of NEAs, and a few that have been imaged from close range by spacecraft. Their sizes are thus known from direct observations. In addition, a dozen or so have passed between us and a star, the size then being obtained from the accurately known rate at which the asteroid moves across the sky, and the length of time for which the starlight is blocked. For the great majority of asteroids, sizes have to be obtained by indirect means.

An important indirect method depends on the measurement of the flux density of the reflected solar radiation that we receive from the body. **Flux density F** is a general term defined as the power of the electromagnetic radiation incident on unit area of a receiving surface. Our receiving surface will be perpendicular to the direction to the asteroid, and F spans the wavelength range of the whole solar spectrum.

Let us assume that the asteroid is in opposition, so that the Sun, Earth, and asteroid are near enough in a straight line. In this case the asteroid is seen from the Earth at what is called zero phase angle, as in Figure 3.4(a), and the flux density received by reflection is labelled $F_r(0)$. It can be shown that

$$F_r(0) = kpA \tag{3.1}$$

where k is a combination of known factors involving the Sun and the distance to the asteroid, A is the projected area of the asteroid in our direction (Figure 3.4(a)), and p is a quantity called the **geometrical albedo**. This is the ratio $F_r(0)/F_L(0)$, where $F_L(0)$ is the flux density we *would* have received from a flat Lambertian surface perpendicular to the direction to the Sun and Earth, and with an area equal to the projected area of the asteroid (Figure 3.4(b)). A Lambertian surface is perfectly diffuse (the opposite of a mirror) and reflects 100% of the radiation incident upon it.

☐ Is the form of equation (3.1) reasonable?

It is reasonable that $F_r(0)$ increases with p and also with A.

If we know p, and if we have measured $F_r(0)$, then A can be obtained from equation (3.1), and the mean radius can then be estimated. The value of p depends on the composition and roughness of the surface. Plausible surfaces have values of p ranging from about 2% to about

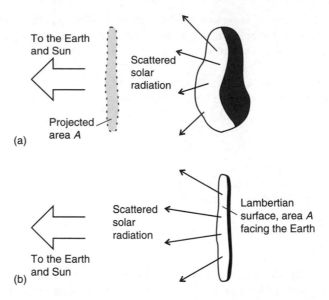

Figure 3.4 (a) An asteroid in opposition. (b) A flat Lambertian surface with the same projected area as the asteroid.

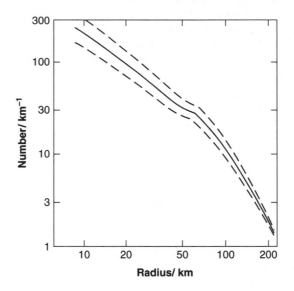

Figure 3.5 The number of asteroids with mean radii per km interval of radius. The dashed lines indicate uncertainties.

40%, and so, unless we can reduce this range, the size of the body can only be obtained to an order of magnitude. Comparison of $F_r(0)$ with the flux density F_e of the infrared radiation *emitted* by the asteroid by virtue of its temperature has provided estimates of p, as have studies of the polarisation of the reflected solar radiation. The details will not concern us.

Indirect techniques have provided nearly all the size data in Figure 3.5. The data have not been extended to sizes much below 10 km radius because at smaller sizes there must be a significant proportion of undiscovered asteroids, the proportion increasing with decreasing size. Even with this cut-off, the general increase in the number of bodies with diminishing size is clear. The general shape of the curve is consistent with an initial population modified by mutual fragmentation as a result of collisions. You have seen that it is thought that during the formation of the Solar System, there were several embryos between Mars and Jupiter, plus a host of smaller bodies, and that further growth was halted by the formation of Jupiter, whose gravitational influence 'stirred up' the asteroid orbits. This increased collision speeds to the point where net growth was replaced by net fragmentation. Note that this is *net* fragmentation – some reassembly of fragments is to be expected to create rubble piles. And so it continues today.

Question 3.2

Even though the number of asteroids increases as size decreases (Figure 3.5), it does *not* follow that the host of tiny asteroids doubtless awaiting discovery will add significantly to the mass of those already discovered. Explain why this is so.

3.1.4 Asteroid Shapes and Surface Features

The term 'radius' in Figure 3.5 has to be interpreted with care. The self-gravitational forces inside an asteroid result in an increase of pressure with depth. If the pressure exceeds the strength of the solid materials in the interior, then the materials yield, and a roughly spherical body results, flattened by rotation at high rotation rates. The internal pressures decrease as the size of the body decreases, and there comes a point where the materials do not yield, and so the body need not be even approximately spherical. For an asteroid made of silicates and iron, the critical radius is about 300 km. Therefore, below this size 'radius' means an average distance from the surface to the centre of the asteroid – the body is not necessarily anywhere near spherical.

▢ Are any asteroids certain to be (very nearly) spherical?

From Table 1.3 you can see that only Ceres is significantly larger than the critical radius. Therefore, we can be confident of a near spherical shape, and this is found to be so. Note that if a body is significantly non-spherical we use the word 'across' rather than 'radius'. In some cases two or three dimensions represent the body fairly well.

A few asteroids have had their shapes determined directly, either from images or from the way that the light we receive from them varies when a planet passes between the asteroid and us. Hubble Space Telescope (HST) images of Ceres showed that it is nearly spherical, with an equatorial radius of 487 km and a polar radius of 455 km, a difference of only about 30 km. Significant departures from spherical form occur at smaller sizes, also as expected. Figure 3.6 and Plate 10 show images of asteroids that display their non-spherical shapes. A few other main belt asteroids have been imaged, including Vesta, Gaspra, and Ida. The image of Vesta in Figure 3.6(a) was obtained from Earth orbit by the HST (this is a model, based on the image). The tiny asteroid Gaspra (Plate 10) was imaged in October 1991 by the Galileo spacecraft en route to Jupiter, and in August 1993 the same spacecraft imaged the larger asteroid Ida (Figure 3.6(b)) and a previously unknown tiny satellite, named Dactyl (beyond the edge of the frame). Outside the main belt, but within the asteroid belt, Mathilde (Figure 3.6(c)) was imaged in 1997 by the spacecraft NEAR. Toutatis is one of a few NEAs that have been imaged by

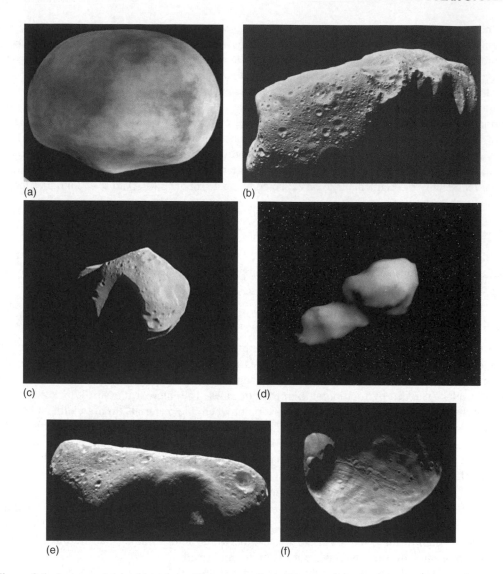

(a)   (b)   (c)   (d)   (e)   (f)

Figure 3.6   Images of asteroid surfaces. The mean radius of Vesta and the longest dimension of the others are given. (a) An image-based model of Vesta, 256 km. (AURA/STScI, NASA, PRC97-27, P Thomas and B Zellner) (b) Ida, 53 km. (NASA/NSSDC P42964) (c) Mathilde 59 km. (The Johns Hopkins University, Applied Physics Laboratory) (d) Toutatis, 4.6 km. (NASA/JPL P46256, S J Ostro and S Hudson) (e) Eros, 33 km. (NASA/JPL-CalTech, PIA03141) (f) Phobos, 26 km. (NASA/NSSDC 357A64)

Earth-based radar. The image of Toutatis in Figure 3.6(d) was obtained in December 1992 when it passed the Earth at a range of only 4×10^6 km. Eros, another NEA, was orbited by NEAR in 2001, which made a landing on 12 February 2001. Figure 3.6(e) is an image of Eros from NEAR when it was in orbit. If, as is thought, the Martian satellites Phobos and Deimos are captured asteroids then these also belong to the list of those imaged. Figure 3.6(f) is a Viking Orbiter image of Phobos; Deimos is broadly similar.

Many asteroids have had their shapes determined indirectly. All asteroids rotate, almost all with periods in the range 4–16 hours, and as they rotate, the observed flux densities F_e and $F_r(0)$ vary with time. There are two possible contributory factors: surface features, and changes in the projected area resulting from non-spherical form. From the variations in F_e and $F_r(0)$ these two factors can be separated. The outcome, as expected, is that large departures from spherical form are common among asteroids. When repeated imaging of an asteroid is possible, the rotation period can be obtained directly. For example, the HST has discovered a dark patch on Ceres, and repeated observations have yielded a rotation period of 9.1 hours.

The rotational data also show a broad range of axial inclinations. These, and the range of rotational periods, are consistent with models in which small bodies frequently collide. Irregular shapes are expected from collisional fragmentation, though in some cases a small asteroid might retain the irregular form it acquired at its origin. The heavily cratered surfaces of asteroids (Figure 3.6) also bear witness to collisions, in this case with smaller bodies. Irregular forms can also arise from the subsequent sublimation of volatile components. Conversely, small bodies can become more spherical through erosion by dust impact. Surfaces exposed to repeated impacts are expected to acquire thin layers of dust, and observations of light reflected and IR radiation emitted indicate that a thin dust cover is indeed a common feature.

3.1.5 Asteroid Masses, Densities, and Overall Composition

A few asteroids have had their masses measured with useful precision. Close approaches of small asteroids to Ceres, Pallas, and Vesta have yielded mean densities of $2100\,kg\,m^{-3}$, $2710\,kg\,m^{-3}$, and $3440\,kg\,m^{-3}$ respectively. The orbit of Ida's tiny satellite Dactyl indicates a density for Ida of $(2600 \pm 500)\,kg\,m^{-3}$. The NEAR mission to Eros in 2001 obtained $(2500 \pm 800)\,kg\,m^{-3}$. Itokawa, another NEA, has been orbited by the Japanese spacecraft Hayabusa since September 2005, and has obtained a density of about $2500\,kg\,m^{-3}$. All of these densities are consistent with high proportions of rocky materials and, except for Vesta, low proportions of the much denser metallic iron. By contrast, the effect of Mathilde on the path of NEAR corresponds to a density of only $(1300 \pm 200)\,kg\,m^{-3}$. Such a low density could be due to hydrated substances in abundance, but analysis of the light reflected from the surface indicates otherwise. Therefore, the low density might indicate high porosity, such as could arise from disruption and (partial) reassembly as a rubble pile. High porosity helps to explain how Mathilde could have survived the large impacts that produced its heavily cratered surface – porosity cushions impacts, thus preventing disruption. A low density has also been obtained for Sylvia, which has two small satellites. The value is about $1200\,kg\,m^{-3}$ which might mean that Sylvia is also a collision-created rubble pile, its satellites being fragments from the collision. Many tens of asteroids are known to have satellites, and therefore each of these asteroids could have its density determined.

Rocky materials thus seem to dominate the few NEAs and asteroids in the asteroid belt that have had their densities measured with useful precision. From such a small sample we certainly cannot infer that all the asteroids are mainly rocky, not even the NEAs and the asteroid belt members, far less so the Trojans and Centaurs. Indeed, surface compositions vary hugely, and these lead to inferences about the interiors, as you will see in the next section.

Question 3.3

(a) Why have gravitational forces failed to make the bodies in Figure 3.6 more spherical?
(b) If a small asteroid were spherical, what would this tell us about its possible histories?

3.1.6 Asteroid Classes and Surface Composition

The surface composition of an asteroid can be inferred from a combination of various types of data. These include ratios of the reflectance in different wavelength bands, derived from flux density measurements. If the bands are very narrow, numerous, and contiguous the measurement is called **spectrometry**, otherwise the measurement is called **photometry**. Further data include the geometrical albedo, and the polarisation induced in reflected solar radiation.

A first step towards compositional determination is to use the observational data to divide asteroids into different classes. This makes the problem more manageable – if we can obtain the (surface) composition of one member of a class, then this is probably similar to that of the other members of the class. Various classification schemes have been proposed. The one we shall describe was devised by the American astronomer David J Tholen in 1983, and it is widely used. It is based on the narrow band reflectances at eight wavelengths in the range 0.3–1.1 μm, plus the geometrical albedo. Fourteen classes are recognised, and Figure 3.7 shows the mean reflectance spectrum of each class, plus an indication of the albedo (high, medium, or low). Note that the classes E, M, and P are distinguished only by albedo. Some classes are represented by very few members. For example, until recently the V class was represented only by Vesta

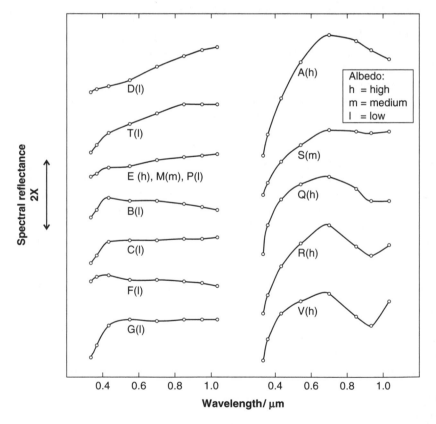

Figure 3.7 Reflectance spectra and geometrical albedos of the 14 Tholen asteroid classes. Note that the reflectances show spectral *shape* on a logarithmic scale, not absolute values, and that the spectra are offset vertically by arbitrary amounts.

(hence the 'V'), and even now only a handful have been added, all of them very small, and in orbits that suggest they are collisional fragments of Vesta. Classes R and Q also contain only a handful of members.

About 80% of classified asteroids fall into the S class, and about 15% into the C class. Members of the C class have geometrical albedos in the approximate range 2–7%, so they are very dark. Ceres, the largest asteroid, is C class, with an albedo of 7.3%.

❏ From Figure 3.7, how would you characterise the *colour* of S and C class asteroids?

At visible wavelengths the reflectances of C class asteroids do not vary much with wavelength, so they are rather grey. The slightly greater reflectances at longer wavelengths gives them a hint of red. S class asteroids are distinctly red, and they have higher albedos, about 7–20%. In Figure 3.6, Ida is S class (as is Gaspra, Plate 10), whereas Mathilde is C class. Eros is also S class, as are most of the NEAs, including Itokawa. The surface composition of Itokawa has been obtained by the IR spectrometer on Hayabusa. It found it to be mainly the silicates olivine and pyroxene (Table 2.3), plus possibly some plagioclase (another silicate) and iron.

By comparing reflectance spectra with laboratory spectra of various substances, and with the aid of albedo and other observational data on asteroids, including radar reflectance, likely surface materials can be identified. The outcome is that class M asteroids match alloys of iron with a few per cent nickel, mixed with little or no silicates. Class S match mixtures of similar iron–nickel alloys with appreciable proportions of silicates. Class C match a type of meteorite called the carbonaceous chondrite, of which more later, but which consist of silicates mixed with hydrated minerals, plus small quantities of iron–nickel alloy, carbon, and **organic compounds**. These are compounds of carbon and hydrogen, often with other elements. Carbon and organic compounds are collectively called **carbonaceous materials**, and they are mainly responsible for the low albedos of class C. Classes P and D are broadly like class C, but correspond to material that is richer in carbonaceous materials. Class V match some subclasses of a type of meteorite called the achondrite, silicate meteorites of which, again, more later.

Note that these mineral matches are to the surface of the asteroid – it could be very different in its interior. Note also that there could be (rare) cases where the matches are merely coincidences, and that the surface composition of the asteroid is quite different from that of the corresponding type of meteorite.

Asteroid classes across the asteroid belt, and asteroid differentiation

Figure 3.8 shows the distribution with heliocentric distance of the five most populous classes. Fractions are shown, such that at each distance the fractions of all asteroids (not just those shown) sum to one. Some idea of the actual numbers can be obtained by comparing Figure 3.8 with Figure 3.1, though it must be noted that Figure 3.1 shows *observed* asteroids, whereas in Figure 3.8 an attempt has been made to correct for various observational biases: for example, that a greater proportion of high-albedo asteroids must have been discovered than of low-albedo asteroids.

It is clear that the distributions differ from one class to another. *If* the mineralogical interpretations outlined above are correct, then the broadest trend is that mixtures of silicates and iron–nickel predominate in the inner belt (class S), and that carbonaceous materials and hydrated minerals become increasingly predominant as heliocentric distance increases (classes C, P, D).

An explanation of these trends is that the materials now in the outer belt formed there, where the cooler conditions allowed condensation of the more volatile substances, such as carbonaceous materials and hydrated minerals. In the warmer inner belt this was not possible, so we get

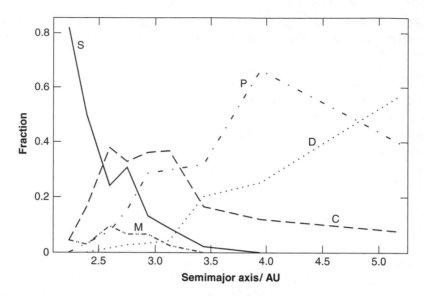

Figure 3.8 The distribution in the asteroid belt of the five most populous classes of asteroid.

only silicates and iron–nickel alloy. This distinction could have been enhanced during the T Tauri phase of the proto-Sun, when the solar wind would have heated the asteroids by magnetic induction, i.e. by the heating from electric currents induced in the asteroids by the action of the magnetic fields entrained in the wind. The heating decreased with heliocentric distance, so asteroids in the inner belt would have been heated more than those in the outer belt. This explanation requires that there has been only limited migration of the different classes across the asteroid belt, and that differences across the belt never became obscured by migration into the belt of planetesimals that formed elsewhere in the Solar System. An alternative explanation places more weight on the loss of volatile materials from the inner belt throughout Solar System history, in which case substantial inward migration of C class asteroids with subsequent modification is a possibility.

Class M asteroids, which have surfaces that are largely or entirely iron–nickel, are presumably iron–nickel throughout – there is no reasonable way of getting such an iron-rich surface and an iron-poor interior. There is no plausible scheme of condensation and accretion in the solar nebula that would give such a silicate-free composition throughout, and it is therefore necessary to assume that internal temperatures in some asteroids rose to the point where the interiors became partially or wholly molten. This allowed a process called **differentiation** to occur, whereby denser substances settled towards the centre of a body, and the less dense substances floated upwards to form a mantle, overlain in turn by a mineralogically distinct crust. The melting could have resulted from heat released by asteroid accretion and collisions, plus heat from the decay of short-lived unstable isotopes, notably the isotope of aluminium ^{26}Al, nearly all of which decayed in a few million years. The asteroid interior would then have cooled, and became solid after a further interval of a few million years.

The temperature rise caused by isotope heating is greater, the larger the body. This is because the mass of the isotopes present is proportional to the volume of the body, whereas heat losses from the body are proportional to its surface area, and the ratio of volume to surface area is greater, the larger the body. The temperature rise from accretion, all other things being equal, is also greater, the larger the body. In a body consisting of mixtures of silicates and iron–nickel alloy (which thus excludes C class), differentiation would have occurred at sizes larger than about 200 km across, and would have resulted in a predominantly iron–nickel core overlain by a mantle and crust largely composed of silicates. There will also be a core–mantle interface consisting of a mixture of iron–nickel alloy and silicates. Collisions can break up these bodies, and fragments of the cores give us chunks of iron–nickel alloy, i.e. class M. Fragments of the mantle and the mantle–core interface could be an important source of S class. The surface properties of Vesta are consistent with the sort of silicates that would form the crust of a fully differentiated body.

The scarcity of M and S classes in the outer belt indicates that differentiation was uncommon there. One explanation is that supplementary heating by T Tauri magnetic induction was too weak at these greater heliocentric distances.

Further discussion of the composition of asteroids is in Section 3.3.4, in relation to meteorites.

Beyond the main belt

The Trojans and Centaurs cannot readily be placed into the asteroid classes outlined above. The Trojans are dark, with albedos in the range 0.03–0.13, similar to the small outer satellites of Jupiter and the other giant planets. Of the small proportion of Trojans that have been classified, most have been placed in D class, and the rest in either C or P.

⊓   In what does this suggest that the surfaces are rich?

This suggests that the surfaces are rich in carbonaceous materials. Trojan spectra are similar to those of the nuclei of short-period comets, and to some of the Centaurs and some of the E–K objects.

There is no spectral evidence for water ice at the surface of any Trojan, though planetesimals at their distance from the Sun would have been icy–rocky, so this could be the typical internal composition. By contrast, there *is* such spectral evidence for some Centaurs, presumably because of their greater average distances from the Sun and the consequent preservation of water ice at their surfaces. The Centaurs otherwise resemble the Trojans, with albedos covering about the same range. The dark surfaces of the Trojans and Centaurs suggest that all the surfaces are rich in carbonaceous materials, further darkened by small impacts that produce dust at their surfaces, and by the bombardment of ions and electrons.

As has been noted, the Centaur Chiron shows evidence of cometary activity. This cannot be driven by the sublimation of water – Chiron is too cold – but it could be driven by CO, CO_2, or NH_3. Moreover, the Centaurs' spectra generally match those of EKOs. This fits with the view noted earlier that the Centaurs are a transitory population between the E–K belt and the short-period comets (except for those Centaurs that are flung out of the Solar System).

Question 3.4

The reflectance spectrum and albedo of the asteroid Eros are shown in Figure 3.9. Explain why it is placed in the S class. Hence deduce its likely surface composition. Where might it have originated?

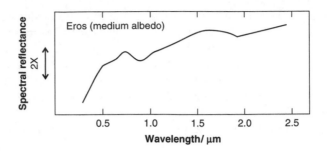

Figure 3.9   The reflectance spectrum (log scale) and albedo of the asteroid Eros.

3.2 Comets and Their Sources

A comet is defined as a body that displays a large, thin atmosphere at some point in its orbit, which can be of gas, of dust, or of both. This is called the coma, and from it develops a very tenuous hydrogen cloud and two tails, as in Plate 22. One of these tails consists of dust, the other of ionised gas. The coma can become as large as Jupiter, the hydrogen cloud larger than the Sun, and the tails as long as a few AU. As seen from the Earth, a comet can be a spectacular sight in the sky for several months. Figure 3.10 illustrates the growth and shrinkage of the tails versus the heliocentric distance as the comet goes around its orbit, indicating that heat from the Sun drives the activity. The source of the coma, cloud, and tails is a solid nucleus, typically a few kilometres across, and this is all that exists of the comet when it is in the outer Solar System. A comet's nucleus must contain sufficient quantities of icy materials to generate the coma, hydrogen cloud, and tails. This indicates that comets formed further out in the solar nebula than asteroids, sufficiently far that solid icy materials were present.

Beyond roughly 10 AU from the Sun, solar radiation is too feeble to create a coma, and this is also the case at smaller distances if there is a devolatilised crust protecting icy materials. We have mentioned that the comets come from the E–K belt and the Oort cloud – more on this shortly. The important point here is that comets are defined by their observed activity. When

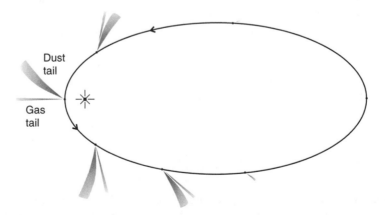

Figure 3.10   The growth and shrinkage of the tails of a comet versus its heliocentric distance.

they are inactive they are usually called something else – EKOs, Oort cloud objects, or, as you will see, dead comets.

3.2.1 The Orbits of Comets

Cometary orbits fall into two main categories, long period and short period. As the names indicate, the categorisation depends on the orbital period. There is no sharp physical division, but the defining orbital period is exact.

Long-period comets (LPCs) have orbital periods in excess of 200 years. In most cases the periods are *greatly* in excess of 200 years, with values extending up to about 10 Ma, and in many cases not measured.

☐ Use Kepler's third law to calculate the semimajor axis a of the orbit of a comet with an orbital period $P = 1$Ma. Express your answer in AU.

From equation (1.3) (Section 1.4.1)

$$a = (P/k)^{2/3}$$

where $k = 1$ year $AU^{-3/2}$. Thus, with $P = 1$ Ma, $a = 10^4$ AU. Such comets are observed only because they have highly eccentric orbits that bring them within a few AU of the Sun. A small number of LPCs arrive on parabolic or hyperbolic orbits, though this could be due to the perturbation of an eccentric elliptical orbit as a comet is on its way towards us. This perturbation can be caused by a close approach to a planet, or by an eruption of gas from the comet. Thus, there is no incontrovertible evidence for any comet having come from interstellar space, though this cannot be ruled out in some cases. If a comet is leaving the inner Solar System on a parabolic or hyperbolic orbit, and if this orbit is not perturbed into an ellipse, the comet will certainly escape.

The huge orbital periods of most LPCs mean that most of them have been observed only once in recorded history. About 1000 different LPCs have been recorded, and about 600 of these have well-known orbits. On average, about half a dozen LPCs are observed per year, and about one per decade becomes noticeable to the unaided eye. A spectacular example was Hale–Bopp in 1997 (Plate 22), and its orbit in the inner Solar System is shown in Figure 3.11(a). One of the factors that determines how spectacular a comet becomes is its proximity to the Earth. Another is its perihelion distance. If this is less than the radius of the Sun (poor comet!) or not much greater than this (Sun-grazing comets), then magnificent tails develop.

The orbital inclinations of the LPCs are randomly distributed over the full range, as are the longitudes of the ascending node and of perihelion (Section 1.4.2). The LPCs thus bombard the inner Solar System from all directions.

Short-period comets (SPCs) are defined as having periods of less than 200 years, and therefore they must have semimajor axes less than 34 AU – comparable with the outermost planets Neptune and Pluto. However, unlike the planets, most of the SPCs are in eccentric orbits, in some cases with $e > 0.9$ (Table 1.4). A few hundred SPCs are known. Of these, roughly half a dozen per year grow bright enough to be visible in a modest telescope.

Many of the SPCs have periods less than 20 years, the majority of these less than 15 years. They move in moderately eccentric orbits with perihelion distances from about 1 AU to a few AU, aphelion distances 4–7 AU, and in nearly all cases with inclinations less than 35° (median value about 11°). Therefore, their aphelia are broadly in the region of Jupiter and so SPCs with periods less than 20 years are often called the **Jupiter family comets** (JFCs). Though all comets

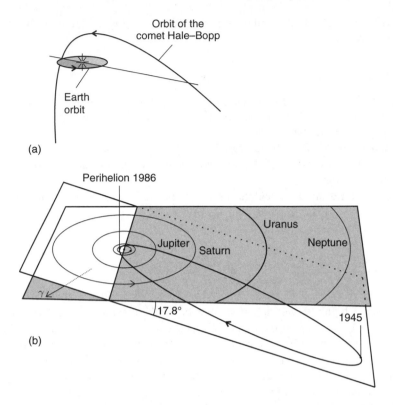

Figure 3.11 (a) The orbit of the long period comet Hale–Bopp in the inner Solar System. (b) The orbit of the short-period comet 1P/Halley.

are prone to orbital changes through gravitational interaction with a planet, this is particularly so for the JFCs. This is because their low-inclination orbits, traversing the planetary zone, give a comparatively high probability of a close encounter with a planet. Such an encounter will result in a drastic alteration of the orbit, with an outcome that can be anything from solar capture to ejection from the Solar System.

Nearly 200 JFCs are known, but survey limitations lead to estimates of the complete population up to a few thousand. In addition, with active lifetimes of the order of a few thousand years, and much longer dynamic lifetimes (before ejection or collision) estimated at about 0.3 Ma, there should be many dead JFCs. Assuming a steady-state population due to resupply, it is estimated that there could be roughly as many dead JFCs as active ones. On the other hand, if devolatilisation usually leads to complete destruction of the nucleus, there would be very few dead JFCs.

The remaining SPCs are typified by the most famous SPC of all, 1P/Halley, and so they are called **Halley family comets** (HFCs). The P denotes a measured period. The members of this small group have periods in the range 15–200 years, and their inclinations are typically larger than for the JFCs, with a few in retrograde orbits. In the 15–20 year period overlap with the JFCs, the distinction between HFCs and JFCs is made on the basis of a comet's orbital elements a, e, and i taken together, in what is called the Tisserand parameter. Its value for the HFCs is distinct from that for JFCs. Consult books on celestial mechanics (Further Reading) for the details.

Halley's comet itself has an orbital period of 76 years and an orbital inclination of 162°, i.e. retrograde (Table 1.4). It has been observed every 76 years or so, as far back as reliable records go – at least as far back as 240 BC. One of its more famous apparitions was in the year of the Battle of Hastings, AD 1066. It is named after the British astronomer Edmond Halley (1656–1742), who noticed that the orbits of the bright comets of 1531, 1607, and 1682 were very similar. He deduced that this was the same comet each time, and he predicted its reappearance in 1758. Halley's comet duly appeared, but alas Halley had died 16 years earlier. It was last at perihelion on 9 February 1986, and in March 1986 it became the first comet to be imaged at close range by a spacecraft. Figure 3.11(b) shows the orbit of Halley's comet.

The number of known HFCs is smaller than the number of known JFCs, but because of their generally higher inclinations and longer periods, it is very likely that a smaller proportion of the HFC population has been discovered than in the case of the JFCs. It is estimated that there could be about twice as many HFCs as JFCs in the complete populations. Like the JFCs, it is assumed that the HFC numbers are roughly in a steady state due to resupply.

Question 3.5

The comet 55P/Tempel–Tuttle has a perihelion distance $q = 0.976\,586\,\text{AU}$, an orbital eccentricity $e = 0.905\,502$, and an orbital inclination $i = 162.49°$. Show that this is an HFC. (Tempel–Tuttle is associated with the Leonid meteor shower, which occurs in November.)

3.2.2 The Coma, Hydrogen Cloud, and Tails of a Comet

The coma that grows when a comet comes typically within 10 AU or so of the Sun is a result of the heating of the nucleus by solar radiation. The coma is a large, tenuous atmosphere, consisting of gases derived from the more volatile constituents of the nucleus, mixed with dust carried aloft by the outgassing. Spectroscopic studies, and measurements made by spacecraft, have shown that the dust in the coma consists of rocky materials like silicates and carbonaceous materials. Such studies also show that except in the innermost part of the coma the gases are predominantly *fragments* of molecules, rather than intact molecules. Such fragmentation is to be expected as a result of the disruptive effect of solar radiation. This is called **photodissociation**, and UV photons are particularly effective. The fragments, and intact molecules, can also be ionised in a process called **photoionisation**, where a solar UV photon ejects an electron.

From the molecular fragments, the parent molecules can be identified. Hydrogen atoms (H), hydroxyl (OH), and oxygen (O) atoms are particularly common, and must have been derived from water molecules (H_2O). Other molecular fragments have been derived from carbon dioxide (CO_2) and carbon monoxide (CO). H_2O, CO_2, and CO have also been detected as intact molecules in the coma. From the relative abundances of molecules and fragments in the coma, it is inferred that the predominant volatile constituent of the nucleus is water, typically accounting for over 80% of the mass of the volatile substances. CO and CO_2 are the next most abundant volatiles. Note that these substances would be present in the nucleus as *solids*. They are sublimed from the nucleus to form gases, and are then photodissociated and photoionised. The more volatile the material, the greater the heliocentric distance at which it can sublime. Water sublimes inwards of about 5 AU, and becomes the main driver of activity. CO_2 sublimes further out. CO is the driver well beyond 5 AU, and thus triggers activity in the LPCs as they move inwards, provided that previous journeys through the inner Solar System have not removed all the CO.

The hydrogen cloud is derived from the coma through the photodissociation of the molecular fragment OH. Even though the cloud can greatly exceed the size of the Sun, there is very little mass involved.

The tails can likewise be huge, extending as far as a few AU from the coma, but they are also very tenuous, so again little mass is present. There are two sorts of tail: a tail stretching almost along the line from the Sun to the coma, and a curved tail that points away from the Sun only in the immediate vicinity of the coma (Plate 22 and Figure 3.10). The spectrum of the radiation received from the curved tail shows it to be a solar spectrum modified in a way consistent with scattering from micrometre (μm) sized dust particles. Therefore, this is a dust tail, and it is seen by the solar radiation that it scatters. By contrast, the radiation from the straight tail shows it to consist of electrons and ionised atoms, plus a trace of very fine dust with particle sizes less than $1\,\mu$m. The evidence is a very weak spectrum like that from the dust tail, and strong spectral lines emitted by the ions, consequent upon absorption of solar radiation. This is the ion tail. One common ion is OH^+, produced from OH in the coma by photoionisation. All the ions that have been identified in the ion tail could have been produced by the ionisation of atoms and molecular fragments in the coma.

It is the strikingly different composition of the two tails that causes their separation in space as they stream away from the coma. The ion tail is swept from the coma by the force exerted on the coma ions by the magnetic field in the solar wind as the field moves across the ions. The details are complicated, but the outcome is that the ions are swept in the direction of the solar wind, i.e. radially from the Sun. The wind speed is much higher than the orbital speed of the comet, so in the time it takes an ion to travel from the coma to where the tail is no longer distinguishable, the comet does not move very far. Consequently, the tail is fairly straight. Because of the intricate structure of the solar wind, the ion tail is highly structured, with filaments and knots, and it can temporarily break away from the coma. The trace of submicrometre dust in the ion tail is carried by the ion flow.

The dust tail is driven from the coma through bombardment by the photons that constitute solar electromagnetic radiation. These carry momentum (as you saw in Section 3.1 in relation to the Yarkovsky effect), and the resulting force on a dust particle is called **radiation pressure**. The dust is driven away from the Sun, but only reaches speeds comparable with the orbital speed of the comet.

☐   So, why is the dust tail curved?

This tail is curved because the comet moves appreciably around its orbit in the transit time of the dust to the end of the visible tail.

The smaller the particle, the greater the acceleration caused by radiation pressure. This is because the area-to-mass ratio is greater for smaller particles, as discussed in relation to the Poynting–Robertson effect (Section 2.3.2). As a result, particles in the coma much greater than a few tens of micrometres in size are retained. The greater effect of radiation pressure on smaller particles raises the question of why it is not an important force on the ions in the tail. This is because, when the particle size is less than the predominant wavelength of the photons, the interaction is enfeebled. For solar radiation the predominant wavelength is about $0.5\,\mu$m, which is very much greater than ion radii.

Often, more than two tails are seen. These extra tails are usually ion or dust tails with slightly different properties, such as the thin tail of neutral sodium atoms seen streaming away from the comet Hale–Bopp.

In January 2006 the NASA Stardust mission returned to Earth with a sample of dust from the vicinity of the comet 81P/Wild 2. Analyses of these samples is providing further information on the composition of comets (Section 3.2.3).

3.2.3 The Cometary Nucleus

From their starlike images in all but the largest telescopes it has long been known that cometary nuclei are small. Size can be estimated in a similar indirect manner to asteroids (Section 3.1.3), but the uncertainties are great. Sizes of the order of a few kilometres are typical, extending up to a few tens of kilometres for the largest comets. The composition of the nucleus can be deduced from the composition of the coma, as indicated in Section 3.2.2. The typical nucleus is deduced to consist of ices, predominantly water, mixed with some CO_2 ice and CO ice, plus rocky and carbonaceous materials.

A great variety of other molecules and molecular fragments have been identified in comas and ion tails, implying the existence of small quantities of other icy substances in the nuclei of some comets. Methanol (CH_3OH), methanal (HCHO), and nitrogen (N_2) are usually the most abundant traces inferred to exist. Some traces in some comets indicate that the icy dust grains that formed their nuclei are interstellar material that has not been heated above about 100 K. On the other hand the LPC Hale–Bopp has isotope ratios for C, N, and S that are the same as in the Solar System in general, indicating that Hale–Bopp's icy grains recondensed from a well-mixed solar nebula.

☐ Why does this suggest that Hale–Bopp comes from the Oort cloud?

Solar nebular theories derive the Oort cloud from icy–rocky planetesimals that condensed in the giant region (Section 2.2.6). Therefore, the composition of Oort cloud comets is expected to resemble that of the nebula. An origin in interstellar space is unlikely because interstellar dust grains were modified by evaporation and condensation in the solar nebula (Section 2.2.2). An origin in the inner E–K belt, which was at least partly populated from the giant region (Section 2.2.6), is ruled out by the long period. The low abundance of neon in Hale–Bopp supports a giant region origin. Except in its inner part, the E–K belt would have been cold enough for neon to condense from the solar nebula, and so neon would now be a more significant component of Hale–Bopp.

Spacecraft missions to cometary nuclei

Our knowledge of cometary nuclei received a huge boost in 1986 when five spacecraft made close observations of 1P/Halley. The European Space Agency's Giotto flew closest, sweeping past at a range of only about 600 km from the nucleus in March 1986, obtaining the image in Figure 3.12(a). Halley was then within 1 AU of the Sun, only a few weeks after its perihelion at 0.53 AU. Consequently, its tails were well developed. The peanut-shaped nucleus is 16 km long, and 8×7 km in typical cross-section. It rotates around its long axis with a period of 170 hours, and this axis precesses with a period of 89 hours around an axis inclined at 66° with respect to the long axis.

The mass of the nucleus was estimated from the effect on Halley's orbit of the forces exerted by gas jets erupting from surface vents. The estimated mass is 10^{14} kg, give or take 50%, and therefore the density of the nucleus is only 100–250 kg m^{-3}, considerably less than the 920 kg m^{-3} of water ice at the Earth's surface. Therefore, the nucleus is not so much a block of dirty ice as a fluffy aggregation of small grains. The effect of jets on the orbits of other comets

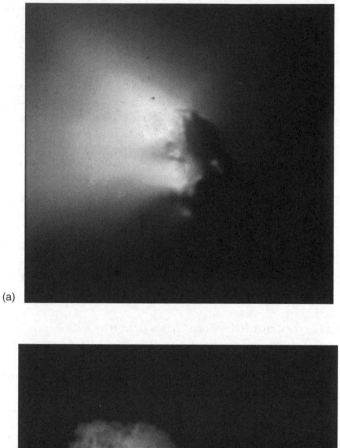

(a)

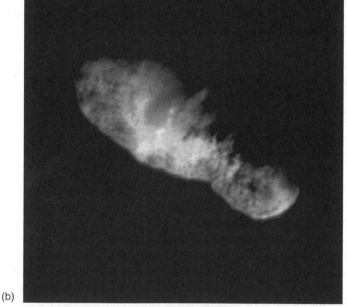

(b)

Figure 3.12 (a) The nucleus of 1P/Halley in March 1986. The long dimension is 16 km, and the Sun is to the left. (ESA 3416 etc. composite. Reproduced by permission of ESA) (b) The nucleus of 15P/Borrelly in September 2001. The long dimension is 8 km. (NASA/JPL PIA03500)

has yielded similar densities, though with much greater uncertainty. The fragility of cometary nuclei is indicated by several that have broken up through their degassing as they approached the Sun, or in the case of Shoemaker–Levy 9 by a close approach to Jupiter.

The spacecraft observations of the coma of Halley's comet, supplemented by ground-based observations, added many details to our knowledge of the composition of the nucleus, but did not change the broad picture very much. It is deduced that Halley's nucleus consists of 80% water, 10% CO, 3.5% CO_2, by numbers of molecules. Definite evidence for methane ice (CH_4) was not obtained, even though this was expected to be relatively abundant. An important detail is evidence that some of the water is probably present in chemical combination with rocky and carbonaceous materials, as water of hydration. Moreover, it seems likely that proportions of the different icy materials are present in what are called **clathrates**, where one material is enclosed in the crystal structure of another. In particular, the rather open crystal structure of water ice can readily enclose molecules of other icy substances, such as CO_2.

The Giotto flyby also confirmed that cometary nuclei can be very dark. Halley has a geometrical albedo of only 3–4%, the result of carbonaceous materials at the surface. Low albedos have since been established for the nuclei of other comets. It seems that, as the icy materials evaporate near the Sun to give the coma and tails, a residue of dust depleted in icy materials concentrates at the surface, where it forms an insulating protective crust over the ice-rich grains beneath. This crust is broken by the vents that spew forth the coma and tail material. Vents tend to switch on when they face the Sun, and switch off when they turn away from the Sun. For 1P/Halley this phenomenon is apparent in Figure 3.12(a).

Vents can explain the transient brightening that some comets exhibit when they are *more* than a few AU from the Sun. Slow evaporation beneath the protective crust would build up the gas pressure to the point where the fluff ruptures, and a vent forms. An interesting example is 29P/Schwassmann–Wachmann 1, which has a nearly circular orbit between Jupiter and Saturn (Table 1.4). Though it is fairly large for a comet – about 40 km across – it would have gone unnoticed but for its outbursts, which occur every year or so. Halley suddenly brightened in 1991, when it was 14 AU from the Sun. This might have been due to a large solar flare which caused shock waves that ruptured the crust.

Since the flybys of 1P/Halley there have been a few more missions to comets. On 22 September 2001 NASA's Deep Space 1 flew past 15P/Borrelly, then 1.36 AU from the Sun and 8 days after perihelion. Deep Space obtained many images, one of which is shown in Figure 3.12(b). You can see that it is a very irregular object, with a longest dimension of 8 km. Its albedo is typically low for a comet's nucleus, only 0.03 and even less in patches – down to 0.007. As for comets in general, this is presumed to be a carbonaceous crust overlying an ice-rich interior. About 90% of its surface is inactive, though a jet on one side makes it visible from Earth.

The NASA Stardust mission in January 2006 returned dust to Earth from the vicinity of JFC 81P/Wild 2. It also imaged the nucleus, a rugged surface, with an albedo of about 3%. There were four to five jets, and circular features that might be impact craters. It is about 5 km across, and has a roughly spherical shape, indicating that it might not be a fragment from a collision. The dust particles contain the sort of materials expected to have been present in the cool outer Solar System, but there are also silicates, including those from the olivine group, that form at high temperatures. These could have been placed in the outer Solar System by jets from the young Sun, or might be pristine interstellar grains, forged by other stars. Further analysis will rule out one of these two possibilities.

Perhaps the most dramatic mission so far was that of NASA's Deep Impact, which, on 4 July 2005, fired a 370 kg copper bullet at $10.2\,km\,s^{-1}$ into 9P/Tempel 1, a $14 \times 4.4 \times 4.4\,km$ JFC, when it was near its 1.51 AU perihelion on 5 July. The goal was to obtain the internal composition. The ESA Rosetta spacecraft made observations before, during, and after the impact. The water content of the ejected dust was measured, and was found, surprisingly, not to be the dominant constituent. Non-icy materials dominate, and though this might be local to the impact site, it is feasible that at least some comets are 'icy dirtballs' rather than 'dirty snowballs'. The activity induced by the impact died after a few days, indicating that such impacts on the crust from meteoroids are not the cause of the longer lived cometary outbursts.

The crust on a JFC's nucleus is estimated to have formed within about 0.1 Ma of its joining this family. For any comet the time taken depends on the accumulated time spent close to the Sun for the surface to devolatilise. But it is possible that comets first arrive with some sort of crust already in place. Prolonged exposure to cosmic rays and UV photons chemically transforms and devolatilises the surface, to form a 1 metre crust in the order of 100 Ma. This is much shorter than the average residence times in the sources of comets – the Oort cloud and the E–K belt (Section 3.2.5). But it is the subsequent growth of the crust that ends activity, and leads to the death of comets.

Question 3.6

In 150–200 words, describe the visual appearance of a comet from when it is about 30 AU from the Sun on its way in, to when it is outgoing at the same distance. Relate the visible changes to events at the nucleus.

3.2.4 The Death of Comets

> When beggars die there are no comets seen: the heavens themselves blaze forth the death of princes.
>
> William Shakespeare (*Julius Caesar*)

Comets die too, because the loss of volatiles is acute within a few AU of the Sun. If perihelion is at 1 AU then the order of 100 perihelion passages will suffice to evaporate all the available ices from a nucleus of typical size, leaving it with a crust so thick that the nucleus no longer has the capability to develop a coma and tails. In some cases this devolatilisation could extend to the centre, in which case the final act of the nucleus is to become dust, perhaps violently. Models indicate that any nuclei smaller than about 1 km across can lose their remaining volatiles sufficiently rapidly to explode. In less extreme cases there is a gentler dissolution to dust. A rapid, jet-driven increase in rotation rate could also disrupt small nuclei. Disruption explains the paucity of small nuclei.

Comets are seen in such last throes of activity – some of the SPCs have very small comas and tails. For example, 133P/Elst–Pizarro shows a very feeble, thin tail. The Infrared Astronomical Satellite (IRAS) that gathered data for nearly the whole of 1983 discovered many Solar System objects with small dusty envelopes. Some of these might be asteroids that never had ices, but are surrounded by fine collisional debris; others might be devolatilised comets.

Elsewhere in the Solar System some members of the low-albedo classes of asteroid, such as the C and D classes, might also be devolatilised comet nuclei. The reflectance spectra and albedos of cometary nuclei resemble those of these classes. For example, Hidalgo (Section 3.1.2) is a D class asteroid with a perihelion of 2.01 AU but with an orbit so eccentric that it has an aphelion of 9.68 AU, unusually distant for an asteroid, and so it is a good candidate for being a comet remnant. Within the asteroid belt there are a few comets that seem to be nearly dead: 133P/Elst–Pizarro orbits in the asteroid main belt, as does the rather more active 2P/Encke, with a period of 3.3 years. Some of the small satellites of the giant planets, particularly those in unusual orbits, might also be cometary remnants, captured by the planet.

Support for the view that some asteroids are dead comet nuclei comes from the Tisserand parameter (Section 3.2.1). Its value is distinct from asteroid values, except for some of the asteroids with albedos around 4%, a value similar to comet nuclei.

A devolatilised comet nucleus has lost not only ices, but also a proportion of its dust, perhaps even all of it. The inner Solar System is pervaded by dust, much of it cometary. The average density of the dusty medium is about $10^{-17}\,\mathrm{kg\,m^{-3}}$, but it is greater along the orbits of comets. (It is even greater in the asteroid belt where dust from asteroid collisions makes a large additional contribution.)

The most dramatic termination of a comet's life is when it collides with another body. As well as collisions with the Sun (Section 3.2.1), collisions with planets also occur. One such was seen in July 1994 – the collision with Jupiter of D/1993 F2 Shoemaker–Levy 9, or rather its fragments. There must have been many other collisions with the planets, including the Earth. Some SPCs have orbits that resemble the Amor and Apollo asteroids, and it is thought that some of these are devolatilised cometary nuclei. The collision of one such nucleus with the Earth might account for the huge explosion in 1908 in the Tunguska River area of central Siberia, though a small asteroid proper is another possibility (Section 3.1.2). There is archaeological evidence for earlier impacts, and in the future the Earth must surely collect further comets. Calculations show that dead JFCs could account for up to 50% of the NEAs.

Question 3.7

As well as perihelion distance, what other orbital property influences the mass of volatile material lost by a comet per orbit? Justify your answer.

Question 3.8

In Section 3.2.1 an estimate of the active lifetime of a JFC was given. Reconcile this with the statement in this section that a comet survives the order of 100 perihelion passages before it becomes inactive.

3.2.5 The Sources of Comets

By now, you will have gathered that there are two sources of comets, the Oort cloud and the E–K belt. These source populations are not called comets – that name is reserved for bodies in orbits such that a coma develops along some part of it. In the sources the bodies are called Oort cloud members and EKOs. In Sections 1.2.3 and 2.2.6 brief descriptions of the Oort cloud and the E–K belt were given. The Oort cloud is a spherical shell of icy–rocky bodies extending from about 10^3 to 10^5 AU. The E–K belt extends inwards towards the Sun to around the orbit of Pluto, and has a flatter distribution, the objects having low to modest inclinations.

As well as being the sources of comets, the Oort cloud and the E–K belt are interesting in their own right, so we describe each of these populations now, and how the comets come from them.

3.2.6 The Oort Cloud

The origin of the Oort cloud was described in Section 2.2.6. You have seen that the members of the cloud are thought to be icy–rocky planetesimals flung out by the giant planets.

❒ According to the giant planet migration model in Section 2.2.5, which region was a particularly copious source of Oort cloud members?

The Uranus–Neptune region would have been a particularly copious source of icy–rocky planetesimals flung outwards. Many were ejected not quite hard enough to escape into interstellar space. The cloud is a thick spherical shell of 10^{12}–10^{13} bodies greater than 1 km across, 10^3–10^5 AU from the Sun (Figure 3.13). In spite of the huge number of comets in the cloud, the total mass is estimated to be only of the order of 10^{25} kg, about the same as the Earth's mass. The cloud is too far away to be observed directly. Its existence has long been inferred from the LPCs.

You have seen that the orbits of the LPCs have aphelia far beyond the planets, and that the aphelia lie in all directions from the Sun. This led the Estonian astronomer Ernst Julius Öpik (1893–1985) to suggest in 1932 that there was a huge cloud of comets surrounding the Solar System, but so far away that only those members with perihelia less than a few AU became visible, through the growth of coma and tails. In 1950 this idea was developed by the Dutch astronomer Jan Hendrick Oort (1900–1992). The Oort cloud is sometimes known as the Öpik-Oort cloud.

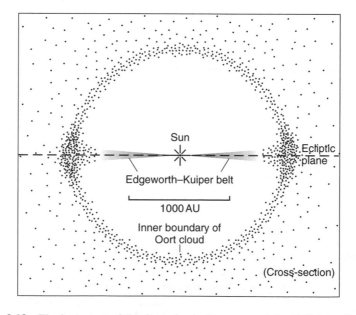

Figure 3.13 The inner part of the Oort cloud of comets, and the E–K belt of comets.

The outer reaches of the Oort cloud are at a significant fraction of the distances between neighbouring stars in the solar neighbourhood – currently the nearest star (Proxima Centauri) is 2.7×10^5 AU away. The stars are in motion with respect to each other, so it is to be expected that from time to time a passing star will perturb the cloud. As a result, some members are drawn out of the Solar System, whilst others have their perihelion distances greatly reduced, so that near perihelion a coma and tails are developed, and a new LPC is observed. Giant molecular clouds in the interstellar medium can have similar effects to stars, as can the Galactic tide (Section 2.2.6). These perturbations on the outer Oort cloud explain the highly eccentric, large, randomly oriented orbits of the LPCs, and the frequency with which these comets are observed. Bodies reaching us from beyond the Solar System might constitute a small proportion of the LPCs.

The HFCs are thought to be LPCs, mainly from the inner Oort cloud, that have had their orbits reduced through interactions with the giant planets.

3.2.7 The E–K Belt

In Section 2.2.6 you saw that the E–K belt is thought to be a mixture of icy–rocky planetesimals composed of a population left over in the giant planet region that was then scattered further out by the giants, and a population formed directly from the solar nebula.

❒ How does the giant planet migration model explain why the space within about 40 AU of the Sun is largely devoid of EKOs?

This was cleared by the 3:2 mmr with Neptune during its outward migration.

Regardless of how it was emplaced, the E–K belt is now thought to be the source of the SPCs. It used to be thought that the SPCs were LPCs that had had their orbits perturbed by the giant planets. However, detailed simulations failed to produce an essential feature of the orbits of the SPCs – namely, orbital inclinations predominantly less than $35°$. By contrast, it is easy to produce this feature from a source population already in low-inclination orbits. Because the SPCs have active lifetimes of a few thousand years before devolatilisation, the population of active SPCs needs to be resupplied. A reservoir of millions of bodies is needed to meet the required rate, and the resupply would occur in two stages. First, an orbit is modified by gravitational perturbations, partly by the outer planets, but particularly by the larger members of the belt itself, of order 10^3 km across. Orbital changes can also result from collisions between EKOs. These result in fragmentation. Second, if the new orbit is such that the object can approach a giant planet, a possible outcome is that the orbit is further modified into one typical of a SPC.

The E–K belt satisfies the requirement for a source population in fairly low-inclination orbits. Its existence was first proposed in 1943, long before EKOs began to be discovered. The idea came from the Anglo-Irish astronomer Kenneth Essex Edgeworth (1880–1972), and eight years later from the Dutch–American astronomer Gerard Peter Kuiper (1905–1973) (which is why it is sometimes called the Kuiper belt). The first EKO was discovered in 1992, and has the name 1992 QB_1 (QB_1 identifies when it was discovered in 1992). It is about 200 km across and occupies an orbit with a semimajor axis of 43.8 AU, an eccentricity of 0.088, and an inclination of $2.2°$.

Over 1000 EKOs are presently known, and their numbers are steadily rising. Very many more await discovery as surveys are extended. Those in the inner E–K belt can presently be detected down to the order of 10 km across, depending on albedo. For a fixed albedo, the brightness decreases as r^{-4}, where r is the distance from the Sun to the EKO – this is a factor r^{-2} for the decrease in solar radiation, and another factor of r^{-2} for the (approximate) distance of the EKO from our telescopes. Therefore, as r increases the population is increasingly undersampled. The estimates of the total population differ widely. One estimate is of at least 10^5 objects greater than 100 km across out to about 50 AU. Thus, given that 50 AU is not the outer boundary, the total population will exceed 10^5, probably by a huge factor for such sizes, and vastly more for sizes greater than 1 km across. The total mass could approach an Earth mass, though other estimates are about a tenth of this, or even less. Figure 3.13 shows the E–K belt blending into the Oort cloud. This is conjectural.

The population of EKOs is divided into three subpopulations: the classical EKOs, the resonant EKOs, and the scattered disc EKOs.

Classical EKOs

These are defined to have perihelion distances $q > 35$ AU, semimajor axes a in the approximate range 40–50 AU, and low eccentricities e, around 0.1. They also have low inclinations i, though this might be an observational selection effect, most searches concentrating near the ecliptic plane. Over 600 are known, accounting for nearly two-thirds of the presently known EKOs. There seems to be a sharp outer edge, which they might have inherited from their birth, or because more distant ones were trimmed off in a close encounter with a star early in Solar System history.

Resonant EKOs

These are the EKOs that have been found in mmrs with Neptune, mostly in the 3:2 resonance, but also a few in the 4:3, 5:3, and 2:1 resonances. The resonances have generally produced larger e and i values than in the classical population.

❑ What are the semimajor axes of these four resonances?

From equation (1.3), $a_{res} = a_N (P_{res}/P_N)^{2/3}$ where $a_N = 30.1$ AU. Thus, with $P_{res}/P_N =$ 1.33, 1.50, 1.67, and 2.00 for the 4:3, 3:2, 5:3, and 2:1 resonances, we get 36.4 AU, 39.4 AU, 42.3 AU, and 47.8 AU respectively. You should recognise 39.4 AU as close to Pluto's current semimajor axis (it varies slightly) of 39.8 AU. The EKOs in this resonance are thus called Plutinos, and over 100 are known, though it is estimated that roughly 1500 larger than 100 km across await discovery.

Recall that the Plutinos are thought to have been pushed there as Neptune migrated outwards. Some Plutinos, and Pluto, have perihelion distances less than 30 AU and so cross Neptune's orbit. Like Pluto, the position of each Plutino in its orbit is such as to avoid a close encounter – a configuration maintained by the 3:2 resonance. If this were not so, the Plutino would not be there!

Scattered disc EKOs

The scattered disc EKOs (SDOs) are characterised by eccentricities greater than those of the classical EKOs, the dividing line being somewhat arbitrary, but 0.25 is in the midst of the various proposals. Values up to 0.9 have been observed, corresponding to aphelia of several hundred AU. Such extremes might be due to a stellar encounter. SDOs also have a greater range of inclinations than the classical objects, extending above 20°. Their semimajor axes are predominantly greater than 35 AU, extending to at least 120 AU. A few hundred SDOs are known, though our searches are very incomplete, and so a far greater number surely await discovery.

The SDOs with perihelia less than about 35 AU could well have been classical EKOs that have been perturbed by Neptune. Those with greater perihelia could be increasingly primordial as the perihelion distance increases, i.e. they could be icy–rocky planetesimals scattered by the giant planets in their migration phase, with the outward migration of Uranus and Neptune making the largest contribution. One theoretical estimate is that about 30 000 planetesimals greater than 100 km across were scattered outwards. This, and other estimates, foretell a cornucopia of discoveries.

The origin of the SDOs and the classical EKOs seems not to be very different. Both could be mixtures of a primordial population and a scattered population. It is not fully understood why their orbital characteristics are somewhat different.

Physical properties of EKOs

Albedos have been obtained for a few EKOs. Among the larger EKOs Pluto has a geometric albedo p varying from 0.5 to 0.7 across its surface, and its satellite Charon 0.38. Varuna, about 40% of Pluto's radius, is dark, with $p \sim 0.07$, but Eris (which HST images show has a 20% greater radius than Pluto) is bright, with $p \sim 0.9$. The albedos of other EKOs mostly lie within the range 0.04–0.4. It is likely that the higher the albedo, the more recently the object has been collisionally resurfaced with fresh icy materials. The colours of the EKOs show significant diversity, from neutral grey through various degrees of redness, uncorrelated with brightness or orbit. Spectra have been obtained for only very few. Some show water-ice features, others do not. Surface temperatures are 50–60 K in the inner E–K belt, depending on the distance from the Sun and the proportion of solar radiation absorbed (see equation (9.8)). Internal temperatures in the larger EKOs could be considerably greater, as you will see in Section 5.2.2.

The mass of Eris will soon be determined from the orbit of its satellite Dysnomia, discovered in 2005 by the Keck telescopes. We will then be able to calculate its density and hence its composition will be constrained.

As well as supplying the SPCs, an EKO could also account, as you have seen, for Neptune's large satellite Triton, which has a peculiar orbit (Section 2.3.1) and resembles Pluto. It could have been captured from the belt, as could some of the small icy–rocky satellites.

Question 3.9

Discuss which feature(s) of the orbits of HFCs indicate an inner Oort cloud origin for most of them, rather than an origin in the E–K belt.

3.3 Meteorites

Meteorites are samples of extraterrestrial material that we find on Earth, and that have come from other bodies in the Solar System, particularly the asteroids but also Mars and the Moon. Well over 30000 have been collected. They are of enormous importance in establishing the chronology of events in the Solar System, the nature of those events, and the composition of the Sun plus its family, as you will see.

3.3.1 Meteors, Meteorites, and Micrometeorites

You have probably seen a 'shooting star', a bright streak of light that flashed across the sky for a second or so before disappearing. You might even have been lucky enough to see a spectacularly bright example, called a fireball, or a bolide if it explodes. These phenomena are caused by **meteors**, small bodies that have entered the Earth's atmosphere at great speed, mostly in the range $10–70\,km\,s^{-1}$. Sometimes the sonic boom produced by the supersonic speed of the body can be heard. They ionise the atmosphere as they travel, and their surfaces become very hot. The streak of light is the glow from the ionisation. In space, the parent body of a meteor is typically less than a few millimetres across.

☐ What are such bodies called?

Such bodies are called micrometeoroids, or dust if smaller than about 0.01 mm (Section 3.1).

Most meteors vaporise completely at altitudes above 60 km. The larger ones, greater than a few tenths of a metre across, usually reach the ground, often fragmenting in the atmosphere or on impact. As you have seen, a fragment, or the whole object from which a set of fragments came, is called a meteorite. Meteorites that are seen to fall are, unsurprisingly, called falls, and there can be no doubt that a fall came from the sky. For a handful of falls there are sufficient observations of the path through the atmosphere for accurate orbits to have been obtained. These orbits resemble those of the NEAs, suggesting an ultimate origin in the asteroid belt.

Only about 1 in 20 of the collected meteorites have been seen to fall. The rest have been found on the Earth's surface some time later. Naturally, these are called finds. You might wonder why a rock on the ground should be thought to have fallen there from the sky. One indicator is a fusion crust on its surface (Plate 25(a)). This is evidence of high-speed travel through the atmosphere. Some of the meteorite burns off in a process called ablation, and the fusion crust is the millimetre or so layer of heat-modified material overlying almost pristine material underneath. However, though this indicates that a rock has arrived at a location via a rapid passage through the atmosphere, it does not establish that it came from interplanetary space. This can be learned from detailed study of its structure and composition, a topic for Section 3.3.2.

Deserts and the Antarctic ice sheets are particularly good places to find meteorites, because small rocky bodies on the surface stand out. Also, in the Antarctic, ice flows concentrate meteorites into glaciers, where subsequent sublimation of ice exposes long-buried meteorites.

A typical unfragmented meteorite is of the order of 10 centimetres across, and has a mass of a few kilograms. Bigger parent bodies tend to fragment, unless they are predominantly iron (Section 3.3.2). For example, the known fragments of the Murchison meteorite seen to fall near the town of Murchison in Australia in 1969 amount to about 500 kg. A particularly massive meteorite was observed to fall near the town of Allende in Mexico, also in 1969. Fragments amounting to over 2000 kg have been recovered. More recently, in 2003, the Park Forest meteorite was observed to break up over the area of this name near Chicago, USA. Many fragments, each a few kilograms, have been recovered. It is estimated that the parent body had a mass of 10 000–25 000 kg. This was the eighth meteorite to have had its orbit accurately determined. The larger the meteorites, the rarer they are. A meteorite of the mass of Murchison, or larger, will arrive at the Earth's surface roughly once a month, but most of these land in the oceans, or in remote areas where they go undiscovered.

Smaller meteorites are more common. The really small ones, a few millimetres or less across, are placed in a separate category called **micrometeorites**. One type is found in abundance in ocean sediments, where their nature is recognised through their spherical form. They are resolidified small bodies that melted in the atmosphere, or resolidified droplets from larger bodies. At sizes below about 0.01 mm, the most common type is a fluffy aggregate of tiny particles, also found in sediments, but also collected by high-flying aircraft. These have traversed the Earth's atmosphere without melting because they are slowed down before they reach their melting temperatures. In many cases those collected might be fragments of larger fluffy aggregates. These dust particles float gently to Earth, and are so common that if you spend a few hours out of doors, even as small a target as you is likely to collect one. Alas! You do not recognise this extraterrestrial mote among all the dust of terrestrial origin that you collect.

Overall, extraterrestrial material is currently entering the Earth's atmosphere at a rate of about 10^8 kg per year, mainly in the form of meteors that completely vaporise.

☐   What is this as a fraction of the Earth's mass?

This is only just over 1 part in 10^{17} of the Earth's mass.

More evidence that meteorites of all classes are of non-terrestrial origin comes from the isotope ratios of certain elements, such as oxygen. The isotope ratios are strikingly different from those found in the Earth's crust, oceans, atmosphere, and Antarctic ice. In most cases the non-terrestrial ratios are consistent with general Solar System values. However, in many meteorites there are tiny refractory grains with very different ratios, indicating that these grains have survived from before the birth of the Solar System. The range of isotope ratios suggests several sources, including condensation in the winds from red giant stars and from the material ejected in supernova explosions. Prominent in these grains are nanometre-sized diamonds, but silicon carbide (SiC), graphite, and corundum (Al_2O_3) are also found.

Question 3.10

Why are most meteorites never found? (Four short reasons will suffice.)

3.3.2   The Structure and Composition of Meteorites

Three main classes of meteorite are defined: stones, stony-irons, and irons. Figure 3.14 shows these classes in the relative numbers in which they occur in *falls*. Finds are excluded because of a strong observational bias that favours irons. As their name suggests, irons are composed almost entirely of iron, and resemble more or less rusty lumps of metal. Stones, as their name suggests, look superficially much like any other stone. Irons thus look much odder than stones, with the result that a much larger fraction of irons are found than of stones. Furthermore, some stones suffer more rapid degradation than irons.

Iron meteorites, as has just been mentioned, consist almost entirely of iron. This is alloyed with a few per cent by mass of the metal nickel, and small quantities of other materials. Naturally occurring terrestrial iron is almost always combined in compounds with non-metals, and so an extraterrestrial origin for irons is at once suspected, particularly given the variety of geological environments in which irons are found. This suspicion can be reinforced by cutting an iron, polishing the fresh surface, and etching it with a mild acid. A pattern emerges like that in Plate 25(b). This is called a Widmanstätten pattern, after the Austrian director of the Imperial Porcelain Works in Vienna, Alois von Widmanstätten (1754–1849), who discovered the pattern in 1808. The pattern arises from adjacent large crystals that differ slightly in nickel content. The large size of the crystals is the result of very slow cooling, 0.5–500 K per Ma, indicating that solidification took place deep inside an asteroid at least a few tens of kilometres across. Such slow cooling in the rare bodies of metallic iron in the Earth's crust is extremely unusual.

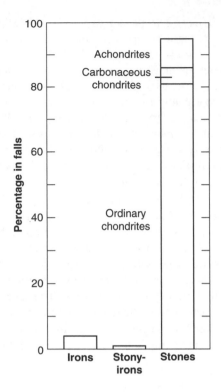

Figure 3.14 The proportions of the three main classes of meteorite in falls, with stones divided into their main subclasses.

Stony meteorites are constituted mostly of various sorts of silicate, though small quantities of iron and nickel are usually present, plus other substances.

☐ What do you think are the main constituents of a stony-iron meteorite?

A **stony-iron meteorite** is a mixture of roughly equal amounts of iron–nickel alloy and silicates, with small quantities of other materials (Plate 25(c)). They are thought to come from transition zones in asteroids that had formed a core of iron and a mantle of silicates, and were then disrupted by a collision.

Stones comprise about 95% of all falls (Figure 3.14) and presumably of all meteorites. The two subclasses are chondrites and achondrites. **Achondrites** are defined on the basis of something that they (and the other two classes) *have not* got, namely chondrules. Their rather uniform silicate compositions indicate that they are from the mantles of asteroids that have cores of iron and transition zones.

Chondrites

The great majority of stones *do* have chondrules (Plate 25(d)), and they are accordingly called **chondrites**. A **chondrule** is a globule of silicates, up to a few millimetres in diameter. They are thought to have been formed by the flash melting of dusty silicate clumps, raising temperatures to greater than about 1500 K, followed by the rapid cooling of the liquid droplets. Flash melting could have been caused by shock waves spreading from the spiral density waves that were present during the formation of the Solar System (Sections 2.1.2 and 2.2.5), or by electrical discharges in the sheet of dust in the solar nebula. However, some chondrules postdate these possible mechanisms (Section 3.3.3). Therefore, impacts between planetesimals or embryos have been evoked. By contrast the silicates outside the chondrules formed by condensation of nebular gas directly to the solid phase. Chondrules are not found in terrestrial rocks.

Ordinary chondrites (OCs) are the most abundant sort of chondrite (Figure 3.14). In the matrix in which the chondrules are embedded there are more silicates, including fractured chondrules, minerals that form at less than $1000\,K$, and 5–15% by mass iron–nickel alloy. The alloy further distinguishes the OCs from terrestrial rocks. The **carbonaceous chondrites (CCs)** are distinguished by a few per cent by mass of carbonaceous materials, and up to about 20% water bound in hydrated minerals. Among the carbonaceous material are many compounds of biological relevance, such as amino acids, which are the building blocks of proteins. There is evidence that a proportion of many of these biomolecules predate the formation of the Solar System. This is from the hydrogen isotope ratio $^2H/^1H$, where ^1H is the common isotope and ^2H, deuterium D, is much rarer. In interstellar molecules this ratio is higher than general Solar System values – the clouds are very cold, which favours incorporation of D into molecules.

The presence of volatile components suggests that the CCs have suffered little heating since they formed. Moreover, they are not fully compacted, indicating that they have never been greatly compressed. These are two of the indicators that CCs have never been in the interiors of bodies more than a hundred kilometres or so across. They are therefore primitive, in that they have been little altered since their formation.

The most primitive of all are the **C1 chondrites**. The matrix is particularly rich in water and in other volatiles. C1s consist of little else but matrix – they are nearly free of chondrules, so presumably predate chondrule formation. Further evidence that the C1s are primitive bodies comes from the relative abundances of the chemical elements in them. Apart from the depletion of hydrogen, helium, and other elements that would have been concentrated in the gas phase of the nebula, the abundances in the C1s are similar to those in the observable part of the Sun. This indicates that these meteorites are not from differentiated bodies, because on fragmentation this would lead to non-solar ratios in each fragment. A particularly well-preserved primitive meteorite is the Tagish Lake meteorite that was seen to fall in Canada on this frozen lake in January 2000, in pieces totalling $56\,000\,kg$. It is intermediate in type between C1 and another primitive subclass CM. Its orbit shows that it came from the outer asteroid belt.

The other CCs also give close composition matches to the Sun, but not as close as do the C1s. Therefore, in C1s it seems we have the least altered samples of the materials that condensed from the solar nebula when the Solar System was forming.

Because the C1s are available for laboratory study, they have been used to refine the relative abundances of all the elements in the Solar System, except those that are very volatile or reside mainly in very volatile compounds.

As well as volatile compounds, CCs also contain irregular white inclusions, typically $10\,mm$ across, that are rich in non-volatile calcium and aluminium minerals such as corundum (Al_2O_3) and perovskite ($CaTiO_3$). Unsurprisingly these are called calcium–aluminium inclusions, CAIs, which are thought to have condensed from the solar nebula. They are rare in the OCs. Radiometric dating (Section 3.3.3) shows that the chondrules generally solidified a few million year after the CAIs, so the melting of CAIs might be a further source of chondrules. Some CAIs show evidence of partial melting.

Question 3.11

In what sort of meteorite would you expect the ratio of carbon to iron to be much the same as that of the Sun? Why are the helium to carbon ratios far smaller in such meteorites than in the Sun?

3.3.3   Dating Meteorites

There are various events in the life of a meteorite that can be dated, but we shall concentrate on two important ones: first, the time that has lapsed since a meteorite, or a component within it, last became chemically separated from its environment, almost always by solidification; and second, the time for which a meteorite was exposed to space rather than protected by some overlying material.

Radiometric dating

Radiometric dating is a powerful technique of wide applicability, as you will see in later chapters. We introduce it here in the context of meteorites.

Imagine that, on chemical isolation, a component in a meteorite contains mineral grains that include, for example, the chemical element rubidium. A small proportion of the rubidium atoms will be of the unstable isotope ^{87}Rb that radioactively decays to form the stable strontium isotope ^{87}Sr:

$$^{87}\text{Rb} \rightarrow \, ^{87}\text{Sr} + \text{e}^- \tag{3.2}$$

where e^- is the electron emitted by the ^{87}Rb nucleus, thus converting it into a ^{87}Sr nucleus. The number of ^{87}Rb nuclei versus time t decays exponentially as

$$N(^{87}\text{Rb}) = N_0(^{87}\text{Rb})\text{e}^{-t/\tau} \tag{3.3}$$

where the zero subscript denotes $t = 0$, and τ is the lifetime of ^{87}Rb, i.e. the time at which $N(^{87}\text{Rb})$ has fallen to $1/e(=36.8\%)$ of its value at $t = 0$. Assume that initially there was no ^{87}Sr in the component, but that it builds up as the ^{87}Rb decays, and that neither of these isotopes escapes from, nor is added to, any of the minerals in the component. The relative quantities of ^{87}Sr and ^{87}Rb in each mineral thus change with time in mirror fashion as in Figure 3.15(a). If, at some time, we measure the ratio $N(^{87}\text{Sr})/N(^{87}\text{Rb})$, then this will tell us how long ago the component became isolated, *provided* that we know the lifetime of ^{87}Rb. Such lifetimes are known, and are usually expressed as the **half-life** $t_{1/2}$ – the time for *half* the atoms to decay. We have $t_{1/2} = 0.693\tau$. For ^{87}Rb, $t_{1/2} = 48\,800$ Ma, with a precision of a few per cent.

☐   What would be the value of $N(^{87}\text{Sr})/N(^{87}\text{Rb})$ after 48 800 Ma, and after twice this time?

After 48 800 Ma, there would be an equal number of the two isotopes, so the ratio would be 1.0. After a further half-life ^{87}Rb will have again halved and $N(^{87}\text{Sr})$ would have increased by half, so the ratio would be 1.5/0.5, i.e. 3. This general method of dating is called **radiometric dating**. Thus, by measuring an isotope ratio, and knowing the half-life of the unstable isotope, we can calculate the time that has elapsed since the component became isolated.

In practice things are more complicated because strontium is likely to be already present in the component on separation. All four of its stable isotopes, including ^{87}Sr, will be there. This isotope builds up as ^{87}Rb decays, but the amounts of the stable isotopes, including ^{86}Sr,

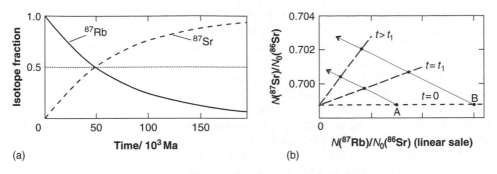

(a)

(b)

Figure 3.15 (a) The principle of radiometric dating using the decay of ^{87}Rb into ^{87}Sr. (b) The Rb–Sr isochron plot for two minerals, A and B. The dashed lines are isochrons at three different times.

are constant. Figure 3.15(b) shows certain abundance ratios that can be measured today, for example, for two mineral grains A and B in a component that differ in their initial endowments of rubidium and strontium. The subscript '0' denotes zero time – the time when the component became isolated. (N_0 (^{86}Sr) does not change.) The arrowed lines show the increase in radiogenic ^{87}Sr relative to ^{86}Sr as ^{87}Rb decays. The time elapsed since isolation is t. The crucial feature is that the slope of each dashed straight line shown is ($e^{t/\tau} - 1$). Thus, knowing τ we can get t from the slope. Because each line in Figure 3.15(b) is for a given value of t, it is called an isochron. Question 3.12 gives you the opportunity to prove that the isochron slope is ($e^{t/\tau} - 1$).

Many radioactive isotopes are used to date meteorites. The ^{87}Rb–^{87}Sr decay has been used here for illustration because the decay to the stable isotope end point is particularly simple (equation (3.2)). In contrast, the decay of ^{238}U to the ^{206}Pb stable isotope end point involves many stages, as does that of ^{235}U to ^{207}Pb. The half-lives of these decays are 4470 Ma and 704 Ma respectively, and are known to higher precision than the ^{87}Rb–^{87}Sr half-life.

The oldest radiometric ages that have been obtained from any body in the Solar System are for the CAIs and chondrules in meteorites, 4570 Ma. This age has been established from ^{238}U–^{206}Pb and other decays. It is taken to be the age of the Solar System. The chondrules are near to 2 Ma younger than the CAIs. To establish such a small age difference between two such large ages use is made of short-lived isotopes. For example, ^{26}Al decays to ^{26}Mg with a half-life of only 0.73 Ma, much faster than the decay of ^{238}U. So, by comparing the lead and magnesium isotope contents of the CAIs and chondrules we can get the age difference with reasonable precision. The details will not concern us. Note that the presence of short-lived isotopes, as inferred from their daughter products, indicates that the CAIs separated within a few million years of the short-lived isotope being created in stars. Furthermore, the CAIs and chondrules could not have survived in isolation for more that a few million year, and so the formation of meteorite parents must have been fairly rapid. This is consistent with the time scale of the formation of planetesimals in Chapter 2. Some separation ages are younger, but very few are less that 1600 Ma. These younger ages are the result of some later melting or vaporisation that reset the radiometric clock.

Space exposure ages

The time for which a meteorite has been exposed to space is obtained from the action of cosmic rays on the parent meteoroid. **Cosmic rays** are atomic particles that pervade interstellar space, moving at speeds close to the speed of light. They are primarily nuclei of the lighter elements, notably hydrogen. When a cosmic ray strikes a solid body it will penetrate up to a metre before it stops, leaving a track, and creating unstable and stable isotopes via nuclear reactions. The quantities of these isotopes increase with the duration of the exposure, and so, by measuring the quantities among the tracks, and knowing the cosmic ray flux in interplanetary space, the cosmic ray exposure age can be calculated.

Many meteorites have exposure ages considerably shorter than their chemical separation (solidification) ages – strong evidence that solid bodies larger than the metre or so cosmic ray penetration depth have been disrupted in space long after they solidified. Most exposure ages are 10–50 Ma, far too long to trace meteorite origins. Some stones have particularly short exposure ages, as little as 0.1 Ma. This is presumably because stony materials are less strong than iron, and are thus more readily broken in collisions and eroded by dust. This gives a constant supply of unexposed material for cosmic rays to lay their tracks in.

Question 3.12

By obtaining an equation for $N(^{87}\text{Sr})/N_0(^{86}\text{Sr})$, show that the isochron slope in Figure 3.15(b) is $(e^{t/\tau} - 1)$. (This needs good facility with algebra.)

Question 3.13

If a certain meteorite is a piece of a larger body, how could it nevertheless have an exposure age far greater than the time ago that it was liberated from the larger body? Why could its calculated exposure age never exceed its solidification age?

3.3.4 The Sources of Meteorites

As well as the Tagish Lake parent's orbit (Section 3.3.2), a clue to the sources of the meteorites is in the few known orbits of the parent meteoroid, which resemble the orbits of the NEAs (Section 3.3.1).

☐ What does this suggest is the source region of these meteorites?

This suggests an ultimate origin in the asteroid belt. The short cosmic ray exposure ages of the stones supports this conclusion, the ages being consistent with the high rate of collisional disruption expected in the asteroid belt, continuously liberating unexposed material, and the relatively short times before many of the meteoroids so generated will collide with the Earth. Many meteorites show evidence of collisional disruption, notably in minerals that have been shocked, and in structures indicating broken fragments that have been cemented together. Sometimes the fragments seem to have come from different bodies, or to have been subject to different processes. To get a meteoroid from an orbit within the asteroid belt into a near-Earth orbit, it is usually necessary for its orbit to be perturbed by Jupiter, or sometimes Mars, when the meteorite encounters an mmr. Such encounters continually occur because of orbital migration caused by the Yarkovsky effect (Section 3.1). Of the more than 30 000 meteorites known, very nearly all seem to be asteroid fragments.

Further support for the view that meteorites are derived from the asteroids comes from comparing the reflectance spectra of the various classes of asteroid with those of the various classes of meteorite. As noted in Section 3.1.6, a clear correspondence exists between the CCs and the abundant class C asteroids. The CCs are presumably collisional fragments of an asteroid that never became sufficiently heated to lose its carbonaceous materials and hydrated minerals, and was far too cool to differentiate. The parent asteroid might itself have been a fragment of a larger unheated body. In the outer belt we see class C asteroids in abundance, indicating that this is where the large asteroids avoided differentiation, perhaps because of weaker magnetic induction heating (Section 3.1.6), and lower proportions of rocky materials and iron, which would give less accretional and radioactive heating. From their position in the outer belt there could well be a low probability of transfer to a near-Earth orbit, which would explain why class C asteroids are common, but the corresponding meteorites, the CCs, are rare.

There is also a clear correspondence between irons and the rare class M asteroids. As pointed out in Section 3.1.6, early in Solar System history the larger asteroids (a few hundred kilometres across, or larger) could have become warm enough to differentiate fully or partially. Figure 3.16 shows the resulting layered structure in the partially differentiated case. The Widmanstätten pattern in irons is indicative of the slow cooling that would occur in the iron core of a large asteroid. Fragmentation of the asteroid can expose the core, which itself could subsequently be fragmented. The core, or its fragments, are the class M asteroids, and the smaller fragments are the parent meteoroids of the iron meteorites. A complication is that the magnesium-rich silicate called enstatite could be mixed with iron–nickel without betraying its presence. Therefore, some class M asteroids might be a mixture of iron–nickel with this type of silicate. Radiometric dating of irons indicates that the parent asteroid formed, in the main, early in Solar System history, just 5–10 Ma after the CAIs.

Stony-irons show some correspondence with S class asteroids.

☐ What is a possible origin of such asteroids?

These asteroids could come from the interface between the iron core and the silicate mantle of (partially) differentiated asteroids, where silicates and iron are mixed.

There are only a few asteroids that match the achondrites. The largest achondrite subgroup comprises the howardites, eucrites, and the diogenites, called the HED subgroup. These are

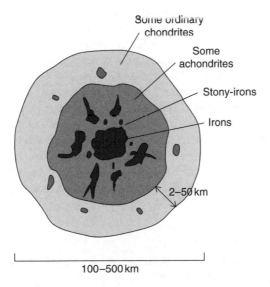

Figure 3.16 A partially differentiated asteroid, showing regions from where various sorts of meteorite could originate.

composed of silicates like feldspar and pyroxene (Tables 2.3 and 6.1). In an asteroid these would be produced by the melting of parent silicates, notably olivine and pyroxene (Table 6.1) followed by differentiation, with the new silicates rising to the top, where they constitute basalts (= feldspar + pyroxene), and the metallic iron sinking to form a core. This requires an asteroid more than a few hundred kilometres across, a necessary (but not sufficient) condition that differentiation is (nearly) complete, so that the achondrite silicates are at the surface and the metallic iron is in a core. The HEDs show a good spectral match with the rare class V asteroids, which includes Vesta, 256 km mean radius, and a handful of small asteroids, presumably collision fragments. HST images of Vesta show a big impact crater (Figure 3.6(a)) that could have supplied a huge number of HEDs, a view supported by the few known HED orbits being similar to that of Vesta. The high density of Vesta (Section 3.1.5) is consistent with a considerable iron core. Radiometric dating of the HEDs indicates core formation within 4 Ma of CAI formation.

Some of the achondrites that are not HEDs could have come from the interiors of partially differentiated asteroids (Figure 3.16) that were collisionally disrupted. The rare basalt meteorite (NWA011) might be from the asteroid Magnya which seems to have a basalt surface, in which case Magnya is a differentiated body.

Ordinary chondrites

The most common class of meteorite is the ordinary chondrite, OC (Figure 3.14), in which the silicates are composed largely of pyroxene and olivine, and (excluding volatile substances) with elemental composition similar to the Sun. This indicates that they originate from undifferentiated asteroidal material. In spite of their abundance, it was only in 1993 that an asteroid was discovered that provided a good spectral match. This is Boznemcova, and it is only 7 km across. Other candidate asteroids are the Q class, though these are few in number. A particularly promising candidate is the S class asteroid Hebe, with a semimajor axis of 2.43 AU that places it in the inner main belt. It has a mean radius of about 90 km, and in 1996 its surface spectrum was shown to match that of the H-type subclass of OCs that accounts for about 40% of them. Moreover, Hebe orbits near to the 3:1 resonance with Jupiter (Figure 3.1), and so chips off its surface would readily find their way to the Earth. Some other OCs could originate from the outermost zone of a partially differentiated asteroid (Figure 3.16).

Other S class asteroids could well be copious sources of OCs too. The S class constitutes about 80% of the inner main belt (Figure 3.8), from where there is ready access to the Earth. These asteroids have spectra that in a few cases are a good match to the OCs, but in many cases display only muted and reddened spectral features of pyroxene and olivine. However, it has been shown that space weathering by solar UV radiation, micrometeorite bombardment, and perhaps cosmic rays, darken and redden OC materials in just the right way, and that the pristine surface of an S class asteroid should be a close spectral match to the OC interiors. NEAR's mission to the S class asteroid Eros has shown that it has the same composition as the OCs.

Martian and lunar meteorites

By mid 2006, there were 34 meteorite finds in which oxygen isotope ratios throughout the group are distinctly non-terrestrial, and sufficiently similar to suggest a common origin. Each one contains minerals of volcanic origin with solidification ages in the range 165–1360 Ma (except for one, ALH84001, which has a solidification age of 4500 Ma). We thus seek an extraterrestrial parent body that could have produced molten rock at its surface by volcanic processes 165–1380 Ma ago. It also has to be relatively nearby, and with at most a thin atmosphere so that a huge meteorite impact could throw surface materials into space. Amongst our neighbours, only Venus and Mars could have had volcanic processes so comparatively recently.

☐ Venus has to be ruled out. Why?

Venus has a very thick atmosphere, inhibiting the escape of rocks. Also, the impact on Venus would have to be so violent that either the rocks would be vaporised completely, or they would bear telltale signs of extreme violence, and these are not seen.

Mars is thus the only candidate. That Mars is indeed the parent body is strongly indicated by gases trapped within one of the meteorites, EETA 79001 – in the mid 1980s these were shown to have a composition similar to the Martian atmosphere, and unlike any other plausible source. Recently, the Mars Exploration Rover, *Opportunity*, found a rock with a mineral composition very similar to EETA 79001. Other meteorites in this group have now also been shown to have Martian characteristics. The Martian meteorites provide us with important information about Mars, as you will see in later chapters.

What sort of impacts on Mars are required to provide the Martian meteorites? Computer models show that an impact that would produce a crater about 3 km across would eject millions of bits of the Martian crust into space large enough to constitute meteoroids rather than dust, and with negligible impact melting. After the heavy bombardment, which ended about 3900 Ma ago, a Martian crater about 3 km across would have been created at average intervals of 0.2 Ma, leading to an estimate of a few meteorites per year landing on the Earth, certainly enough to account for the small sample that has been found. But at least half of the Martian crust had formed by about 4000 Ma ago, so why are the meteorite ages predominantly much younger? One explanation is that the older Martian crust, having been exposed longer to meteorite bombardment, has developed a thick coat of loose rubble and dust (regolith) that has cushioned the larger impacts. Additionally, or alternatively, the widespread presence of sediments on the older terrain could provide a cushion.

Nearly 100 meteorites from the Moon have also been found, a largely undisputed origin because of the compositional similarities with the lunar surface samples that have been returned to Earth by lunar expeditions. Further material from the Moon might be some of the many tektites found on Earth. These are rounded glassy objects, typically 10 mm across, with a presumed volcanic or impact melt origin.

The Moon is very much nearer to us than Mars. It is therefore a puzzle why there are not far more lunar meteorites than Martian ones – models predict a ratio of about 100:1.

Question 3.14

State a possible origin of the stony-iron meteorites, and the likely origins of the OCs, and hence account for the broad differences in their compositions.

3.3.5 The Sources of Micrometeorites

Most micrometeorites are derived from bodies that in space must have been less than a few millimetres across. The great majority of bodies of this size vaporise completely in the Earth's atmosphere and account for most of the meteors. Therefore, if we can find the source(s) of the meteors we will have found the source(s) of the micrometeorites.

If you were to go out on a clear dark night, then on most days in the year you would see on average about 10 meteors per hour. On or around a few dates, the same each year, the hourly rates are considerably greater. These enhanced rates are called **meteor showers**. Just how much greater the hourly rate becomes in a shower varies from year to year, but in exceptional years rates of the order of 10^5 meteors per hour are observed; these are called meteor storms. Observations show that the meteors in a shower very nearly share a common orbit, and for many showers this orbit is the same as the orbit of a known comet. In other cases the orbit indicates an asteroidal source. There are 19 major showers. Table 3.1 lists the six that are usually the strongest, with their dates and associated comet or asteroid.

Figure 3.17 shows how a comet gives rise to a meteor shower (the case of an asteroidal source is similar). Rocky particles are lost by the comet and initially do not get far away. They are estimated to vary in size from submicrometre dust particles, to loose aggregates up to several millimetres across, and sometimes far larger. With each perihelion passage of the comet the debris accumulates, and various perturbations gradually spread it along and to each side of the orbit. The debris moves around the orbit, and when the Earth is at or near the comet's orbit at the same time as the debris, a shower results. The year-to-year variations are the result of a non-uniform distribution of debris along the orbit. The cometary origin of many showers is further supported by estimates of the particle densities, obtained from the rate at which the Earth's atmosphere slows them down. Values in the range $10-1000\,\mathrm{kg\,m^{-3}}$ are obtained, suggesting loose aggregates of dust particles of the sort that comets could yield.

Micrometeorites are also loose aggregates, indicating that they are comet debris that has survived atmospheric entry. This possibility is strongly supported by the composition of the micrometeorites, which is in accord with remote observations of comets and with *in situ* measurements made by Giotto on dust lost by Halley's comet. Micrometeorite composition is something like that of the CCs, though sufficiently different to indicate a source other than class C asteroids. It therefore seems that most meteors, and hence most micrometeorites, originate from the rocky component of comets.

Of the meteors that do not belong to showers, most are thought to be comet debris no longer concentrated along the parent comet's orbit. A few meteors have entry speeds that are so high ($> 72\,\mathrm{m\,s^{-1}}$) that they might have come from beyond the Solar System. This interpretation is supported by the greater fluxes of fast meteors when the Earth is at points in its orbit when it

Table 3.1 The six strongest meteor showers

Shower name^a	Date range^b	Associated parent
Quadrantids	01–06 Jan	96P/Macholtz 1
Eta Aquarids	01–08 May	1P/Halley (HFC)
Perseids	25 July–18 Aug	109P/Swift–Tuttle (HFC)
Orionids	16–26 Oct	1P/Halley (HFC)
Leonids	15–19 Nov	55P/Tempel–Tuttle (HFC)
Geminids	07–15 Dec	Phaeton (an Apollo asteroid)

^{a} Each name is derived from the constellation from which the shower appears to emanate.
^{b} These are the approximate dates each year on which the shower is greatest.

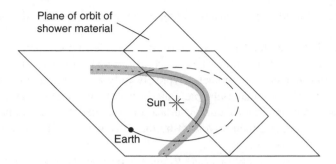

Plane of orbit of
shower material

Sun

Earth

Figure 3.17 Meteor showers and the orbit of a comet.

is either travelling in the same direction through the Galaxy as the Solar System as a whole, or travelling towards nearby massive stars.

Question 3.15

Give two plausible reasons for why some meteor showers are in orbits in which no comet has been seen.

3.4 Summary of Chapter 3

The asteroids are small bodies, the great majority being confined to the space between Mars and Jupiter in the asteroid belt. Their orbits are prograde but on average somewhat more eccentric and more inclined than the orbits of the major planets. They have a total mass of the order of 10^{22} kg, and there is the order of 10^9 bodies greater than 1 km across. Ceres, with a radius of 479 km, is by far the largest asteroid, containing the order of 10% of the total mass.

Beyond the asteroid belt there are near-Earth asteroids, the Trojan asteroids which are near the L_4 and L_5 points of Jupiter, and the Centaurs, with perihelia outwards from Jupiter's orbit and semimajor axes less than that of Neptune.

The asteroids nearer than Jupiter are thought to be the remnants of material in the space between Mars and Jupiter that the gravitational field of Jupiter prevented from forming into a major planet. Throughout Solar System history collisions in the asteroid belt have been common, and so the population has evolved considerably. There has also been a net loss of material, the present mass being only about 0.1% of the original mass.

An asteroid is categorised according to its reflectance spectrum and geometrical albedo. There are 14 Tholen classes (Figure 3.7), with about 80% of classified asteroids falling into the S class and about 15% into the C class. Class C dominates the outer asteroid belt and Class S the inner belt. A class C asteroid is thought to consist of an undifferentiated mixture of silicates, iron–nickel alloy, hydrated minerals, and carbonaceous materials. A class S asteroid has a surface that consists of mixtures of silicates with iron–nickel alloy. The population variations across the belt, and those seen in other classes of asteroid, could be the result of lower temperatures and weaker heating at greater heliocentric distances in the solar nebula.

Comets are small bodies that are distinct from asteroids in that they develop a coma, a hydrogen cloud, and tails when within 10 AU or so of the Sun. Studies of these huge structures indicate that the solid nucleus of a comet – typically only a few kilometres across – contains a significant proportion of icy materials, particularly water, in a loose aggregation with rocky and carbonaceous materials. The volatile materials are liberated by solar radiation to form the coma and hydrogen cloud, and then driven off by the solar wind and by solar radiation to form the tails.

The comets have a wide variety of orbits. Long-period comets have orbital periods greater than 200 years and enter the inner Solar System from all directions. About 1000 have been recorded. It is inferred that they are a very small sample of a cloud of $10^{12} - 10^{13}$ bodies greater than 1 km across, $10^3 - 10^5$ AU from the Sun, called the Oort cloud. This cloud, with a present day mass of about 10^{25} kg, is thought to consist of icy–rocky planetesimals ejected into large orbits from the giant planet region during the formation of the giants, and during any subsequent giant planet migration.

Short-period comets have orbital periods less than 200 years. Most of them have periods less than 20 years and orbital inclinations less than $35°$. These are the Jupiter family comets. A few hundred are known. It is inferred that most of them (the Jupiter family comets) are a sample of the Edgeworth–Kuiper (E–K) belt, planetesimals left over because of ineffective accretion in the solar nebula beyond the giant planets, plus a proportion scattered out by the giant planets, particularly Uranus and Neptune. The remaining short-period comets, typically in higher inclination orbits, are the Halley family comets. These might in some cases be samples of the inner Oort cloud. The Centaurs could be the larger members of a population in transition from the E–K belt to the family of short-period comets.

Meteorites are small rocky bodies that survive passage through the Earth's atmosphere to reach the Earth's surface. The three classes are stones, stony-irons, and irons. Stones account for about 95% of the meteorites observed to fall to Earth. Most of them contain silicate chondrules that define the subclass called chondrites, the remainder being achondrites. About 6% of the chondrites are carbonaceous chondrites, primitive bodies that contain not only silicates, but also hydrated minerals and carbonaceous material. Radiometric dating shows that the oldest components of meteorites – the calcium–aluminium inclusions and the chondrules – solidified 4570 Ma ago. These are the oldest ages obtained in the Solar System, and are taken to be the Solar System's age. A small number of meteorites have come from Mars and the Moon. Micrometeorites are largely derived from the rocky component of comets.

The composition of the particularly primitive C1 chondrites matches that of the observable part of the Sun, except for elements that are volatile, or reside predominantly in volatile compounds. These are greatly undersampled in meteorites. The C1s have enabled the relative abundances of many elements in the Solar System to be significantly refined.

Spectral reflectances of asteroids and meteorites show matches between

- the carbonaceous chondrites and class C asteroids;
- the iron meteorites and class M asteroids, which are the iron–nickel cores of (partially) differentiated asteroids;

- most achondrites and the rare V class asteroids, which are from the silicate crust of fully differentiated asteroids, of which Vesta is the most prodigious parent;
- the ordinary chondrites and S class asteroids.

A least some stony-irons are thought to come from the core–silicate mantle interfaces of (partially) differentiated asteroids.

4 Interiors of Planets and Satellites: The Observational and Theoretical Basis

Our understanding of planetary and satellite interiors is considerable, but will always be limited by their inaccessibility – we have to rely on *external* observations. These are made by telescopes on the Earth or in Earth orbit, and by instruments on spacecraft in the vicinity of the planetary body. For Mars, Venus, the Moon, Titan, and of course the Earth, we also have observations made at the surface of the body. For the Earth and the Moon we have additionally sampled materials from below the surface, though for the Moon only the upper metre or so has been sampled. Even in the case of the Earth, the deepest samples are from only about 100 km – less than 2% of the distance to the centre, and brought to us by volcanoes. Nevertheless, good models of planetary and satellite interiors have been developed.

The basic features of a model are a specification of the composition, temperature, pressure, and density, at all points within the interior. Planets and large satellites are close to being spherically symmetrical, so, at least as a first step, this reduces to a specification of properties versus radius from the centre, or, equivalently, versus depth from the surface. Not all of the features of a model are independent. For example, the density of a substance depends on its pressure and temperature. Some of the relationships between various features are poorly known.

A model will embody certain physical principles. For example, if the material at some depth is neither rising nor falling then the net force on it in the radial direction must be zero. With such principles the model is then used, with initial depth profiles of the various features, to derive properties that are observed externally, and the depth profiles are varied until an acceptable level of agreement with the actual observations is obtained.

We shall now look at the main types of external observations that are available for modelling planetary and satellite interiors. Table 4.1 lists many of the spacecraft missions that have made particularly important contributions.

4.1 Gravitational Field Data

4.1.1 Mean Density

The mean density of a body is an important indicator of bulk composition. It is said to be an important *constraint* because it places useful limits on the range of possibilities.

Discovering the Solar System, Second Edition Barrie W. Jones
© 2007 John Wiley & Sons, Ltd

Table 4.1 Some missions of planetary exploration by spacecraft

Object	Spacecraft		
	Name	Encounter date(s)^a	Some mission features
Mercury	Mariner 10	Mar 74, Sept 74, March 75	The only mission – flybys
Venus	Venera 4^b	Oct 67	First penetration of atmosphere
	Venera 7^b	Dec 70	First lander
	Pioneer Venus	Dec 78–Oct 92	Orbiter – first global radar mapping
	Magellan	Aug 90–Oct 94	Orbiter – radar mapping
	Venus Express^c	Apr 06–	Orbiter – atmospheric studies
Earth	Explorer 1	Jan 1958	First satellite to yield scientific data
Moon	Luna 3^b	Oct 1959	First images of lunar far side
	Apollo 11	July 1969	First manned landing
	Clementine	Mar–Apr 94	First polar orbiter
	Lunar Prospector	Jan 98–July 99	Polar orbiter
	SMART 1^c	Nov 04–Sept 06	Orbiter and collider
Mars	Mariner 9	Nov 71–Oct 72	Orbiter; first global survey
	Viking 1 Orbiter	June 76–Aug 80	Orbiter; second global survey
	and Lander	July 76–Nov 82	First soft lander
	Viking 2 Orbiter	Aug 76–Jul 78	Orbiter; third global survey
	and Lander	Sept 76–Apr 80	Second soft lander
	Mars Global Surveyor	Sept 97–Nov 06	First polar orbiter
	Mars Pathfinder	July–Sept 1997	Lander and first rover
	Mars Odyssey	Oct 01–	Orbiter; fourth global survey
	Mars Exploration Rovers	Jan 04–	Rovers *Spirit* and *Opportunity*
	Mars Express^c	Dec 03–	Orbiter, fifth global survey
	Mars Reconnaissance	Mar 06–	Orbiter, sixth global survey
Jovian system	Pioneer 10	Dec 73	First flyby
	Pioneer 11	Dec 74	Second flyby
	Voyager 1	Mar 79	Third flyby
	Voyager 2	July 79	Fourth flyby
	Galileo Orbiter	Dec 95–Sept 03	First orbiter
	Galileo probe	Dec 95	The only probe
	Cassini–Huygens	Dec 00–Jan 01	Fifth flyby
Saturnian system	Pioneer 11	Sept 79	First flyby
	Voyager 1	Nov 80	Second flyby
	Voyager 2	Aug 81	Third flyby
	Cassini	July 04	Orbiter
	Huygens	Jan 05	Landed on Titan
Uranian system	Voyager 2	Jan 86	The only mission – flyby

(continued)

Table 4.1 (Continued)

Object	Spacecraft		
	Name	Encounter date(s)^a	Some mission features
Neptunian system	Voyager 2	Aug 89	The only mission – flyby
Asteroids	Galileo	Oct 91	Images of Gaspra
	Galileo	Aug 93	Images of Ida
	NEAR	June 97	Images of Mathilde
		Feb 01	Landed on Eros
	Hayabusa^d	June 2005	First asteroid sample return, Itokawa
Comets	Giotto	Mar 86	First comet nucleus image (1P/Halley)
		July 92	Close approach to 26P/Grigg–Skjellerup
	Deep Space 1	Sept 01	Images of 15P/Borrelly
	Stardust	Jan 04	Return of dust from 81P/Wild 2
	Deep Impact	July 05	Bullet penetration of 9P/Tempel 1
	Rosetta	2014	First orbiter and lander, 67P/Churyumov–Gerasimenko

^{a} For orbiters and landers a range of dates is given if the encounter spanned more than a month. Payloads are from NASA, except:
^{b} from the USSR,
^{c} from the European Space Agency,
^{d} from Japan.

If the volume of a body is V and its mass is M, then the **mean density** is given by

$$\rho_m = M/V \tag{4.1}$$

For a sphere, $V = (4/3)\pi R^3$ where R is the radius. Planets and satellites are not perfectly spherical; Saturn, for example, is clearly flattened by its rotation (Plate 16(a)). We can, however, measure the shapes of planetary bodies sufficiently accurately for the actual volumes to be obtained.

There are various ways of measuring the mass M. You have seen in Section 1.4.4 how the mass of the Sun can be obtained from the semimajor axis and period of the orbit of a planet. The same procedure can be applied to get the mass of a planet itself, from the orbit of a satellite or spacecraft around it. If the mass of the satellite or spacecraft is m, then from Newton's laws of motion and gravity we get an equation like equation (1.6) (Section 1.4.5), which can be rearranged as

$$M + m = \frac{4\pi^2 a^3}{GP^2} \tag{4.2}$$

where a is the semimajor axis of the orbit of m with respect to M, P is the orbital period, and G is the gravitational constant. If m is much less than M, it can be omitted from equation (4.2), and we get the value of M without knowing the value of m.

If m is not negligible in comparison with M, e.g. the Moon in comparison with the Earth, then we can still get M by finding the centre of mass of the system. It is the centre of mass that moves in an elliptical orbit around the Sun, and M and m are each in orbit around the centre of mass, as you saw in Section 1.4.5. The position of the centre of mass can be determined from these orbital motions, and the ratio of M and m is then given by an equation like equation (1.7) (Section 1.4.5)

$$M/m = r_m/r_M \qquad (4.3)$$

where r_M and r_m are the distances of M and m from the centre of mass at any instant. From equations (4.2) and (4.3) we can obtain both masses, and we can then use equation (4.1) to get the mean density of each body in turn.

The two bodies do not have to be in orbit around each other. Though the equations are different, the masses can be obtained from the change in their trajectories as they pass each other. If one body has a much smaller mass than the other, then only the trajectory of the smaller body will change appreciably. This is the case, for example, when a spacecraft passes near a planet.

Table 4.2 lists the mean densities, and other data, for the planets and larger satellites. For a wider range of bodies, Figure 4.1 shows the mean density along with the radius. Radius and density are more indicative of composition than density alone, as you can see if we consider two bodies with the same mean density but with greatly different radii. You might think that, with equal mean densities, the two bodies could have the same composition. This is not so. The internal pressures in the larger and thus more massive body will be much greater than in the less massive body.

☐ What effect does this have on the mean densities?

This results in greater compression in the more massive body resulting in a greater density. The hypothetical **uncompressed mean density** of the more massive body must be lower than that of the other body, and therefore the more massive body must contain a higher proportion of intrinsically lower density substances

In Figure 4.1 some distinct groups can be recognised. On the basis of size alone the planets can be divided into four giants, four terrestrial planets, and Pluto. With the mean densities added, this broad division is reinforced by a strong indication of compositional differences between the groups. Consider the giants. It is clear that these bodies are so large that the compression is considerable. If the compression were slight their mean densities would be lower than those in Figure 4.1 by a large factor. The mean densities of the other bodies would be much less reduced, and so the already clear distinction between the giants and the rest would sharpen. The giant planets thus form a quite distinct group.

For the remaining bodies in Figure 4.1, the uncompressed mean densities of Venus and the Earth would be about 20% less than the values shown, with smaller reductions for the rest. Therefore, the groupings among the non-giants are not sharply defined. The bodies in the centre right region, namely the four terrestrial planets plus the Moon, Io, and Europa, constitute the **terrestrial bodies**. The remaining large satellites, namely Ganymede, Callisto, Titan, and Triton, group with Pluto to constitute the **icy–rocky bodies**, a term that reflects their composition. The other bodies in Figure 4.1 are the two largest asteroids, namely Ceres and Pallas, and nearly all of the intermediate-sized satellites. These satellites have mean densities comparable with those of the icy–rocky bodies, whilst those of the two asteroids are more comparable with the terrestrial bodies, indicating broad compositional affinities in each case.

In Section 3.1.4 you saw that there is a threshold size below which a body could have an irregular shape. Bodies large enough to be approximately spherical are often called **planetary bodies**.

Table 4.2 Some physical properties of the planets and larger satellites[a]

Object	Equatorial radius[b]/ km	Flattening	Mass/ 10^{20} kg	Mean density/ $kg\,m^{-3}$	Sidereal rotation period[c]/ days	J_2^d	C/MR_e^2	Mag. dipole moment[e]/ Am^2	Energy outflow[f]/10^{-12} $W\,kg^{-1}$
Mercury	2440	~ 0	3302	5430	58.646	0.00006	?	4×10^{19}	?
Venus	6052	~ 0	48690	5240	243.019	$\sim 10^{-7}$	?	$< 8.4 \times 10^{17}$	6
Earth	6378	0.00336	59740	5520	0.9973	0.001901	0.3308	7.90×10^{22}	5.9
Mars	3396	0.00649	6419	3940	1.0260	0.001959	0.365	$< 2 \times 10^{18}$	3.3
Moon	1738	~ 0	735	3340	27.3217	0.0002037	0.394	$< 1 \times 10^{17}$	15
Io	1822	~ 0	893	3530	1.769	0.00186	0.378	$\leq 4 \times 10^{20}$	A few 100
Europa	1561	~ 0	480	3010	3.551	0.000389	0.347	~ 0	?
Ganymede	2631	~ 0	1482	1940	7.155	0.000127	0.311	1.4×10^{20}	?
Callisto	2410	~ 0	1076	1830	16.689	0.000034	0.358	$< 2 \times 10^{18}$	?
Titan	2575	~ 0	1346	1880	15.945	?	Small	Very small	?
Triton	1353	~ 0	214	2070	5.877	?	?	?	?
Pluto	1153	?	130	2100	6.3872	?	?	?	?
Jupiter	71490	0.0649	18988000	1330	0.4135	0.014697	0.264	1.54×10^{27}	168
Saturn	60270	0.098	5684000	690	0.4440	0.016332	0.21	4.6×10^{25}	130
Uranus	25560	0.0229	868400	1270	0.7183	0.003516	0.23	3.8×10^{24}	3.7
Neptune	24765	0.0171	1024500	1640	0.6712	0.00354	0.29	2.0×10^{24}	33

[a] A '?' denotes that no useful value for the quantity has been obtained, and '~ 0' indicates an (unknown) extremely small value.

[b] For the giants the equatorial radius is to the atmospheric altitude where the pressure is 1 bar (10^5 Pa).

[c] For Jupiter and Saturn the rotation period is for the deep interior (Section 11.1.3).

[d] For the giants J_2, J_4, and J_6 are used in preference to C/MR_e^2 to constrain radial variations of density.

[e] To convert to $T\,m^3$ divide these values by 10^7.

[f] This assumes that Venus has much the same energy outflow as the Earth.

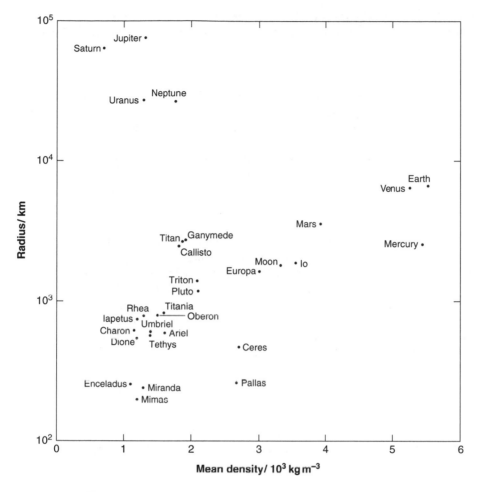

Figure 4.1 Radii and mean densities of Solar System bodies.

The threshold radius is very roughly 300 km, and therefore among the bodies in Figure 4.1, Enceladus, Mimas, Miranda, and Pallas could be irregular.

Question 4.1

Use Figure 4.1 to argue that the giant planets divide into two subgroups on the basis of size and composition.

4.1.2 Radial Variations of Density: Gravitational Coefficients

You have seen that to obtain the mean density of a body we must first measure its mass. This measurement, however it is made, relies on the gravitational force that the body exerts on some *other* body. From more detailed measurement of this force we can go further, and obtain information on the variation of the density from point to point in the body.

Newton's law of gravity (Section 1.4.4) tells us that the magnitude (size) of the gravitational force that a body of mass M exerts on a body of mass m at a distance r from M is given by

$$F = GMm/r^2 \tag{4.4}$$

This equation underpins equation (4.2), and requires either that M and m are point masses or that they are spherically symmetrical, i.e. the density varies only with radius. In this latter case r is measured from the centres of the bodies, and we have to be outside them for the equation to apply. The direction of the force on m is towards the centre of M.

☐ By measuring F and using equation (4.4) can we learn anything about how the density inside a spherically symmetrical body varies with radius?

For a spherically symmetrical body we can learn nothing from external gravity measurements about the variation of density with radius – all possible variations of density with radius for a given total mass M would give the same external force. Fortunately, each Solar System body exerts a gravitational force that is not quite that given by equation (4.4), and therefore we *can* learn something about the variation of density with radius.

It is useful to introduce the concept of **gravitational field** – this will allow us to ignore the mass m, which is of no interest in itself. At any point in space there will be a gravitational force that is the sum of the forces produced by all the surrounding bodies, with due regard to their different directions (in what is called a vector sum). Thus we are not restricted to F as given by equation (4.4). For this general gravitational force F acting on a body of mass m and negligible size, the magnitude g of the gravitational field at the point where m is located is *defined* as the force per unit mass

$$g = F/m \tag{4.5}$$

The direction of g is the same as the direction of F. We can now refer to the gravitational field of M without specifying the mass of the body used to measure it. For example, from equations (4.4) and (4.5) it follows that

$$g = GM/r^2 \tag{4.6}$$

and so g is independent of m. Like equation (4.4), equation (4.6) applies if M is a point mass or if it is spherically symmetrical. From Newton's second law, g is seen to be the acceleration of m.

Because the gravitational forces exerted by planetary bodies do not conform to equation (4.4), it follows that their gravitational fields do not conform to equation (4.6). Instead, the magnitude of the field is given by

$$g = GM/r^2 + \text{extra terms} \tag{4.7}$$

The extra terms at some point of measurement P depend on the position of P. Figure 4.2(a) shows P at some distance r from the centre of a planetary body, with the direction to P specified in terms of the angles θ and ϕ. For planetary bodies the extra terms are small, and the largest is negative, $-J_2 \, (GMR^2/r^4)f(\theta)$, where $f(\theta)$ varies with θ though not with ϕ or r. J_2 (**'jay two'**) is the **gravitational coefficient** of this term. It measures the strength of the extra term, and a particular planetary body will have a particular value of J_2. Note the rapid decrease with r, as $1/r^4$. This means that at large distances this extra term becomes negligible, but that at

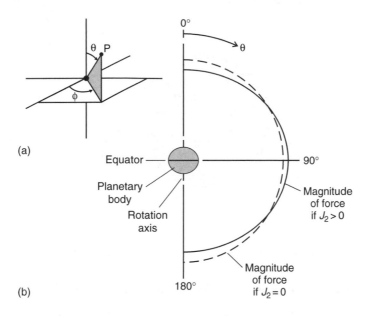

(a)

(b)

Figure 4.2 (a) Specifying the position of a point P. (b) The variation of gravitational field with θ at a fixed distance r from the centre of a planetary body with $J_2 = 0$ and $J_2 > 0$.

sufficiently small distances it is appreciable. The other extra terms decrease no less rapidly with r.

For most planetary bodies the main source of the J_2 term is the rotation of the body. This flattens the sphere as in Figure 4.2(b), and it is the rotational flattening of the body that generates this extra term. For a small degree of flattening the planetary body is an oblate spheroid (Figure 1.13(a)) – note that it is symmetrical around the rotation axis. The **rotational flattening** (also called the oblateness) is quantified by $f = (R_e - R_p)/R_e$, where R_e is the equatorial radius and R_p is the polar radius. Figure 4.2(b) also shows how, at some fixed value of r, the magnitude of the total field g varies with direction when J_2 is the only extra term. The field is symmetrical around $\theta = 0$ (the rotation axis) and it is a maximum at $\theta = 90°$. The larger the value of J_2, the greater the variation of g with θ. With $J_2 > 0$ the field now points towards the centre of the body only at $\theta = 0$, 90°, and 180° – this is not apparent in Figure 4.2(b) because magnitudes alone are shown, not directions.

Regardless of its origin, the J_2 term can be deduced from the gravitational field as mapped by small bodies in the vicinity of the planetary body, such as natural satellites and spacecraft. J_2 provides a constraint on the variation of density with radius, in that only certain variations are consistent with the measured value of J_2. However, a more powerful constraint is obtained if J_2 is combined with other data to obtain a property of the body called its polar moment of inertia. This is a topic for the next section.

Question 4.2

By considering a doubling in the distance r from a planetary body, show that the gravitational field becomes more closely like that of a spherically symmetrical body.

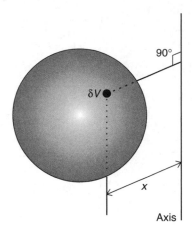

Figure 4.3 The basis of calculating a moment of inertia.

4.1.3 Radial Variations of Density: The Polar Moment of Inertia

A moment of inertia is a quantity involving a body and some axis. Any axis can be used, and the corresponding moment of inertia calculated. If we imagine the body divided up into small equal volumes δV, as in Figure 4.3, then the moment of inertia for the chosen axis is the sum of $(\rho \delta V)x^2$ over the whole body, where x is the perpendicular distance from δV to the axis, and ρ is the density in the volume δV. ($\rho \delta V$ is the mass δM in δV.) The moment of inertia therefore depends on the choice of axis. For a rotating planetary body the rotation axis is a natural choice, in which case we have the **polar moment of inertia** C. C depends on the variation of density from place to place in the body, which is why it is a powerful constraint.

C can be obtained from the precession of the rotation axis (Section 1.5.1). Precession is caused by the gravitational torque on the non-spherical mass distribution of the rotating body. For the planets the torque is provided by any large satellites and by the Sun. In order to calculate C, the precession period has to be known, and also the mass M of the rotating body, its equatorial radius R_e, its sidereal rotation period T, and J_2. So far it has been possible to apply this precession method with useful accuracy only to the Earth, the Moon, and Mars. The details are beyond our scope – see books on celestial mechanics in Further Reading.

To obtain C for the other planetary bodies we have to assume that the interiors are in **hydrostatic equilibrium**, i.e. that the interior has responded to rotation as if it had no shear strength – as if it were a fluid. Solid materials in planetary interiors can achieve such equilibrium in times *very* short compared with the age of the Solar System because of the huge internal pressures that overwhelm the shear strength of the solids. For a planetary body in hydrostatic equilibrium, the value of C can be calculated from M, R_e, J_2, and T – again the details will not concern us. If J_2 is unknown, then the rotational flattening f can be used instead. In fact, J_2 is preferred to f because any departures from hydrostatic equilibrium are likely to be confined to the outer layer of the body, which influences f more than J_2. Furthermore, terms additional to the J_2 term can reveal departures from hydrostatic equilibrium, and these can be used to adjust J_2 so that C can be calculated more accurately.

Table 4.2 lists the values of $C/(MR_e^2)$ for those planetary bodies for which C is known. The division of C by MR_e^2 is useful, because it gives an indication at a glance of the degree of concentration of mass towards the centre. For a hollow spherical shell $C/(MR_e^2)$ is 0.667 (2/3 exactly); for a sphere with the same density throughout it is 0.4 exactly; and for a sphere with its mass concentrated entirely at its centre it is zero.

The giant planets are so rotationally flattened that J_2 is comparatively large, and further terms in equation (4.7) are then also significant. Their strength is measured by the gravitational coefficients J_4 and J_6 (odd-numbered terms like J_3 and J_5 are zero because of the northern–southern hemisphere symmetry of the giant planets). In this case J_2, J_4, and J_6 are in themselves a useful constraint on radial variations of density, and for the giant planets are often used in preference to C.

Question 4.3

Show that $C/(MR^2) = 1$ exactly for a hollow, cylindrical tube, with a radius R and a mass M, for rotation around the axis running along its longitudinal axis.

4.1.4 Love Numbers

The measurement of the gravitational field near a body also enables us to obtain the **Love numbers** of the body, named after the British geophysicist Augustus E H Love (1863–1940). They are derived from the distortion of the surface and interior of a body caused by the tidal forces exerted by other bodies. These tidal distortions can be sensed by their effect on spacecraft orbits. For our purposes we need only note that their values give us a measure of how much the interior of a planetary body deforms under tidal stresses. From this, we can deduce, for example, to what extent the interior is liquid, liquids being much more deformable than solids.

4.1.5 Local Mass Distribution, and Isostasy

Very close to the surface of a planetary body the gravitational field is sensitive not only to radial variations of density, but also to *local* mass distributions. For the modelling of planetary interiors one of the most important things we can learn from such data is whether the planet is in **isostatic equilibrium**. A familiar example of isostatic equilibrium is a block of ice floating on water, as in Figure 4.4(a) and (b). The ice is less dense than the water and in equilibrium floats with its base at a certain depth such that in any vertical column above this depth, within the ice and beneath it there is the same amount of mass. This is where the term 'isostatic' comes from – 'equal standing'. If the ice is raised and released there is a net downward force on it, and if it is pushed down and released there is a net upward force on it. Because liquid water is very fluid the ice quickly regains its equilibrium level, and isostatic equilibrium is restored rapidly. The minimum depth above which there is equal mass in each column is called the depth of compensation.

Imagine measuring the gravitational field at a fixed altitude above the water surface. If the ice is in isostatic equilibrium then the field changes very little over the block, because of the 'equal standing'. Thus, in spite of the mountain of ice, gravity hardly varies across it. By contrast, if there is no isostatic equilibrium the amount of mass in a vertical column changes as we cross the ice, and so the variations in the field are larger. By measuring the field as we cross the ice we can determine whether the ice is in isostatic equilibrium.

❑ How does the field vary if the ice is below its equilibrium level?

If the ice is below its equilibrium level there is a deficit of mass in its vicinity and so the field is slightly weaker than to each side. If it is above its equilibrium level there is an excess of mass and the field is slightly stronger.

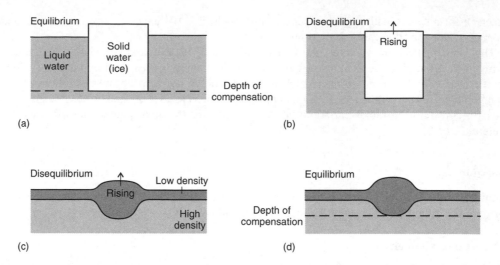

Figure 4.4 Isostatic equilibrium. (a) Equilibrium in ice floating on water. (b) Disequilibrium with ice and water. (c) Disequilibrium at a planetary surface. (d) Equilibrium at a planetary surface.

A planetary body typically has a crust of less dense solids of non-uniform thickness on top of a mantle of more dense solids. For the crust to be in isostatic equilibrium the underlying mantle has to be able to flow in response to departures from isostasy, and the upper layer must be able to deform to take up the equilibrium shape. This is illustrated in Figure 4.4(c) and (d). Given sufficient time, isostasy will be achieved. The greater the plasticity of the interior and the greater the flexibility of the surface layer, the shorter the time required. The plasticity of the interior increases with temperature and pressure, and depends on composition. The flexibility of the surface layer also increases with temperature and depends on composition, and it also increases as its thickness *de*creases. It is quite possible for the adjustment time to be many thousands of years, or longer, and so departures from isostasy are to be expected, and are observed.

4.2 Magnetic Field Data

Magnetic fields are caused by electric currents. Therefore, if a planetary body has a strong magnetic field there must be large electric currents within it, and this can tell us a lot about the interior.

Whereas mass is the source of gravitational field, electric charge in motion – electric current – is the source of **magnetic field**. For magnetic field, the basic source is an electric current loop, and a circular one is shown in Figure 4.5(a). At distances that are large compared with the distance across the loop, the magnetic field is independent of the shape of the loop and the field is then called a **magnetic dipole field**. It has the form shown schematically in Figure 4.5(b), where the loop is presented edge on. The lines show the direction of the magnetic field, and their spacing is a qualitative indication of the magnitude of the field. Note that the field is rotationally symmetrical around the magnetic axis, which is perpendicular to the loop and passes through its centre. The distributions of currents in a planet are much more complex than a loop. Whereas the field sufficiently distant from the currents is a dipole field, at closer range it has a somewhat different form, depending on the configuration of the currents, and it is called a poloidal field.

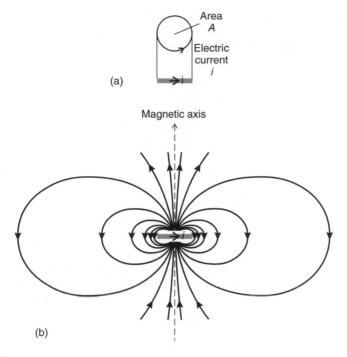

Figure 4.5 (a) A circular electric current loop. (b) Dipole field due to a current loop.

☐ The basic source of gravitational field is the point mass. Describe the equivalent drawing to Figure 4.5(b) in this case.

The gravitational field lines of a point mass would be radial, and would point inwards to denote attraction. The increase in the spacing of radial lines as distance from the mass increases is a qualitative representation of the decrease in the size of the force with distance.

At any point of measurement at a given distance and in a given direction, the magnitude of the magnetic field is proportional to $i \times A$, where i is the electric current circulating around a loop of area A that would generate the same dipole field as that observed. The magnitude of the **magnetic dipole moment** is defined as

$$\mu = i \times A \tag{4.8}$$

The *direction* of the magnetic dipole moment is the same as the direction of the magnetic axis.

The mechanism for sustaining magnetic fields in planetary bodies is not fully understood. The general idea is that the process starts with an electrically conducting fluid in motion. In planetary bodies this will be due to thermal convection. The presence of a small external magnetic field will set up electric currents in the moving liquid that generate further magnetic fields. This influences the liquid motion in such a way that the disordered electric currents are enhanced and coordinated so that the field strength is increased until there is as much energy in the magnetic field as in the liquid motion. If the rotation of the planetary body is sufficiently rapid then the liquid motions are coordinated such that a strong poloidal field is established. Once the field is established, it is probably unnecessary to have a small external magnetic field to sustain it. Thus we have the current sustaining the magnetic field that sustains the current in the moving liquid. This is called a **self-exciting dynamo,** by analogy with the familiar manufactured device that operates in the same basic way. The energy in the field comes from the kinetic energy in the liquid flow. In a manufactured dynamo the kinetic energy comes from the rotation of a coil of wire.

The fluid motions tend to decline through viscous dissipation, and the currents themselves tend to diminish through the electrical heating they cause. In the Earth, where the magnetic field originates in a convective outer core of liquid iron, these losses would eliminate the magnetic field in the very short time of about 10000 years. The losses need to be offset by internal energy sources. In planetary interiors convection can be sustained by deep-seated heat sources, as you will see in Chapter 5.

The crucial role of rotation leads to a fundamental theorem, the **Cowling theorem,** which states that the field and the rotation cannot have the same symmetry (Thomas George Cowling, British physicist, 1906–1990). This means that the rotation and magnetic axes cannot coincide – there must be an angle between them. Most dynamo models require this condition.

The currents inside a planetary body are distributed throughout the conducting region. Consequently the field at, and not far above, the surface of the body is very different from that of a simple loop. The differences can be used to constrain the distribution of the electric currents in the interior, somewhat as the extra terms in the gravitational field can be used to constrain point-to-point variations in density. Nevertheless, the existence of a dipole field at sufficiently great distances means that we can specify the strength of the source of the field by the corresponding size of a magnetic dipole moment. The known values are given in Table 4.2. From a large dipole moment we infer the presence of a considerable body of electrically conducting fluid in motion within the planetary body.

Figure 4.6 shows the magnetic axes of the five planetary bodies that have large magnetic dipole moments – the Earth and the four giant planets. Where the magnetic axis passes through the surface of the body we have the magnetic poles. The magnetic equatorial plane (Figure 4.6) is perpendicular to the magnetic axis, and they intersect at a point that is the centre of the far field. You can see that this point is displaced from the centre of the planetary body, and so the currents are not symmetrically distributed around its centre. The magnetic axis does not coincide with the rotation axis – except for Saturn, where the angle between the axes is close to zero. Saturn thus violates the Cowling theorem, which is worrying, though this might be a transitory circumstance, planetary magnetic fields being variable.

Magnetic fields are subject to small variations in magnitude and in the direction of the magnetic axis – those in Figure 4.6 are for the present time. For planets with rocky surfaces these variations can be traced through rocks that retain the imprint of past magnetisation by the field. One way in which this can happen is through the solidification of a magnetised molten rock.

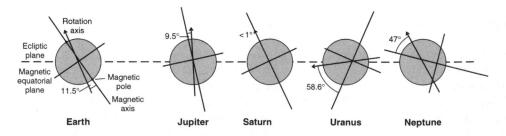

Figure 4.6 The magnetic axes of the five planetary bodies that have large dipole moments.

This remanent magnetism can be used to trace the history of the field, particularly if the time of solidification has been radiometrically dated. For the Earth it is known that the direction of the magnetic axis has varied erratically, though the inclination with respect to the rotation axis has never been very great. In addition, the field has reversed in direction many times. Each reversal takes about 0.01 Ma, and during it the field declines in strength to zero, and then grows in the opposite direction. During the last 80 Ma these magnetic reversals have occurred at intervals of 0.1–1 Ma; the last time this happened was about 0.7 Ma ago. Earlier, there were periods when the reversals were considerably less frequent. Intervals between reversals that exceed 20 Ma are called superchrons. For example, one superchron lasted from 118 Ma to 83 Ma ago. Numerical simulations of the dynamo theory as applied to the Earth's interior do exhibit magnetic variations, including reversals. In the case of the Earth this long-term variation in the frequency of reversals is probably due to the effects on convection of fluctuations in temperature at the top of the liquid iron core where it meets the rocky mantle. The reversals themselves might owe much to chaos – large effects from small causes.

Other sources of magnetic fields, from electric currents in the upper atmosphere to permanently magnetised rocks at and below the surface, are weak, and are sufficiently disorganised on a planetary scale that their net contribution to the dipole field is zero. Note that in planetary interiors there is also a field from a current configuration that does not generate a dipole field at any range. This is called a toroidal field. It has no substantial external manifestation, and so will not be discussed further.

Question 4.4

Making use of what you learned about the planets in Chapter 1, outline reasonable hypotheses based on the self-exciting dynamo mechanism to explain why the Earth and the giant planets have large magnetic dipole moments, but the other planets do not.

4.3 Seismic Wave Data

4.3.1 Seismic Waves

A **seismic wave** travels through a planetary body, and bears information about the conditions along its path. Seismic waves are mechanical waves in that the wave proceeds via the direct push of an atom or molecule on its neighbours. Other mechanical waves include sound waves in the air and the ripples on the surface of a pond. There are several types of seismic wave. Rayleigh and Love waves, rather like water ripples, can exist only at an abrupt change of density, such as at the surface of a planet. But it is two other types that have been of particular value in elucidating the structure of planetary interiors. These types are illustrated in Figure 4.7. In a P wave the oscillatory motion of the substance is longitudinal – to and fro in the direction in which the wave progresses, as in a sound wave in the atmosphere. In an S wave the oscillatory motion is transverse – perpendicular to the progression. S waves require the medium to have shear strength, so they do not travel in liquids or in gases, and they are strongly attenuated in soft, plastic materials.

Seismic waves can be generated by surface impacts, by surface explosions, and by the sudden yielding of interior rocks in which stresses have built up. These stresses can result from the tidal action of external bodies, or from internal heating. Considerable surface motion occurs in the vicinity of the source, and in the case of the Earth this constitutes an earthquake. Source sizes

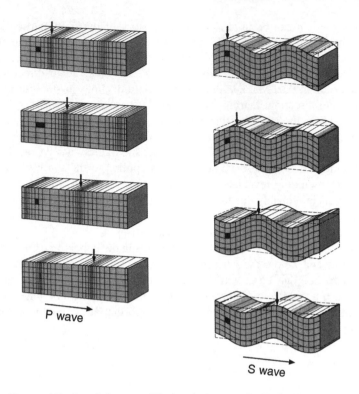

P wave

S wave

Figure 4.7 P and S waves. The vertical arrows indicate a maximum.

are typically only a few kilometres, and seismic waves travel from the source in all directions. For a planet with a liquid core and a solid mantle some typical paths are shown in Figure 4.8. The curvature of the paths is a consequence of the gradual increase in wave speed with depth. You can see that the path bends away from the direction in which the speed is increasing.

The importance of P and S waves is that they can reach a considerable depth in the planetary body. For example, in Figure 4.8 the P wave that enters the liquid core passes near the centre of the body and reaches the surface on the far side. If it is detected there, and if we know the source position and the time that the wave set off, we can calculate the average speed of P waves along the path.

☐ If P waves set off at time t_1 from a shallow depth, pass through the centre of a planet with a radius R, and are detected at t_2 on the opposite surface of the planet, what is their average speed?

The average speed is the total distance divided by the total time: $2R/(t_2 - t_1)$. If average speeds are determined for many different paths, then we can deduce the P wave speed at all points in the interior, and not just as an average along a path. A similar feat can be performed for S waves. What can we learn from these data?

The P wave speed in a material depends on just two of its properties – its density ρ and its axial modulus a. The axial modulus is one of several elastic moduli of a material. In general an elastic modulus is the stress to which a material is exposed, divided by the resulting strain. For P waves the relevant stress is the rapid application of a force along the direction of wave motion, and the resulting strain is the compression or rarefaction along the wave direction with

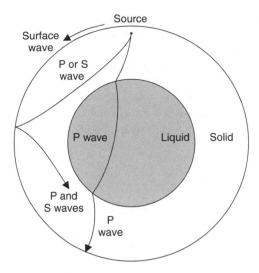

Figure 4.8 Seismic wave paths in a hypothetical planet. (In reality, the P and S waves would follow different paths in the solid material.)

no expansion or contraction of the material perpendicular to this direction (Figure 4.7). In this case the relevant modulus is called the axial modulus, and the P wave speed is given by

$$v_P = \sqrt{\frac{a}{\rho}} \tag{4.9}$$

It is beyond our scope to derive this equation, but you can see that it depends on a and ρ in a reasonable way. Thus, if the material is less deformable then a is larger and it passes the wave along more rapidly. On the other hand, if it is denser then it is more sluggish in its response, and the wave travels less rapidly.

For S waves the speed is given by

$$v_S = \sqrt{\frac{r}{\rho}} \tag{4.10}$$

where r is a different modulus, called the rigidity (or shear) modulus. This is a rapidly applied shear stress divided by the resulting shear strain (Figure 4.7).

Equations (4.9) and (4.10) show that if we know the P and S wave speeds at a point in a material then we know something about the density and the elastic moduli there. Materials differ in these properties so we seem to have valuable information on composition. Unfortunately, because we have just *two* pieces of information – S wave speed and P wave speed – but *three* unknowns – density and two elastic moduli, we cannot deduce the density and the elastic moduli. Therefore, we have only a weak constraint on composition. Tighter constraints are obtained by combining seismic wave speeds with gravitational and other data. The outcome is that the density versus depth is strongly constrained, even deducible with sufficient data, and this provides a strong constraint on composition.

Seismic data alone do, however, provide other types of information on the interior. Depth ranges that are liquid, and therefore inaccessible to S waves, can be identified through the absence of S waves over that range. Also, if the S and P wave speeds change suddenly at a certain depth, this is a strong indication of a change in the minerals present, either in their chemical composition or in their crystal form.

For a body that is fluid throughout, seismic wave speeds can be obtained from the Doppler effect of the wave motions on visible spectral lines in the radiation from the body's outer regions (Section 2.1.2). In the case of the Sun such studies constitute helioseismology, a well-established technique that has yielded much information about the solar interior. Developments are under way to apply a similar technique to spectral lines in the outer atmospheres of the giant planets.

4.3.2 Planetary Seismic Wave Data

To obtain the P and S wave speeds versus depth requires a large network of seismic observatories well spread over the surface of a planetary body, with data from a large number of seismic sources. This has only been achieved for the Earth, and Figure 4.9(a) shows the P and S wave speeds versus depth in our planet.

☐ What is the reason for the absence of S waves from 1215 km to 3470 km from the centre? A liquid shell is clearly indicated by this absence of S waves. At all depths above and below the shell S waves do exist, and so there are no other *extensive* regions that are liquid. (Note that S waves are generated in a solid inner core beneath the liquid shell by P waves that traverse the shell.)

Layering of the Earth is also indicated by large changes in the wave speeds over very short depth ranges, and these generally indicate changes in composition at these boundaries. Between these boundaries there is a general increase of speed with depth. Given that the density in planetary interiors increases with depth, this would seem to be contrary to equations (4.9) and (4.10) which show a *decrease* in speed for an increase in density. However, it is also the case that the elastic moduli of most substances increase with an increase in density (though decrease with an increase in temperature), so the outcome is not obvious in advance. The seismic data are so extensive for the Earth that small changes have been detected in the wave speed profiles, and some of these correspond to slow convective motions within the various layers.

Figure 4.9(b) shows the P and S wave speeds versus depth in the Moon, the only other body for which there are seismic data at present. These are from seismometers set up at the landing sites of Apollos 11, 12, 14–16 (Table 4.1). The large uncertainties shown in the speeds are the combined result of only having these five seismic stations, their being located all on the near side of the Moon, some not working very well, and 'moonquakes' being much less frequent and on average weaker than in the Earth. The speeds in Figure 4.9(b) are averaged over profiles obtained by several seismologists. Every profile shows an increase in P and S speed with depth, as does the average, but the averaging has obscured evidence in most profiles of small, but sharp, changes at a few depths.

The data in Figure (4.9) will be further interpreted in Sections 5.1.1 and 5.2.1.

Question 4.5

Sketch the variation of seismic wave speeds versus depth in an imaginary planet that has both of the following properties:

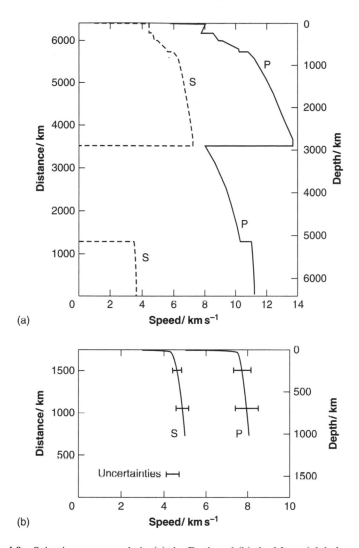

Figure 4.9 Seismic wave speeds in (a) the Earth and (b) the Moon (global averages).

(1) an entirely fluid interior;
(2) a change in composition half way to the centre, with the material just below the boundary having twice the axial modulus of the material just above it and a density 20% greater.

4.4 Composition and Properties of Accessible Materials

4.4.1 Surface Materials

Even though the only planetary materials available for direct compositional analysis are from the surface or near surface, the composition of such materials can give us a useful indication of the composition, temperature, and density deeper down. For example, the Earth's surface consists of rocks with a density of about $2700\,kg\,m^{-3}$ on the continents and about $2900\,kg\,m^{-3}$ in the ocean basins. (In many regions these rocks are covered by a very thin veneer of soil or sediment, but that is of no relevance here.) By contrast, the Earth has a mean density of $5520\,\mathrm{kg\,m^{-3}}$, and it is not possible for the internal pressures to be high enough to squeeze the surface rocks to such a high mean value. Therefore, the interior *must* consist largely of intrinsically denser materials.

As well as densities, we can also constrain the range of minerals that could comprise the interior. This is done by choosing a mineral mix such that when it is subjected to various geological processes that produce surface rocks, it would produce those seen. We shall look at this more closely in Chapters 5–8.

For the giant planets there is no solid surface, and they are all covered in atmospheres that blend into deep outer envelopes. Therefore, determination of the composition of these atmospheres is of particular importance in building models of their interiors.

4.4.2   Elements, Compounds, Affinities

A powerful constraint on composition is provided by observations, not of the body in question, but of other bodies, notably the Sun and meteorites. The relative abundances of the elements in the Solar System are obtained mainly from the composition of the Sun outside its core (Section 1.3) and from the composition of the least altered meteorites, the C1 chondrites (Section 3.3.2). Table 1.5 shows the relative abundances of the 15 most abundant chemical elements. It is to these elements, and to compounds containing them, that we must look to account for most of the mass of a planetary body. In doing so, we must take account of the density of the substance. Table 4.3 lists the densities of some important substances, and distinguishes the more dense from the less dense.

Table 4.3   Densities of some important substances

Substance	Chemical formula	Density^a/kgm^{-3}
Iron–nickel	Fe plus $\sim 6\%$ Ni by mass	7925
Troilite	FeS	4740
Corundum	Al_2O_3	3965
Forsterite^b	Mg_2SiO_4	3270
Silicon carbide	SiC	3217
Diopside^c	$CaMgSi_2O_6$	3200
Alkali feldspars	$(Na,\ K)AlSi_3O_8$	2700
Quartz (silica)	SiO_2	2600
Hydrated minerals^d	$X(H_2O)_n$ or $X(OH)_n$	≤ 2000 (or so)
Water (liquid)	H_2O	998
Carbon dioxide^e	CO_2	1.98
Carbon monoxide^e	CO	1.25
Helium^e	He	0.166
Hydrogen^f	H_2	0.08987

^{a} At 293 K and 1.01×10^5 Pa unless otherwise indicated.
^{b} A particular form of olivine, $(Mg,\ Fe)_2SiO_4$.
^{c} A particular form of pyroxene, $(Ca,\ Fe,\ Mg)_2Si_2O_6$.
^{d} X can be a molecule of a variety of rocky minerals, and n is greater than or equal to one.
^{e} As a gas.
^{f} As a gas, at 273 K.

For example, suppose that we are building a model of the Earth on the basis of its mean density. In spite of their abundance we can rule out the two lightest elements hydrogen and helium as important constituents – the Earth is *far* too dense to contain much of these intrinsically low-density substances. But when it comes to choosing between iron–nickel ($7925\,kg\,m^{-3}$) and zinc ($7140\,kg\,m^{-3}$) for the dominant constituent of a dense central core, density alone is a poor guide – both substances are denser than the mean density of the Earth. There are many grounds for choosing iron, but among them is the relative abundances of the elements in the Solar System – iron is about 1000 times more abundant than zinc.

Of course, most elements will be present as compounds, and of particular importance are abundant rocky materials, notably silicates, and abundant icy materials, notably water.

☐ Do silicates and water include abundant elements?

Silicates are based on the abundant elements oxygen and silicon (Section 2.2.2), and water (H_2O) is a compound of oxygen with the most abundant element of all, hydrogen. Rocky materials are intrinsically denser than icy materials, and icy materials are intrinsically denser than hydrogen and helium.

Chemical affinities are also important. These express the tendency for certain elements to occur together. Elements can be grouped on the basis of these affinities, and three groups of particular relevance to planetary interiors are the siderophiles, chalcophiles, and lithophiles. The **siderophiles** ('iron-lovers') comprise iron, and various metallic elements that tend to be present with metallic iron, such as nickel. The **chalcophiles** tend to form compounds with sulphur. Zinc is one example. The **lithophiles** ('rock-lovers') tend to be found in silicates and oxides – magnesium and aluminium are important examples. Iron is so abundant that it occurs in sulphides, oxides, and silicates, as well as in elemental form.

To illustrate how chemical affinities can be used, consider a planetary body with a surface that is heavily depleted in iron. It can then be argued that iron has been concentrated into the interior, perhaps into a central core of metallic iron. Support for this possibility would be a surface depletion in the siderophilic element nickel, and an enrichment in the lithophilic element aluminium.

4.4.3 Equations of State, and Phase Diagrams

In deciding whether a substance could be a significant component at some depth in the interior of a planetary body, we need to know its density at the pressure and temperature at that depth. The equilibrium relationship between the density of a substance and its pressure and temperature is known as the **equation of state** of the substance. This equation not only allows us to calculate the density at some depth, but also enables our models to achieve consistency. Thus, if we specify a substance at a certain depth, we are only free to specify two of density, pressure, and temperature – the equation of state determines the third.

For the great majority of plausible substances for the interiors of most planetary bodies, pressure is far more important than temperature as a determinant of density. Though the temperatures at all but shallow depths in the interior are far higher than at the surface, the thermal expansion of rocky (and icy) materials is slight, resulting in only a slight decrease in density compared with that at the surface. By contrast, the pressures at such depths squeeze the materials so much that their densities increase far more than that required to offset the thermal expansion. So, to a first approximation we can ignore temperature and concentrate on pressure.

Pressures in the interior

To find the pressures in the interior we use the fact that if a body is neither contracting nor expanding then the pressure at any radius must be just right to support the overlying weight. Consider a thin spherical shell of material of inner radius r and thickness δr, where δr is small compared with r, as in Figure 4.10. To a first approximation we can assume that the planetary body is close enough to spherical symmetry so that we can regard its density as varying only with radius. In this case it is remarkable but true that the mass above the shell exerts zero net gravitational force at all points in the shell. The net gravitational force on the shell due to the rest of the body is thus the gravitational force exerted on it by the inner sphere of radius r. The net force is thus downwards, towards the centre. There is also a downward force on the shell due to the pressure at $r + \delta r$, and an upward force due to the pressure at r. In equilibrium the total downward force must equal the upward force, and so the pressure at r must exceed that at $r + \delta r$. It follows that *pressure increases with depth at all values of r*. This is an important general result.

If p is the magnitude of the pressure at r, and $p - \delta p$ the pressure at $r + \delta r$, then the net upward force F_{up} on the material in the shell has a magnitude $A \times \delta p$, where A is the surface area of a sphere of radius r. Thus

$$F_{up} = 4\pi r^2 \times \delta p$$

The magnitude of the gravitational force on the shell is obtained from Newton's law of gravity, equation (1.5).

☐ For a spherically symmetrical body of radius r, where can all of its mass be regarded as concentrated when calculating the gravitational force it exerts?

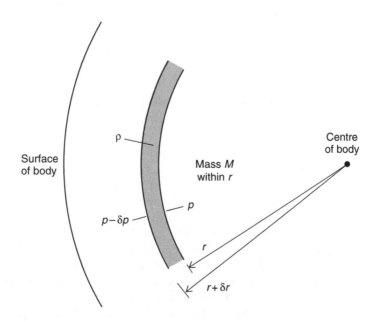

Figure 4.10   The balance of forces on a spherical shell in a planetary interior.

It exerts a gravitational force as if all its mass were concentrated at its centre (Section 1.4.4). We can thus use equation (1.5) as it is, with M as the mass of the inner sphere and m the shell mass. Therefore, the gravitational force has a magnitude GMm/r^2 where G is the gravitational constant. The mass of the shell is its density times its volume, i.e. $\rho(4\pi r^2 \times \delta r)$, and so

$$F_{\text{down}} = -GM \times \rho(4\pi r^2 \times \delta r)/r^2$$

where the minus sign indicates that F_{down} is in the opposite direction to F_{up}. In equilibrium F_{up} must equal F_{down}. Thus, from the equations for these forces, and after tidying up the algebra, we obtain

$$\delta p = -\frac{GM}{r^2}\rho \times \delta r \tag{4.11}$$

The minus sign shows that p decreases as r increases. Equation (4.11) is called the hydrostatic equation. Planetary rotation adds a small term to the right hand side of the equation, but we can ignore it for our purposes.

The central pressure is obtained by applying equation (4.11) step by step from the surface to the centre, calculating the increase in pressure as we proceed. To obtain an exact value we clearly need to know the density versus depth, yet this is one of the things that we *do not* know until our modelling is complete. However, even without this knowledge the useful rough and ready result can be established from equation (4.11) that the central pressure p_c is approximately given by

$$p_c \approx \frac{2\pi G}{3}\rho_m^2 R^2 \tag{4.12}$$

where ρ_m is the mean density of the planetary body, and R is the radius of its surface.

Equation (4.12) indicates how huge the pressures are deep in the interior of a planetary body. In the case of the Earth the equation yields a value of 1.7×10^{11} Pa – over a million times greater than atmospheric pressure at the Earth's surface. For most substances the equation of state is not very well known at such high pressures. But we can still deduce whether a substance is a plausible ingredient. For example, molecular hydrogen (H_2) and atomic helium require pressures in the Earth *far* in excess of 1.7×10^{11} Pa to become even as dense as liquid water at the Earth's surface ($1000 \, \text{kg m}^{-3}$). The same is true of atomic hydrogen. Therefore, these potentially abundant substances can account for no more than a tiny fraction of the Earth's mass. Water, at comparable pressures, is much denser than hydrogen and helium, but still well short of the Earth's mean density. Thus, neither can water, H_2O, a compound of the first and third most abundant elements, account for much of the Earth's mass.

Equation (4.12) also shows how much lower the interior pressures are in smaller, less dense bodies. Thus, at the centre of the Moon, equation (4.12) gives 4.7×10^9 Pa, nearly 40 times less than at the centre of the Earth. Moreover, the higher the pressure given by equation (4.12), the more this approximate equation underestimates the true pressure – by about a factor of 2 in the Earth's case (see Table 5.1).

The general conclusion is that a given substance will have a markedly different density in the interiors of different planets. For comparing compositions, it is therefore useful to obtain the uncompressed mean density. Unlike the mean density it is not directly measurable, but must be calculated from equations of state, assuming a composition. Equations of state are measured in the laboratory up to pressures that fall short of those in the centres of the larger planetary bodies. For higher pressures it is necessary to do the best one can with theoretical equations or reasonable estimates.

Question 4.6

Calculate an approximate value for the pressure at the centre of Jupiter. What can you say about the pressure at shallower depths?

Phase diagrams and equations of state

The pressure and temperature in the interior determine whether a substance there is liquid or solid, and this is of importance to the behaviour of the interior. You have already seen this to be the case with regard to magnetic fields and the passage of seismic waves, and more examples will arise shortly. If a substance were to be a gas in the interior it would be unlikely to be retained, but would escape to the surface. We therefore need to know the ranges of pressure and temperature over which a substance will be liquid, solid, or gas. This information is provided by a **phase diagram**, 'phase' being a generic term for solids, liquids, and gases.

Figure 4.11 shows the typical form of a phase diagram. It consists of the boundaries between the phases. If the substance is in equilibrium, and its pressure and temperature are *anywhere* within the area to the left of the solid + gas or solid + liquid boundaries, it will be solid. Unsurprisingly, the solid phase is confined to the lower temperatures, particularly at low pressure. The liquid phase requires a minimum pressure and temperature, marked by point Tr in Figure 4.11. At each temperature up to Cr the gas phase can exist only up to a certain pressure. At temperatures above Cr the gas phase becomes difficult to distinguish from a liquid, hence the absence of a distinct boundary there. Tr and Cr are, respectively, the triple point and the critical point of the substance.

☐ What do you think is the origin of the term 'triple point'?

The triple point is so named because only at this point do all three phases coexist. By contrast, on each phase boundary just two phases coexist – the phases on each side of the boundary.

Numerical values have not been attached to the axes in Figure 4.11 because the pressures and temperatures of the phase boundaries vary from substance to substance. Note that the density varies across the phase diagram. The density at a specified temperature and pressure is provided by the equation of state of the substance, and so a phase diagram contains only a subset of the information encapsulated in the equation of state.

It is instructive to take a substance on three imaginary journeys across the phase diagram, each one at constant temperature and starting in the gas phase. If we start at T_1 and increase the pressure, the substance gets denser, and it then meets the phase boundary at A, where a further increase in pressure causes a huge increase in density as it solidifies. Within the solid phase there might be structural changes from one crystal form to another, which might give small density changes. If we now reverse the track the substance sublimes at A. If we start at T_2, and increase the pressure, then at B the gas condenses, with a huge increase in density, and it becomes a liquid. At C there is a much smaller increase in density as the substance solidifies. Reversing the track, the substance melts at C and then vaporises at B. Finally, if we start at T_3, higher than the temperature of the critical point, and increase the pressure, then the density goes on increasing to liquid-like values with no sudden changes in density such as occur across phase boundaries.

The pressure and temperature of the triple point is a better measure of whether a substance is volatile or refractory than is the sublimation temperature used in Section 2.2.2, which clearly depends on pressure. The lower the triple point values, the more volatile the substance; the higher the triple point values, the more refractory the substance. For water the triple point values are 6.1 Pa and 273.16 K, whereas for iron the values are many billions of Pa and thousands of

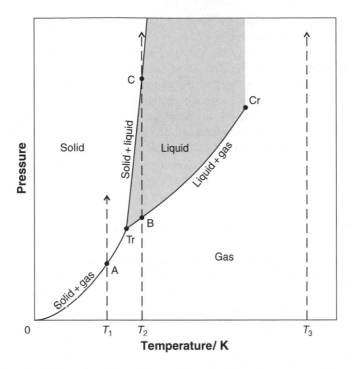

Figure 4.11 The typical form of a phase diagram. For *very* few substances, notably water, the solid + liquid phase boundary slopes to the left, and not to the right as here.

K. Water is volatile, as are all icy materials, whereas rocky materials, which include iron and other metals, are more or less refractory. Carbonaceous materials are intermediate.

In later chapters you will meet phase diagrams, not only in relation to the interiors of specific bodies, but in the context of atmospheres too.

Question 4.7

Water is unusual in that the solid–liquid phase boundary slopes to the *left* from the triple point, and not to the right as in Figure 4.11. Describe the phase changes that occur when water in the gas phase at 273.10 K is gradually compressed at this temperature, to pressures greatly in excess of 6.1 Pa.

4.5 Energy Sources, Energy Losses, and Interior Temperatures

Interior temperatures are an important property of a planetary body. For example, along with composition, they largely determine whether there are any liquids in the interior, and consequently whether a powerful magnetic field is possible. Temperatures in the past can determine whether core formation has occurred, and whether other compositional divisions have become established. Temperatures also determine the dynamics of the interior, such as the rate of any convection. And the thermal history of a planetary body helps to determine the nature of the surface and of any atmosphere, as you will see in later chapters. The temperatures inside a body at any particular moment in its history depend on the accumulated effects of the energy sources and the energy losses at all earlier times.

It is important to distinguish between the internal energy of a body, its temperature, and heat. The **internal energy** consists of kinetic and potential energy. The kinetic energy is almost entirely in the random energy of motion of the particles (atoms and molecules) that constitute the body (the motion of the body as a whole is excluded). The potential energy is the energy of interaction between the particles, be this gravitational or electrical. It generally increases as the mean separation between the particles increases. The temperature of the body is proportional to the random kinetic energy per particle. **Heat** is a particular form of energy transfer, i.e. transfer that is *random* at a *microscopic* level. For example, when a body at a certain temperature is in contact with a body at a lower temperature there is a net transfer of random energy from the higher to the lower temperature region through the random collisions of the microscopic particles at the interface. 'Heat' is often used interchangeably with 'energy'. This is avoided here unless the language would otherwise sound peculiar. **Power** is the rate of energy transfer, by heat, or by any other means.

If a region is a net receiver of heat its temperature might rise, and if it is a net loser its temperature might fall. Temperature changes can also be caused without heating, by any other process that causes a rise in the random kinetic energy part of the internal energy. One example is the temperature rise when two bodies collide. However, heat transfer does not always give a temperature change. For example, heat fed into a solid at its melting point does not cause a rise in temperature but a change in phase from solid to liquid. When a liquid solidifies at its melting point it gives out heat with no change in temperature. Heat that produces no temperature change is called **latent heat**. It is the energy that has to be supplied either to increase the average separation between the particles against their chemical attraction for each other, or (e.g. water) to break chemical bonds between the molecules. In either case there is an increase in the potential energy component of the internal energy.

It is important to stress that the present-day interior temperatures do not depend on the *present* energy sources and losses, but on their accumulated effect right back to the formation of the body. A familiar example of this dependence on history is the heating of water in an electric kettle. The temperature of the water at any instant does not depend on the flow of electric current and the energy losses at that instant, but on their earlier values. We cannot therefore separate the present from the past when we consider energy sources and losses in a planetary body.

4.5.1 Energy Sources

Energy sources no longer active: primordial energy sources

You saw in Chapter 2 that planetary bodies are thought to have been built up from planetesimals and other bodies in a process called accretion. This led to temperature rises as the kinetic energy of the infalling material was partially converted to internal energy of the surface materials at the point of impact. For the giant planets there was also a significant infall of nebular gas. The kinetic energy of the infall, solid or gas, was derived from gravitational accelerations, and so this accretional energy, as it is called, came from the conversion of gravitational energy to internal energy.

If the rate of accretion was low, much of the internal energy would have been lost by IR radiation from the surface to space, so the interior temperatures would not have risen much. But if the rate of accretion was high, the hot surface would have been buried, leading to raised temperatures in the whole interior. For a given mass of impactor, the energy transfer is greater, the more speed the impactor picks up as it is accelerated by the gravitational field of the body to its point of impact on the surface. This leads to accretional energy per unit mass being roughly proportional to $\rho_m R^2$, where ρ_m and R were the then current mean density and radius respectively of the planetary body. Therefore, in large bodies, rapid accretion will lead to extensive interior melting.

Accretion is a primordial source of energy, in that accretion is not continuing today on an important scale. But the *effect* of accretional energy lives on, as you will see later. Another primordial source is any magnetic induction during the T Tauri phase of the proto-Sun (Section 3.1.6). This could have given considerable temperature rises in bodies of asteroid dimensions.

The other primordial source of importance was short-lived radioactive isotopes. Of particular importance was ^{26}Al, which decays to ^{26}Mg. Studies of isotope ratios in meteorites indicate that this isotope accounted for a significant proportion of the aluminium present in the Solar System at its birth; ^{26}Al has a half-life of only 0.73 Ma, and so there are about 62 half-lives in just the first 1% of Solar System history.

❑ By what factor did the initial quantity of ^{26}Al decrease during this time?

With a halving of the quantity every half-life the decrease in 46 Ma was by a factor of 2^{-62} of its initial quantity, i.e. 2.2×10^{-19}. Thus ^{26}Al is long gone. But the shortness of its half-life, and the estimates of a modest initial abundance, indicate that the initial rate of energy release was high. If the isotope was distributed throughout the body, or concentrated at its centre, most of this energy would have been contained, and this **radiogenic heating** would have given considerable temperature rises in bodies of all but the smallest size. If the isotope was concentrated in the surface, then a greater proportion of its energy would have been lost to space. The loss of a greater proportion of energy when an energy source is concentrated at the surface occurs with other energy sources too.

The source of ^{26}Al, and of other short-lived isotopes, is thought to have been supernovae – massive stars ending their lives in huge explosions that generated the isotopes and implanted them in the interstellar medium from which the Solar System was born.

Differentiation

Primordial sources could have raised the temperatures of the interior of a planetary body to the point where an important process occurred that is itself an energy source, and in some bodies it might still be active today. This is differentiation (Section 3.1.6) – the separation of layers of different density. For differentiation to occur, the chemical forces binding different substances together must be weaker than the gravitational forces that tend to separate the denser substances downwards. For example, if the Earth formed with a homogeneous composition throughout (homogeneous accretion) and subsequently underwent partial or total melting, then most of the iron would drain downwards to form a core, and most of the other materials would float upwards to form a predominantly silicate mantle. This downward separation of denser materials would have converted gravitational energy into internal energy, with a consequent rise in temperature. There would have been no such energy source if the iron core formed first and the silicate mantle was acquired later (heterogeneous accretion).

Differentiation can also result from the *cooling* of the interior. For example, if two liquids are already mixed and the temperature falls, the liquids can unmix, the denser settling towards the centre of the planetary body.

Meteorites provide observational evidence that differentiation has occurred. For example, irons and achondrites are readily interpreted as fragments of asteroids that differentiated to form iron cores and silicate mantles, irons being fragments of the core and achondrites fragments of the mantle (Section 3.3.4). The Widmanstätten pattern in irons (Section 3.3.2) is just what is expected from differentiation in which a body of iron melted and then formed a core, which then cooled slowly under the insulating mantle of the asteroid. Stony-irons could come from the core–mantle interface. By contrast, the chondrites seem to represent the sort of material that was around in the asteroid belt before any differentiation occurred.

Energy sources that can still be active

Ongoing differentiation is one possible 'live' source of energy in planetary interiors. Another is latent heat released when a liquid solidifies.

A further type of active source is radiogenic heating. Of particular and widespread importance are the four long-lived radioactive isotopes ^{235}U, ^{238}U, ^{40}K, ^{232}Th. Table 4.4 lists their half-lives and an estimate of the power they would have been releasing 4600 Ma ago in a typical rocky material. The same data are also included for ^{26}Al. You can see that, unlike ^{26}Al, their half-lives are comparable with or exceed the 4600 Ma age of the Solar System, and so these isotopes can have provided energy throughout the life of the Solar System, though at a declining rate as they decay. They are distributed non-uniformly in the interiors of many planetary bodies. For example, in the Earth there is a significant concentration into the outer 30 km or so, a result of earlier partial melting at greater depths.

The other 'live' source of importance in some bodies is tidal energy.

☐ Recall what a tide is, and how a tide is caused.

In Section 1.4.5 you saw that a tide is a distortion produced by differential gravitational forces. Now that you have met the concept of gravitational field (equation (4.5)) it is better to think of this as a differential field, which we can call a tidal field. Figure 4.12 shows the elongation of the whole Earth caused by the tidal field of the Moon. The rotation of the Earth carries the elongation slightly ahead of the line connecting the centres of the Earth and Moon, but of crucial importance to tidal heating is that this rotation also moves material into and out of the elongation, as illustrated by the point P in Figure 4.12, which is fixed on the Earth's surface. From the viewpoint of the Earth's interior, the elongation sweeps like a wave through it – literally, a tidal wave. This flexing increases the interior temperatures in the Earth.

A second, smaller tidal effect is due to the variation in distance between the Earth and the Moon. As the Moon moves around the Earth in its elliptical orbit, the size of the tidal elongation varies, being least when the Moon is nearest to the Earth (at perigee), and greatest when it is furthest (at apogee). This is another way that the Earth's interior is flexed.

The Sun produces similar types of effect, but even though the Sun is far more massive than the Moon it is much further away. Therefore, the solar tidal field across the Earth is about

Table 4.4 Radioactive isotopes that are important energy sources

Isotope	^{26}Al	^{235}U	^{40}K	^{238}U	^{232}Th
Half-life/ Ma	0.73	710	1300	4500	13 900
Power ^{a}/$10^{-12}Wkg^{-1}$	10^4	3	30	2	1

^{a} This is per kg of 'typical' rocky material 4600 Ma ago. The values are only illustrative.

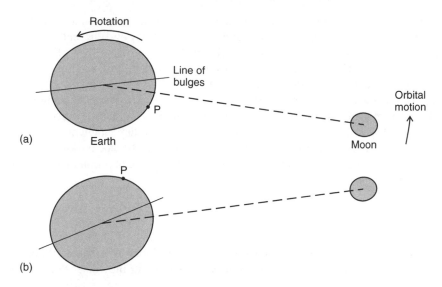

Figure 4.12 Tides (greatly exaggerated) in the Earth due to the Moon (a) at some instant and (b) about 6 hours later, with the orbital motion of the Moon exaggerated five times. P is fixed to the Earth's surface.

46% of that due to the Moon. Overall, the rate at which heat is being generated by tides in the Earth today is about 30 times less than the rate at which heat is being released from long-lived radioactive isotopes. By contrast, in Io, the innermost large satellite of Jupiter, tidal energy is by far the main internal energy source.

For any two bodies, with masses M and m, the power W_{tidal} of tidal heating of either of them is proportional to various properties of the two bodies and their orbit, including, among other properties, those in the following expression:

$$W_{tidal} \propto (Mm)^2 Re/a^6 \tag{4.13}$$

where R is the radius of the body whose tidal heating is being evaluated, e is the eccentricity of their orbits, and a is the semimajor axis of the orbit of one with respect to the other. Note how rapidly tidal heating decreases as a increases.

The ultimate source of tidal energy is the rotational energy of the bodies concerned, and the energy in their orbital motions.

☐ Is it then possible for tidal heating to be associated with changes in rotation rates, and changes in orbital motion?

This is not only possible, but *must* be the case – the internal energy gained through tides must equal a loss of kinetic energy in rotation or orbital motion.

Solar radiation

A small planetary body might for a long time have had no significant energy source except for solar radiation. In this case its interior will be at a fairly uniform temperature determined by a steady state in which the rate at which the body absorbs solar radiation is equal to the rate at which it emits radiation to space. Departures from uniformity will arise from diurnal and seasonal variations in insolation, coupled with thermal inertia. The mean surface temperature will equal the interior value. At a given distance from the Sun the exact value of the temperature will depend on the fraction of the incident solar radiation that the body absorbs, and on the efficiency with which it emits its own radiation. For typical planetary materials at 1 AU from the Sun, the value will be about 270 K. Most planetary bodies have far higher interior temperatures, so solar radiation is not an important energy source in such cases.

High interior temperatures do not necessarily mean that the rate of energy input from sources other than solar radiation *greatly* exceeds the solar input. A modest source confined to a small core would give high core temperatures and a temperature gradient to the surface. If the surface layer is a good thermal insulator, the interior can be hot throughout much of its volume. Temperature gradients are themselves important, because they can drive processes such as convection (Section 4.5.2).

Question 4.8

Draw graphs showing qualitatively how the rate of energy input to a planetary body from the various energy sources outlined above might have varied during the 4600 Ma of Solar System history.

Question 4.9

(a) Use equation (4.6) to show that the difference in the magnitude of the gravitational field across a body of radius R due to a body with a mass M a distance r away is approximately $4GMR/r^3$ if r is much greater than R. (It is the difference along the direction to M that is required.)
(b) Hence show that, for the Earth, the solar tidal field is 46% of the lunar tidal field.

4.5.2 Energy Losses and Transfers

As well as gaining internal energy from various energy sources, planetary bodies also lose internal energy. In the near vacuum of interplanetary space, energy is not conducted or convected away from the body, and so it is by emitting radiation that the internal energy is lost. For an atmosphereless body this radiation is from the surface. If there is a substantial atmosphere then a proportion, perhaps all, of the radiation is from the atmosphere. Overall, the radiation is somewhat like that from an ideal thermal source (Section 1.1.1), and so the wavelengths depend on the temperature of the source, as indicated in Figure 4.13. For planetary bodies the radiation is predominantly at IR wavelengths.

In order to be radiated away, the internal energy must first reach the surface or atmosphere, and there are several ways in which this happens, as follows.

Radiative transfer

This process was outlined in Section 1.1.3, and can be thought of as an outward diffusion of photons. The rate of energy transfer increases very rapidly as temperature increases, and as the opacity of the interior decreases. Planetary interiors are too cool and too opaque for radiative transfer to be important, except perhaps in certain zones deep in the interiors of Jupiter and Saturn.

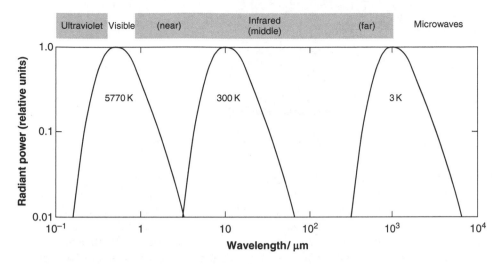

Figure 4.13 The radiation spectra of ideal thermal sources at various temperatures. Note that the lower temperature spectra have been scaled upwards to a peak radiant power of 1.

Thermal conduction

Thermal conduction is a process by which heat is transferred through the direct contact of two regions that are at different temperatures. The more energetic particles in the higher temperature region lose some of their random energy of motion to the less energetic particles in the lower temperature region, just as a fast-moving ball transfers some of its energy of motion to a slower moving ball with which it collides. There is thus a flow of random motion – of heat – from the higher to the lower temperature region. This flow will cool the region of higher temperature unless there is energy generation within it sufficient to offset (or exceed) the heat loss.

Clearly, conduction must play a role in transporting internal energy from the interior of a planetary body to its exterior. Whether it is significant depends on the effectiveness of the two remaining mechanisms, convection and advection.

Convection

Convection differs from conduction in that energy is transported by bulk flows, rather than on an atomic scale. It is a process of energy transfer in a substance heated from below, and requires that the substance can flow. The temperature of the heated substance increases, and it therefore expands, becomes buoyant, and ascends, displacing the overlying cooler material downwards, where it in turn is heated. The rising material loses energy to its surroundings, loses its buoyancy, and descends. This sets up the cycle exemplified in Figure 4.14 by the everyday case of heating a saucepan of liquid, where two so-called convection cells are shown in cross-section.

To obtain a closer look at convection in general, regardless of whether it is in a saucepan, or in the atmosphere or interior of a planetary body, imagine a small parcel of material being swapped with a parcel immediately above it. The hydrostatic equation (equation (4.11)) shows that the raised parcel is now in a lower pressure environment. It quickly responds by expanding until its pressure is the same as the pressure at its new level. This expansion alone causes a decrease in the temperature of the parcel – no heat has had time to flow out of it, and the

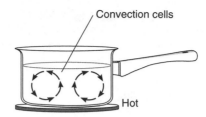

Figure 4.14 Convection in a pan of porridge, showing two convection cells.

decrease is solely a consequence of the expansion. The lower parcel will be compressed, and this will raise its temperature, again with no flow of heat into it. If no heat has flowed into or out of a parcel, then it has undergone what is called an **adiabatic process**. Note that an adiabatic process does not rule out the transfer of energy, only of heat. Thus, energy is transferred through the expansions and compressions that take place. Real processes are not strictly adiabatic – there will be some heat transfer by conduction and by radiation. Nevertheless, an adiabatic process is a very useful idealisation.

The crucial question is whether the parcel's new temperature is greater or less than that of its new surroundings. Let us focus on a parcel that moves upwards. If its new temperature is greater than that of its surroundings, then, the pressures being the same, its density must be lower than its surroundings. It is therefore buoyant, so it will continue to move upwards, and convection will start.

☐ What will happen if its temperature is lower than its surroundings,?

Its density will then be higher than its surroundings, and it will sink back to its point of origin – there is no convection. Corresponding conclusions apply to a parcel that moves downwards.

Whether the temperature of a raised parcel is less than that of its surroundings depends on how rapidly the temperature of the surroundings *decreases* with *increasing* distance from the centre – this is the temperature gradient. The critical value is the adiabatic gradient. This is shown in Figure 4.15 along with two other gradients, A and B. If the actual temperature gradient is greater than the adiabatic value, as in case B in Figure 4.15 (see figure caption), then a rising parcel, which approximately follows the adiabatic gradient, is always hotter than its surroundings and it will continue to rise. Thus, convection occurs. If the actual gradient is less than the adiabatic value, as in case A in Figure 4.15, then a rising parcel is always cooler than its surroundings and it will sink back, and convection will not occur. The value of the adiabatic gradient depends on the gravitational field: the greater the field, the greater the gradient. It also depends on various thermodynamic properties of the material of the parcel, though the details will not concern us here.

If a substance can flow there is a strong tendency for the actual temperature gradient to become equal to the adiabatic value. If the actual gradient is greater (B in Figure 4.15), energy is transported at such a high rate that the deeper levels cool, thus reducing the gradient towards the adiabatic value. If the gradient is less than the adiabatic value (A in Figure 4.15) then energy is transported by conduction, and this is at such a low rate that the deeper levels get hotter, thus increasing the gradient, until convection starts.

The existence of convective currents in planetary bodies does *not* make the hydrostatic equation inapplicable, even though this equation requires that the gravitational force on a shell is equal to the pressure gradient force, in which case there can be no convection. The convection currents are quite slow, and therefore the imbalance between the two forces is relatively slight.

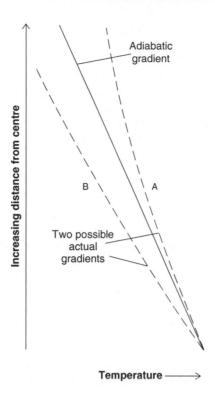

Figure 4.15 The adiabatic temperature gradient and two possible actual gradients. Note that B has the greatest gradient, because in this context the gradient is the rate of decrease of temperature with increasing distance from the centre.

Moreover, averaged over large areas, the rising and sinking motions tend to cancel out, and so the hydrostatic equation applies as a global average.

Convection in solids

It is natural to think of convection as being confined to fluids, and indeed convection does occur in many of the fluid regions of planetary bodies. But it also occurs in some regions that are solid! Solid state convection can occur if a solid behaves like a fluid, and this will be the case if the pressure exceeds the yield strength of the solid.

☐ In what sort of planetary bodies are the internal pressures high?

Equation (4.12) shows that high internal pressures will be found in large planetary bodies. Fluid behaviour is thus expected in a large planetary body, particularly if the yield strength is reduced by raised temperatures. The rate of convection will depend on the composition and on the buoyancy. Though the convective motions are extremely slow, e.g. of the order of 0.1 m per year in the Earth, the associated rate of outward heat transport can exceed that of conduction. In the case of the Earth, convection is particularly well developed and is the dominant process of energy loss from the interior to the crust. Note that solid state convection is far too slow for it to give rise to magnetic field generation.

There are three so-called styles of convection in rocky or icy–rocky planetary bodies. 'Classical' convection as in Figure 4.14 occurs, for example, in the Earth's mantle and crust in the form of plate tectonics (Section 8.1.2). Stagnant lid convection is where the lithosphere (the crust and the upper mantle) is too stiff to yield, and so convection occurs only at depth. This can result in a rise in internal temperatures to the point where a catastrophic disruption of the surface occurs, releasing heat, followed by restoration of the stagnant lid. This quasi-cyclic process might be occurring in Venus (Section 8.2.7). Finally, there is lithospheric delamination. The lithosphere loses material from its base, and this sinks into the more yielding lower mantle (Section 5.1), its downward motion promoting convection. As well as occurring alone, it can supplement classical convection and stagnant lid convection, as you will see in Chapter 8.

Advection

If the uppermost part of the interior of a planetary body is solid it will be at too low a pressure and temperature to convect. But this does not leave conduction as the only mechanism of outward energy transfer. **Advection** is the process of upward energy flow carried by local regions of upward-moving liquids, such as molten rock, or water. If these erupt onto the surface we have volcanic effusions of various sorts. But the liquids need not reach the surface. For example, if molten rock solidifies at some depth it has still contributed to the outward flow of energy.

Table 4.5 indicates the relative importance of advection, convection, and conduction for a few planetary bodies.

Heat transfer coefficient

Regardless of the energy transfer process, the intrinsic property of the material that determines the rate is encapsulated in a heat transfer coefficient. This is the rate at which heat crosses unit area of a material, per unit thickness, per unit temperature difference across the thickness. For many planetary and smaller bodies the heat transfer coefficients are comparatively low at the surface and in any atmosphere, and so these regions can have a significant effect on the overall rate of energy loss.

Question 4.10

Suppose that a rocky planetary body has a weak energy source concentrated at its centre. The interior temperatures throughout are modest, decreasing with increasing radius, almost the entire decrease being across a thin surface layer.

Table 4.5 Mechanisms of heat reaching the surface regions of some planetary bodies today

	Conduction	Convection[a]	Advection
Venus	Probably unimportant	Perhaps dominant	Perhaps significant
Earth	Minor contribution	Dominant	Minor contribution
Mars	Dominant	Negligible	Minor contribution?
Moon	Dominant	Negligible	Negligible
Io	Minor contribution	Negligible	Dominant
Jupiter	Negligible	Dominant	Negligible
Neptune	Negligible	Dominant	Negligible

[a] Including any solid state convection and any large-scale lithospheric recycling.

(a) Explain why it is likely that the only way in which energy is being transferred up to the surface layer is by conduction.

(b) What can you deduce about the heat transfer coefficient of the thin surface layer compared with that of the rest of the body?

4.5.3 Observational Indicators of Interior Temperatures

If planetary modellers only had the various types of energy gains and losses that *might* be present, then models of the interior temperatures of planetary bodies would be very poorly constrained. Fortunately, there are observational data that are direct indicators of the thermal state. These include

- energy flow from the interior;
- seismic waves and the level of seismic activity;
- the degree of departure from isostatic equilibrium;
- the level and nature of geological activity at the surface, today and in the past;
- the magnetic field today, and evidence for its nature in the past.

The energy flow from the interior of a planetary body gives an indication of present day interior temperatures. In principle the flow can be measured via the IR radiation to space to which it gives rise, this being the way that the energy is ultimately lost by the body. If there is flow from the interior then the rate at which IR radiation is emitted to space would exceed the rate at which radiation is absorbed from the Sun. In practice only for Jupiter, Saturn, and Neptune is this IR excess large enough to have been measured accurately. For some other bodies the present energy flow from the interior has been obtained by estimating the outward energy flow at the surface by conduction, convection, and advection. Values for energy outflows from the interiors of some planetary bodies are given in Table 4.2.

Seismic wave speeds can provide an indirect indication of interior temperatures.

☐ If there is a zone in which S waves do not travel, what can you conclude about the physical conditions in the zone?

A zone free of S waves must have negligible shear strength, so it could be a liquid or be highly plastic. Departures from isostatic equilibrium indicate lower plasticity, and changes in isostasy can give an indication of how dynamic the interior is, as can the level of seismic activity.

Geological activity is another indicator of interior temperatures, which will be elaborated in later chapters. By examining the geological activity on surfaces of different ages it is possible to gain insight into interior temperatures not only today but in the distant past. The present-day magnetic field indicates whether there are liquids in convection, and past magnetic fields leave an imprint in the rocks, enabling us to infer the nature of the field in the past, and hence gain further insight into interior temperatures throughout the history of the body.

4.5.4 Interior Temperatures

It has been noted that the temperatures inside a planetary body at any time t depend on the energy gains and losses at all earlier times. But the gains and losses themselves depend on the temperatures. For example, at time t the temperature profile $T(t, r)$ will help determine whether there is any convection. If there is, then this will increase the energy loss rate at t, which will influence subsequent temperatures. We thus have a complicated interplay, represented in

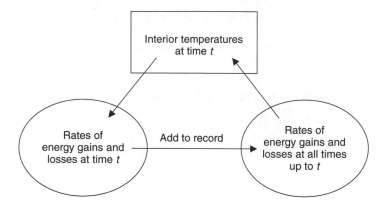

Figure 4.16 The interplay between interior temperatures and the rates of energy gains and losses in a planetary body.

the simplest manner by Figure 4.16. Planetary modellers have to grapple with this interplay until there is self-consistency.

Though the modelling process required to obtain the interior temperatures of a planetary body is intricate, there are three broad features that are readily appreciated. First, energy losses tend to win in the end. Energy sources can slow down the rate of decline of interior temperatures, and can hold interior temperatures steady, even for long periods, by producing energy at the same rate at which energy is lost to space. But most sources of energy either have died already, or are implanting less power as time passes. Sooner or later, interior temperatures will fall, and ultimately all bodies will come to have internal temperatures determined by the balance between solar radiation absorbed and IR radiation emitted to space.

Second, if sufficient time has passed since all energy sources at various depths became active, the temperature decreases monotonically as we move outwards from the centre to the surface. This is because the ultimate energy loss from the body is from its surface. The details of the temperature profile depend on the radial distribution of energy sources and on the heat transfer coefficients at each radius. Only if all energy sources have long been concentrated at the surface will the interior temperature be the same everywhere.

☐ What can give this outcome?

This is the outcome if, for a long time, the only active source has been solar energy. Another way is through the concentration of all radioactive isotopes into the surface.

Finally, a planetary body will lose internal energy faster per unit mass than a larger planetary body, provided that the two bodies have similar heat transfer coefficients, similar interior energy sources per unit mass, and similar spatial distributions of energy sources. This is because the global rate of energy loss increases with the surface area A of the body, whereas the global rate at which energy is released in the interior increases with the mass M of a body. A/M is thus a measure of the rate of loss of internal energy per unit mass, and A/M increases as the size of the body decreases. For a spherical body of radius R

$$\frac{A}{M} = \frac{4\pi R^2}{4\pi R^3 \rho_m /3}$$

where ρ_m is the mean density. This simplifies to

$$\frac{A}{M} = \frac{3}{R\rho_m} \tag{4.14}$$

showing indeed that as R decreases the ratio increases. Thus, in a sufficiently small body, from early in its history, any active energy sources will have been unable to compensate the energy losses, so the interior temperatures will have declined, and the interior will already be at temperatures determined by absorbed solar radiation and emitted IR radiation.

Question 4.11

Describe qualitatively the present-day temperature versus depth in a hypothetical planetary body if its *sole* energy source (apart from solar radiation) has always been each one of the following:

(a) Accretional energy confined to its surface.
(b) Long-lived radioactive isotopes confined to a small central core.
(c) As for (b), but in a body that is a scaled-up version of that in (b).

4.6 Summary of Chapter 4

In order to develop our knowledge about the interior of a planetary body, a model has to be constructed and adjusted until its predictions of external observations match the actual observations as closely as possible.

A model has as its basic features a specification of the composition, temperature, pressure, and density, at all points within the interior. For planetary bodies, variations versus radius from the centre are dominant.

Details of the gravitational field of the body, plus other data such as mass, radius, and rotation period, constrain the density at all points in the interior.

If the body has a large magnetic dipole moment then this is thought to be due to electrically conducting liquid layers that are convecting. If this explanation is correct, then the location of these layers can be deduced.

Seismic waves reveal the presence of liquid layers, the presence of regions that are plastic, and changes in composition. Seismic waves also constrain the composition, particularly if combined with other data.

The composition is also constrained by the composition and properties of accessible materials, and by the relative abundances of the elements in the Solar System.

The temperatures in the interior of a body at any one time depend on the rates of energy gains and the rates of energy losses at all earlier times. The interior temperatures are a factor in determining whether differentiation has occurred, and they also determine the dynamics of the interior and of the surface.

Energy sources no longer active include accretional energy and heat from short-lived radioactive isotopes. Differentiation can be a primordial source of energy, or it can still be an active source. Other active sources include heat from long-lived radioactive isotopes, tidal energy, and solar radiation.

Energy is ultimately lost in the form of infrared radiation emitted to space by the surface or atmosphere of the body. To reach these outer regions energy is conveyed by radiation, conduction, convection (in liquids or solids), and advection. Radiative transfer is negligible (except perhaps in certain zones in Jupiter and Saturn), and the proportions of the overall flow carried by the other three processes vary considerably from body to body.

Some insight into the thermal state of the interior of a body is provided by energy flow measurements, by gravitational and seismic data, by its history of geological activity, and by the magnetic field now and in the past.

There are three broad features of the thermal history of a planetary body:

(1) Except for very early on in Solar System history, interior temperatures have either been roughly constant, or been declining.
(2) The temperatures today increase with depth.
(3) If all other things are equal, the smaller the planetary body the faster it loses internal energy per unit mass.

Table 4.2 summarises some of the observational constraints relevant to interior models of various planetary bodies. Other relevant data are given in Tables 1.5, 4.3–4.5, and in Figures 4.6 and 4.9.

5 Interiors of Planets and Satellites: Models of Individual Bodies

The models outlined here are of the interiors of the planetary bodies as scientists believe them to be today. The evolution of planetary interiors, from the past to the present, and into the future, is largely deferred to later chapters because of the consequences for surfaces and atmospheres.

Recall from the previous chapter that a model of the interior of a planetary body has, as its basic features, a specification of the composition, temperature, pressure, and density at all radii from the centre. From this basic specification, other things follow, such as internal motions, magnetic fields, and so on. The model is arrived at through applying physical principles to calculate the observed properties of the body, and the model is varied until an acceptable level of agreement with the observations is obtained. It is invariably the case that a *range* of models can be made to fit the observational data. The range will be wide if the data are uncertain, or if some data are absent, such as seismic data. Therefore, a model is not unique, though for most bodies certain features are beyond reasonable doubt.

In considering individual planetary bodies, it would be a very lengthy task to relate all the various features of a model to the observational and experimental data that underpin it. Therefore, we shall highlight just a few examples where specific data are strongly suggestive of particular model features.

5.1 The Terrestrial Planets

Figure 5.1 shows models of the interiors of the terrestrial planets that highlight the main compositional divisions within each of them. A common feature is an outer **crust** overlying a **mantle**, which itself overlies a central core. In considering the various layers in Figure 5.1, it is important to realise that it is cross-sections through spherical volumes that are shown. Therefore, it is easy to get the wrong impression of the volume ratios of the different layers. For example, from a superficial look it might seem that the volume of the Earth's core (inner plus outer) is a bit over half the total volume of the Earth – it extends a bit over half way to the Earth's surface. But the volume ratio is actually about one-sixth. A much better impression is obtained from the cutaway drawing in Figure 5.2.

For all the terrestrial planets there is sufficient observational data to indicate that the increase of density with depth is too great for the body to consist solely of the material at the surface, getting denser as the pressure increases with depth. Gravitational data, as outlined in Section 4.1, have

Discovering the Solar System, Second Edition   Barrie W. Jones
© 2007 John Wiley & Sons, Ltd

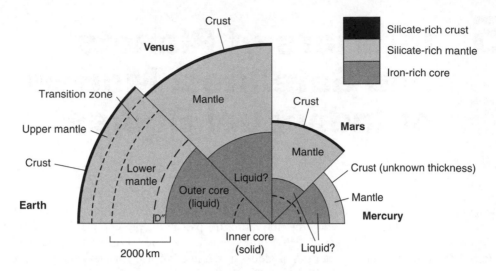

Figure 5.1 Cross-sections through the terrestrial planets.

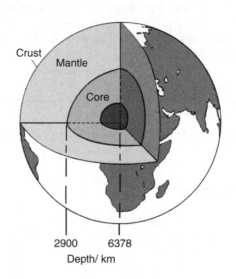

Figure 5.2 Cutaway section through the Earth.

been of particular importance in establishing density profiles. In accord with the observational data, the bodies are shown differentiated, with the intrinsically denser substances lying deeper. Thus, silicates dominate the outer layers, and iron, or iron-rich compounds such as iron sulphide (FeS), dominate the cores. We therefore have rocky materials throughout.

Regardless of the form in which a chemical element is present, over 90% of the mass of each of the terrestrial planets consists of oxygen, iron, silicon, magnesium, and sulphur, though the proportions within this group vary from body to body. These are among the 15 chemically most abundant elements in the Solar System as a whole (Table 1.5). Other abundant elements are less well represented, notably hydrogen, helium, carbon, nitrogen, and neon. This is because the pure element or the common compounds, including icy materials, have densities that are too low even at high pressures to fit the density data. It is their volatility that has led to their scarcity in the terrestrial planets.

Table 5.1 gives temperatures, densities, and pressures at a few depths in the Earth, and Table 5.2 gives the values at the centres of all the terrestrial planets and the Moon. Note that these are model values, and therefore depend on the particular model adopted, so the values are indicative and not definitive. Note also that for a given substance, its equation of state links the three quantities in Table 5.2, and so in principle if we know any two we can calculate the third. However, the equations are not well known, partly because the terrestrial planets consist of mixtures of minerals in somewhat uncertain proportions, and partly because the extreme conditions at great depths are beyond laboratory reach, though it helps that to a fair approximation in terrestrial interiors we can ignore the effect of temperature on density. In the case of the Earth, the density and pressure versus depth are comparatively well known, from seismic and other data. The temperature versus depth is, however, much more poorly constrained.

You can see in Table 5.2 that the central pressures are greater, the larger and denser the body, in accord with equation (4.12) (Section 4.4.3). In Table 5.1 the pressures and temperatures increase with depth.

☐ Could this be otherwise?

In Section 4.4.3 it was shown that pressure must increase with depth, and in Section 4.5.4 it was argued that, given sufficient time, temperature will decrease outwards from the centre.

For a given temperature and pressure, whether a region is liquid depends on the substances present. At a given pressure, the mantle silicates have melting points that are slightly lower than that of pure iron (Table 2.3). However, even a small admixture of other substances into the iron

Table 5.1 Model temperatures, densities[a], and pressures in the Earth

Crust[b]	
T (surface)/ K	288
ρ (ave)/ $kg\,m^{-3}$	2 800
Mantle	
ρ (top)/ $kg\,m^{-3}$	3 300
T (base)/ K	3 300
p (base)/ Pa	1.4×10^{11}
ρ (base)/ $kg\,m^{-3}$	5 400
Core	
ρ (top)/ $kg\,m^{-3}$	9 900
ρ (centre)/ $kg\,m^{-3}$	13 500
T (centre)/ K	5 500
p (centre)/ Pa	3.8×10^{11}

[a] The densities are the estimated values *in situ*, i.e. they are compressed densities.
[b] The crustal values are observed rather than modelled.

Table 5.2 Model densities[a], temperatures, and pressures at the centres of the terrestrial planets and the Moon

	Mercury	Venus	Earth	Mars	Moon
$\rho/\,\mathrm{kg\,m^{-3}}$	9300	12 500	13 500	7000	~ 8000
$T/\,\mathrm{K}$	1370	3 700	5 500	2000	1600
$p/10^9\,\mathrm{Pa}$	44	286	380	43	5.9

[a] The densities are the estimated values *in situ*, i.e. they are compressed densities. In all cases a predominantly iron core is assumed, with Mars having an appreciable proportion of FeS.

reverses the situation, and so in several of the bodies in Figure 5.1 the mantle–core interface is also shown as a solid–liquid interface.

Only at a level of greater detail would there be much dispute about any of the features of Figure 5.1 and the data in Tables 5.1 and 5.2.

5.1.1 The Earth

Observational data pertaining to the Earth's interior are particularly copious, and the model is consequently tightly constrained. As well as an abundance of gravitational and magnetic field data, the composition of the Earth's surface is very well known, as are the details of its high level of geological activity. It is the only planetary body for which we have copious seismic data, far more than for the only other body for which we have any seismic data at all – the Moon. Figure 5.3 shows the P and S wave speeds versus depth in the Earth.

❑   Do the boundaries in the model of the Earth in Figure 5.1 match those in Figure 5.3?

The division of the Earth in the model in Figure 5.1 into a crust, mantle, outer and inner cores is apparent in the seismic data. There is plenty of evidence that at all these boundaries there is a change in composition. The disappearance of seismic S waves in the outer core indicates that it is liquid (Section 4.3.1). Their reappearance at greater depths indicates a solid inner core.

Iron is highly favoured for the core, because of the density indicated at this depth by the seismic and gravitational data.

❑   Among the metals, what is another reason for favouring iron?

Iron is also favoured because of its high relative abundance in the Solar System. Though iron is dominant throughout the core, models that fit the observational data need up to about 4% nickel in the inner core (similar to iron meteorites), though the exact composition of the inner core is uncertain. The density of the outer core needs to be about 10% lower than the inner core, and its melting point needs to be lower too. Various minor constituents in addition to more nickel can achieve this, such as a few per cent of iron sulphide (FeS), or even iron hydrides (FeH_x). These hydrides could contain an amount of hydrogen equivalent to about 100 times that found in the Earth's oceans. The inner core is solid because of its slightly different composition, and because it is at a higher pressure (Figure 4.11).

A liquid iron outer core is consistent with the strong magnetic field of the Earth, for which the dynamo theory requires an electrically conducting fluid in convective motion (Section 4.2). The crystallisation of iron from the outer core on to the inner core is thought to be the main source of energy required at the base of the convecting outer core to maintain the convection. This crystallisation releases latent heat and energy of differentiation (Section 4.5). It also releases

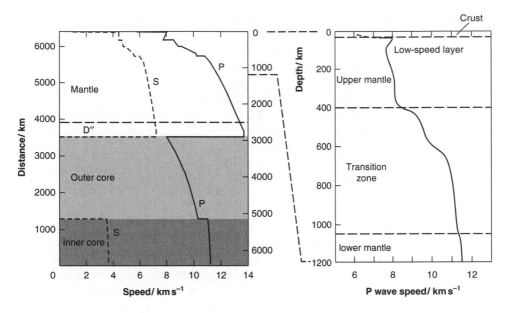

Figure 5.3 P and S wave speeds versus depth in the Earth.

lighter elements that float upwards and this also promotes convection. It is estimated that the solid inner core did not begin to form until roughly 1000 Ma ago. This was when the loss of primordial energy from the core had reduced core temperatures to the point where core differentiation started. At present, the inner core is thought to be growing by about 10 mm in radius per century (≈ 0.1 km per Ma). At earlier times convection in the then totally liquid core could have been sustained by the general outward loss of heat, and this still makes a significant contribution to convection regardless of crystallisation.

In the model the mantle consists of silicates, and from samples of the mantle it is known that the upper mantle consists almost entirely of a rock called **peridotite**. This is largely a mixture of various silicates – about 60% olivine ($(Fe, Mg)_2SiO_4$), about 36% pyroxenes ($(Ca, Fe, Mg)_2Si_2O_6$, where rarely the metals can be Na, Al, or Ti), and 4% other silicates. From a depth of about 400 km to about 1050 km there is a gradual transition zone apparent in the seismic speeds (Figure 5.3) that can be explained by the gradual conversion of the minerals in peridotite into higher density crystal structures that are more stable at high pressures. The effect of such changes is particularly apparent around 410 km and 670 km. Such conversions continue down to the core, though the overall chemical composition remains much the same. The sharp changes in the speeds at an average depth of about 30 km mark the boundary between the continental crust and the mantle; oceanic crust is thinner, 5–10 km. This is a chemical change, though the crust, like the mantle, is dominated by silicates.

Seismic data indicate various structures in the mantle. Of particular interest is the D'' layer that constitutes the lowest 400 km or so. The data are consistent with silicates dominated by $(Mg, Fe)SiO_3$, in various forms, including a mineral called perovskite. Internal layering is also indicated, which could be derived from slabs of the Earth's lithosphere (see below) brought down by mantle convection.

The mantle as a whole is thought to contain compounded hydrogen equal in mass to at least that in the Earth's oceans.

The overall model composition of the Earth follows fairly closely the Solar System relative abundances of the less volatile elements, as seen in the ordinary chondrites. Iron is underrepresented in the mantle and crust because of its concentration in the core. The composition of the crust, which is readily accessible, will be further discussed in Section 8.1.1.

Not far below the crust–mantle division, the seismic data in Figure 5.3 reveal a layer with comparatively low speeds, present in much (but not all) of the upper mantle, and extending over the approximate depth range 50–200 km. This indicates that around this depth the material is particularly plastic. This plasticity is consistent with the isostasy of most of the crust.

The thermal structure of many models suggests that solid state convection is occurring in the mantle. There is shallow convection around the low-speed layer, and larger scale convection probably extending right down to the core, as indicated by the D″ layer. These two types of convective cell are shown schematically in Figure 5.4. The more plastic region of the mantle constitutes the **asthenosphere** (from the Greek *asthenes*, meaning weak). It extends from the base of the lithosphere (see below) to at least a few hundred kilometres, and perhaps right down to the liquid core if that is as far as convection extends. In its upper reaches it seems to incorporate the low-speed layer. Above the asthenosphere, the uppermost part of the mantle and the crust is colder and therefore much tougher, though it is elastic rather than brittle. This is the **lithosphere** (from the Greek *lithos*, meaning rock). Its average thickness is about 95 km, with considerable local variations. Note that at the lithosphere–asthenosphere boundary there is a change in dynamic properties, and *not* a change in composition. In the lithosphere heat transfer is mainly by conduction, whereas in the asthenosphere it is mainly by convection.

Convection in the Earth's interior is manifest at the surface via lithospheric movements that constitute **plate tectonics**. The Earth's lithosphere is divided into a number of plates in motion with respect to each other. At some plate boundaries one plate dives beneath its neighbour; other boundaries correspond to upwelling that creates new lithospheric plates. Plate tectonics is the subject of Section 8.1.2. The important point here is that, regardless of whether convection reaches the core, plates of lithosphere are carried to great depths as part of the

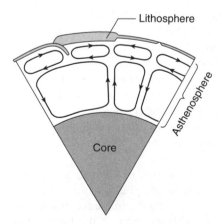

Figure 5.4   Convection in the Earth's asthenosphere and the rigid lithosphere.

internal convection – the lithosphere does not constitute a stagnant lid. Therefore, plate tectonics makes an essential contribution to raising the rate of cooling of the Earth's interior.

There is plenty of evidence that the Earth has had a hot interior ever since its birth about 4600 Ma ago, or not long after. For example, remanent magnetism in radiometrically dated old rocks shows that the Earth's magnetic field extends at least 3500 Ma into the past. Therefore, if it formed undifferentiated it would soon have become warm enough to differentiate. Thermal models indicate that long-lived radioactive isotopes are the dominant source of energy for the Earth *today*. The compositional structure of the Earth, the chemical affinities of these isotopes, and their relative abundances indicate that they are concentrated into the crust and to a lesser extent into the mantle. Heat sources long dead, such as the energy from accretion and from short-lived isotopes, still have an effect, because the Earth's interior temperatures remain a good deal higher than they would be if these extinct sources had never been present.

Question 5.1

At some time in the Earth's distant future its outer core will probably solidify. Describe two observable effects this would have.

5.1.2 Venus

In its size and mean density, Venus is almost the Earth's twin (Figure 4.1). When allowance is made for the lower internal pressures in Venus (because of its slightly smaller mean density and radius) then the uncompressed mean densities are even closer, and these densities are a better basis for comparing compositions. Venus is also our planetary neighbour, and therefore probably had available for its construction much the same sort of materials as the Earth. These few data suggest very strongly that the interior of Venus is not very different from the Earth's interior, and this is reflected in the model in Figure 5.1. Support for this conclusion comes from seven of the Soviet spacecraft that landed on the Venusian surface in the 1970s and 1980s. Six of the craft found the sort of silicates that typify the Earth's ocean basins, and the seventh found silicates of the sort that typify the upper parts of the Earth's continents.

Further support might be expected from seismic, gravitational, and magnetic field data. Unfortunately, we have no seismic data for Venus, and it is so near to having a spherically symmetrical mass distribution that gravitational data provide very weak constraints on the density versus depth (Table 4.2). The spherical symmetry is largely a result of the slow rotation of Venus, the slowest of all the planets, with a sidereal period of 243 days.

☐ Why is this relevant?

There is little rotational flattening, which is the main cause of departure from spherical symmetry for a planetary body.

Evidence of a hot interior is provided by the surface of Venus. Though there is presently no more than modest geological activity, there was a lot more a few hundred million years ago, and the surface displays evidence of a plastic interior (Section 8.2). There is no evidence of plate tectonics and so the lithosphere surely constitutes a stagnant lid on a convecting mantle. In this case, radiogenic heat, produced at roughly the same rate as in the Earth, is escaping only slowly, and the internal temperature must be rising. The dramatic effect of this on the surface is outlined in Section 8.2. That the interior of Venus might well have been kept warmer than had plate tectonics developed makes it likely that a solid inner core has yet to develop. Also, a hot mantle could suppress convection in a liquid iron core.

The magnetic field of Venus is extremely weak, and only a very small upper limit exists for the magnetic dipole moment (Table 4.2). This could mean that there is no iron core, or that any iron core is solid. The absence of a liquid iron core would be very surprising in light of Venus's broadly similar size and mass to the Earth.

☐ As well as an electrically conducting liquid, what other requirements are there for an internally generated magnetic field?

Convection currents and sufficiently rapid rotation of the body are also thought to be necessary (Section 4.2). We have already seen that core convection might be absent. Additionally, Venus rotates slowly. Furthermore, the absence of a solid inner core might mean that there would be none of the crystallisation of iron onto an inner core that is thought to help drive convection in the Earth's outer core. Note also that the Earth's solid inner core means that the Earth's conducting fluid is a shell rather than a sphere, and it turns out to be easier to obtain a large magnetic dipole moment from a shell. Thus, in spite of a very weak dipole field, we can retain a liquid iron core in our models of Venus.

Overall, taking everything that we know about the interior of Venus, it seems beyond reasonable doubt that it has not only a hot interior, but a liquid iron core and a silicate mantle of broadly similar sizes and compositions to those of the Earth. At a more detailed level of modelling there is far more uncertainty.

5.1.3 Mercury

Observational data on Mercury are sparse. We know the mass and the radius, and so the mean density can be calculated, and it is between that of Venus and the Earth. However, Mercury is a good deal smaller in radius, and so the internal pressures are considerably less. The uncompressed mean density of Mercury is not much less than the mean planetary value of $5430\,\mathrm{kg\,m^{-3}}$, whereas the mean densities of Venus and the Earth are reduced to $4000–4500\,\mathrm{kg\,m^{-3}}$ when uncompressed. Therefore, the uncompressed mean density of Mercury is actually greater than that of the Earth, and is the highest of all the planetary bodies. There is only one abundant, sufficiently dense substance that must predominate – iron. Models of the formation of Mercury indicate temperatures high enough for complete differentiation, in which case Mercury has an iron core and a silicate mantle, and the iron core must account for a large fraction of its volume, as in Figure 5.1. Initially, Mercury might have had a silicate–iron ratio more similar to Venus and the Earth, with some of the silicate mantle removed by a large impact after differentiation.

The surface has not been sampled, but IR spectroscopic data from Earth-based telescopes indicate the expected silicate composition, in particular feldspars that include calcium, and perhaps pyroxene (Tables 4.3, 6.1).

Gravitational or seismic data could confirm the existence of an iron core. Unfortunately, as for Venus, we have no seismic data, and the gravitational data are similarly unhelpful, for the same reason as for Venus – the sidereal rotation period of Mercury is 58.6 days, and such slow rotation has led to inappreciable flattening. The planet is therefore nearly spherically symmetrical, and so we can learn little about the increase of density with depth. We do know, however, that the core pressures are low (Table 5.2), in which case it can be shown that the iron core would not have acquired much oxygen (in the form of iron oxides). Also, sulphur, the other element that could be compounded in abundance in an iron core (as FeS), could well have been lost from

Mercury, because of the high temperatures so close to the Sun. Therefore, the core could be fairly pure iron, and consequently denser. In this case, to meet the mass constraint the mantle would be less dense, such as would be the case if it contained less iron (in the form of FeO).

The surface of Mercury (Section 7.2) indicates no internally driven geological activity since very early in Solar System history – it is covered in impact craters that have accumulated over billions of years, with little sign of removal. This lack of activity implies that the lithosphere has long been thick, and this is consistent with the high rate of cooling expected from Mercury's small size. However, a thick, immobile lithosphere would then reduce the cooling rate, which raises the question – is at least some of the iron core still liquid? In this case, Mercury could have a non-negligible magnetic dipole moment.

In fact, this is the case (Table 4.2), which indicates that at least some of the iron core is liquid. Thermal models indicate that, with a large iron core cooling slowly by conduction through a largely lithospheric mantle, and with a modest abundance of long-lived radioactive isotopes, the outer part of the core might still be molten, provided that the core contains a small proportion of lighter elements, e.g. sulphur, to reduce the solidification temperature – a core freezing time of about 500 Ma is deduced for a pure iron–nickel core. The inner core, being at a greater pressure and possibly with a slightly different composition, is predicted to be solid, as in the Earth. But even with a liquid shell, Mercury's slow rotation counts against a strong magnetic field being generated. The rather small dipole moment could well be consistent with slow rotation. A less likely possibility is that the observed field is due to iron-rich surface rocks magnetised early in Mercury's history, perhaps by a strong field in an iron core that had not then solidified. In that early time Mercury would also have been rotating faster – its rotation has since been slowed by tidal interactions with the Sun.

The existence of a liquid core early in Mercury's history is borne out by the surface topography, which indicates contraction early on, such as would result from the solidification of at least some of the core, but not necessarily all of it.

Question 5.2

State and justify your expectations regarding the existence and extent of an asthenosphere in Venus and in Mercury.

5.1.4 Mars

Mars is somewhat larger than Mercury but has the much smaller uncompressed mean density of 3700–3800 kg m^{-3}, which is also significantly smaller than the uncompressed value for the Earth (4000–4500 kg m^{-3}). The surface has been sampled, and from this and other data it is clear that iron-rich silicates are common in the crust. Nevertheless, the low uncompressed mean density indicates that, overall, iron is less abundant in Mars than in the Earth.

Seismic data for Mars are very limited, confined to measurements attempted in 1976–1980 by the Viking 2 Lander. No seismic activity was detected, though the winds were stronger than expected, and would have masked all but the strongest seismic events by terrestrial standards.

Evidence for a core intrinsically denser than the rest of the planet is provided by the value of C/MR_e^2, given in Table 4.2.

❒ What is the value of C/MR_e^2 for a homogeneous sphere?

For a homogeneous sphere the value is 0.4 (Section 4.1.3). The Martian value of 0.365 is sufficiently smaller to be consistent with a small, iron-rich core. The value of C/MR_e^2 has been obtained from the rotation period T, the value of the gravitational coefficient J_2, and the rotation axis precession period (0.1711 Ma), as outlined in Section 4.1.3.

The core is thought to consist of a mixture of iron (with a few per cent of nickel) plus less dense materials such as FeS and perhaps magnetite (Fe_3O_4), though core pressures might be too low for the latter to be a significant component. Such mixtures arise from thermal models of Mars indicating that differentiation might have been less complete than in the Earth. Incomplete differentiation is also consistent with the iron-rich surface silicates and with models than indicate a higher content of FeO in the mantle than in the case of Earth. Isotopes in Martian meteorites suggest core formation within a few tens of million years of the formation of Mars.

Evidence that at least part of the core is liquid is provided by the Love numbers of Mars (Section 4.1.4), derived from the Martian tide raised by the Sun, as observed by Mars Global Surveyor. Even though this tide is only a few millimetres in amplitude at the surface, the associated Love number is too large for a wholly solid interior. Further support for a (partly) liquid core comes from thermal models of Mars. These indicate that a thick lithosphere formed early in Martian history, which is consistent with the gravitational data on a regional scale and with the surface features (Section 7.3). Such a lithosphere could well have acted as a stagnant lid. With a stagnant lid, the core could be warm enough to be (partly) liquid provided that it contains lighter elements, such as sulphur in the form of FeS. It is possible that there is a solid inner core consisting of purer iron–nickel, possibly half the radius of the complete core.

Only an upper limit exists for the magnetic dipole moment (Table 4.2), and this is less than 0.003% that of the Earth. Given the rapid rotation of Mars, if indeed part of the core is liquid, this indicates that it is not convecting. One possibility is that there is no solid inner core forming. Such a completely liquid core is possible if the melting point has been sufficiently lowered by other ingredients. If sulphur accounts for more than about 15% of the core, it would be entirely liquid today. However, such a liquid core seems to have generated a magnetic field in the past. The evidence comes from weak remanent magnetism detected in crustal rocks by Mars Global Surveyor, notably in the more ancient areas of Mars, indicating that Mars had a magnetic field that died perhaps as long ago as 4150 Ma. The preferred explanation for this demise is a rather low sulphur content and the consequent early rapid growth of an inner core, followed by slower growth, so that we have since had a non- or weakly convecting liquid shell. The lack of correlation of the remanent magnetism with topography suggests a magnetised layer deeply buried, presumably including iron-rich minerals.

Figure 5.1 shows an iron-rich core with a radius near the middle of the 1300–1800 km range of sizes that are all consistent with a mantle of composition suggested by the Martian meteorites and with the measured value of C/MR_e^2 (Table 4.2). Variations in the detailed composition of the core contribute to the range of radii.

The modelled mantles are a few per cent more dense than the Earth's mantle. A greater proportion of FeO (ferrous oxide) remaining in the Martian mantle can account for this, The crust is also iron rich, with Fe_2O_3 (ferric oxide), which gives Mars its red tint. If FeS were a significant proportion of the core then the mantle would be correspondingly depleted in sulphur. There is some evidence of such depletion from the Martian meteorites. Though they are samples of the crust, they can be used to make inferences about the composition of the mantle, because the crust was derived from the mantle. One model has the upper mantle dominated by olivine, underlain by a denser mix. If the core is sufficiently small, then the pressures at the base of the mantle could be sufficiently high to create a thin shell of the even denser mineral perovskite, as at the base of Earth's mantle.

Question 5.3

Explain why we would know much less about the interior of Mars if the planet rotated very slowly.

5.2 Planetary Satellites, Pluto, EKOs

The large planetary satellites, Pluto, and the large EKOs fall into two broad groups. The Moon, Io, and Europa are terrestrial bodies in that they are predominantly rocky in composition. Ganymede, Callisto, Titan, Triton, Pluto, and the large EKOs (and presumably the small ones too) differ from the terrestrial bodies in that a substantial proportion of the mass of each of them consists of icy materials, so they are classified as icy–rocky bodies.

5.2.1 The Moon

Next to the Earth, the Moon is the most extensively studied planetary body. This is because it is by far the closest to us. It has been visited by numerous spacecraft, and is the only other body on which humans have walked, during the Apollo 11, 12, 14–17 missions, between 1969 and 1972.

The Moon is sufficiently small that its uncompressed mean density will not be much less than its actual mean density of $3340 \, kg \, m^{-3}$. Even so, this is considerably less than the uncompressed mean density of the Earth and even of Mars, strongly indicating that iron is comparatively scarce in the Moon. The value of C/MR_e^2 has been accurately obtained from a combination of J_2 and the Moon's response (as determined by laser ranging) to tidal torques exerted by the Sun and the Earth. The value, 0.394 (Table 4.2), is slightly but significantly smaller than that of a homogeneous sphere. Because of the slight central compression, this indicates that the Moon is not homogeneous, but has a small dense core. If the core is nearly pure iron then it will be of 300–400 km radius, but if it contains a high proportion of iron-rich compounds, such as FeS, then because these are less dense than iron, the core radius will be somewhat larger. Figure 5.5 shows a model of the lunar interior in which the core, of 470 km radius, is assumed to include a fairly high proportion of iron-rich compounds. A pure FeS core would have a radius around 530 km.

Further evidence for an iron-rich core includes the effect of the Moon on the Earth's magnetic field lines, which indicates the presence of a considerable mass of an electric conductor. There is also depletion of siderophile elements in the mare basalts – these would have been carried downwards during core formation. Radiometric dating of lunar minerals places this differentiation at about 25–30 Ma after the impact that formed the Moon.

A rough indication of present-day internal temperatures has been obtained from measurements of the electric currents induced in the Moon by the magnetic field in the solar wind. These electric currents can be measured through the magnetic fields that they themselves generate. For a given solar wind magnetic field, the higher the internal temperatures, the larger the currents. The temperatures seem to increase with depth, reaching 1000 K or so at 300 km, and 2000 K or

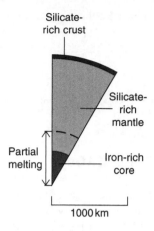

Figure 5.5 A cross-section through the Moon.

so at 1000 km. Further data on internal temperatures have been obtained from Love numbers derived from the 100 mm surface tide at the lunar surface, generated mainly by the Earth. This tide was measured from the round-trip laser travel times to the retroreflectors left by the Apollo missions. This gives the distance at any instant with the required precision. A partially liquid interior is indicated. Such a warm interior is consistent with the stagnant lid provided by the Moon's thick lithosphere.

We also have seismic data. At the Apollo 11, 12, 14–16 sites, seismometers were set up successfully. The Moon is not nearly as seismically active as the Earth, but 'moonquakes' (mostly caused by tidal stresses induced by the Earth, or by surface impacts) have yielded the data in Figure 5.6. These two profiles were seen earlier in Figure 4.9(b) where it was noted that they are each the average of a number of profiles. This has obscured evidence of features visible in most individual profiles. Note especially the large uncertainties, a result of the infrequency of moonquakes and the small number of seismometers, all on one side of the Moon.

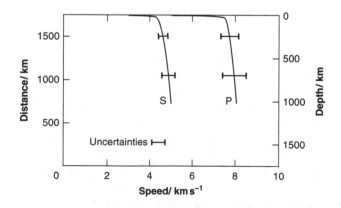

Figure 5.6 P and S wave speeds versus depth in the Moon.

The rapid increase in P and S wave speeds at shallow depths is consistent with a heavily fractured surface layer that becomes less fractured with depth because of the increasing pressure, and is unfractured below a depth of about 20 km. Figure 7.4 shows that there is a small but sharp increase in the speeds at about 40 km (on the near side), widely believed to be due to a change in composition, from the lower density silicates found in abundance at the surface to higher density silicates including the pyroxene and olivine that dominate the Earth's mantle. Samples of the higher density silicates have also been found on the lunar surface, excavated by huge impacts that created large craters. Thus, in addition to the core, there seems also to be differentiation into a crust and mantle. Within the mantle, at a depth of about 500 km, there might be an increase in wave speeds sudden enough to indicate a change in composition – a change in crystal structure in mantle materials would not occur at this depth.

Wave speeds are not shown beyond a depth of about 1000 km. There are too few data from waves that traversed greater depths. This could be due to the deflection of downward-propagating waves from near-surface sources. It is also the case that there have been no moonquakes sufficiently close to the centre of the far side for undeflected waves to have traversed the core. Another possibility is strong attenuation in a plastic and therefore warm medium. It is thus possible that there is partial melting within a radius of about 700 km of the centre. Thus the transition from solid to a partial or total melt could be within the mantle, as shown in Figure 5.5, and not, as in the case of the Earth, at the core–mantle boundary. Thermal models of the Moon, incorporating measured abundances of radioactive isotopes, are consistent with a deep-lying (partially) molten region. These models also indicate that the lithosphere became as much as 60–100 km thick within a few hundred million years of lunar formation. The ancient cratered surface supports this conclusion, because a thin lithosphere would have allowed remoulding of the surface.

The lunar magnetic dipole moment is too small to measure, so we only have an upper limit, about 10^{-6} that of the Earth. Clearly, if any iron-rich core is (partially) liquid, then either it is not convecting today, or the rotation of the Moon is too slow for a field to be generated – the sidereal rotation period is 27.3 days. However, weak magnetic fields have been detected in the surface rocks. In some cases they seem to have been produced by the effects of impacts, but in other cases exposure to a magnetic field 3500–4000 Ma ago seems to be required. Clearly, one possible source of such a field is the lunar interior, indicating that there might have been a time when the liquid proportion of any iron-rich core was more extensive or more convective. The lunar rotation would also have been more rapid then – it has since been slowed through tidal interactions with the Earth. However, it is very likely that the magnetic dipole moment was always considerably smaller than that of the Earth.

There is a lot of observational evidence that the Moon has a whole-body composition broadly similar to that of the Earth's mantle. This is consistent with the theory of the origin of the Moon outlined in Section 2.2.4, in which a differentiated embryo 10–15% of Earth's mass collided with the Earth and ejected the material from which the Moon formed, in as little as a year or so, giving the Moon a warm birth. In this theory, very little of the iron-rich cores of the Earth and the embryo would have been ejected, and so the Moon formed largely from material derived from the mantles of the two bodies. The embryo would have had a similar composition to the Earth, having been formed in the same region of the Solar System, and this is borne out by the oxygen isotope ratios $^{18}O/^{16}O$ and $^{17}O/^{16}O$, which are similar in the Earth and the Moon. The same is true of chromium isotopes.

Therefore this theory explains the broad similarity between the Moon's overall composition and the composition of the Earth's mantle, and the corresponding global depletions in the Moon of iron and the siderophile elements (with additional depletion in the lunar mantle due to lunar core formation). It also explains the Moon's observed depletion in volatile and moderately volatile substances, and the corresponding enrichment in refractory substances – the material sprayed out by the embryo collision would have been hot enough to lose a high proportion of its volatile materials to space. The anorthosite composition of the lunar crust (Section 7.1.6) is also explained by the impact origin – the final stages of lunar accretion would have involved a heavy bombardment that would have melted the surface thus allowing these low-density silicates to rise to the top.

Finally, from the present 5° inclination of the Moon's orbit with respect to the ecliptic plane, it can be inferred that this inclination was initially about 10° with respect to the Earth's *equatorial* plane. This can be shown to be consistent with an impact origin. From all this evidence, the embryo collision theory is highly favoured for the origin of our satellite.

Question 5.4

Outline the factors that are relevant to explanations of why the interiors of the Moon and Mars are cooler than that of the Earth. You should refer to the general principles in Chapter 4, rather than try to come to any firm conclusions.

5.2.2 Large Icy–Rocky Bodies: Titan, Triton, Pluto, and EKOs

The remaining large satellites and Pluto are all in the outer Solar System, and so we have to consider the possibility that icy materials are substantial components. Whether this is the case is indicated by their mean densities, and the values for all but Io and Europa are sufficiently low to indicate that they are icy–rocky bodies, with icy materials accounting for roughly half the mass in the iciest of them. In the solar nebula the icy/rocky mass ratio is thought to have been about 3:1, and so it seems that the greater volatility of icy materials has somehow led to their depletion in these bodies, and to an even greater depletion in Io and Europa. Hydrogen and helium are excluded as significant contributors to the mass because only in the giant planets are pressures sufficient to compress these substances to appreciable densities.

Water was undoubtedly the most abundant icy material in the solar nebula, and it condensed in and beyond the Jupiter region. Beyond the Jupiter region the more volatile ices condensed, notably ammonia (NH_3), methane (CH_4), carbon dioxide (CO_2), carbon monoxide (CO), and molecular nitrogen (N_2). If chemical equilibrium was attained, then at the low temperatures of the outer solar nebula, nitrogen would have been present largely as NH_3 and $NH_3.H_2O$, and carbon as CH_4 and $CH_4.7H_2O$. However, the low densities in the outer Solar System disfavoured chemical equilibrium. In the interstellar medium nitrogen is present mainly as N_2 and carbon as CO, and so the outer Solar System could also have had appreciable quantities of CO, N_2, and $N_2.6H_2O$. Near to where a giant planet was forming, the higher densities and raised temperatures might have promoted a shift towards chemical equilibrium. Sufficiently close to a protogiant many icy materials would have been unable to condense.

☐ Why not?

A protogiant was luminous, and so the temperatures in the protosatellite disc (Section 2.3.1) would have been too high.

Recipes for icy–rocky bodies thus have a range of icy ingredients available. The proportions depend on the volatility of the different ices. Water is not only the most abundant icy material, but also the least volatile, and it is stable over a wide range of conditions. Thus, water is expected to be the most common icy material in the interiors of all of the icy–rocky planetary bodies. The more volatile icy materials are expected to be more abundant, the further the body is from the Sun, and they could be concentrated towards the surfaces. This expectation is borne out by what is known of the surface compositions.

The large icy–rocky bodies are Ganymede, Callisto, Titan, Triton, and Pluto. Ganymede and Callisto are two of the four Galilean satellites of Jupiter, and these four will be considered together in the next section. Therefore we start with Titan, the largest satellite of Saturn, and second only to Ganymede in size among the planetary satellites of the Solar System.

Titan

Titan has a mean density of only $1880\,kg\,m^{-3}$. This is consistent with 52% of its mass being silicates, and the remainder being predominantly water (as ice) with about 15% NH_3. Some CH_4 should also be present. NH_3 and CH_4 are probably caged at a molecular level within surrounding water molecules, in what you have seen are called clathrates. Titan has a substantial atmosphere, largely of N_2 (probably derived from NH_3) but with a few per cent of CH_4, and other gases. Differentiation seems likely given the calculated accretional energy and radiogenic heating. The magnetometer on the Cassini Orbiter (Table 4.1) has failed to detect a magnetic field of internal origin – the magnetic field at the equator is at least 10 000 times less than that at the Earth's equator. The Orbiter, the Huygens Lander, and Earth-based IR observations indicate a surface dominated by water ice, but with ices of CH_4 and other hydrocarbons also present in abundance. The surface appears young, perhaps no more than 300 Ma in some places.

Figure 5.7 shows one possible model of Titan's interior, in which it is assumed that differentiation has proceeded to the point where an iron-rich core has separated from the silicates. This is overlain by a silicate mantle depleted somewhat in siderophile elements, overlain in turn by a mantle of icy materials, predominantly water. If an iron-rich core has not separated, then the silicate mantle would have a 100–200 km larger radius. Any iron core could be (partially) molten, but is not generating a detectable magnetic field, perhaps because convection is not vigorous enough, and because Titan rotates rather slowly, in 15.9 days.

The structure of the icy mantle depends on the temperature and pressure versus depth, which are unknown. In Figure 5.7 a layer of liquid is shown over the depth range 75–375 km, which is possible if the temperatures are in the approximate range 220–250 K and the water is mixed with various recipes of other icy materials. Salts would also be dissolved in the water, in which case they would dissociate into ions, making the liquid electrically conducting. This could be detected by the magnetic field that would be generated by the passage of this conducting liquid through Saturn's magnetic field, and would be distinct in its spatial form from the magnetic field that has been ruled out by Cassini. However, so far, the only magnetic effect of Titan on Saturn's magnetic field is due to Titan's atmosphere.

Triton, Pluto, and EKOs

Triton is the largest satellite of Neptune. Triton is in a *retrograde* orbit, suggesting that is was therefore captured by Neptune after Triton formed (Section 2.3.1). The tidal energy released upon such capture would probably have melted Triton, in which case it is certainly differentiated (Figure 5.7), though it is probably solid throughout now, except perhaps for partial melting in the lower part of the icy mantle. Some of the mantle is probably an asthenosphere, perhaps

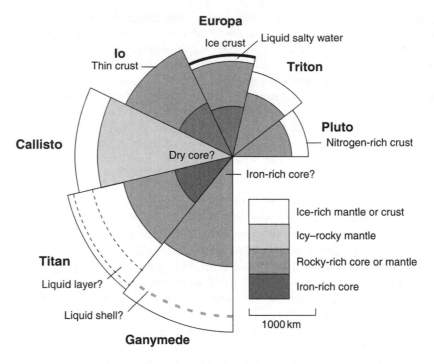

Figure 5.7 Cross-sections of Pluto and the large satellites in the outer Solar System.

most of it. Ices of N_2, CH_4, CO, and CO_2 have been detected on the surface. Observations, particularly from the Voyager 2 spacecraft in 1989, reveal ongoing volcanism involving the very volatile N_2. This is thought to be a lasting consequence of the primordial tidal energy released during capture, and of subsequent radiogenic heating. Volcanism involving icy materials is called cryovolcanism (from the Greek *kruos*, meaning icy cold, or frost).

Distant Pluto is small, and has never been visited by a spacecraft. Consequently, we do not know much about its interior. Its mass has been obtained from the orbit of its satellite Charon, which gives the sum of the masses of Pluto and Charon (equation (4.2)). Charon is not that much smaller than Pluto so we cannot assume its mass is negligible. The individual masses of Pluto and Charon have therefore been obtained by measuring the position of the centre of mass, and using equation (4.3) to obtain the ratio of the masses. The resulting mean density for Pluto is $2030\,\mathrm{kg\,m^{-3}}$. The use of its tiny satellites Nix or Hydra would give a more direct determination.

Pluto is thought to be differentiated into a partially hydrated rocky core, an icy mantle mainly of water ice, and an icy crust that spectrometric observations show to be mostly solid N_2, with a few per cent of CH_4 and some CO. If Pluto was not formed ready differentiated, then radiogenic heating plus the tidal heating due to Charon should have been sufficient to cause differentiation. Tidal heating is promoted by the proximity of Charon, only $19\,570\,\mathrm{km}$ from Pluto, and by the comparatively large mass of Charon – about 10% the mass of Pluto. Today, Pluto is expected to be solid throughout, given its small size, though the lower icy mantle might constitute an asthenosphere.

Table 5.3 Model pressures at the centres of Pluto and the large satellites of the giant planets, plus some central densities and temperatures[a]

	Io	Europa	Ganymede	Callisto	Titan	Triton	Pluto[b]
$\rho^c/\,\mathrm{kg\,m^{-3}}$	8000	4000–8000	5000–8000	2300	5000–8000	?	2800–3500
$T/\,\mathrm{K}$	> 1000	~ 1000	~ 1000	~ 260	~ 1400	?	100–600
$p/\,10^9\,\mathrm{Pa}$	9.3	3.2	3.6	2.7	3.3	2	1.1–1.4

[a] A '?' denotes no useful value is available.
[b] If differentiated.
[c] The densities are the estimated values *in situ*, i.e. they are (slightly) compressed densities. The large ranges for Europa, Ganymede, and Titan correspond to various possible compositions for the cores.

Figure 5.7 shows a model of Pluto's interior, along with models of Titan, Triton, and the four Galilean satellites. Table 5.3 gives further data from typical models.

The EKO Eris, which is slightly larger than Pluto, has a small satellite. Its orbit will enable Eris's mass to be determined and hence its mean density. Several other EKOs have companions, which have yielded densities around $200\,\mathrm{kg\,m^{-3}}$, indicating a loosely consolidated structure, rather like the nuclei of comets, as expected.

Question 5.5

How do solar nebular theories account for Pluto being an icy–rocky body, and being so small?

5.2.3 The Galilean Satellites of Jupiter

Much has been learned about the Galilean satellites from the flybys of Voyagers 1 and 2 and from the Galileo Orbiter (Table 4.1). Figure 5.7 shows interior models, Table 5.3 gives central temperatures, densities, and pressures, and Figure 1.6 shows their orbits.

Io

Io is the innermost of the Galilean satellites of Jupiter. It is slightly larger than the Moon, and somewhat more dense. The mean density, $3530\,\mathrm{kg\,m^{-3}}$, indicates a predominantly silicate plus iron/FeS composition, and the value of C/MR_e^2 of 0.378 is sufficiently less than the uniform-sphere value of 0.4 to require concentration of denser materials into a core. The model in Figure 5.7 has a predominantly iron core extending about half way to the surface, thus accounting for about one-eighth of the volume. The Galileo Orbiter obtained inconclusive evidence for a magnetic dipole moment, somewhat larger than that of Mercury, so the core could be (partly) molten. This is also indicated by Io's density. This is too low for any reasonable proportion of iron or FeS, unless a proportion is liquid.

For the core to be partly molten the interior of Io would have to be hot, and there is dramatic evidence that it is – Io is the most volcanically active of all the planetary bodies in the Solar System (Plate 12)! The mantle, throughout most of its volume, is thought to be an asthenosphere in which solid state convection is occurring, and is overlain by a thin lithosphere that includes a silicate crust rich in sulphur. Partial melting within the asthenosphere would provide the observed volcanic outflows, which are observed to consist of silicates, sulphur, and sulphur dioxide (SO_2).

It came as a surprise to most astronomers that such a small world has a sufficiently hot interior to be so volcanically active.

☐ Why was this surprising?

Equation (4.13) (Section 4.5.4) indicates that small worlds lose energy rapidly, so should now be cool to considerable depth. Indeed, the Moon is comparable in size and mass with Io yet has little or no present-day volcanic activity, and nor does Mercury, which is larger and more massive than Io. Moreover, volcanism is an efficient way for a body to lose heat. We therefore need to find a considerable interior energy source for Io. There is radiogenic heating, but this is far from sufficient. The only plausible additional source is tidal energy.

In Section 4.5.1 you saw that tidal energy will be generated in a body by its rotation, which sweeps the material of the body through the tidal elongation. In the case of Io the rotation period is the same as its orbital period – synchronous rotation – and this is itself a result of tidal effects. Were the orbit circular, Io would keep the same face to Jupiter, as in Figure 5.8(a), and material would not be swept through the tidal elongation – the dot is fixed with respect to the surface. There would then be no tidal heating by this means. Moreover, in a circular orbit Io's distance from Jupiter would be constant, so there would also be no tidal energy generated through variations in the Jupiter–Io distance.

In fact, the orbit of Io is not quite circular. Therefore, there *is* tidal input through its distance from Jupiter varying. Moreover, there is now a contribution from Io's rotation. The rate of rotation is constant, but the rate of motion around the orbit varies, just as the rate of motion of a planet around its elliptical orbit varies in accord with Kepler's second law. Therefore, as seen from Jupiter, Io seems to swing to and fro as it orbits the planet, as shown in Figure 5.8(b), and so material now oscillates through the tidal elongation. Were it not for mean motion resonances with Europa and Ganymede – the orbital periods of Io, Europa, Ganymede are in the ratios 1:2:4 – Io's orbit would be much more circular, and tidal heating would be reduced (equation (4.13)).

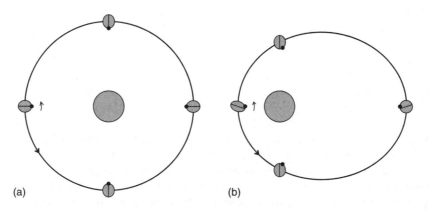

(a) (b)

Figure 5.8 Synchronous rotation (a) in a circular orbit and (b) in an elliptical orbit. The dot is fixed with respect to the surface of the small body.

Question 5.6

The Moon is in synchronous rotation in its orbit around the Earth. With the aid of Tables 1.1 and 1.2 and the answer to Question 4.9, show that the tidal field across the Moon is only 0.004 times that across Io. In relation to tidal heating, why must this factor more than offset the greater eccentricity of the lunar orbit than that of Io?

Europa

Europa, another Moon-sized world, is the next nearest Galilean satellite to Jupiter. Its surface is covered in water ice, though its mean density, $3010\,kg\,m^{-3}$, is only slightly less than that of the Moon, indicating a predominantly silicate composition.

☐ What does the value of C/MR_e^2 in Table 4.2 indicate?

The value of C/MR_e^2 is less than that of Io, which can be met by a greater concentration of denser materials as depth increases. If there is a dry rocky core of uniform composition (mainly silicates plus FeS), overlain by a water mantle, ice, or ice plus liquid, then one model meets the observational constraints with a mantle 150 km thick. A dry core is indicated by thermal models. However, it is likely that the rocky core itself is divided into two zones, the inner zone being iron-rich, perhaps with FeS and/or iron oxides. The Galileo Orbiter obtained inconclusive evidence regarding a magnetic dipole moment. Various detailed compositions, with or without an iron-rich core, lead to a range of models, but there is general agreement that Europa's water mantle is 100–200 km thick.

The same tidal input operates as in Io, though the rate at which energy is released in the shell of water and in the interior is roughly 20 times less, because of Europa's greater distance from Jupiter, which more than offsets the larger eccentricity of Europa's orbit (see equation(4.13) and the answer to Question 5.6). We thus expect Europa to have a cooler interior, though ongoing tidal and radiogenic heating, supplemented by solar radiation and residual primordial heat, should be sufficient to liquefy the lower part of the water shell. There might even be a modest amount of volcanic activity on the ocean floor. The very smooth icy surface indicates an icy crust that could be as little as a few kilometres thick, and though this could be underlain by icy slush rather than by liquid oceans, most astronomers believe that there are widespread oceans. Temperatures throughout most of the mantle are too low for pure water to be liquid. However, salts will be present, notably NaCl, which lower the freezing temperature, and this can be lowered still further by the presence of NH_3, which combines with H_2O to form $NH_3.H_2O$.

Ganymede and Callisto

The two outer Galilean satellites Ganymede and Callisto have surfaces of water ice, with a thin veneer of silicates covering much of Callisto. Their mean densities, $1940\,kg\,m^{-3}$ and $1830\,kg\,m^{-3}$ respectively, require a substantial proportion of ice – unlike Io and Europa these are icy–rocky bodies. The conditions in which Ganymede and Callisto formed, and their observed surface composition, mean that the icy component is predominantly water, probably salty, plus a few percent of NH_3 in Callisto, rather less in Ganymede. $NH_3.H_2O$ is also expected to be present.

☐ From Figure 5.7, what is the proportion of the volume of Ganymede that is water ice?

If Ganymede's water is essentially all in its icy mantle, then the proportion that is *not* water ice is the cube of the radius of the silicate-rich mantle divided by the cube of Ganymede's radius. From Figure 5.7 this ratio is 0.24. Therefore, the proportion of the volume that is water ice is

0.76. However, the proportion that is water by mass is about 50%, because water is much less dense than silicates and iron-rich substances.

The low value of 0.311 for C/MR_e^2 indicates that the silicates and iron-rich materials in Ganymede are concentrated into the core shown in Figure 5.7, perhaps with further concentration of iron-rich materials into an inner core. Such extensive differentiation is to be expected in such a massive body, the most massive in the Solar System. Ganymede has a magnetic dipole moment about three times that of Mercury, which suggests that the core is at least partly molten and convecting vigorously. Vigorous convection requires greater energy input than at present. In the past this might have been driven by greater tides resulting from a more eccentric orbit. It is likely that the orbits of all the Galilean satellites have evolved, and it is possible that Ganymede's eccentricity was substantially higher within the last 1000 Ma or so. The surface of Ganymede reveals relatively recent and extensive geological activity, consistent with a warm interior, and there is also evidence for differentiation at some time in the past (Section 7.4.2). However, it is not clear whether convection driven at some earlier time could still be occurring. Another explanation of the magnetic field is that there is a shell of magnetite, Fe_3O_4, that became permanently magnetized by Jupiter's magnetic field when the shell solidified, perhaps 2000 Ma ago.

A third explanation of the magnetic field is indicated by surface features (Section 7.4.2) that suggest a liquid layer a few kilometres thick at a depth of about 170 km. This would be kept liquid as a result of radiogenic and tidal heating. It would be salty, and therefore electrically conducting.

For Callisto, the value of 0.358 for C/MR_e^2 is significantly less than the value of 0.38 that would correspond to self-compression of an undifferentiated ice–rock mixture. However, it is not small enough to indicate full differentiation, and so there is only a modest degree of concentration of rocky materials towards the centre. Any core is predominantly rocky–icy, though a small core free of icy materials cannot be ruled out. The surface of Callisto suggests that it has always had a cooler interior than Ganymede, and this is consistent with limited differentiation. A likely reason is negligible tidal heating, a result of the low eccentricity of Callisto's orbit, the lack of orbital resonances, and its greater distance from Jupiter. Further factors are the smaller mass ratio of rocky to icy materials in Callisto, resulting in less radiogenic heating, and the possibility of solid state convection in the rocky–icy core that hastened the cooling of the interior. The absence of such a core in Ganymede would have helped promote its more complete differentiation.

❑ Would you expect Callisto to have a magnetic dipole moment?

If Callisto has a cool interior then the magnetic dipole moment should be zero. Callisto is sufficiently far from Jupiter that the Jovian magnetic field is weak, allowing very weak local fields to be detected. This has allowed the Galileo Orbiter to place a very low upper limit on any magnetic dipole moment, entirely consistent with Callisto's other properties. Callisto does, however, disturb Jupiter's magnetic field, indicating the presence of electrically conducting liquids. A liquid shell of water rich in NH_3, $NH_3.H_2O$, and salts, thus lowering the freezing temperature, could be present in the icy mantle. This is also an explanation of the absence of chaotic impact features diametrically opposite the impact basin Valhalla (Section 7.4.2).

The Galilean satellites as a group

Among the Galilean satellites, the proportion of icy materials increases as we go out from Jupiter, from zero for Io, to over 50% by mass for Callisto. Unless these satellites formed further out, and migrated inwards, the explanation is that Io formed too close to proto-Jupiter ever to have had an icy component. Europa acquired a small amount of ice, and Ganymede and Callisto substantial amounts. Water ice would have dominated – the solar nebula in the Jovian region would have been too warm to allow as much of the more volatile ices to condense, or remain condensed.

Increasing distance from Jupiter also correlates with the inferred thermal histories, with greater cooling from Io to Europa, to Ganymede, to Callisto. This trend is largely because of the decrease in tidal energy input with increasing distance from Jupiter.

5.2.4 Small Satellites

There remain the host of small satellites, the asteroids, and the comets. The interiors of the asteroids and the comets were discussed briefly in Chapter 3, and we shall say no more about them here, but concentrate on the small satellites.

The smallest body we have so far considered is Pluto, which has a radius of 1153 km. Next down in size is Titania, the largest satellite of Uranus. This has a radius of 789 km so this is a real step down, and is a convenient point at which to break off discussion of satellite interiors in any detail. Most of the small satellites that orbit the giants will have a composition roughly the same as the large icy–rocky satellites – by mass about half icy materials and half rocky materials. The densities of some of the innermost satellites of Saturn have been estimated from their effect on the rings and on other satellites, and the values are very low – less than the density of water ice ($917 \, \text{kg m}^{-3}$ at 273 K and 10^5 Pa, denser at lower temperatures, but not very sensitive to pressure).

☐ What does this indicate?

Such low densities indicate quite a high degree of porosity. Satellites very close to Jupiter have been warmed, either tidally or by the luminosity of the young giant, to the extent that they are very probably depleted in icy materials. One of these, Amalthea, mean radius 84 km, has had its mass measured by the Galileo Orbiter. This gives a density of about $860 \, \text{Kg m}^{-3}$, which for an ice-free body indicates high porosity, to the extent that it is probably a rocky rubble pile, loosely reassembled after collisional disruption. Its highly non-spherical shape is consistent with this.

Just five small satellites are known not to orbit giants. These are the Martian satellites Phobos and Deimos, rocky bodies that are probably captured asteroids, Pluto's icy–rocky satellite Charon, and its two tiny satellites Nix and Hydra. Tidal heating is likely to have ensured that Charon is fully differentiated. The smallest satellites of all might not have differentiated even if they have been warmed. This is because in very small bodies gravity is relatively weak, and so the chemical forces between constituents can override the gravitational trend towards differentiation.

The surfaces of some of the small satellites are more remarkable than their interiors, so we shall return to them in Chapter 7.

5.3 The Giant Planets

In arriving at the four giant planets we move up an order of magnitude or so in size and mass, with Jupiter, Saturn, Uranus, and Neptune having masses of $317.8 m_{\text{E}}$, $95.1 m_{\text{E}}$, $14.5 m_{\text{E}}$, and $17.1 m_{\text{E}}$ respectively, where m_{E} is the mass of the Earth. They are also worlds that are very different in many other ways. Figure 5.9 shows compositional models of the interiors of Jupiter, Saturn, Uranus, and Neptune that are consistent with the observational data. Some variation is possible and so the models shown are broadly indicative rather than definitive. Much of the

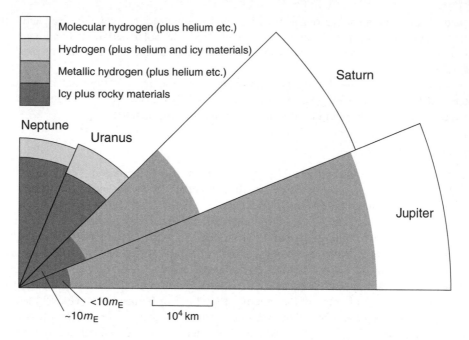

Figure 5.9 Cross-sections of the giant planets (in their equatorial planes).

reason for this is that the equations of state of candidate materials, notably hydrogen and helium, are poorly known at the high pressures encountered in the deep interiors. Another difficulty is that the gravitational coefficient J_2 (Section 4.1.2) is rather insensitive to the density profile within about a third of the distance from the centre. The other coefficients, J_4, J_6, etc., penetrate even less far.

The interiors of the giant planets undoubtedly consist largely of hydrogen, helium, and icy materials, with rocky materials making up only a small fraction of the mass. The dominance of hydrogen, helium, and icy materials is required by the low mean densities of the giants (Table 4.2), which range from a 'high' of $1640\,kg\,m^{-3}$ for Neptune to an astonishing low of $690\,kg\,m^{-3}$ for Saturn. These low densities, several times less than typical densities of rocky materials, are even more remarkable given the large size of the giants.

☐   Why is size relevant?

Equation (4.12) shows that large radii lead to large pressures, and large pressures increase the density of a substance. For Jupiter and Saturn the only models that fit their mean densities consist largely of the intrinsically least dense materials – hydrogen and helium. Uranus and Neptune, being smaller, must have icy materials as the main constituent. On the basis of the relative abundances of the elements in the Solar System, water must be the dominant icy material in the giant planets.

In the models of Jupiter and Saturn, the outer regions are hydrogen and helium, with small quantities of other substances. In the case of Uranus and Neptune, even though the outer envelope is mainly hydrogen and helium, there are substantial proportions of icy materials as well. The outermost regions are directly observable, and of course the models match the observations. The compositional layering in the models is strongly indicated by the gravitational data, though the boundaries between certain layers seem to be fuzzy. There is, however, a concentration of icy–rocky materials towards the centres, though perhaps rather weakly in the case of Jupiter.

☐ Which theory of the origin of the giant planets does this favour?

The core-accretion theory predicts that the giants started as kernels of icy–rocky materials that grew massive enough to capture hydrogen–helium envelopes and icy–rocky planetesimals (Section 2.2.5). Some central concentration of icy–rocky materials is thus expected.

Table 5.4 lists the temperatures, densities, and pressures at a few depths in the giants – again these are broadly indicative, not definitive. High interior temperatures are indicated by the observed IR excesses from Jupiter, Saturn, and Neptune (Section 4.5.3). The corresponding outward flows of energy from the interiors are given in Table 4.2. For all the giants, hot interiors today are predicted by thermal models. Furthermore, the large magnetic dipole moment of each giant planet indicates extensive, convective fluid regions in the interior that are electrically conducting. Models predict that convection is occurring throughout much of the interior of each giant. The low IR excess from Uranus can be explained by a current lack of convection over some range of depths, perhaps due to a composition gradient.

It seems certain that the giants are fluid throughout, in which case hydrostatic behaviour is expected, and the gravitational fields are indeed those expected of rotating hydrostatic bodies.

It is clear that the giant planets can be divided into two pairs: Jupiter and Saturn; Uranus and Neptune (Question 4.1, Section 4.1.1). Indeed, some people used to restrict the term 'giant' to Jupiter and Saturn, and call Uranus and Neptune 'subgiants'. We examine each pair in turn.

5.3.1 Jupiter and Saturn

These two giant planets are most like the Sun in their composition, though whereas in the Sun the heavy elements (those other than hydrogen and helium) account for only about 2% of the total mass, in most models of Jupiter these elements account for 5–10% of the mass, and for a roughly three times greater proportion in Saturn. Therefore, the actual mass of the heavy elements is about the same in the two planets.

Table 5.4 Model temperatures, densities^a, and pressures in the giant planets

	Jupiter	Saturn	Uranus	Neptune
Envelope				
$T(10^5 Pa)/$ K	165	135	76	72
T (base)/ K	~ 6500	~ 6000	~ 2000	~ 2000
p (base)/ Pa	$\sim 2 \times 10^{11}$	$\sim 2 \times 10^{11}$	$\sim 1 \times 10^{10}$	$\sim 1 \times 10^{10}$
ρ (base)/ kgm^{-3}	0.8×10^3	0.8×10^3	?	?
Centre				
ρ (centre)/ kgm^{-3}	2.3×10^4	1.9×10^4	$\sim 10^4$	$\sim 10^4$
T (centre)/ K	$1.5\text{–}2.1 \times 10^4$	$0.85\text{–}1.0 \times 10^4$	~ 8000	~ 8000
p (centre)/ Pa	$\sim 4 \times 10^{12}$	$\sim 1 \times 10^{12}$	$\sim 0.8 \times 10^{12}$	$\sim 0.8 \times 10^{12}$

^{a} The densities are the estimated values *in situ*, i.e. they are compressed densities.

Jupiter

To explore the Jovian interior, we shall take an imaginary journey to its centre. We start in the atmosphere, which is observed to consist by mass, of about 23% helium, nearly all of the rest being hydrogen. Hydrogen is in the molecular form H_2, and helium, an unreactive inert gas, is in atomic form He. This helium mass fraction is significantly less than the primordial solar value of around 27% (Section 1.1.3). Though the Sun and Jupiter formed from the same nebula, it is not difficult to understand why the helium fractions differ. Some downward settling of helium would occur, with the result that there is a greater fraction deeper down. Settling has also occurred in the Sun – in the Sun's atmosphere the helium mass fraction today is about 24%, by coincidence the same as in the Jovian atmosphere. The helium fraction in the metallic mantle (see below) in Jovian models is about 27%, and so, with nearly all of the mass of Jupiter in this mantle, the overall helium mass fraction in Jupiter is about that of the Sun at birth (and outside the fusion core today).

Beneath the Jovian atmosphere we pass quickly through various cloud layers, the temperature, pressure, and density gradually increasing as we descend. At several thousand km below the cloud tops the density is several hundred m^{-3}. This is not much less than the $1000\,kg\,m^{-3}$ density of liquid water at the Earth's surface. Therefore, we have arrived in what we can regard as a hydrogen–helium ocean – and yet we crossed no surface. Figure 5.10 shows the phase diagram of molecular hydrogen (the presence of helium will modify it somewhat, but the story is still the same in all its essentials). Our journey took us along the path shown as a dashed line.

▢ Why was there no sudden transition from the relatively low-density gas phase to the much higher density liquid phase?

The path stayed well to the right of the critical point (Cr), and so the density gradually increased.

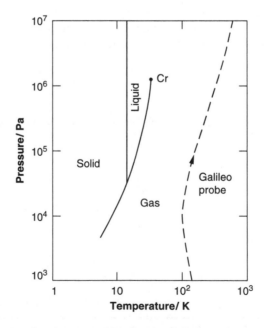

Figure 5.10 Phase diagram of molecular hydrogen, with Jovian conditions represented by the Galileo probe measurements.

This absence of a surface makes the Jovian radius a matter of definition. It is defined as the radius at which the atmospheric pressure is 10^5 Pa (standard atmospheric pressure at the Earth's surface is 1.01×10^5 Pa). The same definition is adopted for all the giant planets. In the case of Jupiter, 10^5 Pa is not far below the top of the upper cloud deck.

We do, however, encounter a surface of sorts deeper down, albeit with only about a 10% increase in density across it. This is the transition between liquid molecular hydrogen (plus atomic helium) and liquid *metallic* **hydrogen** (plus atomic helium). At pressures around 2×10^{11} Pa, the density of molecular hydrogen is about $800 \, \text{kg m}^{-3}$, sufficiently high that each of the hydrogen atoms in the H_2 molecule is attracted to atoms in neighbouring molecules as much as to its molecular partner. Therefore, the H_2 molecules break up. Moreover, the single electron that orbits the nucleus of each hydrogen atom will also become equally attracted to neighbouring atoms, and so the atoms break up too. The hydrogen will then be a 'gas' of electrons moving in a 'sea' of hydrogen nuclei. Many of the characteristic properties of metals arise from the existence of 'electron gases' within them, so the term 'metallic hydrogen' is appropriate. Metallic hydrogen was once just a theoretical prediction, but then in 1996 sufficiently high pressures were produced in the laboratory for metallic hydrogen to appear as expected. In Jupiter the transition to the metallic form takes place over a range of pressures, and divides the interior into a metallic hydrogen mantle and an overlying molecular hydrogen envelope. The range of depths over which the transition takes place is unknown. Figure 5.9 shows the top of the transition. Beneath, there might be a narrow layer of inhomogeneous composition.

One of the properties that an electron gas gives to a metal is high electrical conductivity. Thus, electric currents in the metallic hydrogen mantle are an obvious source of the planet's large magnetic dipole moment. The volume of metallic hydrogen, the hot convective interior, and the rapid spin of Jupiter are all consistent with this conclusion. A small but significant contribution to the field could arise from liquid iron and other conductors much deeper down. The detailed configuration of the field close to Jupiter is consistent with currents in the metallic hydrogen mantle.

Gravitational data indicate an increase in density towards the centre, but it is possible that this is due to self-compression in the metallic hydrogen mantle. Models consistent with the data have cores of rocky and icy materials ranging from 0 to $10m_E$. To satisfy the constraints, there is a complementary range from $42m_E$ down to little more than $10m_E$ in the amount of rocky and icy materials in the whole planet. This broad range of values results from various factors, including uncertainties in the equations of state of hydrogen and helium, the insensitivity of the gravitational coefficients to deep structures, the uncertainty in the width of the transition zone between molecular and metallic hydrogen, and so on. If the icy–rocky core really has a low mass, this could be due to erosion of the core by the high temperatures at the base of the mantle, which is not much less than the central temperature given in Table 5.4.

☐ If icy and rocky materials account for $20m_E$ in the whole of Jupiter, what is the heavy element mass fraction?

Almost all the mass of icy and rocky materials consists of elements other than hydrogen and helium, i.e. the heavy elements. Jupiter's mass is $317.8m_E$, and so the fraction is 6.3%. This is nearly four times the solar mass fraction of about 1.6%. The modelled enrichment of Jupiter by a factor of a few is consistent with (but not the same as) the measured enrichment of the atmosphere (Section 11.1.2). The enrichment is thought to be primordial, from a kernel composed of rocky–icy materials, with a contribution from the subsequent capture of planetesimals. In the gravitational instability model of formation, the planetesimal capture rate after the formation of Jupiter would probably have been too low to give such a high heavy element fraction.

Thermal models show that the present-day high temperatures of the Jovian interior can be accounted for by two dominant energy sources – energy from accretion, plus energy from differentiation when the icy–rocky kernel acquired more mass to become the icy–rocky core. An active but small energy source is the settling of helium through the metallic hydrogen mantle. The two major energy sources became inactive about 4500 Ma ago, yet the central temperature of Jupiter is still about 2×10^4 K. The reason is the large size of Jupiter. This has two consequences. First, there would have been a huge amount of accretional energy per unit mass as Jupiter formed, and so Jupiter would have become extremely hot (Section 4.5.1). Second, Jupiter has a comparatively low ratio of surface area to mass, giving a low rate of internal energy loss (Section 4.5.4) in spite of the efficient outward transfer of energy by convection.

Though convection is expected to occur throughout most of the Jovian interior, calculations also indicate that over the depth range in which the temperatures are 1200–3000 K, energy might be transported outwards by radiation rather than by convection. This would lead to *cooler* interiors than in Table 5.4, and to a deeper transition to the metallic hydrogen phase than in Figure 5.9. This is also the case for Saturn.

Saturn

In many ways, the interior of Saturn is similar to Jupiter, as Figure 5.9 shows. Saturn is a bit smaller, and less dense, leading to lower internal pressures. These lower pressures mean that the transition to metallic hydrogen occurs much nearer the centre, so whereas most of the hydrogen in Jupiter is in metallic form, most of that in Saturn is in molecular form. As with Jupiter, it is the top of the transition that is shown, should this occur over a (thin) shell. As in Jupiter, it is the metallic hydrogen in models of Saturn that can account for the planet's magnetic field. With the two planets rotating at about the same rate, and with comparable internal activity, the lower mass of metallic hydrogen in Saturn is predicted to lead to a smaller magnetic dipole moment.

☐   Is this the case?

Table 4.2 shows that the magnetic dipole moment of Saturn is about 30 times less than that of Jupiter.

The smaller size of Saturn also means that its internal temperatures after formation would have been lower than in Jupiter, and it should have cooled more rapidly. Therefore, the present-day high internal temperatures, indicated by the IR excess, cannot be accounted for solely by energy of accretion and by past differentiation as the icy–rocky kernel acquired more mass to become the icy–rocky core. An additional source of energy is needed, and this is thought to be the ongoing separation of helium from metallic hydrogen (which plays at most a small role in Jupiter). Initially, the helium in the metallic hydrogen mantle was thoroughly mixed at the atomic level, and random thermal motion prevented any settling of the helium atoms downwards – a tendency arising from the greater mass of the helium atom. As the interior cooled, the miscibility of helium in metallic hydrogen fell, and Saturn is estimated to have reached the point about 2000 Ma ago where helium began to form small liquid droplets. These could not be held by random thermal motions in uniform concentration throughout the metallic hydrogen, so downward separation of the helium droplets began. This is essentially the same process as the separation of oil from vinegar in salad dressing. Convection is believed to have slowed the settling rate, but an outer core of helium is forming, and as it does so energy of differentiation is released. An additional source of energy might be continuing growth of an icy–rocky core.

The removal of helium from the metallic hydrogen would result in helium diffusing down from the molecular hydrogen envelope, so we would expect the observable atmosphere to be depleted in helium compared with Jupiter. And indeed it is! In Jupiter the outer atmosphere is observed to consist of about 23% helium by mass, whereas for Saturn the value is about 20% helium. The greater extent of downward separation of helium in Saturn is reflected in models by larger increases in the helium mass fractions with depth. Below the molecular hydrogen envelope, which models indicate has a helium mass fraction not very different from the atmospheric value, the same models give the metallic mantle about 30%. Thus, overall, Saturn, like Jupiter, has about the same helium mass fraction as the Sun at its birth.

The lower internal pressures make the equations of state for hydrogen and helium less uncertain than in the case of Jupiter. This exposes strong evidence for a significant rocky–icy core in Saturn, and thus further evidence against the gravitational instability model of formation. It is likely that the core mass is not greater than 10–$20m_E$, depending on the extent to which heavy elements reside outside the core. The core mass could be reduced by several m_E if, as is quite possible, helium separation has produced a nearly pure helium shell around the core. The mass of heavy elements in Saturn is roughly the same as that in Jupiter.

Question 5.7

As Jupiter's interior cools, a certain energy source will become increasingly powerful. State what this source is, and why it is triggered by cooling. What effect could this source have on Jupiter's subsequent internal temperatures?

5.3.2 Uranus and Neptune

We have noted that the mean densities of Uranus and Neptune show that they are much less dominated by hydrogen and helium than are Jupiter and Saturn. The equations of state of the icy–rocky materials that dominate the interiors are collectively less well known than those of hydrogen and helium, and the range of possible models is thus larger. All models predict that these planets, like Jupiter and Saturn, are fluid throughout.

Uranus and Neptune are smaller and less massive than Jupiter and Saturn, and the overall composition of the models can be obtained, very roughly, by stripping away a good deal of the hydrogen and helium from Jupiter or Saturn.

☐ How is this feature explained in core-accretion theories?

The kernels of Uranus and Neptune formed considerably more slowly than those of Jupiter and Saturn, and so there was less time to capture hydrogen and helium before the proto-Sun's T Tauri phase drove away the nebular gas (Section 2.2.5). In typical models of Uranus and Neptune, hydrogen and helium account for 5–15% of the mass of each planet.

The mass ratio of icy to rocky materials, derived from solar elemental abundances, is about 3, which means that for these planets 60–75% by mass is icy and 20–25% rocky. The greater density of Neptune, $1640 \, \text{kg m}^{-3}$ versus $1270 \, \text{kg m}^{-3}$ for Uranus, in bodies of similar size, indicates that either Neptune has a greater proportion of icy–rocky materials, and/or the *rocky* proportion of the icy–rocky materials is higher in Neptune. Both possibilities are consistent with models in which Neptune formed closer to the Sun than Uranus did, with subsequent outward migration of both placing Neptune further out (Section 2.2.5). In the icy component, H_2O, CH_4, and NH_3 must be the major ingredients. This mix will be rich in ions at depths where the pressures and temperatures are high enough.

The atmospheres where the composition can be observed consist by mass of about 65% hydrogen and about 23% helium – not very much less than the fractions estimated for the young Sun. As in Jupiter and Saturn, the accessible hydrogen is in molecular form and helium is in atomic form. The rest of the atmosphere is icy gases, enriched above what would be derived from solar abundances by the capture of at least $0.1 m_E$ of planetesimals after the atmosphere was in place. In typical models, such as those in Figure 5.9, the hydrogen–helium–icy composition continues in this envelope down to an icy–rocky mantle, possibly with a rocky (but fluid) inner core. The internal pressures are too low for metallic hydrogen, and so there is no metallic hydrogen mantle.

The IR excess of Neptune indicates a hot, convective interior – convective throughout most of its volume. An interior hot enough to be liquid is also implied by Neptune's large magnetic dipole moment. The detailed configuration of the field indicates that the electric currents are located in a thin outer shell of the icy–rocky mantle. As noted above, mixtures of H_2O, CH_4, and NH_3 can become ionised, and thus highly electrically conducting. The predicted convective interior and the observed rapid rotation of the planet complete the requirements of the dynamo theory. For Neptune to have high internal temperatures today there needs to be an active energy source. This is thought to be differentiation, though uncertainties about the internal distribution of the various icy and rocky materials make the details obscure.

Uranus rotates only slightly slower than Neptune and has nearly double Neptune's magnetic dipole moment. As with Neptune, the detailed configuration of the field indicates that the electric currents are located in a thin outer shell of the icy–rocky mantle.

The IR excess of Uranus is barely detectable, corresponding to a power outflow per unit mass about nine times less than that of Neptune. However, the two planets are so similar in so many ways that it is thought that the internal temperatures are roughly the same, and that some process is greatly reducing the rate at which the energy in Uranus is transported to the upper atmosphere, where it would be radiated away to space. It was mentioned near the beginning of Section 5.3 that the suppression of convection over some range of depths, due to a composition gradient, might be the reason for the low rate of transfer. This could be caused by a significant change in the abundance of heavier molecules over some modest range of depths. In the case of Neptune, any such region is likely to be at a greater depth, and consequently less effective. Such zones are not ruled out by the gravitational data.

Question 5.8

It is believed that radioactive isotopes are only a minor energy source (per unit mass) in the giant planets. Why is this believed to be the case?

5.4 Magnetospheres

This chapter concludes with a short account of a particular consequence of a planetary body having a substantial magnetic field. The magnetic field of the planet then interacts with the solar wind to create what is called a magnetosphere around the planet. Very complex mechanisms are involved in magnetospheres, and so the approach here will be qualitative and brief.

5.4.1 An Idealised Magnetosphere

Figure 5.11 shows a comparatively simple magnetosphere that will introduce the essential features. To the left, the solar wind is flowing in interplanetary space, and it is undisturbed by the planet. The magnetic field in the wind in this particular case is perpendicular to the flow, as also, for simplicity, is the magnetic axis of the planetary field. If the field in the wind were static then the interplanetary field would simply be the sum of the undisturbed wind field and the undisturbed planetary field. But the wind field is entrained in the wind – this is because the wind is a plasma, i.e. it is sufficiently ionised for copious electric currents to flow in it. Entrainment means that the wind carries the magnetic field along with it. As a result, the interaction of the wind with the planetary magnetic field gives a different outcome which we shall now explore.

The solar wind 'sweeps' interplanetary space clean of the planetary field except in the vicinity of the planet. On the upwind side of the planet (to the left in Figure 5.11) there is a roughly hemispherical boundary outside of which there is only the interplanetary field. On the downwind side the boundary stretches out into a long magnetotail. Within the boundary the planetary field near the planet is as if there were no wind field, but the planetary field gets more and more distorted as the boundary is approached. The boundary is called the magnetopause, and the volume it encloses is called the **magnetosphere**. The 'sphere' part of the name is to be interpreted as the planet's magnetic sphere of influence, rather than as a description of the shape of the boundary.

There are three sorts of magnetic field lines: there are those that start and end on the planetary surface, distorted though they may be; there are those that never encounter the planet; and there are those that start on the planet but, in effect, connect with the wind field and therefore never return to the planet – these are called reconnected lines.

The size of the magnetosphere is characterised as the distance from the centre of the planet to the upwind magnetopause. This distance is proportional to $\mu^{1/3}/(n^{1/6}v^{1/3})$, where μ is the magnitude of the magnetic dipole moment of the planet, v is the wind speed, and n is the number density (number per unit volume) of the charged particles in the solar wind (mainly electrons and protons). Though v does not vary much with the heliocentric distance, n diminishes as this

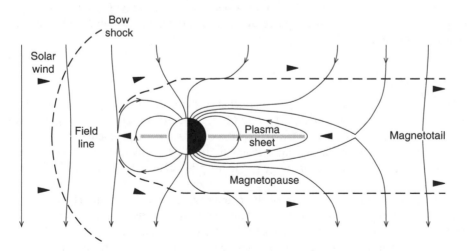

Figure 5.11 An idealised magnetosphere, with magnetic field lines.

distance increases. Also, n varies with solar activity, and so the size of the magnetosphere also varies, being a maximum when n is a minimum.

Because the solar wind approaches the magnetopause at the high speeds around several hundred kilometres a second, it gets a rude shock at a boundary outside the magnetopause, called, appropriately, the bow shock. (The 'bow' is by analogy with a related phenomenon created on the surface of water as the bow of a boat moves through the surface at speeds greater than the speed of the surface ripples.) Between the bow shock and the magnetopause the solar wind is rapidly decelerated, the flow becomes turbulent and the wind plasma is strongly heated. The wind flows around the magnetopause, with very little of the plasma entering the magnetosphere. The region between the bow shock and the magnetopause is called the magnetosheath.

The small fraction of the solar wind plasma that enters the magnetosphere is only one source of magnetospheric plasma. Another is cosmic rays (Section 3.3.3). These highly energetic charged atomic particles readily cross the magnetopause, and though most of them pass out again, a small proportion is trapped. Furthermore, a small fraction of the cosmic rays collides with the atmosphere of the planet, or with its surface if it has no atmosphere. This results in the ejection of particles, and these include neutrons that decay into protons and electrons, many of which are then trapped in the magnetosphere. Yet another source of plasma is a slow leak of particles from the planet's upper atmosphere, both in the form of plasma and in the form of neutral atoms that subsequently become ionised.

Magnetospheric plasma is not uniformly distributed, but becomes concentrated towards the plane of the magnetic equator, where it constitutes the plasma sheet (Figure 5.11). Belts and toruses of plasma surrounding a planet can also occur.

Though the magnetospheric plasma is being added to all the time, there are also losses, outwards to interplanetary space, and inwards to the planet. Among the latter are energetic ions and electrons that reach the upper atmosphere and excite atoms there. The resulting emission of optical radiation is called an **aurora**. Aurorae are concentrated in a ring around each magnetic pole. Large fluxes of energetic electrons plunging into the upper atmosphere generate radio waves with wavelengths of the order of 10–100 metres. Such decametric radiation emanates from the Earth and from the giant planets, and as early as 1955 indicated that Jupiter has a powerful magnetic field. Aurorae and decametric radiation are intermittent phenomena, depending on the strength of the solar wind. Other radio waves, with wavelengths of the order of 0.1–1 metre, are generated in the magnetosphere by electrons travelling at high speeds. This is called **synchrotron emission**.

5.4.2 Real Magnetospheres

In the Solar System, the magnetic dipole moments of the Earth and of the giant planets are far larger than those of any other planetary body (Table 4.2), and correspondingly they have the most extensive magnetospheres. Their magnetic axes are not perpendicular to the solar wind flow, nor on the whole is the solar wind's magnetic field. Nevertheless, the general form of the magnetosphere in each case is roughly as in Figure 5.11, and there is also a plasma sheet and plasma belts.

Figure 5.12 shows the typical form of the magnetosphere of the Earth. There are two main plasma belts around the Earth – the **Van Allen radiation belts**, named after the American physicist James Alfred Van Allen (1914–2006) who discovered them in 1958. The inner belt consists largely of protons and electrons. These come from the solar wind, and also from the Earth's upper atmosphere partly through the action of cosmic rays. The outer belt is more

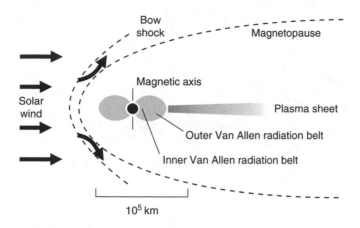

Figure 5.12 The Earth's magnetosphere.

tenuous, and the particles are less energetic. It is populated largely by the solar wind. Within the inner belt there is a third belt in which cosmic rays are prevalent.

There is also a plasma sheet (Figure 5.12). This has a low density, and is hot, the temperatures being $1-5 \times 10^7$K. It is fed largely by the solar wind.

☐ What are the main constituents of the plasma sheet?

Being a sample of the solar wind, its main constituents are electrons and protons. Energetic electrons from this sheet find their way by various means into the upper atmosphere, particularly in a ring around the magnetic poles, where reconnected magnetic field lines intersect the ionosphere (Section 9.2.2). There the electrons can give rise to decametric radiation, and (along with other charged particles) also make a significant contribution to aurorae – the aurora borealis (Plate 26) in the northern hemisphere, and the aurora australis in the southern hemisphere. When the solar wind is strong, as at times of high solar activity, the ring widens and aurorae are then seen further from the Earth's magnetic poles, down to about 70° or so in magnetic latitude. The magnetic axis is tilted by about 11.5° with respect to the rotation axis (Figure 4.6), and so the corresponding geographical latitude depends on longitude. The auroral displays are at an altitude of only about 80–300 km, and so the tropics are not the place to go to see aurorae!

Figure 5.13 shows Jupiter's magnetosphere.

☐ Why is Jupiter's magnetosphere bigger than that of the Earth?

It is bigger because Jupiter is further from the Sun, and so the number density of charged particles in the solar wind is smaller (see the expression in Section 5.4.1), and because the magnetic dipole moment of Jupiter is 20 000 times larger than that of the Earth (Table 4.2). When the solar wind is particularly weak the upwind magnetopause can be about 100 Jovian radii from Jupiter. If we could see such a magnetosphere from Earth with Jupiter at opposition the magnetopause would be appear like a disc with an angular diameter about 2.6 times that of the full Moon. The magnetotail can extend beyond the orbit of Saturn.

The Jovian magnetosphere is particularly rich in plasma – in the denser regions the human body would quickly receive a lethal dose of ions. This richness is a result of copious internal sources, notably the volcanoes of Io, but also the Jovian upper atmosphere and the surfaces of Jupiter's satellites and ring particles. Ions and electrons are ejected from the satellites and rings by cosmic rays. As well as being sources of plasma, the satellites and rings also remove plasma

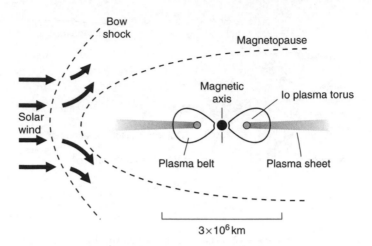

Figure 5.13   The Jovian magnetosphere.

particles that collide with them. The solar wind is not an important source of the magnetospheric plasma, except near the magnetopause and far out in the magnetotail. It can then be shown that the plasma sheet must be a result of leakage from the plasma belt. Leakage occurs preferentially at the magnetic equator, where the magnetic containment is weakest. This domination of internal sources of plasma is a result of electric fields in the magnetosphere plus the rapid rotation of Jupiter – the details are beyond our scope. Jupiter also displays aurorae, with the same basic cause as for the Earth.

Beyond Jupiter there are three more planets with extensive magnetospheres: Saturn, Uranus, and Neptune. The magnetosphere of Saturn is intermediate between that of the Earth and Jupiter in extent and plasma content. Sources and sinks of plasma include the rings and the satellites. Saturn also displays aurorae. Compared with the Voyager encounters in 1980 and 1981, Cassini in 2005 found no detectable changes in Saturn's internal magnetic field. The magnetosphere had changed in extent somewhat, but in line with the variable solar wind.

The considerable magnetospheres of Uranus and Neptune have some peculiarities arising from the large angles between their magnetic and rotation axes (Figure 4.6), and in the case of Uranus from its large axial inclination, but in terms of the above discussion no new major phenomena are encountered.

Question 5.9

Outline the consequences for the Earth's magnetosphere

(a) if the Earth had no atmosphere;
(b) if the speed and number density of charged particles in the solar wind were reduced.

5.5 Summary of Chapter 5

The terrestrial bodies are dominated by silicates, and by iron-rich compounds, including iron itself (see Figures 5.1, 5.5, and Io and Europa in 5.7). They are all differentiated, with iron-rich cores and silicate mantles. Europa has a thin icy crust, mainly water ice, underlain by an ocean of salty liquid water or slush, whereas the other terrestrial bodies have rocky crusts. The Earth's interior is by far the best known, and it is clear that there is an outer core mainly of iron that is hot enough to be liquid. This outer core is the source of the Earth's powerful magnetic field. All but the uppermost part of the Earth's mantle is an asthenosphere, and it is probably undergoing solid state convection. It is thought that the other terrestrial bodies also have warm interiors (Tables 5.2 and 5.3).

Internal temperatures in all of the terrestrial bodies are raised above the values they would have in equilibrium with solar radiation. In all cases, except Io and Europa, this is almost entirely through the effects of primordial energy sources and heat from long-lived radioactive isotopes. The temperatures decrease as the size of the body decreases, with the exception of Io, which, in spite of its small size, has an interior hot enough for silicate volcanism. This is because Io has a dominant tidal component in its internal energy sources. Europa is less tidally heated, being further from Jupiter, but this makes a contribution crucial to sustaining a salty ocean/slush.

Pluto, and the remaining large satellites (Figure 5.7), differ from the terrestrial bodies in having a much larger proportion of icy materials – they are icy–rocky bodies. They are thought to be differentiated into icy-rich mantles, and rocky-rich cores. Ganymede, Callisto, Titan, and perhaps Triton might well be liquid over some depth range. Eris is a bit larger than Pluto, and is icy–rocky, but as yet we know nothing of its interior.

The remaining satellites have Titania as their largest member. Most of them consist of roughly equal masses of icy and rocky materials. The very smallest satellites are unlikely to be differentiated.

The four giant planets (Figure 5.9) are dominated by hydrogen, helium, and icy materials, the proportion of hydrogen and helium being considerably greater in Jupiter and Saturn than in Uranus and Neptune. For all four bodies the hydrogen–helium ratio for the whole body is thought to be similar to that in the young Sun.

Jupiter and Saturn are each differentiated into an icy–rocky core, a mantle of metallic hydrogen, and an envelope of molecular hydrogen (H_2), with helium (He) as the next most abundant component in the mantle and envelope. The boundaries between mantle and envelope are fuzzy. Saturn has a substantial icy–rocky core but in Jupiter the core might be very small, even absent, the heavy elements then being more uniformly distributed.

Uranus and Neptune have deep atmospheres, like Jupiter and Saturn, dominated by H_2, with He the next most abundant component, but with more icy materials. These overlie predominantly icy cores. Though models of Uranus and Neptune are poorly constrained, we can be sure that because of the lower pressures there is no metallic hydrogen.

The interiors of all four giants are hot (Table 5.4), and they all have very large magnetic dipole moments, originating in the metallic hydrogen mantles in Jupiter and Saturn, and in icy mantles in Uranus and Neptune. The IR excess of Jupiter is largely accounted for by primordial accretion and early differentiation. A substantial supplement of ongoing differentiation is required to account for the IR excesses of Saturn and Neptune. In Saturn the differentiation is the separation of helium from hydrogen in the metallic phase. This is happening faster in Saturn than in Jupiter because of its lower internal temperatures. In Neptune the nature of the differentiation is unclear. Uranus has a very small IR excess, and if indeed its interior is hot, then the outward energy transfer rate is somehow being reduced, perhaps through the suppression of convection over some range of depths by a vertical composition gradient.

A planetary body with a substantial magnetic field will interact with the solar wind to produce a magnetosphere. Beyond the magnetosphere the solar wind sweeps space clean of the planetary field. Within the magnetosphere the planetary field becomes increasingly dominant as the planet is approached. There will be a variety of plasma in a magnetosphere, much of it concentrated into a plasma sheet and plasma belts or toruses. Interaction of the plasma with the upper atmosphere can produce aurorae.

6 Surfaces of Planets and Satellites: Methods and Processes

When we wonder about other worlds, it is usually their surfaces to which our thoughts first turn. After all, it is the surface of one particular world upon which we live, and there is a fascination with exploring terrestrial landscapes that differ from those of our own region. Even greater is the immediate fascination with the landscapes of other worlds. Such individual landscapes will be explored in the two chapters that follow this one. In this chapter we shall outline some of the methods of investigating surfaces, and the various processes that have made the surfaces as they are. The giant planets do not have surfaces in the generally accepted sense, so they are excluded from these chapters.

6.1 Some Methods of Investigating Surfaces

The surfaces of Solar System bodies are more accessible than their interiors. The detailed surface form is observable, and the physical and chemical nature of the surface can be determined directly if samples can be obtained, or indirectly from space or from the Earth. We can also observe many surface processes in action, such as volcanism.

But a surface, in the sense of being the outermost layer of a body, distinct from the interior, is not entirely accessible. Therefore, all the methods described in Chapter 4 that are used to investigate the interior are also applied to these shallow depths – seismology, gravitational and magnetic field measurements, and so on. At such depths, as well as a global picture, they also reveal regional and local details.

6.1.1 Surface Mapping in Two and Three Dimensions

Ever since the invention of the telescope around 1600, astronomers have had a growing capability to map the surfaces of planets and satellites. At first this was by looking through the telescope, but later various imaging instruments were added. The best images have been obtained from orbiting or flyby spacecraft (Table 4.1), though ground-based telescopes and telescopes in Earth orbit, such as the Hubble Space Telescope, are yielding images that would have astounded

Discovering the Solar System, Second Edition   Barrie W. Jones
© 2007 John Wiley & Sons, Ltd

astronomers of earlier generations. The images are not restricted to visible wavelengths, but are obtained at UV, IR, and radio wavelengths too. Computer processing can extract every last bit of detail.

A two-dimensional image shows variations *across* a surface, though some altitude information can be obtained, e.g. from the lengths of shadows cast by a feature, or from evidence that one feature is partially obscuring another. We can, however, use other means to obtain a more complete and more accurate picture of the three-dimensional form of a surface – the surface topography. One way or another we need to measure the altitude at every point on the surface – we need to perform altimetry. For the Earth, altitudes have been measured over the centuries by traditional surveying methods on land and sea, resulting in relief maps of exquisite detail, though the deep oceans were only mapped in the latter half of the twentieth century.

For other planetary bodies the best altimetry data have come from spacecraft in orbit around the body. Altitudes have been obtained by sending a radar or laser pulse from the spacecraft to the surface, and timing the interval for the echo to arrive back at the spacecraft, as illustrated in Figure 6.1. The distance d of the spacecraft from the surface is then $c\Delta t/2$ where c is the speed of light and Δt is the round-trip time of the pulse. One way to obtain the altitude from d is to measure the corresponding distance r of the spacecraft from the centre of mass of the planetary body. This distance can be obtained from the spacecraft's orbit. The altitude of the point on the surface with respect to the centre of mass is then $(r-d)$. From the orbit we also know from where on the surface the pulse was reflected, and so the topographic map is built up.

Another technique using radar is particularly useful in low orbit. This is **synthetic aperture radar**, in which a series of radar pulses is sent off below or to one side of the spacecraft, as in Figure 6.1. One advantage in looking sideways is that a left–right ambiguity is avoided. Each pulse illuminates a patch of ground that is much larger than in altimetry. The reflected pulse consists of echoes from every point in the patch, and these points can be distinguished by the different round-trip times and by the changes in wavelength due to the Doppler effect (Section 2.1.2). Points ahead of the spacecraft return shortened wavelengths, and points behind the spacecraft return lengthened wavelengths. Each point in a particular patch is illuminated by several successive pulses as the spacecraft moves in its orbit, and this greatly increases the spatial resolution. Synthetic aperture radar provides a three-dimensional image, and so the surface topography is obtained.

Though the distance $(r-d)$ reveals surface morphology, it is not the best way of specifying altitude. It is *differences* in altitude that are important, and to emphasise these differences we need to define a zero of altitude that lies within or close to the observed range. In the case of Venus, the zero chosen is the mean equatorial radius of 6051.9 km. The surface of zero altitude on Venus is thus a sphere.

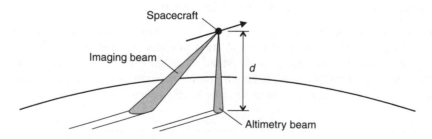

Figure 6.1   Pulse altimetry and synthetic aperture radar.

Though the choice of a sphere seems pretty obvious, it is not sensible for the Earth and many other bodies. This is because these bodies are not as spherical as Venus, but are more flattened by their rotations: the faster the rotation, the greater the flattening; Venus rotates slowly, and so is not appreciably flattened. Consider a rotationally flattened planet in the idealised state of hydrostatic equilibrium, as in Figure 6.2. If a sphere centred on the centre of mass were used as the zero of altitude, the equator would be higher than the poles. Yet there is an important sense in which it is *not* downhill from equator to pole: that is, a plumb line fixed with respect to the rotating surface would hang perpendicular to the surface. Thus, for a real planetary body rotating sufficiently fast to be appreciably flattened, the natural choice of zero altitude is the surface that the body would have were it in hydrostatic equilibrium.

We could calculate the exact shape and size of this ideal surface. It is, however, easier to use any fluids that are widespread at the surface of the planetary body. A fluid flows until it is in hydrostatic equilibrium. In the case of the Earth, we have the oceans ('hydrostatic' derives from 'stationary water'). For zero altitude we could select any depth of given pressure in the water, but the surface is an obvious choice, so mean sea level is defined as having zero altitude on Earth.

☐ Why the 'mean' in mean sea level?

Sea level changes with tides, winds, atmospheric pressure, currents, and the amount of sea ice, so we have to average out these effects. As far as we know, the Earth is the only planetary body with liquid covering most of its surface, but several rotationally flattened bodies are totally covered by a different sort of fluid – an atmosphere. The zero of altitude can then be defined by some value of atmospheric pressure. In the case of the giant planets, the choice is 10^5 Pa – close to mean atmospheric pressure on the Earth's surface.

In the case of Mars, zero altitude has been defined as where the mean atmospheric pressure is 610 Pa – the triple point pressure of water (Section 4.4.3). Recently, a close alternative has become preferred, based on the gravitational field of Mars. From this field a quantity can be obtained called the **gravitational potential**. This is the energy required to take unit mass from a point in space outside a mass M to infinity, assuming no other mass is present. It has the advantage that, unlike the field, it has only a magnitude, and not a direction, which makes it an easier quantity to deal with. For a point outside a spherically symmetric planet, this potential is given by $-GM/r$, where M is the mass of the planet and r is the distance from its centre

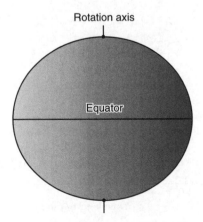

Figure 6.2 A rotationally flattened planetary body in hydrostatic equilibrium.

(compare equation (4.6)). The minus sign is because of the convention to take the gravitational potential to be zero at $r = \infty$; as the distance is reduced, so is the gravitational potential. Departures from spherical symmetry add extra terms, analogous to those in the case of gravitational field. Surfaces exist on which the gravitational potential is everywhere the same, except for local variations. One such equipotential surface is used to define the zero of altitude on Mars, in particular the one that has an average value of r at the equator of 3396.0 km.

Some bodies, such as the Moon, have surfaces that are not closely approximated by a sphere, nor dominated by the effects of rotational flattening; nor do they have an atmosphere, nor a well-known gravitational potential. In this case zero altitude is defined by what is called a reference ellipsoid. This is a distorted sphere characterised by three different radii mutually at right angles.

Question 6.1

Saturn's large satellite Titan rotates slowly and has a massive atmosphere. How could you define its zero of altitude?

6.1.2 Analysis of Electromagnetic Radiation Reflected or Emitted by a Surface

The radiation from the surface of a body includes thermal emission and reflected solar radiation. The radiation can by analysed by the techniques of photometry and reflectance spectrometry outlined in Section 3.1.6 in relation to asteroids. Just as for asteroids, each technique yields information about the surface composition and surface roughness. For example, Figure 6.3 shows the reflectance spectra of Europa and of coarse-grained water ice. The correspondence is close enough to conclude that water ice is the dominant constituent of Europa's surface.

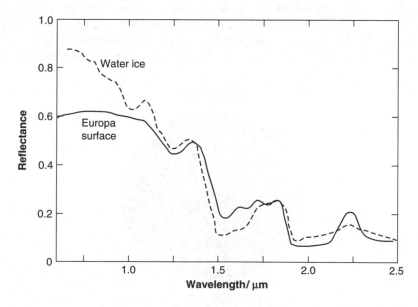

Figure 6.3 Reflectance spectrum of Europa and of coarse-grained water ice.

With radar we provide our own illumination, at wavelengths in the approximate range of 8–700 mm (microwaves). However, radar data are not easy to interpret. The strength of the reflection depends on the angle at which the incident pulse strikes the surface, on the roughness of the surface at the scale of the wavelength used, and on the composition of the surface down to a depth of the order of 10 wavelengths. On the other hand, further information can be obtained if the radar pulse is circularly polarised. Circular polarisation is possible in any wave in which the oscillatory motion is perpendicular to the direction of travel, i.e. if it is a transverse wave. The S wave in Figure 4.7 is transverse, and so too is electromagnetic radiation. A transverse wave is circularly polarised if the oscillations rotate around the direction of travel of the wave to create something like a corkscrew. For radar, if the pulse sent is circularly polarized, then the echo will consist of a component rotating in the same direction as that sent, and a component rotating in the opposite direction. The ratio of the strength of these two reflected components provides further information on texture and composition.

X-ray fluorescence spectrometry reveals the chemical elements that are present. In fluorescence the electrons in an atom are excited (raised in energy) by incoming radiation or particle bombardment. As the electrons return to their initial orbits they emit electromagnetic radiation at wavelengths characteristic of the atom. X-ray wavelengths are particularly useful, and have been observed from spacecraft, such as SMART 1 (Table 4.1), which has detected X-rays from fluorescence caused by the X-rays in solar flares. **Gamma ray fluorescence spectrometry** is an analogous technique, the rays usually coming from the atomic nucleus.

In **neutron spectrometry** the energy spectrum of neutrons emitted by a substance, usually as a result of bombardment, is used to investigate chemical composition.

6.1.3 Sample Analysis

The most direct way of establishing surface composition is to examine samples from known locations. Such samples can also be radiometrically dated. The Earth has, of course, been extensively sampled, and from volcanic activity we even have samples of the upper mantle. For the Moon, we have some meteorites, plus samples brought back to Earth by various missions or analysed *in situ* by landers. In the case of Mars we have *in situ* analyses plus a few tens of meteorites of very probable Martian origin, and for Venus we have *in situ* analysis at four sites. Meteorites and micrometeorites provide samples of asteroids and comets. For all the other bodies in the Solar System we have no samples at all. Samples are analysed by a great battery of chemical and physical techniques, but the details are largely beyond our scope.

Question 6.2

Venus is covered in clouds that are opaque to visible and IR radiation. Describe how a spacecraft above the clouds could determine the surface morphology and composition.

6.2 Processes that Produce the Surfaces of Planetary Bodies

There are many processes that have produced the surfaces of planetary bodies. Here we shall concentrate on those that are widespread and of particular importance. Most of the examples will be for rocky materials, though most of the processes act on icy materials too. From the effect of these processes we can learn much about the surface that has been exposed to them. This, in turn, can tell us much about the interior.

First, we shall see how surfaces are emplaced. Then we shall see how they are subsequently modified.

6.2.1 Differentiation, Melting, Fractional Crystallisation, and Partial Melting

Differentiation of a planetary body was discussed in Section 4.5.1. Here we concentrate on the formation of the crust from the upper mantle. The composition of the crust depends on the composition of the materials from which the planetary body formed.

Consider first those planetary bodies dominated by silicates plus iron or iron-rich compounds. Either the body will have an iron-rich core from the start, or one will start to grow by differentiation as the interior temperatures rise. In either case there will be a silicate mantle more or less rich in iron. In Section 5.1.1 you saw that in the case of the Earth the mantle is predominantly peridotite, which consists of the minerals olivine, $(Mg, Fe)_2SiO_4$, and pyroxene, $(Ca, Fe, Mg)_2Si_2O_6$, where the pyroxene metals can less commonly be Na, Al, or Ti. Table 6.1 lists these minerals, along with some others we shall shortly encounter.

Melting and fractional crystallisation

Consider a peridotite mantle in the late stage of accretion of a planetary body. It is possible that during this stage the uppermost mantle becomes completely molten. A magma ocean is then said to have formed, **magma** being wholly molten material (rocky here, but it could be icy). The ocean will consist of a mixture of substances and, as long as no crystals form, the

Table 6.1 Important igneous rocks and minerals, with their locations in the Earth and Moon as examples

Rock	Mineral content (major components)	Where found in the Earth and the Moon
Peridotite	Pyroxene + olivine	Earth: mantle
Basaltic–gabbroic rocks		
Basalt (extrusive)	Feldspar + pyroxene	Earth: oceanic crust Moon: mare
Gabbro (intrusive)	Feldspar + (Fe, Mg)-rich silicates such as pyroxene	Earth: oceanic crust Moon: mantle (with olivine)
Anorthositic gabbro/anorthosite (intrusive)	Calcium-rich feldspar	Moon: highlands
Granitic–rhyolitic rocks		
Granite (intrusive)	Feldspar + quartz	Earth: upper continental crust
Rhyolite (extrusive)	Feldspar + quartz	Earth: upper continental crust
Andesite	Intermediate between basaltic–gabbroic and granitic–rhyolitic	Earth: continental crust

Pyroxene $(Ca, Fe, Mg)_2Si_2O_6$ (rarely Na, Al or Ti in the brackets).
Olivine $(Mg, Fe)_2SiO_4$.
Feldspar $(K, Na, Ca)AlSi_3O_8$.
Quartz SiO_2 (a particular crystalline form of silica).

random thermal motions keep the magma well mixed. The ocean gradually cools and the first minerals begin to crystallise, with a well-defined chemical composition. Among these will be silicates rich in magnesium and iron, such as olivine. These are dense silicates and so tend to sink downwards. Other minerals that crystallise at the comparatively high temperatures at this stage are calcium–aluminium-rich silicates that are rich also in silicon and oxygen. Feldspars, $(K, Na, Ca)AlSi_3O_8$, are an important example. Such silicates have comparatively low densities and so tend to float upwards. Pyroxenes, $(Ca, Fe, Mg)_2Si_2O_6$, also float upwards. This mixture of feldspars and pyroxenes constitutes **basaltic–gabbroic rocks** (Table 6.1). The process of separation by crystallisation is called **fractional crystallisation**. Note that different elements partition differently between different minerals.

The formation of feldspars, particularly those rich in calcium, is moderated by the effect of pressure. At pressures in excess of about 1.2×10^9 Pa, calcium and aluminium crystallise within denser minerals that do not float upwards. This means that feldspar would not emerge from depths greater than those at which the pressure reaches 1.2×10^9 Pa. Such pressures are reached at depths ranging from 40 km in the Earth to 250 km in the Moon, so the larger terrestrial bodies would have been more vulnerable than the smaller bodies to pressure moderation of feldspar formation.

Complete differentiation of the uppermost mantle might well not occur. Partial differentiation would yield a surface layer rich in feldspar, plus other silicates rich in iron and magnesium, notably pyroxene. Nevertheless, in all cases the formation and subsequent cooling of a magma ocean would result in the surface layer being depleted in iron and magnesium, and enriched in silicon, oxygen, and in metals such as calcium and aluminium. The upper mantle would then consist of peridotite correspondingly somewhat depleted in elements such as calcium and aluminium. A crust–mantle distinction would thus be established, as in Figure 6.4.

Partial melting

Fractional crystallisation is not the only way of creating a crust. This can also occur through the partial melting of a solid outer mantle consequent upon a rise in internal temperatures, or a reduction in pressure. **Partial melting** can occur whenever there is a *mixture* of minerals. As the temperature rises, or pressure falls, there will come a point where a proportion of the mixture will melt, such that the composition of the molten material is different from that of the original solid mixture.

☐ Why does pressure reduction promote melting (see Figure 4.11)?

This is because when the pressure is reduced the melting point of a substance is also reduced (except for water – see Question 4.7).

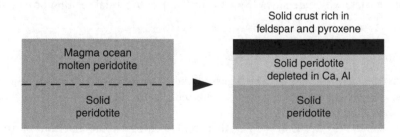

Figure 6.4 Formation of a crust by differentiation in a peridotite mantle.

The onset of partial melting is marked by the emergence of a liquid of distinct composition. For example, consider an upper (solid) mantle with a peridotite composition, subject to (local) warming. The peridotite will typically consist of three components: olivine and two types of pyroxene, namely orthopyroxene and clinopyroxene, that differ in their calcium content. As the temperature rises it reaches what is called the eutectic temperature of the peridotite, which is lower than the melting temperatures of the components. At this temperature a liquid appears of a specific basaltic–gabbroic composition. Regardless of the proportions of olivine and the two sorts of pyroxene in the peridotite, the composition of the liquid (and its eutectic temperature) is fixed until all of one of the three components has melted. Before this occurs, the liquid, being less dense than its surroundings and therefore buoyant, will rise into a cooler environment. It then solidifies to form basaltic–gabbroic rocks – notably richer in silicon and oxygen than peridotite, as pointed out earlier.

In the case of the Earth, the olivine in the upper mantle peridotite will be rich in magnesium. Indeed, the upper mantle contains only about 8% iron by mass. The partial melting does result in some concentration of iron into the melt, and so the Earth's crust is slightly richer in iron than the upper mantle. Note that if the Earth had a magma ocean and an early crust, this would long ago have been recycled into the mantle by plate tectonics (Section 8.1.2), and that the crust today is derived from partial melting of the upper mantle. Though Venus has little by way of plate tectonics it is thought that the whole crust is recycled every few hundred million years, in which case any crust from a magma ocean would also have been recycled (Section 8.2.7). In the case of the Moon, the maria infill is basaltic, having been derived by partial melting of the mantle (Section 7.1.2).

Partial melting also occurs on the icy satellites of the giant planets. For example, in the case of a mixture of water ice with various ammonia hydrate ices (e.g. $NH_3.H_2O$) there is a eutectic temperature of 176 K at a pressure of about 10^5 Pa, where a liquid consisting of two parts H_2O and one part NH_3 emerges. The melting temperatures of H_2O and NH_3 at about 10^5 Pa are 273 K and 195 K respectively.

Once a crust is established, further partial melting can occur within the crust to produce additional differentiation. The outcome is more than one crustal type, and a rich variety of surface rocks. Table 6.1 gives a few examples and, by way of example, where they are found in the Earth and the Moon – more on this in Chapters 7 and 8.

The discovery at the surface of a planetary body of materials that could have resulted from partial or total melting is evidence that such melting has occurred in the past. This indicates that interior temperatures were sufficiently high to cause the melting.

Question 6.3

In one sentence, state why pure silica (SiO_2) is not subject to partial melting or partial crystallisation.

6.2.2 Volcanism and Magmatic Processes

Volcanism covers all processes by which gases, liquids, or solids are expelled from the interior of a planetary body into the atmosphere or onto the surface. On icy worlds the processes are collectively called **cryovolcanism**.

Volcanism and cryovolcanism start with partial melting in the interior, of rocky materials and icy materials respectively. The magma, being less dense on average than the surrounding materials, finds its way to the surface through fissures. En route, rocky magma will dissolve volatile substances such as water, and the magma might partially freeze as it approaches the surface. Volcanism and cryovolcanism lead to further chemical differentiation. For example, **granitic–rhyolitic rocks** (Table 6.1) are richer in silicon and oxygen than basaltic–gabbroic rocks.

The exact composition of the magma, its temperature, and volatile content determine what is called the *style* of volcanism. There are two extreme styles. In explosive volcanism the eruption is very violent, because the magma is viscous and rich in dissolved volatiles. Rocky materials, rock fragments, and ash are erupted at high speeds. By contrast, in effusive volcanism the volatile content is low, or has become low by the time the magma reaches the surface, and the magma also has a low viscosity. Consequently there is a surface flow of molten rock (or icy materials), called **lava**. Viscosity (which depends on composition and temperature) and volatile content are continuous and somewhat independent variables, so there are mixed styles too, including volatile-rich low-viscosity magma that erupts explosively yet leads to lava flows. The details of each style are influenced by the rate of eruption and by the gravitational field. In explosive volcanism the density of any atmosphere and the winds help to determine the spatial distribution of the products.

Given so many different factors, it is not surprising that a wide range of volcanic features is found. Here are some examples. The 'classic' volcanic cone (Figure 6.5(a)) is the result of explosive volcanism where the rate of eruption is modest. At modest eruption rates and low viscosities, a sequence of separate lava flows can create **shield volcanoes** (Figure 6.5(b)), which derive their name from their resemblance to a warrior's shield with a central boss. Less familiar are vast volcanic plains created by low-viscosity lavas flowing at high rates from channels and tubes that radiate out from a vent or fissure.

We need to distinguish volcanic craters from impact craters. Many volcanoes have summit craters (Figure 6.5(a), (b)) and these are called **calderas**. There are also volcanic pits and depressions that do not sit on mountains (Figure 6.5(c)). Some of these forms bear a superficial resemblance to impact craters, but in most cases a more careful morphological examination will reveal their volcanic origin. You can get some sense of this if you compare Figure 6.5 with Figure 6.10.

Sometimes, magma moving up through a fissure does not reach the surface, but spreads out sideways into other fissures and solidifies, often melting adjacent crust and incorporating it into the melt. Erosion can subsequently reveal these **intrusive rocks** as they are called. Lavas and products of explosive volcanism give rise to **extrusive rocks**. Widespread examples of extrusive–intrusive pairs are basaltic–gabbroic rocks, basalt being the extrusive form and

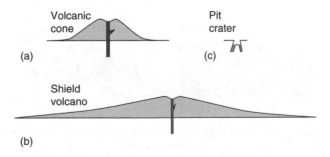

Figure 6.5 (a) Volcanic cone. (b) Shield volcano with summit crater. (c) Volcanic pit.

gabbro the intrusive form, and granitic–rhyolitic rocks, where the extrusive and intrusive forms are, respectively, rhyolite and granite. The extrusive and intrusive forms have much the same mineral content, determined by partial melting. Thus, basalt and gabbro are each mixtures of the minerals feldspar and pyroxene, and rhyolite and granite are each mixtures of feldspar and quartz. The intrusive forms are coarser grained, i.e. have larger crystals. This is because they solidified more slowly in their underground environments than did the rocks extruded onto the surface.

Rocks produced from magma, whether intrusive or extrusive, constitute what are called **igneous rocks**. Table 6.1 lists some important igneous rocks and their corresponding mineral content.

Question 6.4

Why could you conclude from its location that the summit crater in Figure 6.5(b) is unlikely to be of impact origin? How could you confirm your conclusion by surveying the composition of the surface (assume a rocky world), and any layering around the crater perimeter?

6.2.3 Tectonic Processes

Tectonic processes are those that cause relative motion or distortion of the lithosphere. They derive their name from the Greek word for carpenter – *tekton*. Volcanism can be associated with tectonic processes, but this need not be the case. Tectonic features can vary in size from a few kilometres to planetary scale.

Faults are a common tectonic feature. Figure 6.6 illustrates three of the many kinds of fault. The normal fault in Figure 6.6(a) usually arises from stretching of the lithosphere – the lithosphere is in tension and therefore cracks down to some depth, followed by relative vertical motion. Some rift valleys, or **grabens**, are the result of slumping between two parallel normal faults (Figure 6.6(b)). The Great Rift Valley that extends for about 6000 km from Syria to southern Africa is a huge example. It is the longest rift valley on Earth, and has a typical width

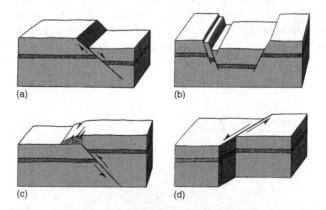

(a)   (b)

(c)   (d)

Figure 6.6   Three kinds of fault. (a) A normal fault. (b) A graben (rift valley), made up of two normal faults. (c) A thrust fault. (d) A strike–slip fault. (From *Moons and Planets, An Introduction to Planetary Science* 3rd Edition, by Hartmann, 1993, Reprinted with permission of Brooks/Cole, a division of Thomson Learning: www.thomson rights.com, Fax 800 730–2215)

Figure 6.7   Part of the eastern wall of the Great Rift Valley. (The author)

of about 50 km. Part of one of its walls is shown in Figure 6.7. Normal faults and grabens are common not only on the Earth, but on other bodies, such as Mercury, the Moon, Venus, Mars, and some of the large satellites of the giant planets.

A thrust fault, or reverse fault (Figure 6.6(c)) usually results from lithospheric compression. A strike–slip fault (Figure 6.6(d)) can arise from compression or tension, but is characterised by motion that is predominantly horizontal and not vertical as in the other examples. Thrust faults and strike–slip faults are seen on the Earth and on many other planetary bodies. Compression can also cause bending and upthrust of the lithosphere, in extreme cases resulting in mountains, e.g. the Maxwell Montes range on Venus (Section 8.2.6).

The Earth seems to be unique in that its lithosphere is divided into a global system of many plates in motion with respect to each other, predominantly laterally. This lateral tectonics produces mountain chains such as the Himalayas in Asia, the Alps in Europe, the Andes in South America, and so on. This dominant aspect of terrestrial tectonics, plate tectonics (Section 5.1.1), is the subject of Section 8.1.2. Other planetary bodies have lithospheres that are not divided into plates, and consequently tectonic motion is predominantly vertical.

The extent to which a lithosphere has experienced tectonic processes depends on its composition and its thermal history. If the lithosphere is thick, or rigid, few tectonic features will be present, and those that are will be mainly due to vertical motion. Conversely, the extent and nature of tectonism provides important information on the thermal evolution of the body. Further discussion of individual planetary bodies is in Chapters 7 and 8.

☐   Would you expect icy lithospheres to be subject to tectonic processes?

Yes indeed, though their expression will reflect differences between the behaviour of ices and rocks. For example, except at very low temperatures, icy materials are more plastic than rocky materials and have much less strength. This will rule out the building of high icy mountains.

6.2.4   Impact Cratering

We now shift emphasis away from processes that emplace surfaces to those that modify them, though there is no sharp division between emplacement and modification. The most ubiquitous process in the Solar System is impact cratering.

You will recall from Chapter 3 that interplanetary space is populated with large numbers of small bodies in orbits that can intersect those of planetary bodies. It is therefore to be expected that the surfaces of planetary bodies will bear the scars of individual impacts. Indeed they do, in the form of impact craters. The lunar craters are a familiar consequence of impact (Figure 6.8(a)), and 174 impact craters, or structures clearly produced by impact, have been identified on Earth, the best known being Barringer Crater in Arizona (Figure 6.8(b)).

An **impact crater** is produced when a projectile strikes the surface of a planetary body with sufficient kinetic energy to excavate a hole. The kinetic energy of the projectile is given by

$$E_k = mv^2/2 \tag{6.1}$$

where m is the mass of the projectile and v is its speed with respect to the surface. The value of v depends on the relative motions of the projectile and the planetary body, and on any changes in projectile speed during the encounter, notably through its acceleration by the planetary body's gravity. Orbital speeds and the acquired speeds are each about $10 \, \text{km s}^{-1}$, to an order of magnitude. Therefore, to an order of magnitude v is $10 \, \text{km s}^{-1}$. In the case of the Earth, the impact speeds are in the range 5–$70 \, \text{km s}^{-1}$. Such high impact speeds mean that even a small mass can excavate a considerable crater.

☐ What is the kinetic energy of a small body with a mass of $10^8 \, \text{kg}$ and with a speed of $15 \, \text{km s}^{-1}$?

From equation (4.1) the kinetic energy is $E_k = 10^8 \, \text{kg} \times (15000 \, \text{m s}^{-1})^2/2 \approx 10^{16}$ joules. This is the energy that would be liberated by the explosion of about 3 million tons of TNT, and it is also about the energy that was required to excavate the 1.2 km diameter Barringer Crater. This crater was excavated by an iron meteorite, and so the radius of the meteor at the Earth's surface would have only been about 15 metres if it had been travelling at $15 \, \text{km s}^{-1}$.

Figure 6.9 illustrates the stages in the formation of an impact crater. In stage 1 (Figure 6.9(a)) the projectile has struck the surface and has penetrated perhaps only 2–3 times its diameter

(a)

Figure 6.8 (a) The lunar impact crater Copernicus, about 90 km in diameter. (NASA/NSSDC AS17-151-23260) (b) Barringer Crater in Arizona, about 1.2 km diameter. (D R Roddy, IcT, The United States Geological Survey)

(b)

Figure 6.8 (Continued)

before being brought to rest. It does not get far because the speed of the projectile at impact exceeds the speed of seismic waves in the surface material, which at most will be about $4\,\mathrm{km\,s^{-1}}$. Therefore, the material ahead of the projectile gets no 'advanced warning' of the impact, and so cannot move away. This leads to the piling up of a sudden and enormous compression – a **shock wave**. The pressures generated exceed the strength of the materials by factors of 10^3–10^4, and so the surface materials and the projectile are highly fractured. At the same time nearly all of the kinetic energy of the projectile goes into heating the projectile and its immediate surroundings. Most of this material is vaporized and this produces a violent explosion. During stage 1 surface materials are ejected at high speed.

In stage 2 (Figure 6.9(b)) shock waves spread out from the site of impact, fracturing and melting subsurface layers and throwing huge quantities of material outwards. The immediate result (stage 3, Figure 6.9(c)) is a hole with a volume greatly exceeding the volume of the projectile itself. The hole is rimmed with the distinguishing characteristic of overturned rock strata. Subsequently, modifications can occur by a variety of processes, as you will see.

The direction of the projectile has little effect on the shape of the crater unless it impacts at grazing incidence – the explosion and the shock waves spread out uniformly from the point of impact. Craters are thus roughly circular, unless produced by a near grazing impact, when the crater will be elongated in the direction in which the projectile was travelling.

The volume of a crater is roughly proportional to the kinetic energy of the projectile. It also depends on the gravitational field of the planetary body, and on the strength and density of its surface layers. For example, for a surface of given strength and density, the crater size diminishes as the gravitational field increases. This is because the surface materials are held down more strongly. Also, there is a big difference between rocky surfaces and icy surfaces, partly because of the lower densities of icy materials, and partly because of the lower strength of icy materials, unless they are far below their melting points. For a projectile of specific kinetic energy and a specific surface gravitational field, a crater produced in an icy surface will be roughly double the diameter of one produced in a rocky surface.

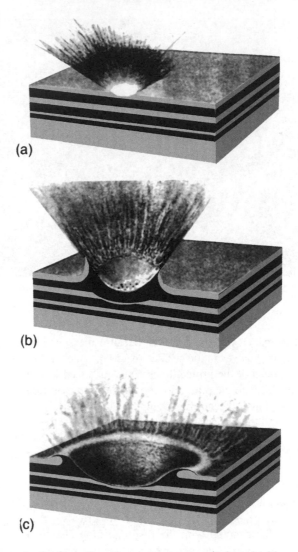

(a)

(b)

(c)

Figure 6.9 Three stages in the formation of an impact crater. (Adapted with permission from Kluwer, from Figure 3.1 of *Planetary Landscapes*, R. Greeley, Chapman and Hall, 1994)

The morphology of a crater depends on its size. The reasons are complicated, but the outcome is illustrated in Figure 6.10, where the diameter ranges are for the Moon. Small craters are simple, bowl-shaped depressions (Figure 6.10(a)). The bowl is flatter at larger diameters, and at yet larger sizes a central peak will be present (Figure 6.10(b)), probably the combined result of floor rebound immediately after the excavation is complete and the reflection of shock waves from any deep interfaces. At about this size there might also be slumping of the crater walls soon after the impact, forming terraces. In yet larger craters the central peak is replaced by a cluster of peaks, or by a ring of peaks, called a peak ring (Figure 6.10(c)). The largest craters are complex multi-ringed structures, called multi-ring basins (Figure 6.10(d)), that are probably the result of a combination of slumping and waves of surface motion. The larger craters can

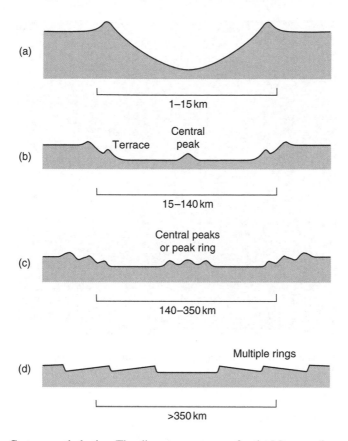

Figure 6.10　Crater morphologies. The diameter ranges are for the Moon, and are approximate.

become flooded by lavas, though this might not happen for millions of years. The larger craters are also modified by isostatic adjustment (Section 4.1.5).

▢　What effect do you think this will have?

A crater is a deficit of material, so isostatic adjustment will cause uplift within the crater at the surface (and horizontal motion deeper down). This reduces the depth-to-diameter ratio to 1/20 or less. Craters of all sizes are continuously subject to erosion and partial infill by some material or other.

The actual size ranges of the different types of crater depend on the gravitational field and on the strength and density of the surface layers, and so vary from body to body. For example, on the Earth, which has a rocky surface like the Moon but a larger gravitational field, central peaks occur in craters as small as 6 km diameter. On the icy surface of Ganymede, which has a comparable gravitational field with the Moon, they occur in craters down to about 10 km in diameter.

As well as the crater itself, there will also be evidence of the ejected surface material, notably rays, ejecta blankets, and secondary craters, i.e. craters produced by the impact of ejecta thrown out by the primary impact (Figure 6.11). These features, like the crater itself, are subject to modification through erosion and through deposition of material.

Figure 6.11 Rays, secondary craters, and an ejecta blanket from the lunar crater Euler (27 km in diameter). (NASA/NSSDC AS17-2923)

6.2.5 Craters as Chronometers

Impact craters can be used to determine the age of a surface. The older a surface, the greater the number of impacts per unit area it will have accumulated. Therefore, a heavily cratered surface on a planetary body must be older than a lightly cratered surface. Figure 6.12 shows a heavily cratered region on the Moon adjacent to a much more lightly cratered region. Clearly the latter is the younger surface.

For the Moon as a whole, Figure 6.13 shows the number densities of craters of different sizes, i.e. the number of craters per unit area in defined ranges of diameters. These densities are known with good precision. Two graphs are shown, the one averaged over the lunar highlands, the other averaged over the lunar maria, which are the smooth, dark areas in Plate 7. ('Maria' is the plural of 'mare' (ma-ray), Latin for 'sea', though we now know that there are no seas on the Moon.) The graphs clearly show that for the Moon as a whole the maria are younger than the highlands. Regional studies show that even the oldest mare is younger than the youngest highland area. Both graphs show that the smaller the craters, the greater their number density. Therefore, the smaller the kinetic energy of a projectile, the more numerous they are. Among the great range of projectile masses, impact speed is uncorrelated with mass and so, broadly speaking, Figure 6.13 shows that massive projectiles in the Solar System have been fewer in number than less massive ones. This is in accord with our understanding of the size distribution of planetesimals (Section 2.2.3) and with the subsequent evolution of the remnant population.

The shapes of the curves in Figure 6.13 are sufficiently similar to indicate that the same population of projectiles is responsible for both. These shapes are consistent with impacts by projectiles with a distribution of sizes that would be expected if they came from the main belt of

Figure 6.12 Lunar craters in the highlands and the adjacent Oceanus Procellarum. The frame is about 160 km wide. (NASA/NSSDC AS15-2483)

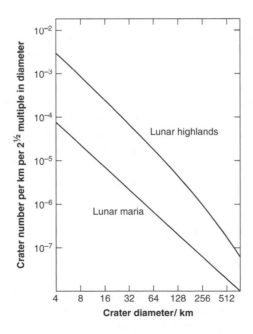

Figure 6.13 Number densities of lunar craters of different sizes, averaged over the lunar highlands and over the lunar maria.

asteroids. That this has long been the case is indicated by data from lunar surfaces of different radiometric ages (see below), which show that the size distribution has not changed much over the past 4000 Ma. A significant cometary contribution cannot be ruled out, particularly if the size distribution of comets has always resembled that of the main belt asteroids – the data on comets are too sparse to tell. Bombardment on all the terrestrial planets seems to have been caused by the same type of population.

In what *sense* is the age of a surface indicated by craters? On a given planetary body, one surface will have a lower crater density than another because of some resurfacing event that obliterated some or all of the existing craters. An obvious example is a flood of lava. Another is melting of the surface. In such cases, the age of the surface indicated by its crater density is the time in the past when such resurfacing occurred. Craters can also be obliterated by various erosional or depositional processes. These will obliterate small craters much more rapidly than large ones. This is a different sort of resurfacing, continuous in time rather than concentrated near to a particular time, and it is important to avoid confusing its effects with those that result from lava floods and the like. For example, suppose that erosion is more powerful near the poles of a planet than near its equator. The crater density at the poles, particularly for small craters, will consequently be lower than at the equator for surfaces of equal age.

☐   How could this be misinterpreted?

This could be misinterpreted to mean that the polar regions had been resurfaced by lava or by melting more recently than surfaces elsewhere. It is therefore common to exclude from counts those craters with diameters of less than a few kilometres.

Further complications arise from the need to avoid counting craters of volcanic origin and secondary craters. Fortunately, most volcanic and secondary craters have morphologies and spatial distributions that betray their origin. Again it is wise to exclude small craters, many of which will be secondaries.

A final complication is the phenomenon of saturation. A surface becomes saturated with impact craters when further impacts, on average, obliterate as many craters as they create. It is then not possible to distinguish between older and younger saturated surfaces on the basis of impact crater densities alone. The most heavily cratered parts of the lunar highlands are saturated, as might be regions on some of the icy–rocky satellites, though definitively saturated surfaces in the Solar System are rare. One measure of saturation is the degree of randomness of the spatial distribution of craters. Simulations show that as an area approaches saturation, the random distribution that characterises subsaturated areas becomes more uniform as sparsely cratered subareas acquire more, and subareas near to saturation are little changed.

With care, crater densities can be used to place the various surfaces on a single planetary body in the order of their age since some widespread resurfacing. This can also be done for the surfaces on *separate* planetary bodies, provided that we know of any differences in

- the bombardment history of the two bodies;
- their surface properties;
- rates of erosion and deposition.

Unfortunately, such differences are rarely well known, particularly the first. The bombardment history depends (among other things) on the local flux of potential projectiles. Differences between local fluxes are poorly known. For example, Mars is close to the asteroid belt, and so will surely have been more heavily bombarded than the Moon. But estimates of exactly how much more are highly uncertain. One estimate is that a (unsaturated) surface of given age on Mars has received twice the number of impacts as a surface of similar age on the Moon. If this is the case then we can estimate (with appreciable uncertainty) whether the surface on Mars is older or younger than a particular surface on the Moon.

To obtain *absolute* ages we need to know in absolute terms the cratering history of at least one body, and how to apply the data to other bodies. Only for the Moon do we have the absolute ages for surfaces with *widely* different crater densities. These ages have been obtained by radiometric dating, and they have enabled us to deduce the rate of impact cratering on the Moon to about 4000 Ma ago. The crater densities and corresponding absolute surface ages are shown in Figure 6.14 for the total number of craters larger than 4 km in diameter. (Smaller craters are excluded to avoid the problems noted earlier.) Terrestrial craters are few because much of the surface is young, and because erosion and deposition are very active. The limited data for the Earth have been used to help get the curve at young ages (after correction for Earth–Moon differences). The uncertainties in the data are not large, and on the y axis in Figure 6.14 correspond to a factor, very roughly, of 1.5 either way, not huge on a logarithmic scale. The curve for times preceding 4000 Ma is shown dashed, because it is an extrapolation, older lunar surfaces being saturated.

Figure 6.14 also shows the impact rate in the Earth–Moon system inferred from the crater densities. These densities are not nearly as well known as the crisp line in Figure 6.14 indicates, and so the impact rate is correspondingly uncertain. Nevertheless, it is certain that at the earliest times the rate was very high and then declined rapidly. Before about 4200 Ma ago two possibilities are shown. In the one there is a monotonic decrease in bombardment, but in the other there is a peak. In both cases the overall rapid decline presumably reflects the final stages of mopping up of interplanetary debris left over from the formation of the Solar System. Different models of these final stages lead to different impact rates at the earliest times. You have seen that the period from the end of accretion about 4500 Ma ago to about 3900 Ma ago

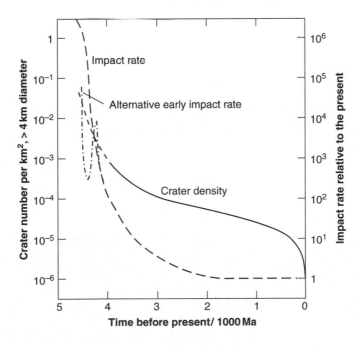

Figure 6.14 Crater densities versus surface age in the Earth–Moon system, and an inferred impact rate.

is called the heavy bombardment. Any peak during the latter part of this time is the *late* heavy bombardment. Not shown in Figure 6.14, because it is not quantified, is evidence of a peak in impact rate on the Moon at 3900 Ma, lasting about 100 Ma, perhaps caused by a comet crashing into the asteroid belt. This peak would be the last gasp of any late heavy bombardment.

In applying the data in Figure 6.14 to other bodies we have the same sources of uncertainty that we encounter in placing surfaces in age order, particularly the differences in local fluxes of projectiles through the ages. This leads to some uncertainty in placing absolute ages on various surfaces on Mercury, Venus, and Mars. The only feature that presumably has roughly the same age everywhere in the Solar System is the decline of the heavy bombardment. Therefore, any near-saturated surface is likely to be older than about 3900 Ma. This reveals an interesting difference between the inner Solar System and the outer Solar System, with the orbit of Jupiter as a rough dividing line. On the most densely cratered surfaces, the relationships between the number densities and crater size are rather different in these two regions, which indicates that the sources of heavy bombardment projectiles were also different. This also seems to have been the case subsequently, and this makes for considerable uncertainty in placing absolute ages on outer Solar System surfaces that postdate the heavy bombardment. The difference in the sources is presumably because the asteroid main belt cannot act as a significant source of projectiles in this far-flung region.

Question 6.5

A plain called Chryse Planitia on Mars has a crater density of 2.2×10^{-4} per km^2 for craters greater than 4 km diameter.

(a) How many such craters are there in a typical area of $10^5 \, km^2$ in Chryse Planitia?
(b) Assuming that data for the Moon can be applied to Mars, estimate how long ago this area was resurfaced. Why is your value an upper limit to the age?

6.2.6 Gradation

Gradation covers three sets of processes:

(1) those by which a surface is disintegrated or eroded;
(2) those that transport the loosened or liberated material;
(3) those that deposit it, perhaps in a different mineralogical form.

The gravitational field of the planetary body tends to cause settling to the lowest available altitude, and therefore gradation tends to level off a landscape.

Disintegration and erosion

You have already met impact cratering. Additionally, disintegration of surface materials can be caused, for example, by seismic waves, expansion and contraction in the day–night thermal cycle, and the expansion and contraction of water in pores and crevices of rocks as it changes from liquid to ice and back again. Material can be loosened by chemical reactions at the surface, particularly if there is an atmosphere, and by the alteration of surface materials through UV irradiation. Material can also be made subject to disintegration by the removal of adjacent material. Erosion is caused by the impacts of micrometeorites and raindrops, by wind-borne dust, and by the action of rivers and glaciers. Note that in this last case water need not be the agent. For example, you will see that on Mars, as well as water, CO_2 ice plays a role, and on Titan CH_4 acts analogously to water on Earth.

Transport – mass wasting

Mass wasting is the downslope movement of loosened materials under the influence of gravity. Material can be loosened by all the processes that cause disintegration and erosion. Once sufficiently loosened, gravity does the rest. The distinctive feature of mass wasting is that transportation is downhill directly from where the material lay to its new location – there is no long-range transportation. One of many possible outcomes is shown in Figure 6.15, where mass wasting has produced the fan-shaped deposits.

Transport – aeolian processes

Whereas mass wasting can occur on the surface of *any* planetary body, aeolian processes require an atmosphere in which winds blow. Wind moves solid particles such as dust and sand, and it can move them over large distances before they are deposited. Small particles tend to be carried in the wind, and larger particles tend to bounce or creep along the surface. Wind-borne particles cause further erosion – Figure 6.16. Wind can also sculpt a sandy surface to create sand dunes, and these can creep over the landscape like slow waves. Accumulations of small particles are called sediments, and deposition from winds is one way of producing them.

Transport – evaporation, sublimation, condensation, precipitation

The more volatile constituents of a surface can be transported by becoming a gas, through either evaporation from a liquid, or sublimation from a solid, followed by motion through the atmosphere through winds or diffusion, to places where the gas condenses, either directly on the surface as frost, or in the atmosphere from where it precipitates. On Earth water is the volatile constituent.

☐ What is it on Mars and Titan?

On Mars it is water and CO_2, and on Titan CH_4 (plus other hydrocarbons).

Transport – processes involving liquids

One of the more familiar results of liquid transport is the terrestrial system of river valleys resulting from surface drainage. The very existence of such systems is testimony to the large quantities of material eroded and moved downstream by the surface run-off of liquid water. Motion of groundwater near the surface and of water deeper underground also results in erosion and transportation. Valley glaciers and ice caps remove huge quantities of material, but rather more sedately. When glaciers and ice caps recede, characteristic landforms are left behind, such as valleys with U-shaped cross-sections.

Lakes and oceans also erode materials, transport, and deposit them, as can be seen in landforms like cliffs and beaches, and in extensive deposits of sediments, particularly on ocean floors. Sediments deposited in water can subsequently be exposed if the oceans recede, or if a shallow lake dries up.

Water moves material not only mechanically, but also chemically by dissolving some or all of the minerals in rocks. Limestone, which is mainly calcium carbonate ($CaCO_3$) and

Figure 6.15 An example of mass wasting, in Wyoming. (T A Jagger, The United States Geological Survey)

Figure 6.16   An effect of wind erosion in Chile. (K Segerstrom, The United States Geological Survey)

dolomite $(CaMg(CO_3)_2)$, is particularly susceptible, especially if the water contains dissolved CO_2. The removal of limestone from under the Earth's surface has produced many magnificent underground caverns.

Deposition

Many transport processes are inextricably linked with deposition, including *all* of the processes outlined above, from the short-range deposition in mass wasting to the long-range deposition of rivers. Deposits borne by fluids and by atmospheres are called sediments, and often form sedimentary rocks, as outlined in the next section.

Question 6.6

List the gradational processes to which a planetary surface is subject if the planet has an atmosphere but no surface liquids.

6.2.7   Formation of Sedimentary Rocks

A sediment, however formed, and wherever it lies, can form a **sedimentary rock**. Chemical cementation of the solid particles is an important part of the process, though modest pressure helps through the consolidation it produces. On the Earth, shale is a particularly abundant sedimentary rock, making up an estimated 4% of the upper 6 km of the Earth's crust. It consists of very fine grains derived from consolidated sediments. These sediments consisted of clays and silts, dominated by clay minerals in both cases but distinguished on the basis of size – clay particles are defined as those smaller than 1/256 mm across, and silt particles as those in the range 1/256-1/16 mm.

Clay minerals are a common outcome of chemical modification of a sediment by water. One of the commonest processes of clay formation on Earth is the chemical decomposition of feldspar, $(K, Na, Ca)AlSi_3O_8$, by water, but other minerals are modified too. Clay minerals are also derived from insoluble minerals in limestone. One example of many clay minerals is montmorillonite. This has a large molecule with the formula $(Al, Mg)_8(Si_4O_{10})_3(OH)_{10}.12H_2O$, which exemplifies the effect of water on silicates not only through the attached water molecules H_2O but also through the presence of the water molecule fragment OH (hydroxyl).

Limestone is another type of sedimentary rock, consisting mainly of cemented carbonate particles. A **carbonate** is a compound that contains the chemical unit CO_3. A common terrestrial example is calcium carbonate, $CaCO_3$.

6.2.8 Formation of Metamorphic Rocks

In addition to igneous and sedimentary rocks, there is just one other major rock type – metamorphic rock. A **metamorphic rock** is an igneous or sedimentary rock that has been modified but not completely remelted. Metamorphosis can result from any combination of raised pressure, raised temperature, or a change in the chemical environment. For example, if shales are subject to a combination of raised pressure and temperature, then slate is a possible outcome. If granite is subjected to high pressure and temperature then granite-gneiss ('nice') is formed, with a prominent banded structure.

Metamorphic rocks can result from magmatism, which raises the temperature of rock next to the magma, and might include modification of the rock by water. The cycling of rock through the interior can also result in metamorphism, through high pressure as well as high temperature. On the Earth this cycling is almost entirely a consequence of plate tectonics (Section 8.1.2), but on other bodies it might be achieved in some other way.

Regardless of how metamorphic rocks are produced, this usually happens beneath the surface, and therefore they have to be transported upwards to be seen.

Figure 6.17 summarises the processes described in Section 6.2. The three open arrows indicate that the processes apply to all three types of rock in the triangle. 'Subduction' denotes any process by which crustal materials are carried downwards, usually by (plate) tectonics. 'Uplift' is self-explanatory. Table 6.2 lists the planets and large satellites along with the processes that dominate their surfaces *today*.

6.3 Summary of Chapter 6

Surfaces are investigated through mapping that establishes the topography. Mapping can be performed using images, altimetry, and synthetic aperture radar. For bodies that are rotationally flattened, the zero of altitude can be defined by a surface of hydrostatic equilibrium, such as mean sea level on Earth. The gravitational potential is used on some bodies. For irregular bodies, a reference ellipsoid is used.

The composition and other characteristics of a surface are investigated by photometry and reflectance spectrometry, through the use of radar, by X-ray and gamma ray fluorescence spectrometry, and by neutron spectrometry Direct analysis of surface samples has only been

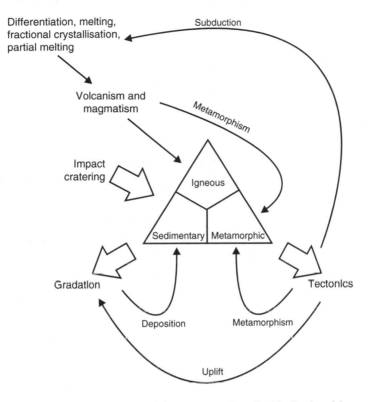

Figure 6.17 A summary of the processes described in Section 6.2.

Table 6.2 Dominant surface processes *today* in planets and large satellites

Planetary body	Dominant surface processes today
Mercury	Impact cratering; dry gradation^a
Venus	$V + M^b$; tectonics; gradation; metamorphic rock formation?
Earth	Plate tectonics; other tectonics; $V + M$; gradation; $S + M^c$
Mars	Gradation; impact cratering; tectonics?
Pluto	Impact cratering; dry gradation; cryo-V+M?
Moon	Impact cratering; dry gradation
Io	V+M; gradation; impact cratering
Europa	Cryo-V+M; cryotectonics; impact cratering; gradation
Ganymede	Impact cratering; dry gradation
Callisto	Impact cratering; dry gradation
Titan	Cryo-V+M; cryotectonics; gradation; cryo-S+M?

^{a} Dry gradation is gradation that does not involve fluids (liquids or gases).
^{b} Volcanic and magmatic processes, including the formation of igneous rocks.
^{c} S denotes sedimentary rock formation, and M denotes metamorphic rock formation.

achieved for the Earth, the Moon, Venus, and Mars. Meteorites and micrometeorites provide samples of asteroids and comets, and also of the Moon and Mars.

There are several different types of process that create and modify the surface of a planetary body:

- melting, fractional crystallisation, and partial melting;
- impact cratering – in which bodies from interplanetary space impact the surface;
- (cryo)magmatism and (cryo)volcanism – all processes by which gases, liquids, or solids are expelled from the interior;
- tectonic processes – all processes that cause relative motion or distortion of the lithosphere;
- gradation – all processes by which material is eroded from a surface, and transported and deposited elsewhere, sometimes with chemical modification, notably the production of clays;
- the formation of sedimentary and metamorphic rocks.

Impact craters can be used to place surfaces on a given body in a sequence of ages: the greater the crater density, the older the surface, i.e. the further into the past it was last resurfaced by lava or by melting. Care has to be taken to allow for gradational effects, and to exclude volcanic craters and secondary impact craters. Absolute ages of surfaces with widely different crater densities have been obtained for the Moon. In comparing one body with another we have to allow for any differences in the bombardment history of the two bodies, their surface properties, and in the rates of gradation.

7 Surfaces of Planets and Satellites: Weakly Active Surfaces

Except for the Sun and the giant planets, which have no surfaces in the generally accepted sense, the surfaces of all bodies in the Solar System are still subject to impact cratering and gradation, but only some of them now experience (cryo)volcanism or tectonic processes to any significant extent, and in this sense are active. This suggests an obvious grouping of the surfaces into two sorts, and this is the basis on which Chapters 7 and 8 are organised. Thus, Chapter 7 describes surfaces that are now weakly active or inactive, and Chapter 8 describes those that are much more strongly active. The bodies with active surfaces are the Earth and Venus – the two largest terrestrial planets – and some of the larger satellites. The bodies with weakly active surfaces range in size from Mars down to the smallest bodies in the Solar System, though we shall concentrate on the larger ones. Broadly speaking, a body has a weakly active or inactive surface because its interior has cooled to the point where its lithosphere is now too thick to allow (cryo)volcanism or tectonic processes to occur.

7.1 The Moon

The Moon is the only body on which we can see surface features with the unaided eye (Plate 7). The dark areas are the maria (singular, mare). These are relatively smooth, and they lie amidst more rugged highland terrain that constitutes most of the lunar surface. The highlands reach up to 16 km above the lowest lying regions, and are dominated by impact craters.

The Moon is in synchronous rotation around the Earth. Therefore one side – the near side – always faces towards the Earth, and the other side – the far side – always faces away from the Earth. However, this does not mean that one face of the Moon never sees the Sun. When we see the Moon as less than full then part of the near side is in darkness, and consequently part of the far side must be in sunlight. With no atmosphere to moderate surface temperature, the temperatures at the equator reach about 400 K at lunar noon only to plunge to about 100 K at lunar midnight. At any point on the lunar surface the average time between successive noons – the lunar 'day' – is 29.53 days. A moment's consideration will show that this is also the average interval between successive new Moons, an interval which is called the synodic month.

With the Moon in synchronous rotation we would see just 50% of the lunar surface if every one of the following conditions was met (you do not need to dwell on these):

Discovering the Solar System, Second Edition   Barrie W. Jones
© 2007 John Wiley & Sons, Ltd

Figure 7.1 Lunar far side (left half) and near side (right half), imaged by the Galileo spacecraft in 1990 en route to Jupiter. (NASA/JPL P37327)

- its orbit around the Earth were circular;
- its rotation axis were perpendicular to the plane of its orbit around the Earth;
- the lunar orbit did not precess about an axis perpendicular to the ecliptic plane;
- the radius of the Earth were negligible (so we would always observe the Moon from the same vantage point as the Earth rotates).

None of these conditions is met, and as a result, from any point on the Earth's surface, the Moon appears to rock slowly to and fro (east–west) and nod up and down (north–south). These are called geometric librations, and they allow us to see 59% of the lunar surface over a period of about 30 years. In addition, because of tidal forces, the lunar rotation rate oscillates, and consequently there is a physical libration that allows us to see a tiny bit more. Nevertheless, about 40% of the lunar surface was hidden from us until 1959, when the spacecraft Lunik III provided us with our first images of the far side. A more recent image of part of the far side is shown in Figure 7.1.

A further consequence of the eccentricity of the Moon's orbit around the Earth and the tilt of its rotation axis, plus the Earth's orbital eccentricity and other effects, is that the lunar 'day' varies slightly in length.

7.1.1 Impact Basins and Maria

Impact craters are the dominant lunar landform. Much of the highlands is near to saturation, indicating great age. The largest impact craters are the impact basins, many of which have subsequently been partially filled to create the maria, which constitute about 17% of the lunar surface. The maria are concentrated on the near side (Figure 7.1). The largest impact basin of all, the South Pole–Aitken Basin, is on the far side near the South Pole. It has a maximum depth of 8.2 km below the reference ellipsoid that defines zero altitude, and a diameter of 2250 km, making it the largest known impact basin in the Solar System. There is very little infill. The impact might have excavated the upper mantle, and image analysis is consistent with the presence of silicates richer in iron and magnesium than those in the crust.

There is plenty of evidence that the maria are partially filled impact basins. The circular nature of the mare boundaries (Plate 7) and arc-shaped mountain ranges within the maria surely indicate a multi-ring impact basin beneath. Further evidence is provided by gravitational field measurements which show that coinciding with many maria there is excess mass – a mass concentration called a **mascon**. One possible cause is the upward bowing of relatively dense mantle immediately after the impact. The bowing would have greatly reduced the mass deficit after excavation, but would still leave a depression. Any subsequent infill of the basin would create the mascon, provided that the lithosphere had by then become too rigid to achieve isostatic adjustment (Section 4.1.5). In some cases, the upward bowing could have been excessive, owing to rebound held in place by a rigid lithosphere, which would create a mascon before any infill.

Studies of impact basins, filled and unfilled, indicate that the Moon has long had a thick lithosphere that has prevented isostatic equilibrium from being achieved.

7.1.2   The Nature of the Mare Infill

There is a good deal of evidence that the mare infill is mainly lava, and not debris from later impacts, nor migrating dust. For example, mare samples have a basaltic composition, in sharp contrast to the pyroxene-poor composition of the surrounding anorthositic highlands (Table 6.1 and Section 7.1.6). Also, shallow channels, called sinuous rilles, snake across the maria – these could be the remnants of lava supply channels or collapsed lava tubes. Some linear rilles might be graben created by the extraction of subsurface magma. More obvious signs, such as volcanoes, are represented by only a handful of small features, and so it is presumed that the upwelling of the lava was mainly through fissures that the lava itself buried – there are terrestrial examples. Had the lava erupted all in one go, the resulting fluid would have filled the basin to the level required by isostatic adjustment and there would be no mascon. It is therefore necessary to suppose that the infill was in a series of sheets, each a few tens of metres thick. There is evidence for such sheets on Mare Imbrium, in the form of scarps on the surface (Figure 7.2).

It is widely believed that the mare lavas are derived from the partial melting of the mantle a few hundred kilometres beneath the lunar surface, the melting being aided by the release of pressure as the mantle material ascended. Detailed studies of large basins of different ages suggest that this infill was available more readily early in lunar history than later, and this indicates that the lithosphere gradually thickened as the interior cooled, making it more difficult for lava to be released after later large impacts. Even so, the magma always took some time to reach the surface. Radiometric dating shows that several hundred million years separated maria basin formation from maria infill.

☐   What other evidence is there that the maria surfaces are younger than the rest of the lunar surface?

The low density of impact craters on the maria surfaces also indicates relative youth. There is thus plenty of evidence for the delay in mare infill necessary to explain the maria mascons.

Figure 7.2 Part of the lunar Mare Imbrium, showing what are probably thin sheets of lava. The frame is about 60 km across the base of this oblique view. (NASA/NSSDC AS-17-155-23714)

7.1.3 Two Contrasting Hemispheres

On the far side there is less infill of impact basins. The second largest of the far side basins, Mare Orientale, about 900 km across, is in the centre of Figure 7.1. It is only slightly filled, and the gravitational field shows a deficit of mass – a 'negative mascon'. The lack of infill on the far side is a possible outcome of the observed higher altitudes of the basin floors there than on the near side, which could have placed the far side basins beyond the reach of magma. The exception is the South Pole–Aitken Basin, which is so deep that the lack of infill indicates significant regional differences in the properties of the crust and upper mantle.

Another striking difference between the near side and the far side is the crustal thickness. This has a near side mean value of about 40 km, but gravitational and topographic measurements, particularly by the Clementine Orbiter (Table 4.1), have shown that the mean thickness on the far side is about 12 km greater than on the near side. On a regional rather than hemispherical scale, the thickness of the crust varies over the approximate range 20–120 km, though it is much thinner than 20 km in the South Pole–Aitken Basin. These regional variations must be due in part to the transport of crustal materials across the surface of the Moon by giant impacts, and this could also explain the difference on the hemispherical scale. Another regional factor with hemispherical consequences might be major variations in the extent of melting of the lunar exterior. Additional possibilities are that erosion by mantle convection early in lunar history moved crustal material from the near side to the far side, or that mantle convection has reincorporated into the mantle more of the near side crust than the far side crust.

7.1.4 Tectonic Features; Gradation and Weathering

Tectonic features include faults and ridges, largely confined to the maria, and some of the linear rilles – these occur in the highlands and on the maria. These features can be explained by a combination of crustal tension early in lunar history, crustal deformation around impact basins, and cooling of lava.

Gradation is confined to mass wasting and impact-related events. There is no evidence at all that the Moon ever had oceans, lakes, rivers, or an atmosphere of any significance to weather the surface, though there might be water ice in craters near the poles whose floors and some walls have been shaded from sunlight for at least the last 1000 Ma or so, with temperatures persistently around 100 K. The Moon must have been so dry at birth that any water is likely to have been delivered subsequently, by volatile-rich impactors such as comets, or perhaps by solar wind ions.

7.1.5 Localised Water Ice?

In 1996 radar data from the Clementine Orbiter found indirect evidence of ice near the South Pole in the sloping walls of the Shackleton Crater (whose floor cannot be seen from the Earth). In 1998 the orbiter Lunar Prospector, using neutron spectroscopy, found indirect evidence of ice at both poles. In the case of the Moon, low-energy neutrons were detected that have been generated by cosmic ray impacts on lunar surface materials, and lose much of their energy in collisions with hydrogen atoms. Assuming that water is the dominant source of the hydrogen, a few per cent by mass of the top metre or so of the surface in various polar craters would be water ice. Any such water must have migrated there from elsewhere on the Moon, and has survived because the sublimation rate from permanently shadowed areas is very low, though dust protection might be necessary to shield any water from photodissociation by UV radiation from starlight.

Doubt about the existence of near-surface water ice came when the Lunar Prospector mission was ended by crashing the spacecraft into a crater near the South Pole on 31 July 1999. The Hubble Space Telescope and Earth-based telescopes searched unsuccessfully for spectroscopic evidence of water. Further doubt has since been cast by radar data from the 300 m Arecibo radiotelescope on Puerto Rico. Radar provides evidence for large lumps or sheets of ice because it gives a strong reflected signal, not so much from the surface, but throughout a depth equal to several wavelengths of the radar. At a wavelength of 0.7 m, Arecibo has failed to detect ice anywhere, thus ruling out large lumps down to a depth of about 5 m. The positive radar outcome from Clementine could well be due to its observation of a sloping crater wall. A rough surface at a low angle can mimic water ice because it too reflects radar strongly.

To explore the upper metre with greater precision, Arecibo also used a wavelength of 0.12 m, again with a null outcome.

☐ How can this be reconciled with the Lunar Prospector detection of hydrogen to this depth? A possible reason is that the ice is not present in large lumps or sheets but as tiny particles mixed with the dust and rubble, perhaps generated by solar wind ions that gave rise to frosts.

7.1.6 Crustal and Mantle Materials

Lunar samples have been returned to Earth by the six Apollo manned landings and by three Soviet Luna robotic missions, all on the near side, and, except for Apollo 16, all on maria. Samples have also been analysed at the surface by several robotic landers, again on the near

Figure 7.3 A typical view of the lunar surface, dominated by dust with few large rocks. This is at the Apollo 17 landing site, on the edge of Mare Serenitatis. (NASA/NSSDC AS-17-145-22165)

side maria. In all cases the samples comprise small rocks found lying on the surface, pieces chipped off larger rocks, and samples from cylindrical tubes that have penetrated to 2.4 m below the surface. The orbiter SMART-1 carried out a global survey of the elements at the lunar surface by X-ray fluorescence spectrometry (Section 6.1.2).

The surface of the Moon is covered in fine dust, called lunar fines or 'soil'. Figure 7.3 is a typical view of the dusty surface. The fines are a complicated mixture of silicate rock fragments and glassy particles with much the same composition at the nine sites from which it has been sampled. On the maria the fines have albedos of only 5–8%, whereas in the highlands the values are in the range 9–12%, which is why the highlands look brighter. There are four sources of the fines: recondensed minerals that were melted or vaporised by impacts; fine ejecta; surface rocks fractured by micrometeorites; and dust infall from space. As well as the fines there are lots of small rocks, but few large ones. These small rocks are mainly breccias ('bretchy-ars') – the result of pressure and temperature welding of rock fragments by impacts. Fines plus breccias dominate throughout the length of the core samples. The whole assemblage of fines and pieces of rock is called the **regolith**.

What lies beneath the regolith? Seismic data from landers, all on the near side, are shown in Figure 7.4, and indicate that beneath the regolith there is broken rock that becomes fully compacted at a depth of about 15 km (perhaps a bit deeper).

☐ What does the sharp increase in speed at the greater depth of 40 km indicate?

The sharp increase in speed at about 40 km (at least on the near side) is widely interpreted as a change in composition at a crust–mantle interface.

We turn now to the mineralogy of the lunar rocks. The highland rocks are dominated by anorthosite, an intrusive igneous rock consisting largely of the mineral feldspar (Table 6.1), and

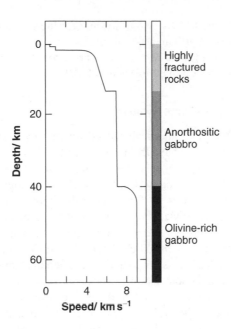

Figure 7.4 Seismic wave speeds versus depth for the outer part of the Moon.

in particular the calcium-rich variety plagioclase feldspar. Anorthosite in this quantity could be produced by strong differentiation in a widespread magma ocean created by impact melting in the later stages of lunar accretion. To account for the anorthosite the ocean would have needed to be a few hundred kilometres deep, with a peridotite composition similar to that of the Earth's mantle, though depleted in the more volatile elements and modified by the formation of a small iron-rich lunar core. Fractional crystallisation would have established a lunar crust rich in anorthosite, underlain by a mantle rich in olivine. It is this change in composition that is held to be responsible for the increase in seismic speed at about 40 km in Figure 7.4. The seismic speeds are consistent with anorthosite above this level and olivine-rich gabbro below it. It is possible that the heat-producing radioactive isotopes of uranium, thorium, and potassium are concentrated into the crust, and this would help explain why the Moon's interior seems to have cooled early in its history (Section 4.5.1).

The mare rocks differ from the highland rocks through being dominated by the extrusive igneous rock basalt – just the sort of rock that would result from partial differentiation of the mantle after crust formation. Though the maria cover about 17% of the lunar surface, the mare infill is at most only a kilometre or so deep, and so the basalts comprise less than 1% of a lunar crust of mean thickness of roughly 40–50 km.

Taking the crust plus mantle as a whole, analyses show that, compared with the Earth, the lunar crust plus mantle is enriched in refractory compounds but heavily depleted in iron and iron-rich compounds and in siderophile elements such as magnesium. The lunar crust plus mantle is also heavily depleted in volatiles such as water, carbon dioxide, and hydrocarbons (compounds of carbon and hydrogen), and in the more volatile silicates, such as those rich in potassium. Metallic iron particles in the lunar rocks indicate that the lunar surface has *never* been exposed to oxygen-rich volatiles to much extent – otherwise they would have been dissociated and oxidised all the metallic iron.

☐ Does any theory of the origin of the Moon account for these differences in crust plus mantle compositions between the Earth and the Moon?

As explained in Section 5.2.1, these differences can be accounted for by the collision theory of the origin of the Moon that was outlined in Section 2.2.4.

7.1.7 Radiometric Dating of Lunar Events

Two types of event have been radiometrically dated for the lunar rock samples. For the mare basalts and many highland rocks we have the age of solidification; for the breccias as well as the solidification ages of component fragments, we also have the time at which the rock fragments were impact welded. In addition, isotopes of hafnium (Hf) and tungsten (W) have been used to establish that the lunar core formed by 25–30 Ma after the formation of the Moon: ^{182}Hf decays into ^{182}W with a half-life of only 9 Ma; ^{182}W is removed by iron, so ^{182}W depletion in rocks is the telltale measurement, but there are many complications that will not concern us.

Turning to crust and mantle events, the oldest dates are solidification ages of 4460 Ma for some highland samples of anorthosite. This is not long after the Moon formed, at about 4500 Ma. Most other highland rocks are of comparable antiquity, though towards the lower end of the age range, about 3800 Ma, the radiometric clocks could well have been reset by the impacts that made the samples available on the mare. The earliest ages are in accord with the great antiquity of the highlands inferred from the high impact crater densities there. Mare basalt solidification ages range from about 3950 Ma to 3150 Ma. Breccia welding ages range down to 3100 Ma ago, except for an age of only about 900 Ma for breccia from the impact that formed the crater Copernicus (Figure 6.8(a)). The fresh-looking crater Tycho is even younger, perhaps as little as 100 Ma.

The older breccias help us to date the formation of the mare basin impacts, and the mare basalt solidification ages help us to date the mare infills. The breccia welding ages range from 4000 Ma for Maria Serenitatis, Nectaris, and Humorum to 3900 Ma for Mare Imbrium. Other basins, such as Mare Tranquillitatis, are inferred to be older, because that have been modified by the earliest dated basins. Other ancient basins have presumably been entirely obliterated. Table 7.1 gives the ages of some lunar basins, along with the mare infill basalt solidification ages. As noted earlier, you can see that infill was delayed for several hundred million years, the most recent, that of Mare Procellarum, being completed about 3200 Ma ago. It is not known what caused such delays. There are no radiometric dates for the lunar far side, but Mare Orientale, from ejecta relationships, seems to be a bit younger than Mare Imbrium, and so has an estimated age of 3800 Ma.

Table 7.1 Ages of some lunar basins and mare infill

Basin	Basin age/ Ma	Infill age/ Ma
Tranquillitatis	Before 4000	3600
Fecunditatis	Before 4000	3400
Serenitatis	4000	3800
Crisium	3900	3400
Imbrium	3900	3300
Procellarum	?	3200
Orientale	About 3800	?

'?' denotes an uncertain value.

A small fraction of the lunar highlands is somewhat depleted in craters. This is largely due to blanketing by ejecta from huge impacts, though in some regions there might be ancient lava flows, over 4200 Ma old, since broken up by impacts and mixed with ejecta. Some features here and there on the Moon are consistent with more recent volcanism. For example, crater densities on parts of Mare Imbrium are so low that volcanism might have persisted here until about 2500 Ma ago. Even today there is the rare observation of possible gas or ash emission in tiny quantities. However, these transient lunar phenomena (TLPs) have many possible causes, e.g. meteor strikes and moonquakes – volcanism is not generally required. Two possible exceptions are an 8 second flash observed by an amateur astronomer in 1953, and a possible venting of lunar gases (on the near side) seen by Lunar Prospector. But overall, from about 4200 Ma ago, lunar volcanism was dominated by lava infill of mare basins, and this was almost entirely over by 3200 Ma.

The lunar cratering rate through lunar history

It was explained in Section 6.2.5 how the radiometric ages of lunar surfaces with different crater densities have been used to deduce the cratering rate throughout its history, with the outcome shown in Figure 6.14. The cratering rate was very high early on, and as a result most of the lunar surface became saturated. There has since been little resurfacing in what we see now as the highland regions. The cratering rate declined steeply as the supply of debris left over from planetary formation diminished, with 3900 Ma ago marking the end of the heavy bombardment, possibly marked by a narrow peak, as outlined in Section 6.2.5. The rate continued to decline until about 2000 Ma ago, since when it has not varied much.

7.1.8 Lunar Evolution

We can outline a plausible lunar evolution in broad terms as follows, on the assumption that just after its formation the Moon had a uniform composition broadly similar to that of the Earth's mantle today, though depleted in volatiles and enriched in refractories.

(1) A small lunar core formed by 25–30 Ma after the Moon formed (at about 4500 Ma), depleting the mantle in siderophile elements.
(2) Impact melting helped create a magma ocean several hundred kilometres deep. A thin skin solidified and gradually thickened, but not before a crust rich in anorthosite formed, overlying a mantle rich in olivine. The crust was in place by about 4400 Ma.
(3) By about 4000 Ma the magma ocean had solidified throughout its depth, with the whole surface nearly saturated with impact craters, including some large basins. Further impact basins formed, but with a rapidly declining impact rate even the youngest (Orientale) is about 3800 Ma old.
(4) By about 4000 Ma ago the heat from radioactive decay had raised temperatures to the point (about 2000 K) where there was an extensive asthenosphere. Isolated pockets of partial melt formed in the rising legs of convection cells, and the asthenosphere migrated deeper as radiogenic heating declined.
(5) The partial melt supplied basalt magma to infill some of the large impact basins, notably on the near side where the basins are at lower altitude. Because of slow, deep convection the magma did not reach the basins until hundreds of million years after they formed.

(6) As the rate of radioactive heat generation subsided, the interior cooled and the lithosphere thickened, to about 300 km by perhaps 3600 Ma ago, and it continued thickening until magma became only very rarely available from about 3200 Ma ago.

Note that the magma ocean *might* not have been planet-wide at any one time. The degree of isostatic compensation varies considerably across the Moon in a manner indicating that the crust and upper mantle might have become rigid at different times in different regions.

Question 7.1

If the lunar highlands had a peridotite composition, how would this modify our view of lunar evolution?

Question 7.2

(a) If the mare infill had been derived from the highlands through gradation, how would the composition of the infill differ from that observed?
(b) Were a large impact basin to be excavated in the Moon today, why is it likely that any subsequent infill would only be through gradation?

7.2 Mercury

Mercury, the planet closest to the Sun, is a small world almost entirely devoid of atmosphere. Its proximity to the Sun means that at perihelion, when it is only 0.31 AU from the Sun, the equatorial temperature at noon is about 725 K, though just before dawn the absence of an atmosphere leads to a frigid 90 K. Mercury rotates three times during two orbits of the Sun, giving a mean solar rotation period (Mercury's solar 'day') of 176 days. It is the tidal force of the Sun that has slowed Mercury's rotation to the point where it is now in this 3:2 resonance.

Mercury is much less well explored than the Moon. It is very difficult to see much on the surface of Mercury with ground-based telescopes, and with a maximum orbital elongation of 28° it is always too close to the Sun to be viewed by the Hubble Space Telescope. The Arecibo radiotelescope is not constrained in this way, and has imaged the hemisphere unseen by Mariner 10, but at low resolution. The only spacecraft to have visited the planet was the flyby mission of Mariner 10, twice in 1974 and again in 1975. Only 46% of the surface was imaged, with a resolution of about 100 m – about the same as that of the Moon in the larger Earth-based telescopes. Earth-based radar has indicated that the altitude range on Mercury is around 4–5 km, but with a large uncertainty.

We have no samples of the Mercurian surface, though the remotely sensed properties of the surface from Mariner 10 and from the Earth are consistent with dusty basaltic silicates everywhere, plus iron sulphide. This makes it rather like the lunar fines.

Mercury is a dark world, though not quite as dark as the Moon, and it is generally more uniform in its albedo, not displaying the maria–highlands contrast of the Moon. The albedo (at visible wavelengths) over much of the surface lies in the range 10–20%.

7.2.1 Mercurian Craters

The surface of Mercury is dominated by impact craters (Plate 4), ranging from small bowls at the limit of resolution up to huge impact basins. Mercurian impact craters are broadly similar in form to those on the Moon, the main differences being attributable to the higher gravitational field at the surface of Mercury, $3.7\,\mathrm{m\,s^{-2}}$, compared with $1.6\,\mathrm{m\,s^{-2}}$ on the Moon.

❐ What effects on the craters should the higher field have?

Among the expected effects of higher gravity are that, for surfaces with similar compositions and compaction, the craters will be smaller for given projectile kinetic energy, and ejecta will be flung less far. Also, the diameter ranges of the various crater morphologies in Figure 6.10 will be different. Though these expectations are borne out, closer inspection reveals further differences between Mercurian and lunar craters, presumably because the two surfaces have somewhat different properties. Also, though the distributions of sizes are not hugely different, such differences as there are remain largely unexplained.

Some craters are fresh, with bright rays of ejects, the brightest features on Mercury. Others craters are degraded. Gradation of Mercury's craters has occurred in various ways. Ejecta from more recent craters partially obscures neighbouring older craters, and many craters show evidence of infill that has created so-called smooth plains – in many cases this infill might be lava. Larger craters show evidence of isostatic adjustment. There has also been mass wasting, presumably a result in part of seismic activity. The paucity of craters greater than about 50 km in diameter suggests that the crust might have been rather soft early in Mercury's lifetime, leading to viscous relaxation. This occurs in any solid substance warm enough to flow slowly under its own weight, and the effect is exaggerated if the lithosphere is thin. The number of large basins (300–1000 km diameter) is much the same as on the Moon, but they are more subdued because of burial and infill.

The largest basin, indeed the largest structure on the surface, is Caloris, a multi-ring basin 1300 km across, with ejecta reaching as far again beyond the outer rim (Figure 7.5(a)). The infill is a mixture of ejecta and impact melt. Caloris also has smooth plains thought to be lava flows, an interpretation supported by the possible presence of lava channels. The smooth plains seem to have been modified by isostatic adjustment, by magma withdrawal, and by tectonic processes. Diametrically opposite the Caloris Basin is a unique region of hilly terrain criss-crossed by linear features (Figure 7.5(b)). On a smaller scale, radar studies from the Earth show this region to be very blocky, or fractured. It is thought to be the result of the focusing by a large and dense planetary core of seismic waves from the Caloris impact. This is further evidence for the large iron core postulated in Section 5.1.3. The Caloris impact was probably the final major impact.

7.2.2 The Highlands and Plains of Mercury

About 40% of the 46% on Mercury's surface imaged by Mariner 10 consists of large craters, basin rim structures, hilly terrain, and terrain with roughly linear topography. These Mercurian highlands constitute the analogue of the lunar highlands, and constitute the heavily cratered terrain. The more heavily cratered areas in Figure 7.6 and in Plate 4 are examples.

❐ What could have caused this type of terrain?

This terrain is considered to be left over from the terminal phase of the heavy bombardment, as in the case of the lunar highlands.

The remaining 60% of the surface imaged by Mariner 10 comprises two types of plain, the smooth plains, already mentioned, and the intercrater plains. The smooth plains constitute a smaller fraction of the surface than the intercrater plains. As well as being found in the Caloris

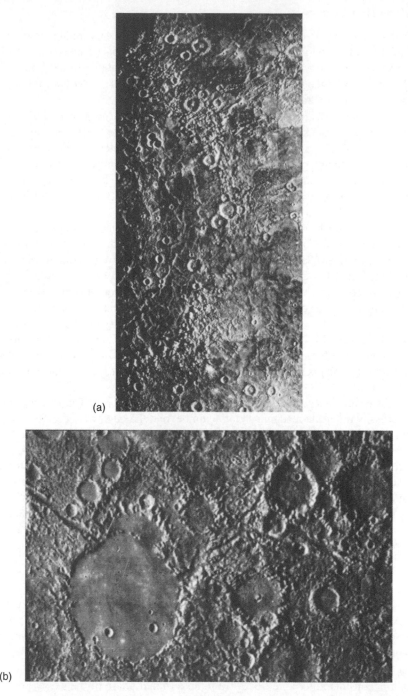

(a)

(b)

Figure 7.5 (a) The Caloris Basin on Mercury. Note the ejecta, the multiple rings, the infill, the smooth plains. The inner ring is about 1300 km in diameter. (NASA/NSSDC AoM F21) (b) Hilly terrain diametrically opposite the Caloris Basin. The largest crater, Petrarch, is about 150 km across. (NASA/NSSDC FDS27370)

Figure 7.6 The surface of Mercury, showing smooth plains (upper half) and heavily cratered terrain. The frame is about 490 km wide. (NASA/NSSDC P15427)

Basin they are also found elsewhere on Mercury, such as in other basins and in large craters (like Petrarch in Figure 7.5(b)). The less cratered half of Figure 7.6 is an example of where the smooth plains are found exterior to impact basins and large craters. Though the name suggests that the smooth plains are flat and sparsely cratered, they are more heavily cratered than the lunar maria, which are the least cratered areas on the Moon. The cratering density is similar on all such plains, suggesting that all are of much the same age. This is estimated (from the lunar cratering record) to be around 3800 Ma, which is when the heavy bombardment was declining. Though ancient, the smooth plains are the youngest surfaces on Mercury, borne out by the relationship of smooth plains to adjacent material, which shows that the smooth plains came later. All of them could be lava flows, a suggestion supported by sinuous lobed ridges that could be the edges of lava sheets. The smooth plains seem to be the Mercurian equivalent of the lunar maria.

The commonest type of surface on Mercury has no lunar equivalent. These are the rolling intercrater plains, so called because they lie between clusters of large craters (greater than about 20 km in diameter). Most are in the highlands, and in low-resolution images seem to have few craters. Many of the craters are secondaries, as indicated by their shallow, elongated forms and tendency to form clusters and chains. The less cratered areas of Plate 4 are intercrater plains. In fact, the crater density on the intercrater plains is much higher than on the smooth plains, the population being dominated by craters smaller than 10 km in diameter. The intercrater plains are therefore older than the smooth plains, with estimated ages of 4000–4200 Ma, which is during the heavy bombardment. The intercrater plains are probably of volcanic origin, rather than a result of erosion and deposition, or impact ejecta mantling. However, you should not envisage volcanoes and volcanic eruptions as creating these plains. Rather, at that early time, Mercury's lithosphere would have been thin, so we have subsurface magma emerging through fractures in the lithosphere, creating widespread flooding.

It is quite possible that plain formation on Mercury was more or less continuous, from some time during the heavy bombardment to when the youngest of the smooth plains were created.

7.2.3 Surface Composition

Earth-based IR spectra have detected feldspars and perhaps pyroxene, which also dominate the lunar highlands (Section 7.1.6), though the detailed composition is different on Mercury. The seven times greater solar radiation and solar wind at Mercury also contribute to surface differences. These differences account for Mercury having a generally higher albedo than the Moon. The feldspar-rich surface supports the view that Mercury is highly differentiated. IR and microwave spectra have failed to detect basalts, which indicates that extrusive volcanism since Mercury acquired a crust has been rare. If so, then loss of internal heat by volcanic extrusion has not occurred, thus helping to sustain high temperatures in the interior.

Mercury, with an axial inclination of zero, has permanently shadowed regions near the poles, like the Moon, that act as cold traps for migrating water. But unlike the Moon, there is good evidence, from Arecibo radar, that in these regions there *is* ice, at least 1 metre thick, under at most a thin layer of dust. How can Mercury have bodies of ice of moderate lateral extent, whereas the Moon might well not? This could be a result of greater cometary bombardment on Mercury, due to its position much closer to the Sun, and its lower water loss rate due to its larger gravitational field.

7.2.4 Other Surface Features on Mercury

A few craters have dark halos, as have a few craters on the Moon. Such craters might be volcanic, but in the case of Mercury better images are required to investigate this possibility. Arecibo radar has seen a dome on the hemisphere unseen by Mariner 10, which might be volcanic.

Scarps are very common, 1–2 km high, and cross all types of terrain. Figure 7.7 shows part of the scarp Discovery Rupes. Ridges are also common. Together with the scarps they suggest planet-wide crustal compression. A 1–5 km decrease in the radius of Mercury could account for them, though some might be the result of more local tectonics and others could be the result of lava flows. Local crustal *tension* is indicated by a few graben and a few strike–slip faults. Mercury continues to cool slowly, so compressional features will grow, equally slowly, though on the basis of lunar cratering rates there has been little volcanic or tectonic activity on Mercury in the past 3000 Ma.

7.2.5 The Evolution of Mercury

On the basis of the limited observations, including those that relate to the interior, a plausible (but not unique) picture of the evolution of Mercury is as follows:

(1) Any early magma ocean would have solidified by 4000 Ma ago, a crust having formed much earlier.
(2) Though we have no radiometric dating for Mercurian surfaces, models indicate that the bombardment history of Mercury has been broadly similar to that of the Moon, and so the heavy bombardment of Mercury would have ended about 3900 Ma ago.
(3) The existence of heavily cratered terrain indicates that a lithosphere formed before the end of the heavy bombardment. It was not very rigid, and therefore allowed some of the craters,

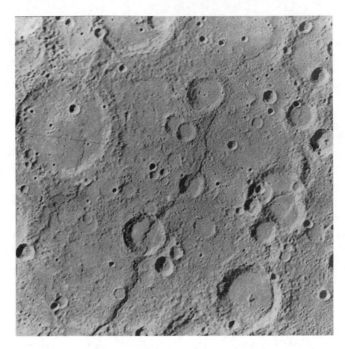

Figure 7.7 The scarp Discovery Rupes on Mercury. The largest crater is Rameau, 55 km in diameter. (NASA FDS528881-4, 27398-9, 27386, 27393, R G Strom)

particularly those around 50 km in diameter, to relax into oblivion, thus contributing to the creation of the intercrater plains.

(4) The intercrater plains have ages of 4000–4200 Ma, and are largely the creation of lava flows.

(5) The formation of the massive iron core and radiogenic heating warmed the interior sufficiently for an asthenosphere to extend up to the comparatively shallow depth of about 50 km, thus making magma available to the surface. Any early expansion of the crust during this heating phase is not seen in surface cracks, so it would have ended before the end of the heavy bombardment.

(6) Crustal contraction created fractures, scarps, and ridges. Lava flows created more intercrater plains. The contraction could have resulted from cooling and shrinkage of an iron core, aided by lithospheric contraction. Fracturing could also have resulted from the stresses produced by the tidal slowing of Mercury's rotation by the Sun. If at this early time there was a molten iron core, then it must still be largely molten today, because the large crustal shrinkage that would have occurred on solidification would be preserved today, but is not seen.

(7) The final stage of heavy impact cratering created Caloris and its associated features, rather as it created Mare Orientale on the Moon.

(8) Subsequently, when the impact rate had declined somewhat, the smooth plains were emplaced, also through lava flows, and have ages around 3800 Ma.

(9) The lithosphere gradually thickened, to about 200 km by 2000 Ma ago, and it continued to thicken thereafter. There might have been little volcanic or tectonic activity on Mercury in the past 3000 Ma.

Question 7.3

Discuss the factors relating to energy gains and losses that are relevant to the conclusion that the lithosphere of Mercury is thicker today than those of the Earth and Venus.

7.3 Mars

Mars is one of the most spectacular sights in the night sky, particularly near its oppositions, which occur every 780 days. In the months before and after, it gleams brightly, like a red eye. It has inspired much literature, mostly based on the notion that Mars is inhabited, usually by hostile beings.

Mars is somewhat larger than Mercury. On average it is 1.5 times as far from the Sun as we are, and therefore under its thin atmosphere its surface is cold, with temperatures rarely above 273 K, and at night plunging to as low as 150 K in the coldest regions. The day on Mars is nearly the same length as on Earth, the mean solar rotation period, the sol, being 24 hours 39 minutes 35.2 seconds, compared with 24 hours for the Earth. Mars takes longer than the Earth to orbit the Sun, so there are nearly 669 sols in a Martian year.

Mars has for a long time been a favourite target of investigation, with its surface markings that exhibit seasonal changes, and the consequent possibility of life, past or present. For nearly 400 years it has been scrutinised through telescopes, and in the Space Age there have been many successful spacecraft missions (Table 4.1). Orbital missions have investigated Mars in a great variety of ways: imaging, mapping, gravity measurements, spectroscopy at a variety of wavelengths, neutron spectroscopy, and so on. Five successful landers have added *in situ* investigations (Section 7.3.7), and Martian meteorites (Section 7.3.8) have provided samples from other sites on the surface.

7.3.1 Albedo Features

From the Earth, the most obvious features on Mars are the white polar caps, the dark markings, and the widespread light-red background (Plate 8). The polar caps consist of ices of carbon dioxide and water. The light-red regions are fairly uniformly covered in dust, whereas the dark regions are more streaky, and are darker because of a higher proportion of dark dust, and because of the exposure of dark underlying terrain.

These albedo features exhibit seasonal changes – a result of Mars's axial inclination of 25.2°. The orbit of Mars is fairly eccentric, with a perihelion distance of 1.38 AU and an aphelion distance of 1.67 AU. Midsummer in the southern hemisphere occurs near perihelion and midwinter near aphelion (the same as for the Earth). Therefore the seasonal changes in the southern hemisphere are more extreme than in the north. The most obvious seasonal change is the cyclical growth and retreat of the polar caps, visible from the Earth even in a modest telescope. We now know that this cycle is the result of carbon dioxide condensing at the poles in the autumn and subliming back into the atmosphere in the spring.

There are also seasonal changes in the shape of the dark areas and in their contrast against the light areas. The agent is atmospheric winds. These mobilise dust, and as a consequence streaks are created downwind of obstacles such as craters – light streaks where light dust has been deposited on dark terrain, and dark streaks where dark dust has been deposited on light terrain. Winds also create dark streaks by scouring away overlying light dust to reveal dark terrain beneath. The dark areas are only weakly correlated with large-scale topographic features, but are strongly correlated with small-scale topography (such as small craters) and with wind strength and direction. It is seasonal changes in the winds that cause the seasonal changes in the dark areas, e.g. by scouring light dust from some areas and depositing it in others. Furthermore, changes in the winds on time scales of decades cause changes in the dark areas on a similar time scale.

The most dramatic manifestation of wind-raised dust is huge dust storms. These are most frequent near to perihelion, and sometimes cover almost the whole planet in yellow-tinted clouds that consist predominantly of the light dust. Spectrometric studies of these clouds and of the light areas of the surface itself indicate basaltic minerals mixed with various clay minerals, notably montmorillonite (Section 6.2.6). The dark material seems to be dominated by basaltic silicates rich in iron and magnesium. The red tint of both the light and the dark areas is the result of iron-rich minerals. It is thought that the light dust is derived from the dark material by various physical and chemical processes.

☐   How are clay minerals produced?

Clay minerals are the result of the aqueous alteration of silicates. Direct evidence for the action of water is a 500 km diameter area near the equator which is rich in haematite (Fe_2O_3). Such a concentrated deposit of haematite suggests that it was formed in a body of liquid water, now long gone. In Section 7.3.6 you will learn a lot more about water on Mars.

7.3.2   The Global View

Spacecraft images show that the surface of Mars consists of two contrasting hemispheres, divided approximately by a great circle inclined at 30° to the equator (Figure 7.8). The mean altitude of the hemisphere north of this line, the northerly hemisphere, is a few kilometres lower than that of the southerly hemisphere, and it is generally much flatter. Why it is so flat and why it is at a lower mean altitude will be discussed shortly. Figure 7.8 shows altitude data in more detail, as the frequency of occurrence of altitudes across the whole surface – the **hypsometric distribution** for Mars. Two peaks are apparent. The one from −2 to −5 km corresponds to the northerly plains and constitutes 34.5% of the surface. The one from 1 to 3 km corresponds to the southerly highlands and constitutes 35% of the surface. The altitude of the boundary between the two hemispheres ranges from −2 to 1 km, and accounts for 25% of the surface. The small proportions at the extremes correspond to high mountains and deep depressions. The zero of altitude is defined from gravity data as outlined in Section 6.1.1.

Gravity data, along with topographical data, have been crucial in crustal models that indicate a mean crustal thickness of about 60 km in the southerly hemisphere and about 35 km in the northerly hemisphere, with uncertainties of 10 km or so. It has been suggested that the thinner crust in the northerly hemisphere is the result of mantle convection. It is less likely to have been caused by multiple large impacts – these would probably have been too few – and one giant impact would have disrupted the planet. Whatever the cause, it is possible that the thinning of the crust was instrumental in producing the lower altitudes in this hemisphere.

The hemispheres also differ in other ways. The southerly hemisphere is dominated by impact craters and impact basins, whereas the northerly hemisphere is dominated by plains that constitute the flattest areas known in the Solar System. In the northerly hemisphere there are also domes, volcanoes, and huge grabens and pits. The scarcity of impact craters in this hemisphere shows that it is the younger one, and this is also indicated by abundant evidence that it has been resurfaced by basaltic lava flows, sediments, and sedimentary rocks. At least some of the sediments could have come from the southerly highlands, which have morphologies indicating

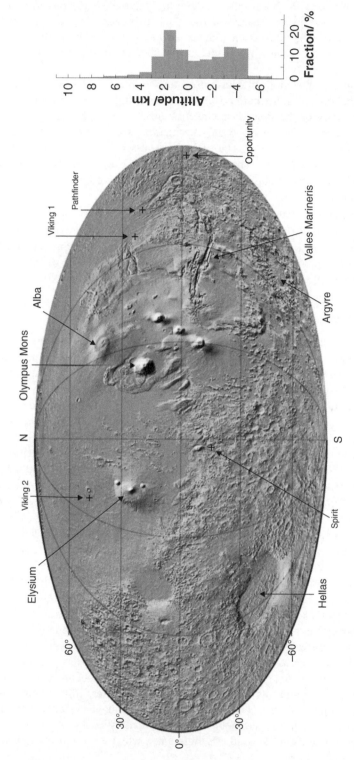

Figure 7.8 A simplified topographic map of Mars, and the Martian hypsometric distribution. Only the main features are shown. Lines of longitude are shown at 60° intervals, with 0° at the right hand edge. (NASA/MOLA Science Team, S Soloman)

the removal of material in the Noachian (see below), to a depth of at least a kilometre. The relative youth of the northerly hemisphere is consistent with a thinner northerly crust, because magma would then have had easier access to the surface.

However, the radar on board Mars Express (Table 4.1) has detected many impact craters and impact basins at a depth of a few kilometres in the northerly hemisphere, indicating that the young plains there are underlain by a far older surface.

Relative ages from crater densities are used to define three epochs:

(1) The Noachian, with near-saturation crater densities, lasts until the heavy bombardment was over. This terrain is found mainly in the southerly hemisphere (the highlands).
(2) The Hesperian, with moderate crater densities, is the time immediately following the Noachian. These areas consist mainly of ridged plains that overlap Noachian terrain.
(3) The Amazonian, with light cratering, immediately follows the Hesperian, and extends right down to the present. This terrain dominates the northerly hemisphere.

Radiometric dating has not been carried out at the Martian surface. We therefore have to obtain absolute ages by adapting to Mars the lunar cratering rate in Figure 6.14, making due allowance for Mars's different gravity, its proximity to the asteroid belt, the different nature of the Martian surface, and the greater degradation on Mars. The end of the Noachian is placed at around 3700 Ma, by which time the bombardment rate on Mars had declined considerably from its peak. The end of the Hesperian is not as well known, with dates ranging from 3300 Ma to 2900 Ma, or even earlier and later.

A pre-Noachian epoch is now recognised, based on huge numbers of craters tens to hundreds of kilometres across that underlie Noachian surfaces. The pre-Noachian surface constitutes the early crust of Mars. It is *defined* to be older than the huge Hellas Basin, which has been dated (using impact craters) at 4080 ± 60 Ma.

Another date yielded by impact craters is that of the formation of the boundary between the two hemispheres, and therefore of the initial formation of the northerly lowlands. Its value, 4120 ± 80 Ma, places the formation in the pre-Noachian. Partially buried craters and basins in the northerly lowlands indicate an underlay of (pre-)Noachian terrain.

Indications of surface composition have come from spectra obtained by Mars orbiters. In the southerly hemisphere, IR spectra indicate the presence of basalts or basaltic materials. In the northerly hemisphere, given the many basaltic volcanic structures there, it was a surprise to find evidence of andesitic signatures over wide areas. It is thought that the basalt, presumed to be widespread, has become altered at many locations by weathering, and distributed by winds. Also, the IR spectra that have provided much of the data from orbit only 'see' 1–2 mm below the surface, and therefore it is possible that the alteration is not deep. Evidence from a few surface locations is outlined in Section 7.3.7.

Let us look more closely at each hemisphere separately.

7.3.3 The Northerly Hemisphere

Figure 7.8 includes the main topographic features of the northerly hemisphere. The plains that constitute a large fraction of this hemisphere are young and thought to be largely volcanic. The numerous volcanic features support this conclusion. Spectrometric evidence from orbiting spacecraft and evidence from surface analyses by the landers is consistent with the presence of iron-rich basaltic materials such as would constitute silicate lava flows. Radar reflectivities and the rapid response of the temperatures of some areas to changes in insolation are consistent with a covering of volcanic ash. This ash could have been carried by the wind from the sites of explosive volcanic eruptions.

Though the northerly plains are rather flat, superimposed on them are two large raised domes, the Elysium and Tharsis regions. The Tharsis region is the more dramatic, a broad dome about 5000 km across, rising to 10 km above the surrounding plains. In the case of Tharsis there is evidence, e.g. from magnetic anomalies, that it was probably in place by the late Noachian. On both domes, substantial lava flows occurred up to a few hundred million years ago. Since then, volcanic activity has been rare, though virtually uncratered hillsides indicate activity perhaps more recently than 3 Ma. Various processes could have contributed to the formation of these domes – convective plumes in the mantle, isostatic adjustment, intrusive and extrusive igneous activity. Different regions of each dome have different ages, indicating protracted formation. Whatever single or combined process created the domes, the gravitational fields in their vicinity show that there is excess mass present near the surface, perhaps because the underlying lithosphere is so thick that it takes hundreds or thousands of million years for isostatic equilibrium to be achieved, or perhaps because the domes are still being borne aloft by convective plumes in the mantle.

Each of these domes is associated with features arising from crustal tension that would result from uplift, or perhaps in some cases from the effect of the intrusion of dykes of magma. Most spectacular is Valles Marineris, a system of cracks and grabens enlarged and modified by slumping, landslides, wind erosion, and magma withdrawal, and perhaps also modified by water flow at and below the surface. It is nearly 4000 km long (Figures 7.8 and 7.9), up to several hundred kilometres across, and with an average depth of 8 km. Mars Global Surveyor (MGS) images were the first to reveal layers in the walls, each layer about 10 km thick. The layers are thought to be lava flows. At the west end of Valles Marineris is Noctis Labyrinthus, consisting of short grabens running in all directions. At its east end is an example of Martian chaotic terrain – jumbled, heavily mass wasted, with wide channels leading downhill.

The domes also bear huge shield volcanoes. One old volcano, Alba, on the Tharsis dome, has a lopsided topography that suggests it predates the Tharsis dome, and was then tilted as the dome grew. At the other extreme, some volcanoes are so unweathered that they could be dormant or recently extinct, and could have contributed lava flows to dome building. The least eroded and youngest volcanoes are four South of Alba, on the Tharsis dome, on what is called

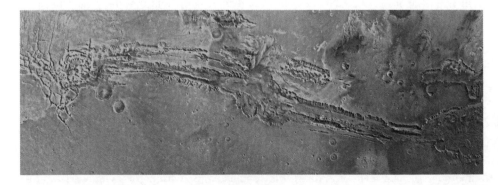

Figure 7.9   Valles Marineris on Mars. It is nearly 4000 km long, extending from a system of grabens to the west (left), called Noctis Labyrinthus, to chaotic terrain to the east. (NASA/USGS PIA00422)

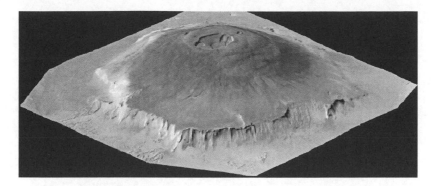

Figure 7.10 Olympus Mons on Mars. The steep scarp is about 550 km across, and the summit is 25 km above the adjacent plains. The vertical scale is exaggerated. (NASA/JPL)

the Tharsis Ridge. These resemble terrestrial shield volcanoes such as Mauna Loa in Hawaii, though the Martian examples are much larger than the terrestrial ones. Olympus Mons is the largest of all (Figures 7.8 and 7.10), 550 km across, and rising to 25 km above the adjacent plains. Lava has flowed from a huge complex of summit calderas and from numerous vents on its flanks. The expected basaltic composition of the lava is borne out by orbital spectrometry. Gravity data, combined with topographic data, give a density of $3100 \pm 200 \, kg \, m^{-3}$, which is consistent with such a composition. The great size of each Martian shield volcano could result from the combined effects of a magma source in a fixed position with respect to the crust, and a thick lithosphere retarding isostatic adjustment.

Some volcanoes seem to have erupted ash rather than lava.

☐ What does the magma need to contain to form ash?

Ash indicates the presence of dissolved volatiles in the magma. Other volcanic features include fissures from which lava emerged, and channels along which lava has flowed. Low-rise lobes are common, consistent with lava flows, though the source vents in many cases are now hidden, as in the case of the lunar maria.

In spite of these extensive indications of volcanism, detailed studies show that the accumulated volcanic activity on Mars is *considerably* less than on the Earth. This might be because crustal formation from a magma ocean concentrated the heat-producing radioactive elements in the crust, depriving the interior of a heat source. This might also be the case for the Moon. Mars, however, is bigger than the Moon, which is why Martian volcanism and tectonic activity have been more widespread and persisted longer.

7.3.4 The Southerly Hemisphere

The southerly hemisphere is dominated by impact craters, much of it near to saturation. The largest crater is the impact basin Hellas, within which is found the lowest elevation on Mars, 6.5 km below zero altitude, and about 8 km below the general altitude of the surrounding highlands Figure 7.8. About half way down to its rather flat floor it is about 1800 km across, making it second only to the Moon's South Pole–Aitken Basin among impact basins in the Solar System. Beyond the main ring of uplifted mountains are fractures, with associated volcanism. Within the Hellas Basin is the plain Hellas Planitia, built of lava and wind-blown dust. The next largest impact basin is Argyre (Figures 7.8 and 7.11), 900 km across, also filled in the manner of Hellas,

Figure 7.11 Part of the Argyre Basin on Mars, visible in the lower left quarter of this image. The basin is about 900 km across. (C J Hamilton and NASA)

as indeed are many smaller craters. Impact crater densities have been used to date Hellas to 4080 ± 60 Ma, and Argyre to about 4040 Ma. The most ancient surviving highland crust is dated at about 4400 Ma.

On the most heavily cratered parts of the southerly hemisphere, many of the craters are heavily weathered, and smaller craters have been totally eradicated. On slightly less cratered and therefore slightly younger terrain the smaller craters still survive, implying a decrease in weathering rate at some time in the distant past, before 3000 Ma ago. A decline in atmospheric mass would account for such a decrease – more on this in Chapter 10.

Impacts have reduced the surface density of the southerly hemisphere. Gravity data yield values in the range 2500–3000 kg m^{-3}. This is too low for solid rock of basaltic composition – Martian meteorites (Section 7.3.8) have densities 3200–3300 kg m^{-3}. A fractured surface to some unknown depth is indicated.

Apart from impact craters, the southerly hemisphere displays a few heavily weathered, almost obliterated shield volcanoes, and lava channels that indicate long-extinct volcanic activity. There are not many such features, and so the activity was not as widespread as in the northerly hemisphere.

The boundary between the hemispheres

The boundary between the two hemispheres is ragged, with hummocky outliers of high ground in the northerly hemisphere, some of them surrounded by landslides. This is called fretted terrain, and it could be the result of retreating scarps, or the weathering of a once-sharp boundary, or the removal of underground magma, or some combination of all three. You have seen that the boundary is dated at 4120 Ma.

7.3.5 The Polar Regions

The seasonal polar caps in the northern and southern hemispheres extend in midwinter to latitudes of about 65 °N and 50–60 °S respectively. In both hemispheres the caps undoubtedly consist largely of CO_2 ice. This is the dominant constituent of the Martian atmosphere, and the temperature of the seasonal caps, about 150 K, is that at which CO_2 would condense (gas to ice) at Martian surface pressures. The temperature remains near to 150 K throughout much of the long winter, sustained by the latent heat released on condensation of CO_2. Some of this condensation occurs at ground level, but in the polar regions of the atmosphere it also forms CO_2 clouds from which powdery snow falls. The depth of the seasonal caps has been estimated from their rates of advance and retreat, a metre or so topping the range of values.

By high summer the seasonal cap has retreated, leaving behind the residual cap (Figure 7.12). The North Pole residual cap reaches temperatures of about 240 K, far too high for it to consist of CO_2. These temperatures, at Martian surface pressures of water vapour, are not far above those on the 'solid + gas' line on the phase diagram of water (Figure 9.9), and so the residual northern cap is presumed to be water ice. Reflectivity measurements by the Galileo Orbiter indicate that dust is also present, in accord with earlier findings. At the South Pole the residual cap seems to be CO_2 (plus dust), though presumably underlain by dusty water ice.

❏   If seasonal changes are more extreme in the southern hemisphere, how could the residual cap be CO_2 when in the northern hemisphere it is water ice?

The South Pole is at the higher altitudes characteristic of the southerly hemisphere, and so is colder than the lower lying North Pole.

The northern cap is generally smooth, but both caps are scarred by wind-scoured pits and valleys. These show that the caps overlie extensive, nearly horizontal layered sediments a few kilometres deep. Each layer is 10–50 m thick and is richer in dust than the residual caps. The layering is thought to arise from a variation in the dust/water-ice ratio from layer to layer, and it is estimated that each layer represents many years of deposition. Much thinner annual layers could exist within those resolved. These layered sediments overlie unlayered and largely ice-free bright dust that is several hundred metres thick near the poles. It extends further towards the equator than the layers, thinning as latitude decreases, and petering out at mid latitudes. Nearer to the poles, and encircling the residual caps, are dark polar collars that consist of dunes of dark grains. The polar deposits are all very lightly cratered, but though Amazonian they could still have ages in excess of a few hundred million years. The pits and valleys extend down to the underlying terrain, which in the north is seen to be the lightly cratered plains typical of the hemisphere. The north cap reaches an altitude of at least 2.5 km above its surroundings. Beneath the polar deposits at the South Pole lies the heavily cratered terrain that is typical of that hemisphere.

The dust has presumably been transported to the polar regions by winds, each dust particle having acted as a nucleus on which CO_2 and water condensed from the atmosphere, the water adding to any water content the grain initially had. The particle with its icy mantle was then

Figure 7.12 Residual Martian cap at the North Pole in October 2006 (see also front cover). The width across the 'Swirly' cap is about 1000 km. Clouds are visible to the left of the cap. (NASA/JPL/Malin Space Science Systems, MOC2–1607)

precipitated at the pole. Subsequently, at least some of the CO_2 content was lost through sublimation. For so much dust to be present at the poles, evidence of denudation elsewhere is a reasonable expectation, and such evidence is found. For example, Valles Marineris seems to have suffered wind erosion, as have various hummocky terrains.

A scenario that can account for many of the polar features has as its central feature switches from net deposition to net erosion. First, the thick unlayered deposit is laid down over many years, perhaps as a set of layers no longer preserved. Net deposition then switches to net erosion, cutting into the deposit. Then there is a switch back to net deposition, resulting in the oldest layers that *are* preserved today. The layered deposits are then partially eroded and this is followed by another deposition episode. Evidence for several layering episodes is provided by what is called an unconformity, a mismatch in orientation between a set of layers and another set above it. Net deposition then switches to net erosion again, to give the pits and valleys that we see today. It is not known whether there has recently been a switch back to net deposition. A major factor in these switches is probably quasi-periodic changes in the axial inclination of Mars – more on this in Chapter 10.

The polar layering is currently (early 2007) being explored by the radar and high-resolution optical imager on board the Mars Reconnaissance Orbiter (MRO, Table 4.1). So far, these instruments have revealed variations in the thickness and composition of the layers at both poles, which will elucidate climate changes. The MRO climate sounder has detected atmospheric pressure variations due to CO_2 snowfall at the winter pole, which makes the major contribution to layer formation.

The MRO spectrometer and imager are detecting gypsum ($CaSO_4.2H_2O$) near the North Pole, and clay minerals elsewhere, indicators of wet conditions at some time in the past. Water-related features are the subject of the next section.

Question 7.4

In Plate 8 there is a variety of albedo and topographic features. In 300–400 words explain the origin of an example of each of them.

7.3.6 Water-related Features

The most intriguing features on Mars are those that seem to require the involvement of liquid water. The intrigue is that liquid water is not evident anywhere on the Martian surface today, and that the present surface conditions would everywhere result in rapid freezing of water. Nevertheless, the features indicating the involvement of liquid water are many and varied.

Unusual ejecta blankets

A small proportion of Martian impact craters have unusual ejecta blankets – Figure 7.13 shows one example. Such blankets are readily explained by the impact-generated melting and surface flow of liquid water with entrained rocky materials, and so they are evidence that the Martian surface contains water in some form. (A less favoured explanation is that such blankets were produced by dry ejecta interacting with the thin atmosphere.) The lack of impact craters on the blankets indicates their youth, and so the water is presumably still there. Older blankets of this sort that would now bear craters are not seen, presumably because such low-relief features are readily eroded away by the atmosphere.

Martian channels

More intriguing are Martian channels. Figure 7.14 shows three sorts of channels, all of which suggest the flow of liquid water. Their locations indicate that their formation was concentrated in the late Noachian or Hesperian. They could, however, form under present conditions, because salts can lower the freezing point of water, and ice can cap slow-moving streams, protecting them from freezing.

Figure 7.14(a) shows an example of an outflow channel. These are of the order of 10^3 km long, 10^2 km wide, and a few kilometres deep, and their morphology strongly suggests the sudden outflow of liquid water, in many cases more than once. Downstream there are features that indicate ice flow as well as liquid flow, presumably because some of the water froze. Lava flows would have generated quite different morphologies.

Figure 7.13 The Martian crater Arandas, with an ejecta blanket suggesting a surface flow of water plus entrained rocky materials. The crater itself is about 28 km across. (NASA/NSSDC P17138)

At the head of the channel in Figure 7.14(a) is an example of Martian chaotic terrain. Such terrain is found at the head of most outflow channels.

☐ What mechanism does this suggest for the formation of chaotic terrain?

A ready explanation is that chaotic terrain is the result of collapse following the removal of the water that created the channel. The outflow channels indicate that the removal was rapid, and this is thought to be due to the release of groundwater from an aquifer that was sealed beneath a permafrost layer until the layer was suddenly disrupted. This disruption could have been caused by an impact, direct or nearby, by fault movement, or by volcanic activity. Another possibility is the bursting forth of water from the aquifer as the pressure of water in it exceeded a critical value. The terracing of channel walls is one indication that there have been repeated floods as the aquifer refilled and emptied. In this case there might be further outflows in the future, though most of the channels are on Hesperian terrain and so were formed in that early epoch.

The remaining outflow channels originate in canyons. Groundwater seepage into these canyons could have created lakes with ice-covered surfaces. Subsurface seepage of water could have eroded and weakened the canyon walls leading to catastrophic break-out of the water beneath the ice, and the formation of the outflow channel.

Most of the outflow channels flow into low-lying plains in the northerly hemisphere and into the Hellas Basin in the southerly hemisphere. Lakes must have formed on these plains, and there are features in the northerly plains that are consistent with this possibility, e.g. possible shorelines and layered sediments. It is even conceivable that much of the northerly hemisphere

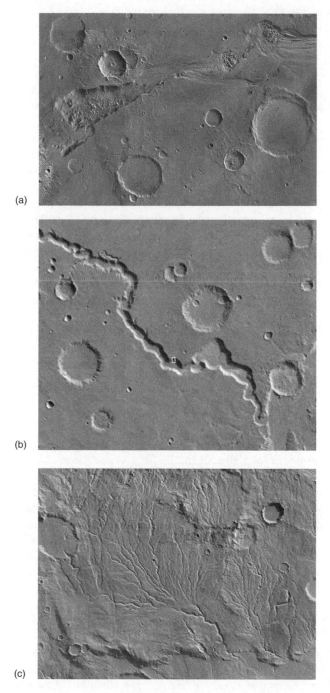

(a)

(b)

(c)

Figure 7.14 (a) The Martian outflow channel at the head of Simud Vallis. The width of this frame is about 300 km. (C J Hamilton and NASA (P16893 is similar)) (b) The fretted channel Nirgal Vallis on Mars. The frame is about 160 km across. (NASA/JPL) (c) A valley network on Mars. The frame is about 130 km across. (C J Hamilton and NASA (from 63A09))

was once covered by an ocean. It is not clear where the lake or ocean water went. It might still be present as ice deposits hidden by dust, or it might have seeped into the groundwater system. At equatorial latitudes under present conditions, surface ice can quickly be lost by sublimation. We shall return to Martian oceans and lakes shortly.

Fretted channels (Figure 7.14(b)) are narrower than outflow channels and are more sinuous. Another distinguishing feature is short, stubby tributaries. They occur along parts of the boundary between the northerly and southerly hemispheres, and stretch hundreds of kilometres into the uplands.

Valley networks (Figure 7.14(c)) consist of channels that are narrower still, with typical widths of 1–10 km and typical depths of 100–200 m, and they have better developed tributary systems. Most of them occur in the southerly hemisphere in areas of Noachian age, particularly at high altitudes and low latitudes. It is widely accepted that the fretted channels and the valley networks were carved by the flow of liquid water. But this leaves open the question of whether the water was supplied by precipitation or by groundwater sapping. The detailed form of the fretted channels and of most of the networks is similar to that of terrestrial systems that were created by sapping, though some Martian networks might have been carved by precipitation. For precipitation it is necessary for atmospheric temperatures and pressures to have been higher in the Noachian than they are today. But if sapping was the sole cause then conditions need not have differed much from the present day. We shall return to Martian atmospheric conditions in Chapter 10.

A minority view is that the fretted channels and valley networks were not caused by liquid water alone, but by subsidence mobilised by groundwater at the base of the debris, in which case it would have been rocky dust and rubble that comprised most of the flow. It is, however, difficult to see how such a process could have produced such long narrow channels.

Martian gullies

Gullies were first imaged by MGS (Table 4.1). A typical example is shown in Figure 7.15, in the south-facing wall of Nirgal Vallis, at the location of the small box in Figure 7.14(b). The frame width is just 2.3 km.

The gullies are about 1 km in length, 10×10 m in section, and are located on steep slopes on young surfaces with ages of about 1–10 Ma. Older gullies have presumably been lost through erosion, which would rather rapidly erase such small features. They are confined to latitudes $30°–70°$ north and south. The gullies typically consist of three sections: an alcove at the top, a tapered channel, and debris at the bottom. The tapering shows that the liquid water was lost rapidly, as would occur by freezing and evaporation/boiling – the slopes are too steep for loss by infiltration. This rapid loss of liquid also explains the short lengths of the gullies. The loss mechanism is consistent with the altitudes at which the gullies are found, altitudes at which the pressure and temperature are below the triple point of water, thus leading to rapid conversion to solid and gas.

All of these features are well explained by a model in which a subsurface aquifer at the altitude of the alcove feeds the gullies with liquid water, though aquifers extending to depths well below the alcove cannot be ruled out. Rarely, snowmelt could provide the liquid, though a counter indication is that gullies are found on slopes of all orientations, not just Sun-facing ones. Also, the alcoves are usually below the top of the slope, again inconsistent with snowmelt.

In the aquifer model the aquifer is created by geothermal heating of ice, trapped within rock layers, at a few hundred metres depth. Increase in aquifer fluid pressure fractures the ground

Figure 7.15 Gullies in the south-facing wall of Nirgal Vallis, in the small box in Figure 7.14(b). Frame width 2.3 km. (NASA/JPL/Malin Space Science Systems)

ice plug and liquid water emerges from the slope. The regional clustering of gullies can be explained by the geothermal heating being localised. The cause of the pressure increase could be a rise in temperature consequent upon the variation in Mars's axial inclination, which alters the amount of solar radiation received at each latitude (see Section 10.2.4). Ice plugs at the required shallow depths would not be stable at equatorial latitudes, which explains the current absence of gullies within about 30° of the equator. If at any earlier times the axial inclination was so high that the polar regions were significantly warmer than today, then gullies should have formed at higher latitudes. If this was so, these gullies have not survived.

Bright new deposits seen in two gullies by the camera on MGS (Table 4.1) suggest that water has transported sediment through these gullies at some time between 1999 and 2005.

Martian oceans and lakes?

An indication that Mars once had oceans and lakes is terraces in an ancient north polar basin and in other basins, notably Hellas, interpreted as shorelines. The laser altimeter on MGS provides support by showing that each terrace is at a gravitational equipotential. Another lake/ocean might have existed to the south-east of Elysium (Figure 7.8). Here, a very flat area, $800 \times 900\,\mathrm{km}$, displays features seen in pack ice on Earth – break-up into plates followed by drift with rotation, and pressure features from ice floe collisions. The paucity of impact craters indicates an age of 3–7 Ma. If volcanic ash quickly covered the ice then the sublimation rate could have been low enough for some water, now frozen, still to be there. Lava flow is less likely – the plates are too big and the area is too flat.

Support for the view that liquid water on Mars was widespread comes from the OMEGA visible and IR spectrometer on Mars Express. This instrument has revealed that Mars is widely coated in hydrated silicates and sulphates that are evaporites. These are several billion year old. Carbonates were also expected to have formed in abundance from atmospheric CO_2 and liquid water. But MGS's thermal emission spectrometer indicates a 2–5% carbonate content in the dust, too little if much of the presumed massive early CO_2 atmosphere was removed by dissolution in liquid water. Such dissolution could have been inhibited by high acidity, due to a high concentration of dissolved iron compounds and sulphates.

At the surface (Section 7.3.7) the Mars Exploration Rover (MER) *Opportunity* found chemical and textural indications that the region around it was formed under water, acidic water at that. The MER *Spirit* crossed into a region where haematite and perhaps sulphates are present, indicating deposition from liquid water.

Global evidence

Mars Odyssey's gamma ray spectrometer has detected hydrogen at shallow depths – below about 0.2 m at the South Pole, gradually deepening to below about 1.2 m at 40°S, the range of latitudes investigated. Neutron spectrometry indicates a large quantity of buried water.

When was Mars wet, and why not now?

Clearly, there is much subsurface water on Mars, presumably largely frozen today, much of it mixed with rock dust. It is also clear that at some earlier time liquid was fairly abundant at or near the surface. This seems to have been in the distant past, in the early Hesperian and Noachian. Most of the features described above indicate this.

☐ What are the exceptions?

The gullies, which are all recent, are an exception, and those outflow channels that have released liquid water as a result of magma heating. Further evidence for more abundant surface water in the past is that in the first 1000 Ma or so of Martian history the rate of fluvial and glacial erosion was about 10–100 times greater than subsequently.

So, why is liquid water now rare near the surface? The usual explanation is that the climate in the Noachian was warmer. This requires a much more substantial, and possibly different, atmosphere as discussed in Section 10.6.3. Most of the atmosphere was gradually lost. This would have included water, some to space and, as temperatures fell, some to the (sub)surface as ice.

An alternative is that Mars was normally in much the same atmospheric state as it is today, except that early in its history, large, infrequent impacts in the heavy bombardment ejected subsurface water ice into the atmosphere as liquid, raining out in about 10 years. This is long enough to carve the valley networks, whether by groundwater sapping or precipitation. The fretted channels presumably formed by sapping. Additionally, major volcanic episodes could have released water from magma.

Question 7.5

State, with reasons, which *one* water-related landform on Mars is least likely to be found on Earth.

7.3.7 Observations at the Martian Surface

In 1976 the Viking Landers became the first two spacecraft to land on the Martian surface and carry out a long and successful series of investigations. They landed in well-separated areas in lowland plains in the northerly hemisphere (Figure 7.8). The view from both sites was broadly similar: a gently undulating, dusty landscape strewn with boulders in the 0.01–1 m size range, most of which were probably ejected from nearby impact craters. The limited photometry that the Lander cameras could perform on the boulders is consistent with basalt. Bedrock might be visible here and there. At the Viking Lander 2 site some of the surface has the same crusty appearance as so-called duricrust on the Earth, where it is the result of dust grains becoming bound together by materials precipitated in pools of water as they evaporate. In the case of Mars, the water could have come from upward percolation, or from the melting of permafrost.

At both sites the dust was sampled and analysed. The relative abundances of chemical elements and isotopes were measured, rather than chemical compounds being detected, and so the mineralogy has to be inferred from the elemental analysis and from sparse data from other observations. Broadly, the data can be matched by dust consisting largely of clays rich in iron and magnesium. On Earth such clays result from the action of water on iron- and magnesium-rich basalts, and on both planets such basalts are to be expected in volcanic lavas. Orbital spectrometry provided evidence that iron- and magnesium-rich clay minerals are common in the bright regions. The dust in general has a high iron content. Oxidation, among other things, produces haematite, and this gives Mars its red appearance, though the reason for the oxidation is unknown. The dust is rich in sulphur, a possible result of sulphates left behind by water evaporation. The action of water is also expected to produce carbonates. The Landers had no means of detecting carbonates.

Evidence for organic compounds was not found. This is consistent with the negative results of the Lander experiments to detect life. But even without the activities of living organisms, organic compounds must continuously be delivered to the surface of Mars

☐ What sorts of bodies deliver organic compounds?

The bodies delivering these compounds are comets and carbonaceous meteorites. The absence of any traces of organic compounds is thought to be the result of their destruction by peroxides produced by solar UV radiation. Such radiation reaches the Martian surface almost unattenuated, because the atmosphere is largely devoid of the ozone that protects the Earth's surface.

In July 1997 Mars Pathfinder landed on Mars near the mouth of the outflow channel Ares Vallis, just inside the northerly hemisphere (Figure 7.8). The view to the west is shown in Figure 7.16. Particularly noticeable are the two peaks on the horizon, about 1 km away. These are 30–35 m tall, and, unsurprisingly, have been dubbed Twin Peaks. Though the most recent outflow from Ares Vallis might have been as long as 2000 Ma ago, the outflow was copious, and the expected evidence of water flow at the Pathfinder site is found in topographical forms such as ridges, troughs, and a distant streamlined island, and also in the boulders – the assortment of types, their size distribution, the roundness of some of them, and the way that some are stacked. One or two boulders, and also the more southerly of Twin Peaks, might be layered, suggesting that they are sedimentary.

As well as evidence of water flow, there is abundant evidence at the Pathfinder site, as elsewhere on Mars, of aeolian processes – dunes, ripples, moats, wind tails, and centimetre-sized scours on rocks. In the thin atmosphere of Mars it takes a very long time to produce such features, and their prominence is further evidence of the low level of the volcanic, tectonic, and fluvial processes that would erase the aeolian features. Further evidence of wind action is

Figure 7.16 Mars, a view from Pathfinder. The two peaks on the horizon – Twin Peaks – are 30–35 m tall and 1–2 km to the west. (NASA/JPL P48984/PIA00765)

air-borne dust. At the Pathfinder site samples were collected and analysed, and a high magnetic content was found, at least partly due to maghemite, an iron oxide slightly depleted in iron compared with magnetite, that on Earth is formed by precipitation from water that is rich in iron compounds. The surface dust was also analysed and found to be much the same as at the two Viking Lander sites. Bright dust was more common than dark dust, the latter predominantly being found where winds would have swept the bright, finer dust away.

A crucial aspect of the Pathfinder mission was a briefcase-sized rover, called *Sojourner*, that made several trips up to 12 m from Pathfinder to measure the relative abundances of the elements in six boulders – the Viking Landers could only perform such analyses on the dust. These analyses were supplemented by photometry in more wavelength bands than the Viking Landers could perform, covering the range 0.44–1.0 μm. The somewhat surprising result is that some of the boulders are rather andesitic in composition, indicating chemical differentiation in basaltic–gabbroic crustal rock (Table 6.1). If so, then rather more extensive differentiation has occurred in the Martian crust than previously thought. However, as noted above, it is possible that the andesitic material is only a veneer produced by weathering.

The most recent landers are the two MERs, *Spirit* and *Opportunity*, which arrived in January 2004. Plate 9 shows a view from each of them, and Figure 7.8 shows their locations. By November 2006 they had spent over 1000 sols on Mars, photographing and analysing the surface (a sol is a day on Mars, just slightly longer than a day on Earth). Each of them carries a panoramic camera and a low-power microscope. They also carry a rock abrasion tool and three spectrometers – IR, X-ray, gamma ray. Unfortunately, they have no means of carrying out radiometric dating. By November 2006 *Spirit* had travelled about 7 km and *Opportunity* about 10 km. What have they found on their travels?

Spirit landed in Gusev Crater, a 150 km diameter impact crater in the southerly hemisphere, not far south of the equator, in a region 3000–4000 Ma old. This crater was selected because the channel system Ma'adim Vallis drains into it from the south, so signatures of a water fill were expected. However, there was no immediate evidence that a lake had ever been present – no sedimentary rocks or deposits, no water-formed minerals (e.g. carbonates). *Spirit* found magnesium-rich basalts dominated by olivine. Presumably, if there were any lake sediments then they would have been buried by lavas, deep enough so that the small craters within Gusev would not have reached them.

Spirit then made a journey towards some hills within Gusev Crater, named the Columbia Hills, 3–4 km from where it landed. En route it saw some layered outcrops that could be sedimentary. On reaching the Columbia Hills it ascended one of the tallest, Husband Hill, though it is only about 100 m above the surrounding crater floor. In this region it found rocks with a wide variety of compositions. Detailed analyses indicate that many are basaltic impact ejecta, subsequently altered aqueously, but only slightly, perhaps by steam and local fumeroles. At least one rock is an unaltered basalt, and one other is a basaltic sandstone cemented by sulphates, though unaltered pyroxene and olivine indicate that the wetting was brief. *Spirit* then visited Home Plate, a bright, low plateau. This is an altered basalt, its association with nearby blocks of lava of similar composition favouring a volcanic origin over an impact one.

Husband Hill (and presumably the rest of the Columbia Hills) seems to be a sample of the ancient basaltic Martian crust, with impacts as the main modifying process. Water has played only a minor role, probably confined to a brief period over 3500 Ma ago. Plains volcanism postdates the formation of the Columbia Hills.

Opportunity landed in Meridiani Planum, a 3000–4000 Ma old plain in the southerly hemisphere just east of Valles Marineris, 2°S, in a crater 20–30 m across, subsequently named Eagle Crater. This site was chosen because it is rich in haematite, indicative of the past presence of liquid water.

Eagle Crater has punctured the dark, sandy basaltic plain to reveal fine layers running half way around the crater wall, resembling sedimentary rocks on the Earth. The action of (salty) water here is supported by rock textures and evaporite mineral compositions. Similar layers were found 700 m away in the 130 m Endurance Crater. This larger crater has penetrated to different material, so *Opportunity* was sent several metres down a slope named Karatepe to examine it. It encountered a knobby texture, in contrast to the finely laminated upper material. The knobby texture can be explained by recrystallisation, the water reaching the level marked by the transition to the upper material. The chemical composition also changes between the two layers. This can be explained by the aqueous removal from the basaltic upper layer of about 55% of its Fe, Mg, and Si.

Opportunity then went on to a highly eroded impact crater called Erebus, about 300 m across, where it found sediments displaying evidence of water action. Finally, it journeyed to the 800 m diameter impact crater Victoria where it is examining the layering in the crater walls.

In this long journey it has scrutinised the Burns Cliff. The lower part is aeolian dunes, truncated by a middle layer that is an aeolian sand sheet that must have eroded the dunes. The upper layer shows the rippled texture typical of that formed on sand by flowing water. This was a habitable area, but only intermittently. At another location, Escher rock has polygonal fracturing consistent with shrinkage due to dehydration of sulphates. Other rocks are erratics, i.e. have come from elsewhere. For example, Bounce Rock is pure pyroxene, ejected from a distant impact crater. It resembles EETA79001, the youngest Martian meteorite (see Section 7.3.8). A meteorite on Mars has also been found – an iron meteorite named Heat Shield Rock.

Of particular interest are the 'blueberries', granules 1–5 mm across. That they formed in water is indicated by a composition with more than 50% haematite, and a morphology consistent with growth in liquid water. These are found in the crater layers and in sedimentary outcrops outside them.

The approach to Victoria Crater is slightly uphill, the blueberries and the surface cracks disappearing. The blueberries reappear higher up. One explanation is the creation of the gentle slope by aeolian erosion, which removed blueberries from the surface, followed by the formation of Victoria Crater, a process that created the blueberries near its rim. Victoria Crater has a crenellated rim, the result of erosion, and its interior contains sand dunes. The crenellation is due to erosion, a conclusion borne out by *Opportunity*'s image of the crater wall, which shows an ejecta blanket overlying barely disrupted rock that could not have been at the edge of the initial crater. The crater wall shows layering that *Opportunity* will explore by working around the crater.

The action of water is much more in evidence in the area explored by *Opportunity* than that explored by *Spirit*. The area is also rich in elements that are mobilised by water. Water certainly seems to have been important in this region.

In summary, exploration at the surface has taught us that the surface composition of Mars, and the textures and morphologies, are consistent with volcanically produced basalts, perhaps with some subsequent modification by partial melting and certainly by weathering, including the action of water. The areas explored were damp, perhaps wet, at some time in the distant past. The absence of organic materials is thought to be due to their destruction by peroxides produced by solar UV radiation. These conclusions are consistent with what we have learned from orbit. Are they also consistent with what we learn from Martian meteorites?

7.3.8 Martian Meteorites

The samples examined by the various landers are not the only samples of the Martian surface that we have scrutinised at close range. In Section 3.3.4 you learned that by mid 2006 there were 34 meteorites that are widely regarded as having come from Mars. The radiometric solidification ages vary from one to the other (more properly called chemical isolation ages). The oldest (ALH84001) has a solidification age of 4500 Ma; for the others it ranges from 1360 Ma to as recent as 165 Ma, indicating prolonged igneous activity. For reasons outlined in Section 3.3.4, Martian surface as old as about 4500 Ma is underrepresented in the meteorites. Even so, with just one so far in the collection, it is deduced that such an ancient crust is more common on Mars than on the Moon.

Martian meteorites are broadly basaltic–gabbroic in composition, and so have presumably crystallised from magma at and below the Martian surface. Basalt predominates in several of them, thus providing further evidence that basalts are common on the Martian surface. Note, however, that the meteorites are so few that their sampling of the crustal composition is poor. In particular, they do not necessarily represent bulk crustal composition – all but one are less than 1360 Ma old, and most are cumulates of various minerals, which makes it hard to infer the composition of parent magmas. They do, however, indicate that the mantle is depleted in siderophile elements and in elements that preferentially enter sulphide melts. This indicates an iron-rich core in which FeS is prominent.

They have also enabled the dates of key events in Martian history to be obtained, through accurate measurements of isotopes in various minerals in Martian meteorites. For example, the concentration of hafnium and tungsten isotopes in meteorite minerals has been used to date Martian core–mantle separation at about 150 Ma after the formation of Mars. The method relies on the decay of ^{182}Hf into ^{182}W, also used to date the Moon's core formation (Section 7.1.7). The decay of the samarium isotope ^{146}Sm into the neodymium isotope ^{142}Nd (half-life 103 Ma) has been used to date the separation of crust to within 100 Ma of Mars's formation. The details will not concern us.

Also of global significance is the evidence that since solidification, the Martian meteorites have been modified by liquid water, up to at least the last few per cent of Martian history. Weathering products in them add to the evidence for clays on Mars. Carbonates are also present. Martian meteorites thus support the idea of liquid water being available near the surface of Mars throughout much of its history.

7.3.9 The Evolution of Mars

A plausible picture of the early history of Mars resembles early lunar history, with differences because of the lower rate of Martian heat loss per unit mass, consequent upon its greater size, and because of its greater abundance of water and other volatiles.

(1) Towards the end of accretion there was a magma ocean created by impact melting (perhaps aided by a thick blanketing atmosphere). This led to the formation, within 100 Ma of the formation of Mars, of a crust of different composition from the mantle, perhaps with a concentration of the heat-producing radioactive elements in the crust. The magma ocean, always shallow because of Mars's low mass, had solidified by the end of the heavy bombardment.

(2) Ancient impact basins became places of crustal weakness that were sites of subsequent volcanic activity that flooded these basins with lava, and in some cases led to later volcanic activity.

(3) Probably before a well-developed lithosphere was established, and around 150 Ma after the formation of Mars, the core formed consisting largely of iron and iron sulphide.

(4) Around 4120 Ma ago (about 450 Ma after the formation of Mars), the northerly/southerly divide was created, as a result of mantle convection. This resulted in a thinner northerly crust, and this made the northerly hemisphere more prone to volcanic resurfacing.

(5) The uplift of the Tharsis dome was complete by the late Noachian (which ended around 3700 Ma ago), producing extensive fracturing. This uplift might have been the result of a convective plume in the mantle. Radiogenic heating increased the temperature of the interior, and the subsequent crustal expansion resulted in further fractures and in volcanic activity, particularly in the northerly hemisphere. There is a similar story for the Elysium dome.

(6) Volcanism persisted throughout much of Martian history, focused in later times on the Tharsis and Elysium regions, perhaps to within the last 3 Ma. Volcanism declined as radiogenic heating declined, so the interior cooled and the lithosphere thickened. The lithosphere, about 80 km thick at 4100 Ma, is at least 120 km thick today, and the asthenosphere might now have vanished completely. There is no evidence for plate tectonics (Section 8.1.2).

(7) Volcanic and tectonic activity might rarely still occur. Impact cratering and degradation are certainly continuing, with the thin Martian atmosphere playing a prominent role.

(8) There is ample evidence that liquid water was more readily available and more persistent in the Noachian and early Hesperian than today. The likely cause is a warmer climate, and perhaps major volcanic episodes or large impacts. Liquid water can occur locally today, as seen in the gullies, and perhaps some outflow channels can still be prone to activity, driven by geothermal heating and the variation in Mars's axial inclination.

Question 7.6

Make a brief case for transferring this section on Mars to the next chapter – on active surfaces.

Question 7.7

Outline how the surface of Mars might have been different if its crust had everywhere been as thin as it is thought to be in the northerly hemisphere.

7.4 Icy Surfaces

Pluto, its satellite Charon, and most of the satellites of the giant planets have icy surfaces. This is revealed by spectroscopy, and by the low mean densities of those bodies for which data are available (Table 1.2). Icy surfaces are expected in the outer regions of the Solar System because of the low temperatures, and you have seen that, because of its high cosmic abundance and low volatility among icy materials, water is expected to be dominant. Beyond Jupiter other ices, such as those of NH_3, CH_4, and N_2, can be significant components of surfaces. The near absence of these icy materials from the satellites of Jupiter is due to the higher temperatures closer to the Sun, and to IR radiation from the formation of this massive planet.

Among the icy surfaces, only those of Europa, Titan, Enceladus, and Triton are active. The rest are, at most, weakly active, and we shall concentrate on the four largest of these. Table 7.2 highlights the distinguishing features of the largest of the remainder – the icy satellites of intermediate size. Not included are the smallest satellites, icy or rocky. Very few of these have in any case had their surfaces explored to any significant extent.

7.4.1 Pluto and Charon

With a radius of 1153 km Pluto is far smaller than the eight planets closer to the Sun, and because of its great distance from the Sun its surface temperatures never exceed about 60 K. Its visual geometric albedo ranges from 0.5 to 0.7 across its surface, consistent with fairly pure icy materials. Earth-based IR spectrometry has identified nitrogen (N_2) ice on the surface, plus smaller quantities of CH_4 ice and a trace of CO ice. Hubble Space Telescope images have revealed a bright polar cap and bright patches elsewhere, presumably consisting of clean ice. The rest of the surface could be ice mixed with rocky materials, or with organic compounds generated by the action of UV radiation on ices. Organic tars called tholins have been detected.

The distribution of albedos over the surface exhibits changes, perhaps due to atmospheric transport of frosts. There is a very thin, slightly hazy atmosphere with a current surface pressure of about 15 Pa. It is dominated by N_2, with some CH_4. CO is also presumed to be present. It is thought that a layer of solid CH_4 about 10 km thick could have been lost to space from the surface over Pluto's lifetime, in which case the surface will be depleted in impact craters up to about 10 km in diameter. No spacecraft has yet visited Pluto so we cannot confirm this prediction.

An icy surface is in accord with the model of the interior in Figure 5.7. Water ice is the main constituent of the icy mantle, but this is expected to be topped by the more volatile ices that have been detected at the surface. Any CO_2 would also be hidden from view. Note that in the low nebular pressures where Pluto formed, N_2 is more likely to harbour nitrogen than NH_3, so

Table 7.2 Distinguishing surface features of the inactive intermediate-sized icy satellites

Object	Orbit semimajor axis/ 10^3 km	Radius/ km	Surface features
Saturn			
Mimas	186	199	Heavily cratered, though craters over 30 km are rare, perhaps due to partial resurfacing before the end of the heavy bombardment. Herschel crater, at 140 km across, and by far the largest, shows long grooves opposite this impact, possibly a result of seismic waves
Tethys	295	530	Heavily cratered, except for a region of early cryovolcanic resurfacing
Dione	377	560	Cratered, though with considerable cryovolcanic resurfacing near the end of the heavy bombardment
Rhea	527	765	Imaged hemisphere dominated by the heavy bombardment. A later, lighter bombardment confined near to the North Pole. Little resurfacing
Iapetus	3561	718	Heavily cratered. In synchronous rotation. The leading hemisphere has a visual albedo of 0.04 ('soot'), the value for the trailing hemisphere is over 0.6. The 'soot' is no more than about 1 km thick, and postdates the craters. Its source is one or both of the outer satellites (by impact erosion) and interplanetary particles
Uranus[a]			
Miranda	130	236	Heavily cratered, plus three regions resurfaced after the heavy bombardment, perhaps due to collisional disruption followed by tectonics and cryovolcanism
Ariel	191	579	Surface cratered after the heavy bombardment with global tectonic features and associated volcanism lasting to about 2000 Ma ago
Umbriel	266	585	Heavily cratered. A particularly low albedo (0.19), perhaps from a surface veneer additional to effects of methane decomposition
Titania	436	789	Similar to Ariel, and perhaps with a similar surface history
Oberon	583	762	Heavily cratered, rather like Umbriel
Neptune[a]			
Proteus	118	210	Poorly imaged. Craters and tectonic features present, plus a large depression
Nereid	5513	170	No good images or other data, but presumably icy. Some evidence of surface markings

[a] Many of the satellites of Uranus and Neptune are darkened by the cosmic ray decomposition of methane in the surface ices.

NH_3 could only exist in the interior. Likewise, CO would be the nebular repository of carbon as opposed to CH_4 (mainly) and CO_2.

Pluto experiences extreme seasons. This is due to its high orbital eccentricity (0.254) and its high axial inclination (123°). The latter is an expected result of the collision between Pluto and another body that is thought to have formed Charon. This collision can also explain the surface cleansing that has given Pluto a significantly higher albedo than many EKOs. Pluto also has a highly variable atmospheric mass. The very volatile N_2 is expected to condense on the surface as Pluto recedes towards aphelion, having passed through perihelion in 1989 in its 251 year orbit.

Pluto's largest satellite, Charon, has a radius of 603 km, about half that of Pluto. With a visual geometric albedo of 0.4 it too seems to have been cleansed by the collision. Spectroscopy indicates that about 70% of its surface is water ice, in a crystalline form indicating slow condensation. The remaining 30% or so has a spectrum matched by a mixture of NH_3 and $NH_3.H_2O$. The NH_3 is probably a thin veneer, derived from the interior: as noted above, N_2 is the stable form at low pressure, such as that at the surface. CH_4 might be absent from the surface of Charon.

Radiogenic and tidal heating might have been appreciable, but it is very unlikely that volcanic or tectonic processes still operate.

❑ Why is this?

They are small bodies, and so cool rapidly.

Pluto and Charon are in mutual synchronous rotation, with a period of 6.387 days.

7.4.2 Ganymede and Callisto

Of the four Galilean satellites of Jupiter, the outer two, Ganymede and Callisto, are no longer tectonically or volcanically active. Ganymede and Callisto have been investigated by a variety of flybys, and from December 1995 to September 2003 by the Galileo Orbiter. The outer regions (Figure 5.7) are dominated by water ice. Hydrated minerals are also present, along with small quantities of so far unidentified substances. These satellites are too close to the Sun for the more volatile ices to be present as more than traces.

Callisto is the third largest satellite in the Solar System, slightly smaller than Titan. It is very heavily cratered (Plate 15), indicating little by way of resurfacing since the heavy bombardment, except for the partial burial of small craters by dark, smooth material. Indeed, among the icy satellites, Callisto is one of the darkest, with a visual geometric albedo of only about 0.2. This might be a consequence of the great age of the surface – loss of ice through sublimation or through bombardment by charged particles in the magnetosphere could have created a surface enriched in the rocky materials that were originally a minor component. A significant proportion of rocky dust in the ice is also indicated by the manner of gradation of some crater walls and by the low radar transparency of the ice.

❑ Where would you expect to see high albedos on Callisto?

If there is purer ice beneath the surface, then young craters and the ejecta from them should have high albedos. This is observed to be the case. There is no evidence of volcanic or tectonic activity.

Large craters, greater than about 60 km diameter, are scarce, and many craters show evidence of viscous relaxation – the effect on large craters is greater than on smaller ones, and as a result many of the oldest larger craters have vanished. Under present conditions on Callisto this would not happen even in 4600 Ma. Therefore, a thinner lithosphere and warmer subsurface conditions are indicated for the past, and this is consistent with thermal models of Callisto. The higher interior temperatures in the distant past were a result of residual heat from accretion and differentiation, tidal slowdown of the rotation (to yield the present synchronous rotation), and greater radiogenic heat from radioactive isotopes in the rocky materials. A thin lithosphere and viscous relaxation in the past are also indicated by several multi-ring basins with flat areas in the centre. Today, a comparable impact would not generate rings around the impact basin, and the central region would be bowl shaped. The absence of lumpy, grooved, and hilly terrain opposite the impact basin Valhalla contrasts with the hilly terrain diametrically opposite the Caloris Basin on Mercury, and indicates a soft shell inside Callisto in the distant past. Indeed, there is some evidence for a soft shell today, in particular an electrically conducting liquid. The evidence, as noted in Section 5.2.3, is the effect Callisto has on Jupiter's magnetic field. The liquid would be water, and the conductivity would be provided by dissolved salts. These, along with NH_3, would also serve to reduce the freezing temperature sufficient for the liquid phase.

The other inactive Galilean satellite, Ganymede, is the largest satellite in the Solar System. It has a radius 1.08 times that of Mercury, though only 45% of Mercury's mass.

☐   What does this indicate about the composition of Ganymede?

This indicates a large proportion of icy materials (water), as in Figure 5.7. Its surface displays two distinct terrains, each accounting for about half the surface, though the two are intermingled (Plate 14). There are regions of heavily cratered darker terrain, and bands of less heavily cratered brighter terrain that here and there makes wedge-shaped incursions into the darker terrain.

On the darker terrain the crater density is about a third of that on Callisto, with a notable scarcity of craters larger than about 100 km in diameter. The craters show evidence of even greater viscous relaxation than on Callisto and this could account for their smaller number, though cryovolcanic resurfacing before the end of the heavy bombardment is another possibility. Though large craters are scarce on Ganymede, there are flat circular features called palimpsests that could be almost fully relaxed impact basins, or impacts that penetrated a thin lithosphere and released slushy ice.

The brighter terrain is criss-crossed by belts of ridges and grooves that suggest several episodes of formation and a consequent range of ages for this type of terrain (Figure 7.17). This conclusion is borne out by the variation in crater densities from place to place on it. Overall, the crater density on the brighter terrain is less than on the darker terrain, indicating that the brighter terrain is younger, though the high-resolution Galileo Orbiter images have confused this simple picture by revealing that one dark area has been reworked relatively recently, and that one bright area has fine grooves plus numerous small impact craters that can be explained by crustal expansion long ago.

The most likely origin of the bright terrain is from the thermal expansion of Ganymede consequent upon differentiation, or expansion upon freezing of an icy mantle. This cracked the lithosphere to form grabens that filled with slushy ice. This ice froze, and the ridges and grooves could be the result of cracks and subsidence, or of cryovolcanism along cracks. In this model the infill is brighter because as a partial melt it is relatively free of dust and rock fragments. This also makes it less dense than the overlying material, which helps it to rise to the surface. Episodes of enhanced tidal heating, with the most recent perhaps about 1000 Ma ago, can explain the range of ages in the bright terrain. Such episodes could arise from orbital evolution (Section 5.2.3). It is even possible, as noted in Section 5.2.3, that there is a liquid salty water shell today, sustained by radioactive and tidal heating. IR spectra have revealed magnesium sulphate ($MgSO_4$) at the surface, which alone would make any liquid shell conducting. This is one way to explain (some of) Ganymede's magnetic dipole moment. Models indicate that a liquid shell would be a few kilometres thick at a depth of about 175 km (shown in Figure 5.7).

Compared with Callisto, Ganymede has clearly been more extensively resurfaced. This is consistent with Ganymede's greater size (slower cooling rate), its slightly greater ratio of rocky to icy materials (greater radiogenic heating), and a possible history of significant tidal heating.

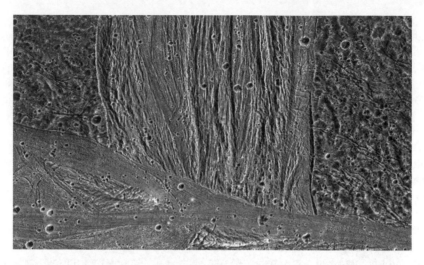

Figure 7.17 Two swathes of brighter terrain on Ganymede, one down the centre and a narrower one across the bottom of this 950 km wide frame. (NASA/JPL P50037)

This is consistent with the clear evidence that Ganymede is differentiated more completely than Callisto.

Question 7.8

State, with justifications, how you would modify the orbit of Pluto and the orbit of Ganymede to cause cryovolcanism on these bodies.

7.5 Summary of Chapter 7

The solid surfaces in the Solar System can be divided into those that today are subject to little other than impact cratering and gradation, and those that are additionally subject to a significant level of volcanic or tectonic activity. Broadly speaking, the larger the body, the more likely it is to have a surface that falls into the latter category, though tidal heating blurs this size-based distinction.

The largest bodies that are no longer subject to volcanic or tectonic activity comprise the Moon, Mercury, and Mars, which have rocky surfaces, and Pluto, Charon, Ganymede, and Callisto, which have icy surfaces. In general, these surfaces are now dominated by impact craters, though they all bear evidence of volcanic and tectonic activity early in their history, the details varying from one body to another. There is no evidence of plate tectonics (Section 8.1.2).

The surface of Mercury has been modified by crustal shrinkage and by viscous relaxation early in its history, and subsequently by lava flows. This has produced a variety of terrains that differ in their crater densities, from heavily cratered terrain, through intercrater plains, to smooth plains.

The Moon has a predominantly anorthositic crust, thought to be derived from a peridotite mantle. The impact basins on the near side have been filled by basalt lavas to form the maria – areas with much lower crater densities than the nearly saturated highlands that dominate much of the Moon. The basins on the far side, perhaps because they are at higher altitudes, are only partially filled. The crust on the far side is generally thicker than that on the near side.

The surface of Mars, like that of the Moon, also has two distinct hemispheres. The southerly hemisphere is dominated by impact craters, and has an abundance of channels and other features that have been created by flowing water. It is an ancient domain. The northerly hemisphere has a lower altitude, and the crust is thinner there. There are few impact craters, but extensive evidence of volcanic and tectonic activity, with some lava flows as recent as a few million years, though mostly much older. Unlike Mercury and the Moon, there are extensive deposits of ices of water and CO_2, particularly at the poles. The observable surface of Mars is largely basaltic, the result of volcanism. Subsequent alteration by various processes has occurred, including those involving water and solar UV radiation, the latter eliminating organic materials. Further crustal differentiation might have taken place, to produce rocks more andesitic in composition, though the andesitic signatures might be largely or wholly due to a thin surface layer produced by the weathering of basalts. Gulleys provide evidence that aquifers can deliver water to the surface today.

Pluto, and many of the larger icy–rocky satellites, have a long history of surface inactivity, and are known or presumed to be heavily impact cratered. Their surfaces are not pure water ice, but contain rocky dust and, beyond Jupiter, other ices, in some cases as a thin veneer. They all show some variation in composition across the surface. Callisto and Ganymede show evidence of viscous relaxation in the distant past. Ganymede also seems to have experienced some resurfacing due to enhanced tidal heating, most recently perhaps about 1000 Ma ago.

Except for Enceladus, the smaller bodies all lack volcanic and tectonic activity today. The surface characteristics of some of them are summarised in Table 7.2.

8 Surfaces of Planets and Satellites: Active Surfaces

In this chapter we discuss the surfaces that are still (cryo)volcanically and tectonically active – the surfaces of the Earth, Venus, Io, Europa, Titan, Enceladus, and Triton.

8.1 The Earth

The Earth is the largest of the terrestrial bodies, slightly larger than Venus, and for obvious reasons its surface is the best known in the Solar System. About 70% of the surface is covered in oceans, but in this chapter, as well as the continents, we are concerned with the solid surface under the oceans, postponing a discussion of the oceans to Chapter 10.

The Earth's surface is both tectonically and volcanically very active, largely because of its high internal temperatures and thin lithosphere. The high temperatures are a consequence of primordial heat plus the heat from long-lived radioactive isotopes. The Earth is unique in that it has a *global* tectonic system called plate tectonics, referred to briefly in Section 5.1.1. This has sculpted the large-scale features, and has given the Earth a predominantly youthful surface almost devoid of impact craters. Plate tectonics will therefore be the focus of this section. To understand plate tectonics we first need to look at the Earth's lithosphere in more detail.

8.1.1 The Earth's Lithosphere

On the basis of rock samples, seismic, gravitational, and other evidence, a typical section through the Earth's lithosphere is known to be as in Figure 8.1. The concentric layering that characterises the structure of the deep interior is absent. Instead, the structure of the lithosphere varies from one region to another. It is mostly 90–100 km thick, and consists of the Earth's crust plus the uppermost part of the mantle. The mantle consists of peridotite, which you will recall is a mixture of iron-and magnesium-rich silicates, notably pyroxene and olivine (Table 6.1). The crust consists of less dense silicates that are not so rich in iron and magnesium. The crust is subdivided into oceanic crust and continental crust. As their names imply, these are found under the oceans and on the continents respectively. However, though continental crust reaches higher altitudes than oceanic crust and thus accounts for most of the dry land, and though oceanic crust

Discovering the Solar System, Second Edition   Barrie W. Jones
© 2007 John Wiley & Sons, Ltd

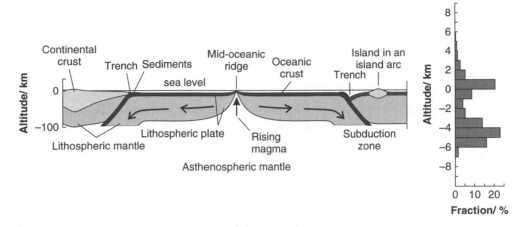

Figure 8.1 The Earth's lithosphere and hypsometric distribution.

with rare exceptions lies beneath the sea, the sea shore is rarely the boundary between the two crustal types. In many regions continental crust only gives way to oceanic crust some distance off shore. Only if the volume of water in the oceans were somewhat reduced would the sea shore predominantly coincide with the boundary.

Continental crust not only reaches higher altitudes than oceanic crust, but also occupies a largely separate range of altitudes. This is shown by the Earth's hypsometric distribution in Figure 8.1. You can see that there are two peaks – it is bimodal. The lower peak corresponds mainly to oceanic crust, and the higher peak mainly to continental crust. Oceanic crust is 5–10 km thick (mostly about 7 km), has an average density of about $2900\,\mathrm{kgm^{-3}}$, and is dominated by basaltic–gabbroic rocks. Continental crust has an average thickness of about 30 km, ranging from about 20 km to about 100 km beneath large mountain ranges. It has an average density of about $2600\,\mathrm{kgm^{-3}}$, and has an overall andesitic composition, which is intermediate between basaltic–gabbroic rocks and granitic–rhyolitic rocks (Table 6.1). In its upper reaches continental crust is more granitic–rhyolitic and in its lower reaches it is more basaltic–gabbroic. Much of the Earth's continental crust is covered by a thin veneer of soil. This is derived from rocks through gradation plus extensive modification by the Earth's biosphere. Much of the oceanic crust is covered in sediments.

⬚   What general processes could give rise to the formation of a crust on the mantle, and the separation of the crust into two types?

Partial melting created the oceanic crust and then fractional crystallisation separated it into two types. These processes continue today.

In Figure 8.1 you can see that crustal elevations are mirrored by deep 'roots'. This is a consequence of isostatic equilibrium, evident in gravitational data, and common over the Earth's surface. Such extensive isostasy implies the existence of a plastic region – an asthenosphere – for which there is much evidence (Section 5.1.1). It underlies the lithosphere. Were oceanic crust as thick as continental crust then, because of the greater density of oceanic crust, isostasy would still lead to it 'floating' lower on the asthenosphere than does the continental crust. The smaller thickness of oceanic crust increases the altitude difference, as does the weight of water lying on top of it.

8.1.2 Plate Tectonics

Most of the large-scale features of the lithosphere, and the volcanic and tectonic processes that mould it, can be explained by the theory of plate tectonics. This theory developed from seeds sown in the early years of the twentieth century, and it had its full flowering and wide acceptance in the 1960s.

According to the theory, the lithosphere of the Earth is divided into fairly rigid lithospheric plates in motion relative to each other. The continents are carried on these plates, and therefore the modern map of the world is transitory. There are seven large plates, as shown in Figure 8.2, and a greater number of smaller ones, only a few of which are shown. Their relative motions define three kinds of boundaries between the plates. The first kind is where two plates are sliding past each other, side by side. This kind is called a conservative margin, and the boundary is a transform fault where the predominant displacement is horizontal. Figure 8.2 shows many examples, e.g. the San Andreas Fault, marked by the line near the Pacific coast of California.

The second kind of plate boundary is the constructive margin. The plates on each side of a constructive margin are continuously created from the upper mantle by partial melting of the mantle beneath mid-oceanic ridges (Figures 8.1 and 8.2). This partial melting is the result of pressure release as material ascends, and it creates a magma of basalt–gabbro which has a lower density than mantle peridotite and therefore rises to form basaltic–gabbroic crust as the upper layer of the plate. This is how oceanic crust is created. The rest of the oceanic plate consists of mantle peridotite somewhat depleted in the elements that are enriched in basalt–gabbro. The elevation of the ridges is largely due to isostasy associated with their higher temperatures and consequent lower densities. The oceanic plates slowly spread away from the ridges. This motion is recorded in the oceanic crust by remanent magnetism, which preserves reversals of the Earth's magnetic field as the crust solidified.

The movement of the oceanic plates away from the ridges has its immediate cause in the descent of oceanic plates into the Earth at the third kind of plate boundary, the destructive margin, to which we now turn.

Destructive margins

Because the Earth is not expanding, plate spreading must lead to plate collisions elsewhere. To consider collisions, it is necessary to distinguish between plates where the crust is oceanic and plates where the crust is continental.

Consider first the collision between two *oceanic* plates, as shown to the right in Figure 8.1. This is the first kind of destructive margin. You can see that one plate dives steeply beneath the other and descends into the Earth – it is subducted. As it descends, it pulls the rest of the plate with it. This is called slab pull. The pull is considerably enhanced by the dehydration of the descending oceanic lithosphere and its conversion into denser minerals, constituting eclogite, a mixture of pyroxene (Table 6.1) and garnets (typically Mg–Al–Fe silicates). The dehydration results in the hydration of the mantle part of the *overriding* plate. This lowers its melting temperature, which results in partial melting to form basaltic magma. The hydration, and other volatiles, lead to explosive volcanism. In the crust of the overriding plate, usually at its base, fractional crystallisation of the basaltic magma can form magmas with higher silica content, and continental crust is born in the form of island arcs (Figure 8.1 and, for example, many islands around the Pacific Ocean).

Consider now the case where continental crust has grown to the point where it occupies a significant fraction of a plate. We can then have the second kind of destructive margin, as shown

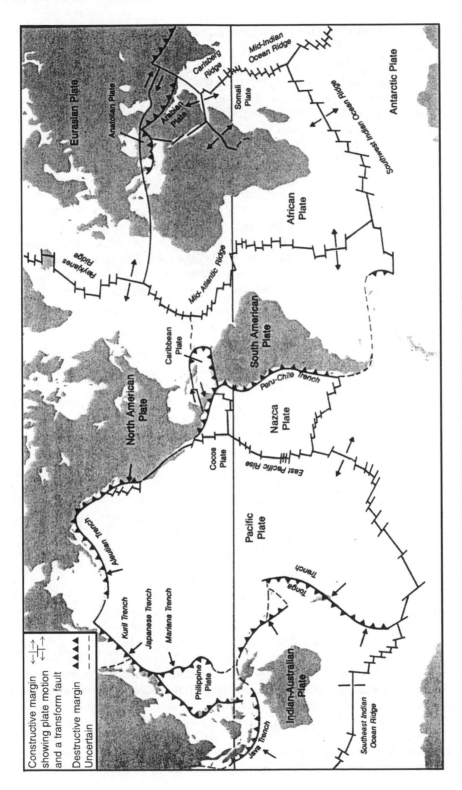

Figure 8.2 The Earth's major lithospheric plates.

to the left in Figure 8.1. The lithospheric plate that bears continental crust will always override the oceanic plate.

❒ Why is this?

Continental crust is less dense than oceanic crust, and thicker. Magmas form from the mantle of the overriding plate as outlined above. As these ascend, they pass through the existing continental crust and partially melt it to form andesitic magmas (andesites – Table 6.1). These are named from the Andes, the mountain chain built largely from andesites (plus upward bending of the overriding plate. It lies alongside the Peru–Chile Trench in Figure 8.2. Andesites are widespread.

The continental crust formed by the process of fractional crystallisation of magma typically consists of two layers, an upper layer particularly rich in silicon and oxygen (granitic–rhyolitic in Table 6.1) and a lower layer richer in magnesium and iron (though remember that iron is not a major component of the Earth's mantle and crust). The lower layer gives the crust strength, because it does not deform plastically at the prevailing temperatures, unlike the upper layer.

The descending plates at both kinds of destructive margins do not melt, but seem to descend to the core–mantle boundary – the mantle, though solid, is plastic enough to allow this. There is seismic evidence for plate fragments in the lower mantle, mentioned in Section 5.1.1. How the sunken plates then evolve is unknown, but presumably they merge with the mantle.

Continental collisions

If the diving plate also carries continental crust, then, as subduction continues, this crust can ultimately meet the continental crust of the other plate. This will uplift any sedimentary deposits and bits of oceanic crust trapped between the two continents. There is also a thickening of the continental crusts from the compression of one or both plates and from reverse faulting. Isostasy ensures that the topography rises, to form mountain ranges such as the Himalayas.

The continental part of some plates is particularly rigid, and is called a craton. An example is India. When the plate carrying India collided with the plate carrying Asia, the Himalayas and the Tibetan highlands were raised as a result of compression of the Asian Plate, where the lithosphere is warm in parts and relatively weak. In contrast, India was relatively undeformed. Other cratons include the continents of Africa, South America, Australia, and Antarctica. Along with India, these are pieces of an ancient supercontinent called Gondwana – this is apparent within the dashed circle in Figure 8.3 below in an early stage of break-up. The rigidity of cratons can have several contributory causes: metamorphism that has removed radiogenic isotopes, resulting in a cooler lithosphere; a dry lower crust (which increases the yield strength) from early partial melting; and an unusually thick lithosphere (up to about 200 km).

A new constructive margin can form within a continent as a result of a rising (solid) plume in the mantle. In its initial stages the plume will create a rift valley, such as the Red Sea, where the associated subduction under the Zagros Mountains of Iran. Therefore, plate motion not only moves continents and joins continents, it can also disrupt them. From the onset of surface rifting to the start of the creation of new oceanic lithosphere takes of the order of 10 Ma. Note that the Great Rift Valley in Africa (Figure 6.7), a splendid example of a graben, is also the result of a plume, but does not seem to be a constructive margin yet – there is no slab pull in operation.

Creation and destruction of continental crust

Destructive margins are the main source of continental crust, through the emplacement of extrusive and intrusive igneous rocks. Since about 2500 Ma ago the rate of creation has averaged

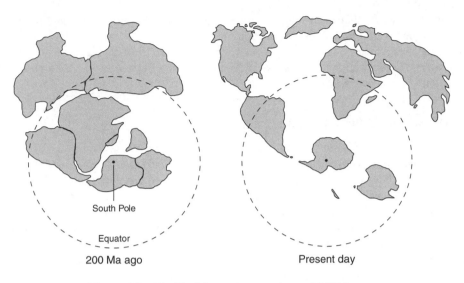

South Pole

Equator

200 Ma ago   Present day

Figure 8.3 The Earth's continents today and 200 Ma ago.

about $1.7\,km^3$ per year. What about its rate of destruction? The volume of *oceanic* crust is kept roughly constant by subduction (Figure 8.1). For subduction to cause significant loss of *continental* crust, continental oceanic sediments would have to be subducted. This happens only to a very limited extent. The sediments are too buoyant and a large proportion is thus left behind where it can rejoin continents or remain on the sea floor. Delamination could also remove continental crust (see below). But overall, the volume of continental crust plus sediments derived from it is thought to have gradually increased over the billennia.

As well as preserving continental crust it is also necessary to maintain its altitude above sea level. It has been estimated that gradation at the rates that we know have been operating over the past few hundred million years can convert all of the continental crust above sea level into ocean sediments in much less than 100 Ma. Therefore, some opposing mechanism must have been operating. Thickening of continental lithosphere in continental collisions helps. Also, as the lithosphere thickens it gets denser. The lower mantle part of the plate reaches densities that exceed that in the upper asthenospheric mantle. It breaks away, and isostatic adjustment causes the remaining lithosphere to rise. This loss of the lower lithosphere is called **delamination**. In some cases the breaking away might extend to the denser lower crust, in which case this contributes to the loss of continental crust.

Plate tectonics in the early Earth

Before about 2500 Ma ago, plate tectonics seems to have operated differently. The evidence includes peculiar geological landforms on ancient continental crust. At that time heat flow from primordial sources was several times greater than today. This resulted in either smaller plates, or a higher rate of spreading from constructive margins. Either way, oceanic crust was warmer and more buoyant. Consequently, it descended at a shallow angle and thus had time to melt partially before it had time to dehydrate and form eclogite. Models, plus geochemical evidence, show that there was no fractional crystallisation of magma at this time, and that therefore the whole continental crust was rather like the upper crust today. It was therefore weak and in collisions between continents high mountain ranges could not be built.

The rather abrupt change at about 2500 Ma might have been caused by a pulse of continental crust formation, perhaps triggered by the decreasing heat flow from primordial sources reaching some threshold.

8.1.3 The Success of Plate Tectonics

You have probably already noted many features of the Earth's surface that can be accounted for by plate tectonics. Thus, it accounts for the two types of crust, oceanic and continental. These arise from the partial melting at constructive and destructive margins. Additionally, it explains the division of continental crust into upper and lower layers, the result of fractional crystallisation. It also accounts for the hypsometric distribution – the bimodal form of this distribution arises from a combination of the greater density of the oceanic crust and the greater thickness of the continental crust, with both crusts in isostatic equilibrium over most of their areas. Plate tectonics also accounts for the greater thickness of the continental crust.

One of the earliest indications of the existence of mobile plates was the fit of the continents, notably on the opposite sides of the Atlantic Ocean. Down the centre of the Atlantic there is a mid-oceanic ridge (Figure 8.2), roughly equidistant from Europe and Africa on one side, and the Americas on the other side. This was explained by the creation of new oceanic plate material and the consequent spreading of oceanic plates away from the ridge, carrying the pre-existing continents further apart. Today there are measurements of the rate at which oceanic plates spread from mid-oceanic ridges – typically a few centimetres per year. For example, the South Atlantic Ocean is presently widening at about 3 cm per year.

☐ The South Atlantic Ocean is about 5000 km wide in the direction of widening. If it has always been widening at about 3 cm per year, how long ago were South America and Africa in contact?

At this rate of widening, the Americas were in contact with Europe and Africa about 200 Ma ago. Figure 8.3 shows an estimate of where all of the continents were at that time. Very much earlier they were somewhat more scattered again.

It is clear that, according to the theory, oceanic crust near a constructive margin is young, and that it is older, the further it is from the margin. This is just what is found by radiometric dating. Oceanic crust near mid-oceanic ridges is less than 1 Ma old, increasing up to about 200 Ma near subduction zones. As expected, most continental crust is older than this, some of it more than 2600 Ma old.

As well as mid-oceanic ridges, many other types of expected landform exist, and in the right places. For example, many mountain ranges border a coastline with a deep ocean trench that has every appearance of a subduction zone. Moreover, the andesitic materials found in such regions are as predicted by plate tectonics (Section 8.1.2). Transform faults also exist, and some of these are in motion, the most notorious being the San Andreas Fault in California.

Earth's volcanic and tectonic activity is concentrated at transform faults, mid-oceanic ridges, the mountain ranges that border ocean trenches, and at island arcs. In plate tectonics these features are near constructive and destructive plate boundaries, just where the theory predicts that volcanic and tectonic activity will be concentrated. Also, in the region of mountain ranges and mid-oceanic ridges, there are departures from isostasy of the sort expected from continuing vertical lithospheric motions at plate boundaries.

The explanatory power of plate tectonics is vast, and you can explore it further in Question 8.1.

Question 8.1

In terms of plate tectonics, account for each of the following features of the Earth's surface: (a) the scarcity of impact craters; (b) a rift valley that crosses Iceland; (c) the Urals (the mountain range east of Russia); (d) seismic activity at shallow depths near mid-oceanic ridges and ocean trenches; (e) the Aleutian Islands (alongside the Aleutian Trench).

8.1.4 The Causes of Plate Motion

The superficial cause of plate motion is slab pull, outlined in Section 8.1.2, the pull arising from the steeply descending part of the plate that has acquired a greater density through its conversion into eclogite. The sideways motion that carries a plate away from a constructive margin is assisted by the slight downhill gradient.

The underlying cause of plate motion is solid state convection in the mantle (recall that the mantle is almost entirely solid). A constructive margin is where motion in the convection cells is upwards, and a destructive margin is where it is downwards. A possible (simplified) pattern of mantle convection was shown in Figure 5.4. The depth range in the mantle that permits convection can be regarded as the asthenosphere.

If there were no convection there would be no plate tectonics. However, the absence of plate tectonics does *not* imply that there is no convection. For example, if all of the continental plates were sufficiently rigid, it is likely that the global system of plates would not permit relative motion – it would lock up. In the case of Venus, where there is evidence for mantle convection, the absence of plate tectonics might be due to a rigid lithosphere, or to some other cause – see Section 8.2.7.

As well as the large-scale convection that drives the plates it is known that there are huge rising columns of hot (but solid) mantle material underneath plates, well away from any margins, called plumes. Therefore, though convection offers an explanation of plate motion, there is no simple relationship between convection patterns and plate boundaries.

Plumes account for some strings of volcanic islands well away from plate boundaries (unlike island arcs). The Hawaiian Islands are one such chain. The sequence of ages when the different volcanoes there became extinct indicates that plate motion over the plume generated the chain, and continues to do so. Plumes are also a possible source of the huge quantities of basalt that have occasionally flooded parts of the Earth. As the plume head rises the pressure decreases, and in the asthenosphere the pressure can decline to the point where extensive partial melting occurs, creating large volumes of basalt that spill onto the Earth's surface to create volcanic plains. Mantle plumes can be detected as 'hot spots'. However, of the 100 or so hot spots, only about 10 seem to be due to mantle plumes, as indicated by seismic imaging (tomography), by uplift of the crust, and by correlation with plate motion. The other 90 or so could be due to cracks in the lithosphere, or mantle heterogeneity, or small-scale mantle convection.

Of course, the deeper question is: what causes the large-scale convection?

☐ Recall from Section 5.1.1 what the cause might be.

Models of the Earth's interior indicate that heat from long-lived radioactive isotopes, plus the residual effect of primordial heat, can sustain an adiabatic temperature gradient in the mantle, and that the temperatures are sufficiently high for solid state convection to occur from the base of the lithosphere to great depth, perhaps as far as the outer core boundary.

Plate cycling is an efficient means of removing heat from the Earth's interior. It accounts for about 70% of the energy that reaches the crust, as hot plates are created at constructive margins and cold plates descend at subduction zones. Plumes probably account for a large proportion of the remaining 30%.

8.1.5 The Evolution of the Earth

A plausible history is as follows. We start nearly 4600 Ma ago, with a composition that was roughly uniform at all but the smallest scales. At the smallest scales, certainly at the level of centimetres and millimetres, there was a mixture of free metal (mainly iron) and silicates. The overall composition was broadly similar to that of the ordinary chondrites.

(1) After a few million years, radiogenic heating, perhaps assisted by energy from accretion and other heat sources, had raised temperatures to the melting point of iron, and so partial melting occurred with the appearance of liquid iron (plus some minor constituents such as nickel). This was denser than the remaining material and so the iron drained downwards, a process aided by the plasticity of the surrounding material. The iron core was thus formed, and heat of differentiation was released. The overlying mantle was thus depleted in iron and siderophile elements, and acquired the composition of peridotite. Core formation took only about 30 Ma.

(2) The final giant impact, at about 4500 Ma, was from a body of broadly terrestrial composition that created the Moon. It contributed the final 10% or so of the Earth's mass.

(3) Impact melting (perhaps assisted by the thermal blanketing of a dense atmosphere) created a magma ocean several hundred kilometres deep. Fractional crystallisation led to the formation of a chemically distinct oceanic type crust of basaltic–gabbroic composition, and a corresponding depletion in the upper mantle of elements that are enriched in the crust. Subsequent impacts in this primeval crust might have led to regional differences that aided the subsequent formation of continental crust. The magma ocean froze throughout its depth before 4000 Ma ago.

(4) Direct evidence of the early heavy bombardment that pervaded the Solar System is found in some of the oldest surviving rocks, 3700–3800 Ma, from Isua in Greenland – isotope ratios indicate that a meteorite component is present. There is also evidence for a late veneer from the particularly primitive C1 chondrites.

(5) Meanwhile, at the surface, plate tectonics was already established, and had destroyed the early crust. New oceanic crust was created and entered the plate tectonic cycle. This slowly increased the volume of continental crust up to about 2500 Ma ago, since when the volume has continued to increase slowly.

(6) At about 2500 Ma there was a change in the way plate tectonics operated, probably caused by a pulse of continental crust formation, perhaps triggered by the heat flow from primordial sources decreasing to some threshold.

(7) Around 1000 Ma ago, as the core lost primordial heat, it cooled to the point where the separation of the solid inner core began to occur. It is growing at about 10 mm per century.

(8) Plate tectonics continues, driven by mantle convection in the asthenosphere, which might extend to the core–mantle boundary. Throughout Earth history it has been responsible for the continuing high level of tectonic and volcanic activity, and thus for much of the sculpting of the Earth's surface.

(9) The Earth's surface is also sculpted through gradation, mainly by water and winds. Impact craters are scarce because of such gradation and all other forms of geological activity. Nevertheless, over 170 large impact craters have been identified. Most of them are heavily

eroded. Impacts continue, and a Tunguska-size impact (Section 3.2.4) occurs every 800–1800 years. The famous impact, 65 Ma ago, that contributed to the demise of the dinosaurs and many other species was far larger, but, thankfully, far rarer.

8.2 Venus

Venus has a volume only 15% smaller than that of the Earth, and its global mean density is 5% less than the terrestrial value. As noted in Section 5.1.2, there is little doubt that its interior broadly resembles that of the Earth, though Venus is not the Earth's twin, particularly in its massive, hot atmosphere and at its surface.

As well as Earth-based radar, Venus has been extensively explored by orbiters and by landers. The landers were from the USSR, and they reached Venus in the 1970s and 1980s. Several of them returned data from the surface, and four of these returned images – a considerable achievement given the mean surface temperature of 740 K. None carried out seismic measurements. The many orbital missions included NASA's Magellan Orbiter that between 1990 and 1992 mapped 98% of the surface using radar altimetry and synthetic aperture radar to 'see' through the 100% cloud cover. The images obtained with synthetic aperture radar have spatial resolutions down to a few tens of metres. The Magellan Orbiter subsequently mapped the gravitational field, in an extension of the mission with the spacecraft in a lower orbit that provided gravitational detail. Currently, Venus is being explored by ESA's Venus Express, which went into orbit in April 2006. Its instruments are exploring the atmosphere (Section 10.3) and also mapping the surface temperature.

8.2.1 Topological Overview

Figure 8.4 shows a highly simplified topographic map of Venus. Venus is nearly spherical, so the zero of altitude is defined as the mean equatorial radius, 6051.5 km. About 80% of the surface of Venus is volcanic plains. The plains are streaked with low, broad ridges, indicating gentle folding and shortening, and networks of wrinkle ridges indicating compressional deformation. Within the plains is highly deformed terrain, called tesserae, lying above zero altitude, and constituting 8% of the surface. Ishtar Terra (Figure 8.4) consists mostly of tesserae and mountain belts, and includes Maxwell Montes that rises to 11 km, the highest peak on Venus. There are over 100 gently sloping volcanoes that resemble the basaltic shield volcanoes on Earth. They constitute 4% of the surface, and have a maximum altitude in Maat Mons, at 9 km (Plate 5). There are numerous small volcanic constructs, including domes, and concentric ringed features, called coronae. Coronae are up to about 1 km high, and mostly less that 300 km across, though a few exceed 1000 km.

Reaching downwards, there are chasms resembling rift valleys, presumed to be the result of tectonic extension – these are called chasmata. They are typically a few kilometres deep, hundreds to thousands of kilometres long, and constitute 8% of the surface. A few of the longest are shown in Figure 8.4, including Diana Chasma, where the lowest altitude on Venus has been recorded, about 4 km below zero. The extreme altitude range on Venus is thus about 15 km, rather less than the Earth's 20 km range.

Figure 8.4 also shows the hypsometric distribution. It is unimodal, whereas the Earth's (Figure 8.4) is bimodal.

☐   What does this indicate?

The bimodal distribution on Earth arises from the existence of two types of crust, continental

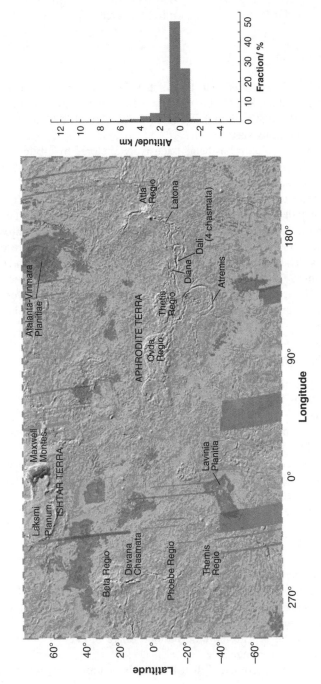

Figure 8.4 A simplified topographic map of Venus, showing only the highest and lowest altitudes: the lighter the shade, the higher the altitude, except for Maxwell Montes. The hypsometric distribution is also shown.

and oceanic. The unimodal distribution on Venus thus indicates that there is one crustal type, or that any other type is scarce. There is much evidence (see below) that the crust is basaltic. One model has a mean crustal thickness of about 70 km. Comparison of the two distributions also shows that the altitudes cluster around the one maximum on Venus more than they do around either maximum on the Earth, and therefore on a large scale Venus is a good deal smoother, probably the result of its higher surface temperature – this reduces rock strength.

8.2.2 Radar Reflectivity

Almost all the imaging of the Venusian surface has been done by orbiting radar, at wavelengths of the order of 10 cm. At such wavelengths the radar reflectivity of the surface correlates with the surface altitude – broadly, the greater the altitude, the higher the reflectivity. This suggests that on a scale of a centimetre or so the lowlands are smooth whereas the highlands are rough. This is thought to result from gradation, in particular the erosion of highland material to form fine dust that collects on the lowlands. The generation of dust presumably owes much to the high atmospheric pressures and temperatures, and to traces of corrosive gases. Further evidence for gradation is wind streaks behind obstacles, and the existence of some dune fields and landslides.

At altitudes above 3.5 km the reflectivity abruptly becomes considerably higher, and this might be due to a thin veneer of various metal chlorides, fluorides, and sulphides. These are somewhat volatile, and therefore condense at the lower surface temperatures in the highlands – the temperature decreases by about 8 K for every kilometre increase in altitude. These substances would be emitted in small but sufficient quantities by the volcanoes of Venus. The very highest altitude summits have low reflectivities, perhaps because of a veneer of materials that are even more volatile.

Radar shows that about 75% of the surface is bare rock, and the four landers that returned images each showed much bare rock, with only patches of gravel and finer material. This is one indication that gradation is not as powerful on Venus as on the Earth. This is not surprising given the lack of any precipitation and the low surface wind speeds – less than about $2\,\mathrm{m\,s^{-1}}$.

8.2.3 Impact Craters and Possible Global Resurfacing

The surface of Venus is lightly covered in impact craters, ranging from simple bowls a few kilometres across, to multi-ringed basins with diameters of more than 100 km (Figure 8.5(a)). Really small craters are absent because Venus has long had a dense atmosphere that has disrupted small projectiles before they reached the ground; any surviving fragments would produce small pits well beyond Magellan's image resolution. The fragments of larger projectiles are presumably responsible for the observed crater clusters and for shallow surface scars.

Many craters have radar-dark halos. These are presumed to be smooth, perhaps because the ejecta blanket contains much fine dust, or perhaps as a result of surface pulverisation by an atmospheric shock wave caused by the projectile – such a shock wave would be very powerful in the dense atmosphere. Flows extend from some craters out to about 150 km. These could be ejecta that was confined near to the surface by the dense atmosphere. Alternatively, or additionally, the flows could be magma released by the impact.

The spatial distribution of impact craters is not very different from random, with a global surface density indicating a mean surface age in the range 500–800 Ma. Moreover, many of the craters have suffered little degradation. One interpretation is that most of the Venusian surface was wiped clean of craters 500–800 Ma ago and that there has been little resurfacing since. Another view is that the resurfacing has been more extended in time, and that 500–800 Ma

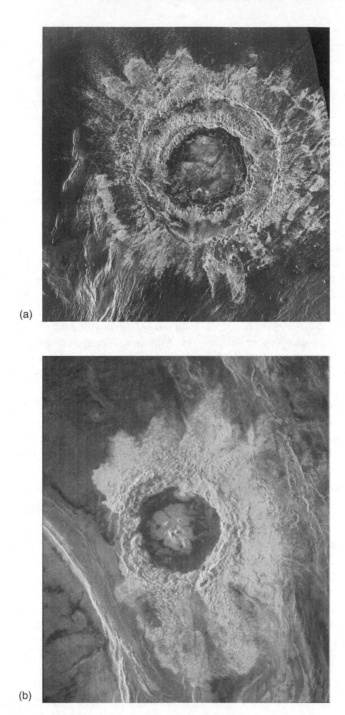

(a)

(b)

Figure 8.5 Radar images of two Venusian impact craters. (a) Meitner, about 150 km across the outer rim. (NASA, part of F-MIDR55S319, R Greeley) (b) Dickinson, 69 km in diameter. (NASA/NSSDC P37916)

is an average retention age for craters. There has, however, been some recent resurfacing, because areas of particularly low crater density correlate with volcanic and tectonic features. The predominant view is that 80–85% of the surface dates to a short period of intense geological activity concentrated at some time between 500 Ma and 800 Ma, followed by much lower activity down to the present. Possible reasons for this history are outlined in Section 8.2.9.

8.2.4 Volcanic Features

Some were briefly mentioned in Section 8.2.1. Volcanic features are abundant on Venus. Particularly widespread are volcanic domes from a few kilometres to several tens of kilometres across. The largest are flat topped and only about 1 km high, and are consequently called pancake domes (Figure 8.6). Domes also occur on Earth, where they are the result of viscous lava oozing from a central vent, and though on Earth they are only up to a few kilometres in diameter, they are thought to have the same origin on both planets. There are also Venusian shield volcanoes (Plate 5), built from low-viscosity lavas. They are up to hundreds of kilometres in diameter, several kilometres high, and tend to be on broad regional rises. The larger shields are the same sort of size as the largest shields on Earth.

Lava floodplains cover about 80% of the surface. Some areas have very few craters and so these areas must be recent. Presumably the lava erupted from fissures, and ran along the channels that are seen to meander for up to thousands of kilometres. Some of these channels now run uphill, presumably as a result of uplift since the lava flowed. The vast floodplains and long channels indicate lava with low viscosity.

☐ What else do these features indicate?

The lava must also have remained liquid for a long time. The high surface temperature of Venus would facilitate longevity and low viscosity, but a distinct composition is also required. One possibility is lava with a high sulphur content. Another is carbonatite lavas. These are rich in igneous (not sedimentary) carbonates and in simple compounds such as sodium chloride. They

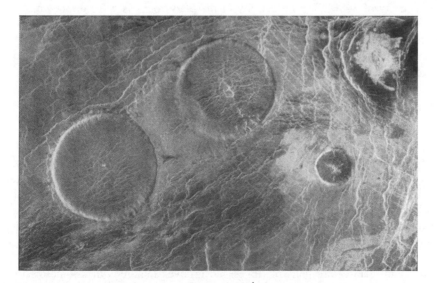

Figure 8.6 Radar image of pancake domes in the Eistla region of Venus. The largest is about 65 km across. (NASA/NSSDC P38388)

have melting points not much above the Venusian surface temperature. Their sources would be near the surface.

Low lava viscosity also helps to account for the rarity of explosive volcanism. A further factor would be any dryness of the Venusian crust and mantle, and the consequent low volatile content of the magma (Section 10.4). For any explosive volcanism that did occur, the high surface atmospheric density would have confined the explosively erupted material close to its source.

Volcanism might continue today. There is indirect evidence in the existence of fresh-looking lava flows, and also in fluctuations in atmospheric sulphur dioxide (SO_2) that could indicate variations in volcanic activity. However, as yet there have been no direct observations of active volcanism on Venus.

8.2.5 Surface Analyses and Surface Images

Direct analyses of surface materials have been carried out by several of the Soviet landers – Veneras 7–10, 13, 14 (between 1970 and 1982), and Vegas 1 and 2 (in 1984). These all landed in regions of volcanic flows. Venera 8 landed 5000 km east of Phoebe Regio (Figure 8.4), and the rest of the Veneras landed on the flanks of Beta Regio. The Vegas landed on Rusalka Planitia, which is on the northern flanks of Aphrodite Terra, a predominantly highland region with several large volcanoes. Gamma and X-ray fluorescence spectrometry indicate that, except at the Venera 8 and 13 sites, the rocks are basaltic, consistent with a basaltic–rhyolitic crust. At the other two sites they are more granitic. This suggests that some division into two crustal types has taken place, though the hypsometric distribution indicates that this has not gone very far.

Veneras 9, 10, 13, and 14 returned images from the Venusian surface, and those from Veneras 13 and 14 are shown in Figure 8.7. Venera 9 landed on the steep slope of a tectonic trough, with talus and platy rock fragments in view. Venera 14, and also Veneras 10 and 13, landed on plains of low-relief outcrops of platy rocks with fine material in small depressions between them. The fines everywhere consist of particles less than 4 mm across, and are neutral in colour, with a reflectivity of 3–5%. The reflectivity of the platy rocks is 5–10%, presumably because they are larger and less fragmented than the fines. The rocks at all four sites are finely layered, perhaps, in some cases, a result of wind deposition followed by consolidation. At the Venera 13 and 14 sites (Figure 8.7) the rocks are porous, either because of weak consolidation, or because they are volcanic tuff. Some rocks could consist of thin lava layers.

8.2.6 Tectonic Features

Tectonic features are abundant on Venus. Some mountain ranges have complex folds and faults resulting from crustal compression and tension. Among these is Ishtar Terra (Figure 8.4), which includes the mountain range Maxwell Montes plus three other ranges and a high plateau. As well as mountain ranges, the highlands also bear rugged tracts of faulted and folded terrain called tesserae. Alpha Regio, a rugged plain with a mean altitude of about 2 km and about 1000 km across, is a major tessera. Some of the fracturing in the tesserae might be due to a process that degrades all elevated features on Venus. This is the plastic flow of rocks at the surface, a result of the high surface temperatures. The process is much more rapid on Venus than on the Earth, but it still takes tens of millions of years to have a significant effect.

Other tectonic features are belts of ridges and grooves hundreds of kilometres long. There are also the chasmata, many of which seem to have formed through crustal tension. One such example is Devana Chasma which, at 3000 km long, is of Africa's Great Rift Valley

Figure 8.7 Images from two of the four Venus landers that returned images. Left, Venera 13; right, Venera 14. (Via NASA/NSSDC, processed by Don P Mitchell and used with his permission)

dimensions, and divides the young volcanic regional rise Beta Regio and also the neighbouring rise Phoebe Regio (Figure 8.4). Other chasmata, such as those in Figure 8.8(a), have profiles that suggest they might be places where one slab of crust is being forced under an adjacent slab, somewhat like subduction on Earth. Another type of tectonic feature is the coronae. These roughly circular plains, typically a few hundred kilometres in diameter but no more than about a kilometre high, are ringed by ridges and grooves. Some of them bear volcanic domes and display volcanic flows (Figure 8.8(b)) – their spatial relationships with other landforms indicate that they were created at different times. The tectonic interpretation is that each one is the combined effect of crustal uplift, perhaps involving rising plumes of hot mantle, plus volcanism, and sagging, in a prolonged and intricate sequence, not always with uplift first. A few coronae might be highly degraded or deformed impact craters.

8.2.7 Tectonic and Volcanic Processes

In spite of the tectonic features just described, Venus lacks landforms that would suggest planet-wide plate tectonics. For example, long sinuous constructive margins like Earth's mid-oceanic ridges seem to be entirely absent. Moreover, the unimodal hypsometric plot indicates that there has been no widespread creation of a second kind of crust, and surface samples suggest that any second crust is at most rare.

If, as seems likely, plate tectonics on Venus is at most a local phenomenon affecting only a small fraction of the surface, then this requires explanation. It could be because the hot surface has made the lithosphere so plastic that widespread tectonic stresses are distributed, with the result that we get many lithospheric fractures that prevent the formation of a few large plates. Alternatively, it might be that the plate system is too rigid to be mobile because it lacks widespread continental-type plates, which on the Earth might give the system a necessary lack of rigidity (except at the cratons). Such rigidity could arise from a lack of water, and indeed the evidence from the surface and atmosphere indicates that Venus is very dry – more on this in Section 10.4. The high crustal temperatures also contribute to the stiffening by precluding 'slippery' minerals like chlorides and serpentine. Observations also indicate a stiff lithosphere,

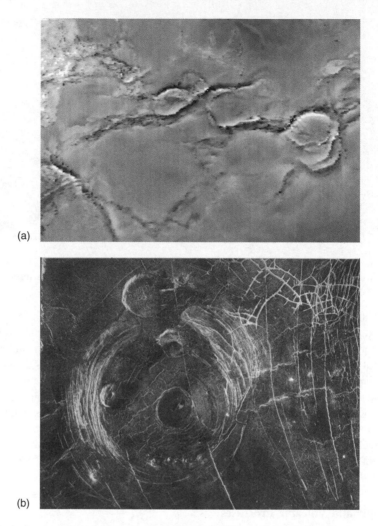

(a)

(b)

Figure 8.8 Radar images of tectonic features on Venus. (a) Diana and Dali Chasmata. The frame is about 4200 km across. (MIT17/12/97, NASA, P G Ford and G H Petengill) (b) A corona, about 150 km across, in Aino Planitia. (NASA/NSSDC P38340)

particularly the correlation of regional gravity with topography. These data also point to little by way of an asthenosphere, at least not at relatively shallow depths. The scarcity of water in the lithospheric materials will also have raised their melting temperatures, so that a subducting plate would have to penetrate deep to melt (partially or wholly). There might be too much friction between a long length of plate and its surroundings for deep penetration to occur. This would further inhibit plate tectonics.

☐ Plate tectonics is absent from some planetary bodies probably because they have *thick* lithospheres. Why is this an unlikely explanation for Venus?

Even though increased melting temperatures, consequent upon dryness, would cause some lithospheric thickening, a really thick lithosphere on Venus is unlikely because there are widespread tectonic features, and tectonism might still be occurring today. Lower estimates of lithospheric thickness of a few tens of kilometres are obtained from a close examination of how the surface has responded to stresses.

With no plate tectonics, how do we explain the tectonic and volcanic features, and how do we explain a possible global resurfacing 500–800 Ma ago?

One explanation is that there is a system of convective hot plumes in the mantle, rather as seems to be the case underneath some of the plates on the Earth. Hot plumes on Venus could give rise to volcanoes that create some new crust, as in Figure 8.9. In this model these are the sites of the coronae. The sequence of events that create the coronae, as outlined in Section 8.2.6, then includes material being uplifted by a plume, then volcanism, and finally sagging as the plume wanes. All stages in this proposed sequence are visible in different coronae. In some cases sagging seems to have preceded uplift. The tesserae result from the *downgoing* part of a convective cycle (Figure 8.9). On this view the tesserae have been created by crustal compression. Some of the folded, faulted highlands might also have originated this way, such as Ishtar Terra (Figure 8.4). However, the topography of Ishtar Terra is also consistent with mantle uplift in this region. Such is the persistent puzzle that is Venus!

Support for the plume model comes from gravitational field measurements. Detailed gravitational analyses reveal that some topographic highs are in isostatic equilibrium due to crustal thickening, whilst others are probably supported dynamically. In the latter case isostatic adjustment makes a contribution to the raised topography through the reduction in density consequent upon the greater temperatures in dynamically supported regions. These interpretations are in accord with the particular types of high terrain that fall into each gravitational category. Thus, many of the candidates for dynamic support, such as Beta Regio, show evidence that they are volcanic constructs that could have been created by a mantle plume that also provides dynamic support. By contrast, many of the candidates for isostatic support, such as Alpha Regio, are not obviously volcanic but look more like regions of crustal thickening such as could arise from crustal compression. The largest of these, Ishtar Terra, is so large that it might not simply be a thicker version of the surrounding crust, but also have lower density, in which case it is a rare example on Venus of crust analogous to continental crust on the Earth.

The occurrence of a particularly large number of plumes about 500–800 Ma ago could possibly have generated enough magma at the surface to account for the widespread resurfacing that might have occurred around that time. An alternative explanation is that the lithosphere gradually thickened until plastic deformation within it decreased its viscosity to that of the underlying convective mantle. At this point there was global subduction of the lithosphere.

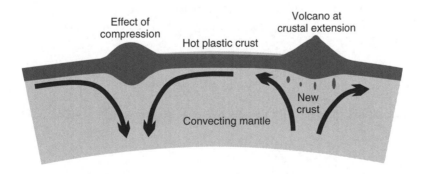

Figure 8.9 Mantle convection and its effects in Venus.

A new lithosphere formed, and the cycle is being repeated. However, there is as yet no general agreement on any one of these or any other explanations, even supposing such widespread resurfacing occurred.

8.2.8 Internal Energy Loss

You have seen that the interior of Venus today is probably not very different from that of the Earth. Therefore the interior energy sources have probably been similar in the two planets. In the case of the Earth, most of the internal energy has been escaping through plate recycling, but this does not seem to occur on Venus, and conduction through the lithosphere is probably weak. How then has Venus been losing its internal energy?

It is possible that mantle plumes can get rid of much of the energy, in which case the coronae are important sites of heat loss. Subduction of lithosphere at certain chasmata could also make a contribution, because the lithosphere is cool and so would absorb internal energy from the interior. The episodic global subduction of the lithosphere that has been proposed to explain global resurfacing could result in enormous, if occasional, heat loss. The new lithosphere then puts a stagnant lid on the interior (Section 4.5.2), and the internal temperatures rise until the next subduction. However, as with many things to do with Venus, there is much uncertainty.

8.2.9 The Evolution of Venus

The interior evolution of Venus has probably been much the same as that of the Earth. The main differences are at the surface. No global system of plate tectonics has developed, and various possible reasons were outlined in Section 8.2.7. Instead, there might be a system of mantle plumes and downwellings that subduct little or no lithosphere but that account for many of the tectonic features. We therefore have predominantly one type of crust, a basaltic–rhyolitic crust derived by partial melting of a peridotite mantle, the crust gradually increasing in volume. Plumes, delamination, and subduction might account for most of the current loss of internal energy.

Episodic global subduction of the lithosphere might occur, and this could account for the global resurfacing that seems to have happened 500–800 Ma ago. The loss of the stagnant lid would release a burst of internal energy. Any extensive resurfacing 500–800 Ma ago could alternatively have been caused by an outburst of mantle plumes at that time that resulted in sufficiently copious production of basaltic lavas to obliterate almost all of the surface. We have no idea what happened on Venus before this extensive resurfacing.

Volcanic activity might continue today, as might tectonic activity, and further resurfacing episodes are possible.

Question 8.2

What is a reasonable explanation for the absence of huge impact basins on Venus?

Question 8.3

Speculate on how the granites found on Venus by Veneras 8 and 13 could have formed.

8.3 Io

Io is the closest of the four Galilean satellites to Jupiter. It is slightly larger than the Moon, and like the Moon it has a rocky surface and a rocky interior. Unlike the Moon it is volcanically active, and highly active at that.

☐ What are the main sources of energy to drive this activity?

The main source is tidal, with a supplement from radiogenic heating (Section 5.2.3). The tides raised by Jupiter are huge, about 100 m at the surface.

The volcanic activity of Io was discovered on images returned during the flyby of Voyager 1 in 1979 and it came as a surprise to most planetary scientists. Voyager 1 revealed not only a surface free of impact craters and dominated by fresh-looking volcanic features, but also nine active volcanoes that made Io the most volcanically active body in the Solar System! Two active volcanoes are visible in Plate 12, one edge on at the limb, the other face on in the centre.

Since Voyager 1, Io has been flown past by Voyager 2 in 1980, and scrutinised by the Galileo Orbiter at various times from December 1995 up to its plunge into Jupiter in September 2003. At the end of 2000 Cassini flew by at 97 million kilometres en route to Saturn.

Well before Voyager 1 it was known that Io had a reddish hue. Voyager 1 showed that this is due to an intricate pattern of red, orange, yellow, white, and black. IR spectrometry from the Galileo Orbiter indicates that the white areas are frosts of sulphur dioxide (SO_2). The other colours could be due to other sulphur oxides and to the various forms of sulphur itself. On the basis of cosmic relative abundances, Io could certainly have been born with sufficient sulphur for a surface layer several kilometres thick, and less than a few tens of metres would since have been lost to space. However, it seems likely that most of Io's sulphur is in an iron-rich core as iron sulphide (FeS), and that at the surface sulphur and its oxides are present only as a thin veneer that colours the surface. This veneer is presumed to have come from the volcanoes.

Over 500 volcanic sites are known, randomly spread over the surface. Activity at any one time is concentrated at a few of these sites. Some sites are shield volcanoes up to a few hundred kilometres across and a few kilometres high, complete with summit calderas, and others are paterae, low-relief volcanoes with vents up to 200 km in diameter. Some caldera walls are so high and steep that sulphur and its oxides would be insufficiently strong to constitute them. Silicates would be sufficiently strong, thus supporting the view that sulphur and its oxides are present only as a thin veneer. The vents switch on and off, and new ones appear, as indicated by changes in the 3 months between the Voyager 1 and 2 flybys, by subsequent observations from the Earth of transient infrared hot spots, and by Galileo Orbiter observations in 1996–2003. The estimated rate of volcanic resurfacing is 100 m depth per Ma, which explains the absence of impact craters.

The observed volcanic activity of Io includes volcanic plumes that rise to hundreds of kilometres (Plate 12). Some shield volcanoes, e.g. Pele, have remained active since the Voyagers. Plumes require some gaseous content to create them, and vaporised sulphur or SO_2 seems likely, as opposed to water as on the Earth. Lava flows are also observed, with temperatures of around 2000 K. The SO_2 frost, caldera wall strength, Galileo Orbiter IR measurements, and thermodynamic arguments all suggest that the dominant constituents of the plumes and of the lava are silicates, with sulphur and SO_2 as minor constituents. Variations in the proportions of sulphur to silicate could explain the different types of flow and plume seen.

Though volcanoes are numerous, a second type of terrain covers most of Io's surface. These are flat blotchy plains (Plate 12). Here and there erosion has exposed layers, each one a few hundred metres thick. A volcanic origin is presumed, involving effusive lava flows hundreds of kilometres long, plus fallout from volcanic plumes.

The third type of terrain is mountains. These are dotted around the surface, though constitute less than 2% of it. The largest are about 200 km across and 10 km high. Most of them do not seem to have a volcanic origin, and the plains sweep around the mountains in a way that shows the mountains to be older. They are too steep and too high to be made of sulphur and its oxides, and though they have a reddish tint, sulphur would be redder, and so they are presumed to be predominantly silicate.

Tectonic features on Io include faults attributed to tidal flexing of the lithosphere, but there is no sign of plate tectonics.

The estimated heat flow from the interior approaches a prodigious $10^{-9}\,\mathrm{W\,kg^{-1}}$, much greater than the $0.006 \times 10^{-9}\,\mathrm{W\,kg^{-1}}$ for the Earth (Table 4.2). Indeed, this is several times more that the estimated input to Io from tidal heating! Much of the outflow is in the radiation from broad regions at temperatures of about 300 K, which though less than the 1000 K or so of the IR hot spots is significantly greater than the 130 K mean surface temperature of Io. Silicate lavas are a possible explanation of the warm regions, in which case Io might be in an atypically active volcanic phase, and the heat flow is then anomalously high.

The evolution of Io

Io presumably formed in a high-temperature environment near to proto-Jupiter. In this case it would have formed depleted in water and in substances with greater volatility. Subsequent radiogenic and tidal heating could soon have completed the devolatilisation. It seems certain that Io is differentiated into a predominately silicate crust 20–30 km thick, overlying a partially molten, predominantly silicate mantle, overlying in turn a core of iron plus FeS, and that there is a thin lithosphere and convective asthenosphere (Figure 5.7). The crust is probably rapidly recycled through being created volcanically and destroyed by melting at the base of the lithosphere.

Question 8.4

If Io were devoid of sulphur, what difference would this make to volcanism on Io and to its surface?

8.4 Icy Surfaces: Europa, Titan, Enceladus, Triton

We end this chapter with a brief discussion of the four bodies with *icy* surfaces that seem to be volcanically or tectonically active today.

8.4.1 Europa

Europa is the second closest of the Galilean satellites to Jupiter, and was particularly well imaged by the Galileo Orbiter in 1996–1997 (Plate 13). It is predominantly a rocky body, but is known from spectroscopy to be entirely covered in fairly pure water ice, plus some salts and perhaps NH_3. The albedo in most areas is high, indicating that the ice has a clean surface. Areas of mottled terrain and dark plains might be where the surface has been darkened by ion implants from external bombardment, or by some non-icy component from interior and exterior sources. The existence of water ice is also indicated by Hubble Space Telescope UV spectra that have detected monatomic oxygen gas (O) in a thin atmosphere. This is very probably derived from water vapour through dissociation by solar UV. Models suggest that most of the oxygen is present as O_2, and that the atmospheric pressure is only 10^{-6} Pa.

The surface has almost no craters larger than a kilometre across. An exception is the crater visible at lower right in Plate 13 in the centre of the bright area. This 26 km diameter crater is called Pwyll (pronounced, roughly, 'Poo-eel'), after a character in Welsh legend. The paucity of impact craters indicates that the surface is nowhere older than about 200 Ma, and much of it must be considerably younger. Though viscous relaxation has played a role in this rejuvenation, there must also have been cryovolcanic resurfacing, perhaps continuing today. This will give a surface with the observed high albedo that will darken with age through ion implantation and dust infall.

The surface is remarkably flat, the tallest features being systems of dark ridges up to about 300 m high. These systems might be the result of thin cracks through which liquid water escaped to form a ridge on each side of the crack. The weight of the ridges produced further, parallel cracks, and the process repeated itself. Subsequent cracks can cross the older ones, to give the observed network of ridge systems, some of which is visible in Plate 13. The ridges gradually disappear through viscous relaxation to leave smooth, dark bands. Ridges display faults, and there are areas where the surface has been broken up, liberating rafts of ice that moved away to be trapped by upwelling freezing water (Figure 8.10). Tidal stresses are probably responsible for disruption of the icy crust, though the patterns indicate that a shift in the rotation axis might also have contributed. As well as the area in Figure 8.10, many other areas on Europa display features consistent with the creation of new icy crust. To keep the surface area constant it is necessary for plates of ice to converge and disappear into the interior. Evidence for this has now been found, solving a long-standing mystery.

The smoothness of the surface, detailed topographical studies, and the ready access that liquid water seems to have to the surface indicate that the icy crust could be as little as a few kilometres

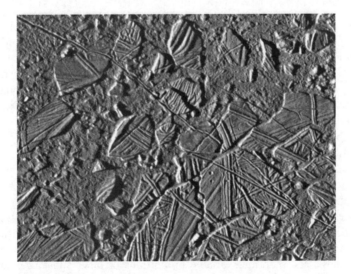

Figure 8.10 The ridged and faulted surface of Europa, including rafts of ice frozen into new positions. The frame width is 42 km. (NASA/JPL P48526)

thick, and that underneath it there is icy slush, perhaps mixed with rock. More dramatically, it is increasingly believed that widespread oceans of liquid water are present. Support for this view might have been provided by Galileo's magnetometer. A possible influence of Europa on Jupiter's magnetic field could be due to a conducting fluid, such as salty water. Salts also lower the freezing point, as does NH_3 : 33% NH_3 in water would lower the freezing point to about 173 K.

The shell of water, ice, and liquid probably has an average thickness of roughly 150 km (Section 5.2.3). Throughout most of its depth the icy shell is plastic, and it thus constitutes an asthenosphere. Though tidal heating today is 20 times or so weaker in Europa than in Io, tidal heating coupled with radiogenic heat, perhaps with the aid of generally raised temperatures from solar radiation and residual primordial heat, could sustain an ocean or an icy slush. Heat from the underlying rocky mantle could be driving convection in the water, and this in turn could have created some of the cracks and ridges. Silicate volcanism in the rocky mantle is possible, extending to the ocean floor, and this could vent material rich in organic compounds that might darken some of the water that reaches the surface. It has even been speculated that life has originated in the oceans of Europa, and this is certainly worth investigation.

A less dramatic, less widely held view is that there is a solid icy asthenosphere, and it is partial melting in its upper reaches and at the base of the icy lithosphere that produces icy magma for cryovolcanism.

That Europa has a shell of water and Io does not was explained in Section 5.2.3 – it comes down to the greater distance of Europa from proto-Jupiter.

Question 8.5

Describe the differences between the surfaces of the Galilean satellites that can be explained by the decrease in tidal heating that occurs with increasing distance from Jupiter.

8.4.2   Titan

Saturn's huge satellite, Titan, larger than Mercury and second only to Ganymede among the planetary satellites, formed from the disc of gas and dust around Saturn. Its low density, $1880 \, kg \, m^{-3}$, indicates a substantial fraction of water. Titan could well be partially or wholly differentiated (Section 5.2.2). If it is differentiated, then it should have an outer mantle of water, wholly or mainly solid, with perhaps about 15% NH_3 by mass, plus smaller quantities of other substances (Figure 5.7). Irradiation of the atmosphere and surface by solar UV radiation, cosmic rays, and charged particles from Saturn's magnetosphere would have dissociated the NH_3 to yield a considerable atmosphere consisting largely of N_2, as observed. In the lower troposphere it also contains 4.7% CH_4 and traces of other carbon compounds.

The CH_4 forms clouds and the carbon compounds form hazes. As a result the surface is obscured at visible wavelengths. Consequently, the images returned by Voyagers 1 and 2 in 1980 and 1981, respectively, revealed nothing of the surface, but only the general orange appearance of the atmospheric aerosols. The atmosphere is, however, less obscuring at certain near-IR wavelengths, and observations in these windows have been made from the Earth's surface and from the Hubble Space Telescope. Water ice is dominant, as expected, but the surface is not uniform in brightness, and therefore variable proportions of other substances are presumed to be there, notably hydrocarbons and NH_3. The global mean surface temperature and pressure ($94 \, K$, $1.5 \times 10^5 Pa$), and the atmospheric composition, had indicated that there might be a surface veneer of CH_4 and ethane (C_2H_6), plus smaller quantities of larger hydrocarbon molecules. The possibility of extensive oceans of hydrocarbons had been diminished by Voyager 2 in its 1981 flyby, and also by Earth-based radar investigations, the atmosphere being transparent at wavelengths greater than about 1 cm. The radar reflectivities were deduced to be consistent with regions of bright ice consisting of mixtures including water, NH_3, and other regions darkened by solid and liquid hydrocarbons.

Our knowledge of Titan has been hugely increased by the NASA–ESA Cassini–Huygens mission that reached Saturn in June 2004. In January 2005 Huygens landed on Titan, from where it transmitted data for about 80 minutes to the Cassini Saturn Orbiter and thence to Earth.

Images and data from Huygens

Figure 8.11 shows a mosaic of three frames of the surface obtained during the descent of Huygens by the descent imager/spectral radiometer (DISR). The width across the top of the rightmost frame is 3.0 km. A river-like drainage system is apparent, flowing downhill to a flatter, darker area bounded by what looks like a coastline, though the area is currently dry. The narrow channels with side feeders at sharp backward angles indicate precipitation run-off, whereas the

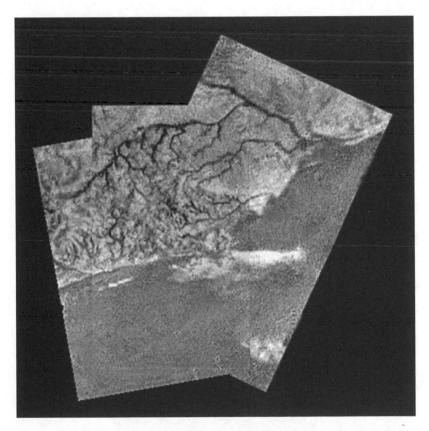

Figure 8.11 A mosaic of three frames of the surface from Huygens' descent imager/spectral radiometer. The width across the top of the rightmost frame is 3.0 km. (NASA/JPL/ESA/University of Arizona PIA07236)

short stubby channels that join the main channel at right angles indicate sapping and subsurface fluid flow. Spectral analysis shows that the darker area into which the channels feed is covered in material that came from the adjacent higher land, thus supporting the view that rivers have flowed to fill a lake. But it was not water that flowed, but a hydrocarbon, quite possibly CH_4. CH_4 can exist as a solid, liquid, or gas at and near Titan's surface, in accord with the surface temperatures and pressures and the phase diagram of CH_4 (Figure 10.17). Consequently, CH_4 plays the same role as water does on the Earth, and at the surface can occur in any of the three phases.

Water on Titan is very hard and strong at the low temperatures. It dominates the surface and plays a somewhat analogous role to rocks on the Earth. Figure 8.12 is a DISR view at Titan's surface, showing pebbles of water ice and hydrocarbon ices. The two just below image centre are about 15 cm and 4 cm across, and about 0.85 m distant. The surface is darker than expected and consists of a mixture of water ice and hydrocarbon ice. The pebbles are rounded, and have a size distribution consistent with transportation by a fluid. Huygens probably landed just within a dark area, possibly a lake bed where the pebbles were carried by hydrocarbon rivers.

Surface impact data as Huygens landed are consistent with both a solid granular material with very low cohesion and a damp surface like sticky 'tar'. Fine grains of ice plus a collection

Figure 8.12 A DISR view at Titan's surface, showing ice pebbles. The two just below image centre are about 15 cm and 4 cm across, and about 0.85 m distant. (ESA/NASA/JPL/University of Arizona PIA07232)

of photochemical products could comprise any tar. The gas chromatograph–mass spectrometer (GCMS) on Huygens detected CH_4 as a gas, but not the next member of this hydrocarbon series, ethane (C_2H_6). This might be because of its lower volatility or because it has drained from the area. NH_3 is also much less volatile than CH_4, so could also be present. Certainly, a surface rich in organic compounds is to be expected, generated from N_2 and CH_4 in the upper atmosphere by solar UV radiation and by energetic electrons from Saturn's radiation belt. These organics filter downwards and condense to form the haze layers (Plate 17) that gradually rain out to the surface.

Mountains, Channels, lakes, and oceans

Among its many instruments, the Cassini Orbiter is equipped with radar, the imaging science subsystem, ISS ($0.2–1.1\,\mu m$), and the visible and infrared mapping spectrometer, VIMS ($0.35–5.1\,\mu m$). These instruments are being used to map the surface, and also to perform compositional studies.

Among the many types of landforms seen is a mountain range about 150 km long, 30 km wide, and 1.5 km high. The range must be as hard as rock, though made of water ice and other icy materials. In the upper reaches there is bright, white material, perhaps CH_4 snow or exposed fresh material. The range could be the result of material welling up to fill gaps opened when two crustal plates were pulled apart.

Many channels have been seen, with broadly similar characteristics to those in Figure 8.11. One notable area is Xanadu, revealed by Cassini's radar to be a mountainous, Australia-sized region, with eastern and western margins laced with networks of narrow channels that lead into darker regions. The liquid that carved the channels was surely CH_4 or/and C_2H_6. The darker regions might be lakes or dry lake beds. In the case of Xanadu the overall observational evidence indicates that it consists of rather porous water ice washed clean of hydrocarbons, which explains its brightness in the near IR.

There is evidence for other lakes, perhaps dry. One of these is near the South Pole, where Cassini saw a dark feature, 234×73 km, with a shore-like boundary, in what was at the time the cloudiest region on Titan. From these white CH_4 clouds, CH_4 rain would fall, and evaporate only slowly at the low temperatures, not far from the triple point (Figure 10.17). Thus, this area is either a lake of liquid CH_4, perhaps with liquid C_2H_6, or a dry lake with dark deposits – or, disappointingly, a broad depression, such as a caldera, filled with dark, solid hydrocarbons. The crucial test would be for a specular reflection of the Sun in the near IR, such as would arise from a liquid surface, but this test has not yet been made on this feature.

❐ Why not make this test at visible wavelengths?

At visible wavelengths the atmospheric haze blocks sunlight from the surface. The best evidence for lakes so far (early 2007) is dark patches at high northerly latitudes, up to 70 km actors. Radar data indicate that these are (predominantly) of liquid CH_4.

The specular reflection test has been made in the search for large oceans. The 10 m Keck II telescope on Mauna Kea examined 27 locations in the near IR. At none of these was a specular reflection observed, thus ruling out large oceans. At the much longer wavelengths of radar beamed from the 300 m Arecibo radiotelescope, specular reflections were observed at 12 of 16 locations, but at these wavelengths a better interpretation, and one consistent with the Keck data, is of very flat *solid* surfaces, though perhaps covered by liquid in the past.

The lingering possibility that the dark equatorial regions were liquid has also been ruled out. Cassini radar has shown that these are indeed seas, but of sand dunes. These consist of particles of unknown composition, presumably formed by erosion during heavy rainfall of CH_4. These particles must be mobilised by the surface winds, which, though generally very light, must occasionally blow stronger.

With no large oceans of CH_4, replenishment of atmospheric CH_4, which constitutes 4.7% of the lower atmosphere, can not be met by evaporation from open bodies of liquid CH_4. Replenishment is necessary because the lifetime of atmospheric CH_4 against disruption by solar UV radiation and fast electrons from Saturn's magnetosphere, is only about 10 Ma. Evaporation from surface/subsurface CH_4 ice must make a contribution. But cryovolcanism could make a much larger one.

Cryovolcanic features, impact craters, and surface age

Evidence for cryovolcanism has been obtained by the Cassini Orbiter. Possible cryovolcanic features include domes, lobate flows, rough terrain, and landforms that resemble volcanoes. 'Smoking guns' have also been seen – east–west cloud streaks hundreds of kilometres long originating at fixed locations on the surface. These could well be due to gas vented by cryovolcanic activity. It is not only CH_4 that cryovolcanism releases, but water vapour and other carbon compounds.

Cryovolcanism would be driven by a combination of residual heat from Titan's accretion, radiogenic heat from the core, and modest tidal heating. Overall, the rate of heat loss is less than 10% of that from the Earth, but this is sufficient for cryovolcanism.

Cryovolcanism can obliterate impact craters. Obliteration will also result from wind/fluid erosion, by the accumulation of organic compounds from the atmosphere, and by tectonic processes. If not obliterated, some craters will be subdued, though this can also result from viscous sagging. Cassini's radar has (so far) mapped a few per cent of the surface at high resolution, and indeed very few craters have been seen. This scarcity, over much of the surface, has been confirmed by Cassini's ISS and VIMS. There are a few as large as a few hundred kilometres, increasing in number as size decreases, with a cut-off at about 20 km because of the shielding effect of Titan's considerable atmosphere, which will cause small projectiles to break up.

The visible surface of Titan is therefore young, much of it only a few hundred million years old, though there might be regions of considerable greater age.

Question 8.6

Explain why water has played a relatively minor role in weathering the surface of Titan.

8.4.3   Enceladus

Enceladus is a medium-sized satellite of Saturn, radius 253 km, with a mean density of $1120 \, kg \, m^{-3}$.

☐   What does such a low density suggest about its interior?

The low density suggests a predominantly icy composition. The surface is certainly icy, and has an albedo of about 0.9, the highest in the Solar System. This suggests that the surface is very fresh. Its light-scattering properties are consistent with the texture of frost. The high albedo keeps the surface of Enceladus cold, with a mean surface temperature of 75 K, lower than any other of Saturn's many satellites.

The first spacecraft to visit Enceladus was Voyager 1, which flew through the Saturnian system in November 1980. In August 1981 it was followed by the flyby of Voyager 2. There was then a long gap until the arrival of Cassini–Huygens in June 2004. Since then, the Cassini Orbiter has made several close approaches to Enceladus, culminating on 14 July 2005, when the spacecraft descended to a minimum altitude of 175 km. Figure 8.13 shows an image taken during the approach, revealing the richly varied landscape first revealed by the Voyagers.

Crater densities vary from almost zero, as, for example, in the lower part of Figure 8.13 (the south polar region is at lower right), to about that of the least cratered areas on the surfaces of the other satellites of Saturn. This indicates that even the oldest terrain postdates the heavy bombardment, and the youngest terrain could be very much younger, and certainly no older than 1000 Ma. Within each type of terrain the craters show various degrees of degradation, due in part to viscous relaxation. The most degraded areas might have been above interior hot spots analogous to mantle plumes in Venus.

There are various types of tectonic grooves and ridges, and these are concentrated in plains. Some of the grooves and ridges seem to be tensional features, others compressional, and yet others might be cryovolcanic fissures and vents. The south polar region is particularly interesting. As well as being virtually devoid of impact craters, it is replete with tectonic features, including the parallel wavy lines at lower right in Figure 8.13 that have been dubbed 'tiger stripes'. These are cracks, about 130 km long and spaced about 40 km apart. They are 'warm', with temperatures up to about 110 K. There are also many boulders present, 10–100 m across. Together, this evidence indicates that this region is particularly young and active.

Figure 8.13 Enceladus, from the Cassini Orbiter. The south polar region is at lower right, and displays the 'tiger stripes' that seem to be the source of cryovolcanism. (NASA/JPL Space Science Institute PIA06249)

Dramatic evidence for activity came from Cassini during its 14 July 2005 flyby, when it imaged a broad fountain of icy particles and vapour reaching an altitude of at least 500 km. The source was the south polar region, notably the 'tiger stripes'. Cassini flew through this fountain, sampled it with its mass spectrometer, and found the vapour to consist (by numbers of molecules) of about 65% water, 20% H_2, and some N_2 or CO.

☐　Why can a mass spectrometer not distinguish between N_2 and CO?

The common isotopes of nitrogen, carbon, and oxygen have atomic masses of 14, 12, and 16 respectively, so the molecular mass of N_2 is the same as that of CO. The composition of the ice at the surface of Enceladus has been measured by Cassini's VIMS, which has revealed it to be near pure water ice. Other ices, such as NH_3 and CO_2 that are surely present in the interior, could be too volatile to exist on the surface.

Such cryovolcanic activity accounts for several other observations. The magnetometer detected oscillations of Saturn's magnetic field near Enceladus at a frequency suggesting ionised water vapour. Allowing for ionisation by solar UV radiation, this shows that a *very* thin atmosphere is present, mainly water vapour. It might be thin, but there is still too much to have been retained by the satellite's weak gravity. Cryovolcanism could sustain the atmosphere. Such activity also explains the resurfacing that has created the high albedo and the areas with few impact craters. The continuing escape of the atmosphere also accounts for Saturn's tenuous E ring (Figure 2.14) which has a maximum density at the orbit of Enceladus. The water-ice particles that constitute the E ring have estimated lifetimes of only about 10 000 years and thus need replenishing.

The presence of cryovolcanism indicates that at least some of the interior is warm enough for water, and other icy materials, to melt. Evidence for (partially) fluid subsurface materials is provided by the altitude range. This is largely confined to 1000 m, low for a body of the size of Enceladus, and indicates interior softness. This is consistent with Cassini magnetometer data, which showed Enceladus to be deflecting and slowing charged particles in Saturn's magnetosphere, as would be expected from liquids made electrically conducting by dissolved salts.

How can a small body like Enceladus sustain cryovolcanism and tectonic activity? The neighbouring satellites are similar in density – Mimas is about the same size as Enceladus, and Tethys is considerably larger, yet neither have signs of current or recent activity. This points to tidal heating for Enceladus, but not for its neighbours. But equation (4.13) indicates that the present eccentricity of the orbit of Enceladus, 0.0001, is too small for sufficient tidal heating today. Moreover, heat from radioactivity in Enceladus's rocky component cannot make up the shortfall. Perhaps, as a result of gravitational interactions between Enceladus and some other satellites, the eccentricity was greater in the past, and this could have led to intermittent periods of greater tidal heating, each lasting 10–100 Ma, with the last period ending relatively recently. This is plausible – at present Enceladus is in a 1:2 mean motion resonance with the larger satellite Dione, radius 560 km, which could raise its eccentricity.

8.4.4　Triton

Much of what we know about Triton, which at 1353 km radius is by far the largest satellite of Neptune, is from the flyby of Voyager 2 in August 1989, supplemented by Earth-based spectroscopy and occultations of stars by Triton. Voyager 2 obtained high-resolution images of about two-thirds of the sunlit hemisphere, with rather more of the southern than of the northern hemisphere being included. Plate 21 is a composite of these images, with southerly latitudes in sunlight.

Icy materials must comprise about a third of Triton's mass and 60% of its volume, probably concentrated into an icy mantle around a rocky core (Figure 5.7). Though water ice is expected to be the dominant icy material, IR spectroscopy reveals that at the surface N_2 ice dominates, and that CH_4 ice and CO ice are also present. These surface ices are presumably fairly thin veneers, derived from the mantle through their greater volatility than water. An indication that N_2 and CH_4 ices are no more than veneers is provided by topographic features such as 1 km high scarps – only water ice is strong enough at the surface temperature of 38 K to support such topography. There are also patches of CO_2 ice, perhaps the result of chemical reactions triggered by impacts. The surface ices, if pure, would be colourless, and so the pinkish appearance of much of the surface (Plate 21) is presumably because cosmic rays and solar UV radiation have produced traces of complex organic compounds.

The surface is overlain by a thin atmosphere of N_2, plus traces of CH_4 and CO. This is clearly consistent with the surface composition. The surface pressure measured by Voyager 2 was 1.4 ± 0.1 Pa. In November 1997 the atmospheric pressure was measured again, this time from the Earth, utilising the transit of a star. This is not very accurate, but placed the pressure in the range 2.0–4.5 Pa, a considerable increase in the 8 years since the flyby of Voyager 2. Possible explanations for this are presented in the section on Triton's atmosphere, Section 10.8.2. One of them is surface activity releasing N_2 into the atmosphere.

Evidence of surface activity

There are three main types of terrain on Triton. High smooth plains are presumed to be the result of a series of flows of icy lavas. At their edges they are seen to overlie the other two types of terrain

☐ What does this imply about the relative ages of the three types?

The high smooth plains must be the youngest, a conclusion borne out by the low density of impact craters on these plains. A similar type of surface is seen beyond the smooth plains, constituting the floors of four depressions 100 km or so across. A rugged area at the centre of each floor indicates a cryovolcanic origin for this infill, in which case the depressions could be calderas.

The second type of terrain is hummocky plains that border the smooth plains on their poleward side. Domes and a single ridge suggest volcanic creation of this type. It is the most cratered terrain on Triton, though only about as much as the lunar maria. No absolute ages can be deduced, but the hummocky plains, though still young among the surfaces of planetary bodies, are at least twice as old as the high smooth plains.

Finally, there is cantaloupe terrain, which is unique to Triton, and can be seen towards the day–night boundary in Plate 21. It gets its name from its resemblance to the skin of a cantaloupe (melon), and consists of a patchwork of approximately 10 km diameter dimples and bumps criss-crossed by grooves and troughs. Some of the grooves and troughs might be graben, and some have ridges running along their centre lines, which might be extrusions of viscous ice. Such extrusions might also account for some of the bumps. Fresh long troughs cut all types of terrain, and might be recent graben. The high smooth plains and hummocky terrain overlap cantaloupe terrain and so it must be the oldest type, yet there are few impact craters – this is a mystery.

In Plate 21 a polar cap is visible extending from the South Pole towards the equator. It generally has a higher albedo than the rest of the surface, reaching 0.89 in its brighter parts (though even the darkest parts of the imaged surface have the rather high albedo of 0.62). Any north polar cap was hidden on the night side. The polar cap presumably consists of the same materials as the rest of the surface veneer. The cap's high albedo would then be the result of the seasonal deposition of pure ices, notably the more volatile N_2, presumably as N_2 frost is sublimed from one pole and deposited at the other.

Within the south polar cap four plumes have been identified in Voyager data. They rise straight up to an altitude of about 8 km, such high altitudes resulting from their buoyancy in the atmosphere. At 8 km the plumes are no longer buoyant and high-speed winds then carry them sideways for 100 km or so. Dark steaks elsewhere in the polar cap indicate deposition from earlier plumes, and some of these are visible in the lower right of Plate 21. The plumes are probably geysers from the sublimation of N_2 below the surface, the N_2 gas breaking through the overlying ice. The sublimation could be the result of heat flow from the interior, or from the solar heating of darkened ice beneath a metre or so surface layer of pure transparent ice. Increased geyser activity could account for the increase in atmospheric pressure between 1989 and 1997.

Though the geysers are the only definite sign of ongoing activity on Triton, much of the surface is comparatively young, and models of the thermal evolution of Triton, in which there was powerful tidal heating early in its life, show that cryovolcanism could be occurring today as a result of the raised temperatures from primordial tidal heating plus radiogenic heat. The presence of nitrogen ice in the outer layers enables these heat sources to drive cryovolcanism because nitrogen is very volatile.

Question 8.7

Why is nitrogen cryovolcanism probably confined to Triton?

8.5 Summary of Chapter 8

The bodies that are still subject to volcanic and tectonic activity comprise the Earth, Venus, and Io, which have rocky surfaces, and Europa, Titan, Enceladus, and Triton, which have icy surfaces. Rocky surfaces experience silicate volcanism and icy surfaces cryovolcanism. Impact craters are scarce as a result of resurfacing.

The Earth seems to be unique in having a global system of plate tectonics, in which the Earth's lithosphere is divided into a number of plates in relative motion. Oceanic crust is created at constructive plate margins from where the crust spreads, and it is destroyed at destructive plate margins where it is subducted into the mantle. At both types of margin there is a great deal of volcanic and seismic activity. Oceanic crust is derived by the partial melting of the mantle, and is basaltic–gabbroic in composition. Continental crust is, on average, andesitic, with a granitic–rhyolitic composition in its upper regions. It is derived, ultimately, from oceanic lithosphere. Continental crust is eroded mainly through the production of ocean sediments by gradation. Plate motion is sustained by solid state convection in the asthenosphere, driven by the residual heat from primordial energy sources plus ongoing radiogenic heating. Plate recycling accounts for about 70% of the energy that reaches the crust, with mantle plumes accounting for most of the rest.

On Venus there is no widespread system of plate tectonics. It is unlikely that this is because of a really thick lithosphere, but more likely because its lithospheric properties differ from those of the Earth. Instead, there might be a system of mantle plumes and downwellings. Sample analyses indicate a predominantly basaltic crustal composition, in accord with the lack of an extensive system of plate tectonics. The predominance of one type of crust is also indicated by the unimodal hypsometric distribution. Volcanic activity has been widespread within the past few hundred million years, and little of the surface of Venus is older than 500–800 Ma. This might be due to episodic global subduction of the lithosphere, which otherwise acts as a stagnant lid. Any extensive resurfacing 500–800 Ma ago could alternatively have been caused by an outburst of mantle plumes at that time that resulted in sufficiently copious production of basaltic lavas to obliterate almost all of the surface. Plumes, delamination, and (limited) subduction might account for most of the loss of internal energy. Any episodic global subduction could make a major if intermittent contribution. Volcanic activity might continue today, as might tectonic activity, and further resurfacing episodes are possible.

Io is the most volcanically active body in the Solar System. Its silicate volcanism depends largely on tidal heating. Europa has a smooth icy surface, the result of cryovolcanism, also sustained largely by tidal heating. Titan has a surface dominated by water ice, but so cold that it behaves like rock on Earth. CH_4 plays a similar role to water on Earth, and in its liquid form must have been responsible for most of the channels and other fluvial features. There is widespread evidence of cryovolcanism, thought to be driven by a combination of residual heat from Titan's accretion, radiogenic heat, and modest tidal heating.

The icy surface of Enceladus has displayed its cryovolcanism most dramatically in the form of a broad fountain of icy particles. Its cryovolcanism needs substantial tidal heating. This does not seem sufficient today because of its low orbital eccentricity, but gravitational interactions with other satellites, notably Dione, could have increased the value in the past, and could do so again in the future. On the icy surface of Triton, dramatically expressed in the form of N_2 geysers, the present cryovolcanic activity could be the result of the residual effects of primordial tidal heating plus radiogenic heating, the presence of the highly volatile nitrogen ice in the outer layers enabling these heat sources to be effective.

9 Atmospheres of Planets and Satellites: General Considerations

The atmospheres of the planetary bodies are richly diverse. This is apparent from Table 9.1, which lists some of the properties of the substantial atmospheres. As you can see, the list includes the main constituents of each atmosphere. This is specified as the **number fraction** of each constituent: that is, the number of molecules of the constituent divided by the number of molecules in the whole atmosphere. For convenience, we shall extend the term 'molecule' to include constituents that are present as single atoms, such as argon in the Earth's atmosphere. In the case of the four giant planets there is no clear distinction between the atmosphere and the interior (Section 5.3), and so zero altitude is defined at the altitude at which the atmospheric pressure is 10^5Pa. The number fractions do, however, apply to the whole atmosphere down to pressures that are greater than this, and for non-condensing components, considerably greater.

The total quantity of an atmosphere can be expressed as the number of kilograms 'standing' on each square metre of the surface of the planetary body. This global mean value is called the **column mass** m_c, and these are listed in Table 9.1. The total, except for the giant planets, mass of an atmosphere is the column mass multiplied by the surface area of the body. The total is a very small fraction of the planetary mass – an atmosphere is a low-density veneer.

Table 9.1 also gives the size of the atmospheric pressure p_s at the surface of the body. This is related to the column mass as follows:

$$p_s = m_c g_s \qquad (9.1)$$

where g_s is the magnitude of the gravitational field at the surface (Section 4.1.2). To understand this relationship note that the right hand side is the mass of atmosphere standing on unit area of the surface, multiplied by the magnitude of the gravitational force per unit mass. The right hand side is therefore the size of the force exerted by the atmosphere per unit area of surface. A force per unit area is pressure, and thus the right hand side is the magnitude of the surface pressure. Strictly, equation (9.1) is an approximation because g_s neglects the effect on p_s of rotation of the body, and the slight decrease of the gravitational field with altitude. In Table 9.1 the surface gravity is g_e, the acceleration of gravity that would be experienced by an object fixed at the equator, and so includes the slight effect of rotation.

Also included in Table 9.1 is the mean surface temperature T_s. In equilibrium, T_s is related to the mean surface pressure p_s and the mean surface density ρ_s via an equation of state

Discovering the Solar System, Second Edition   Barrie W. Jones
© 2007 John Wiley & Sons, Ltd

Table 9.1 Some properties of the substantial planetary atmospheres

Planetary body	Major gases/number fractions	Escape speed[c]/ km s^{-1}	Surface gravity[c]/ m s^{-2}	Mean surface pressure/ 10^5Pa	Column mass[d]/ 10^4kgm^{-2}	Mean surface temperature/ K
Venus	CO_2 0.96, N_2 0.035	10.4	8.38	93	102	740
Earth	N_2 0.77, O_2 0.21, Ar 0.009, H_2O 0.01	11.2	9.78	1.01	1.03	288
Mars	CO_2 0.95, N_2 0.027, Ar 0.016	5.0	3.72	0.0056	0.015	218
Titan	N_2 0.95, CH_4 0.047	2.64	1.35	1.5	11	94
Triton	Mainly N_2, some CH_4	1.45	0.78	$2–4.5 \times 10^{-5}$	$2.6–5.8 \times 10^{-4}$	40
Pluto[a]	Mainly N_2, some CH_4 and CO	1.3	0.8	$\sim 15 \times 10^{-5}$	1.9×10^{-4}	40
Jupiter[b]	H_2 0.863, He 0.135, CH_4 0.0018	59.5	24.7	1	247	165
Saturn[b]	H_2 0.881, He 0.122, CH_4 0.0045	35.5	10.4	1	104	135
Uranus[b]	H_2 0.83, He 0.15, CH_4 0.016	21.3	8.8	1	88	76
Neptune[b]	H_2 0.83, He 0.15, CH_4 0.022	23.5	11.1	1	111	72

[a] Pluto's orbital eccentricity is large and its atmospheric properties vary as it goes around its orbit. The values are for the present, with Pluto not far from perihelion.
[b] The surface quantities for the giant planets are at the altitude where $p = 10^5$Pa. The composition is for the total mass of atmosphere to greater depths.
[c] At the surface, at the equator, and including the effects of rotation.
[d] Obtained by dividing the value in the fifth column by that in the fourth.

(Section 4.4.3). Atmospheric gases are at relatively low densities. Therefore they can be regarded as **perfect gases**, also called ideal gases, and we can then use the perfect gas equation of state. This equation applies locally, and is

$$p = \frac{\rho kT}{m_{av}} \tag{9.2}$$

where k is a universal constant called Boltzmann's constant (Table 1.6), and m_{av} is the mean mass of the molecules in the atmosphere at the point in question. If we apply equation (9.2) at the surface then it allows us to calculate any one of ρ_s, T_s, and p_s from the other two.

☐ In order to do so, what else needs to be known at the surface?

We also need to know m_{av}. This is obtained from the atmospheric composition.

You can see from Table 9.1 that atmospheric compositions and column masses vary considerably from body to body. For the planetary bodies not listed, any atmospheres are far more tenuous than those included. Why atmospheres vary so much is a topic for Chapters 10 and 11, where we examine individual atmospheres in detail. Before then, we shall look at atmospheric processes in general. But first, we take a brief look at how atmospheres are studied.

9.1 Methods of Studying Atmospheres

One way of studying an atmosphere is from instruments within it, and the Earth's atmosphere has been widely studied in this way for centuries. For other atmospheres to be studied in this manner it is necessary to fly a probe through the atmosphere, or to place a lander on the surface. Venus, Mars, Jupiter, Titan, and the Moon have had their atmospheres directly explored by probes or landers. But much of what we know about atmospheres comes from studying them remotely from the Earth's surface, from telescopes in Earth orbit, or from spacecraft that fly past the remote body or go into orbit around it.

Spectrometry

Spectrometry is of great importance in remote studies of atmospheres. As a typical example, Figure 9.1 shows part of the IR spectrum of Mars. This is the spectrum of the thermal radiation *emitted* by the Martian surface and then modified by passage through the Martian atmosphere. The atmosphere is cool and/or thin enough so that it emits more feebly than the surface, and so the surface acts as a bright source of background illumination. Atmospheric molecules absorb the background at various wavelengths, and then emit the absorbed radiation at the same wavelengths, but equally in all directions, so the net effect is removal of some radiation from our line of sight. This results in a set of absorption lines, often narrower than the absorption lines produced by solids or liquids, and therefore readily distinguished from them. A particular molecule absorbs at a characteristic set of wavelengths, and therefore if absorptions at such wavelengths are detected, the presence of the molecule is inferred. The absorption line in Figure 9.1 lies at a wavelength known from laboratory studies to correspond to gaseous CO_2 and so it shows that CO_2 is present as a gas in the Martian atmosphere.

At the wavelengths in Figure 9.1 the solar radiation scattered from the Martian surface is much weaker than the radiation emitted by the surface, and it therefore makes a negligible contribution to the source on which the absorption lines are produced. At shorter wavelengths solar radiation dominates. The atmospheric constituents then make an imprint not only on the fraction scattered by the surface, but also on the solar radiation on its way in to the surface.

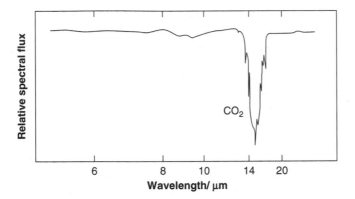

Figure 9.1 Part of the IR spectrum of Mars, with a prominent absorption line due to CO_2. Note that the smooth shape of the spectral flux from an ideal thermal source at 273 K has been subtracted. (Adapted from C.A.Beichman, JPL Publication, 96-22, (1966))

Note that as well as a solid (or liquid) surface, clouds or deep atmospheric layers will do just as well as emitters or reflectors. Note also that, in addition to absorption lines, emission lines from atmospheres can also be observed in suitable circumstances, e.g. if the source region is hotter than the background. The wavelengths of emission lines also indicate composition.

So far we have used only the *wavelengths* of the spectral lines. There is clearly more to spectral lines than that: they have width, area, and shape; absorption lines have depth (Figure 9.1); emission lines have height. These characteristics can be used to infer three further properties of the atmosphere: namely, the number of molecules that produced the line(s), the temperature of the molecules, and their pressure. We shall use the two absorption lines in Figure 9.2 to exemplify how these properties are extracted.

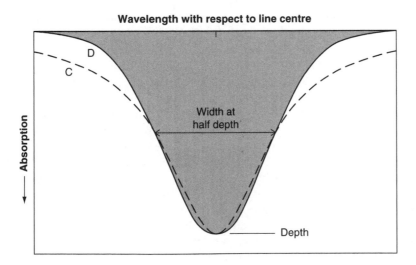

Figure 9.2 Two spectral absorption lines. The depth and width (at half depth) are the same in both cases, but the shapes and areas differ.

The number of molecules that produce a spectral line is the total number along our line of sight through the region in which the spectral line is produced. This total influences the *area* of the line: the greater the total, the greater the area. In Figure 9.2 the area of one of the lines has been shaded. The area of the other line is greater. The area also depends on the average temperature along the line of sight. For example, consider a gas with a copious source of photons beyond it, i.e. a gas against a bright background. A particular spectral absorption line in the gas requires the molecule to be in a particular state before the photon is absorbed. The proportion in this state depends on the temperature of the gas. If the proportion is low, it is *as if* far fewer molecules are present, and the area of the line is consequently reduced.

Temperature also influences the *width* of the line (Figure 9.2). This is through the Doppler effect – the greater the temperature T, the greater the spread in the random velocities of the atoms or molecules, and therefore the greater the Doppler width. At a given temperature, the greater the mass m of the atom or molecule causing the line, the smaller the spread in the random velocities, and so the Doppler width is reduced. Overall, the Doppler width is proportional to $\sqrt{T/m}$. The spreading of the line due to this Doppler broadening gives the line a characteristic shape – line D in Figure 9.2.

❏   Will the Doppler effect change the wavelength of the *centre* of the line?

The directions of motion are random, so at any instant there are about half the particles moving away from us and about half towards us, regardless of the temperature. Therefore, the centre of the line remains at the same wavelength. Only if the source of the line as a whole is moving towards or away from us is there a change in wavelength at the centre of the line.

A spectral line can also be broadened owing to collisions. If, during the absorption or emission of a photon, an atom or molecule collides with another, the photon energy will be changed slightly and hence so will its wavelength. The overall effect is that the width of the line is proportional to $p/\sqrt{T}$, where p is the pressure. This **collisional broadening** (also called pressure broadening) imparts a characteristic shape to a spectral line, different from that in Doppler broadening – line C in Figure 9.2. To show that the line width is proportional to $p/\sqrt{T}$ we need to use three fairly fundamental relationships. Do not worry if you are not familiar with them. They are: that the line width is proportional to the frequency of collisions; that this frequency is proportional to $v\rho$, where v is the average speed of a molecule and ρ is the density; and that v is proportional to $\sqrt{T}$. Accepting these, you can see that the collision frequency and hence the line width is proportional to $\rho\sqrt{T}$. Using the perfect gas equation of state (equation (9.2)), we then obtain $p/\sqrt{T}$ for the factor to which line width is proportional.

Clearly a huge amount of information is embedded in a spectral line. Even more information is embedded in *sets* of lines. For example, the relative area of two different lines from the same region depends on the temperature of the region. The main point is that by examining the details of one or more spectral lines it is possible to establish the total number of molecules of the sort that produce the line(s), and the temperature and pressure of the molecules.

Different absorption lines can originate from different altitudes. Therefore by observing a variety of absorption lines it is possible to estimate altitude variations of composition, pressure, and temperature. Moreover, a *particular* line can be formed over a great range of altitudes, and this is a further influence on the shape of the line. Therefore, a single line can also provide information about conditions at different altitudes. Variations with altitude can also be determined by observing how absorption lines change as the detector line of sight is swept across the disc of the planet. Near the centre of the disc we are looking vertically through the atmosphere and the tenuous upper reaches have less effect than when we are looking near the edge of the disc.

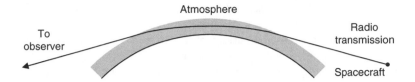

Figure 9.3 The occultation method of studying planetary atmospheres.

Occultation

Another way of investigating atmospheres is to use the radio transmission from an orbiting or flyby spacecraft when the craft passes behind (is occulted by) the planetary body as seen from the Earth. Then, to reach us, the radio waves have to pass through the atmosphere as in Figure 9.3, and they are refracted slightly. The angle of refraction can be measured and the mean refractive index along the path can then be calculated. During the **occultation** the path changes, which allows the refractive index versus altitude to be calculated. The refractive index depends on the mean density, the mean molecular mass, and on certain properties of the candidate molecules that are known from laboratory studies. Radio waves can penetrate clouds with particular ease, and so we can get information from depths that are obscured at other wavelengths. A similar technique at shorter wavelengths uses the occultation of a star by the planetary body.

Clouds

Clouds can be studied spectroscopically, though, as for all condensed matter, the spectral signatures that reveal composition are usually much less distinctive than for gases. As well as spectroscopy, we can measure how the intensity of the solar radiation scattered by the cloud particles varies with the phase angle, i.e. the angle between the incoming solar radiation and our line of sight. This not only provides further information on composition, but also reveals particle shapes. Atmospheric circulation is revealed by following particular cloud features.

Question 9.1

Describe the effect on the spectral line in Figure 9.1 of

(a) a decrease in the number of CO_2 molecules in the Martian atmosphere;
(b) an increase in the temperature of the Martian atmosphere;
(c) an increase in the pressure of the Martian atmosphere.

9.2 General Properties and Processes in Planetary Atmospheres

9.2.1 Global Energy Gains and Losses

In considering global energy gains and losses, we need to include the gains and losses at the planetary body's surface. The giant planets have no surfaces, in that there is no abrupt change in density from an atmosphere to an interior that is opaque to solar radiation. Their 'surfaces' can crudely be regarded as some depth in the atmosphere beyond which solar radiation does not penetrate. For bodies with well-defined surfaces but no atmospheres, it is clear that the references to the atmosphere below are to be ignored.

If we treat the atmosphere plus the surface as a single global entity, then energy can be gained by this entity in two major ways: namely, from solar radiation and by heat flow from the interior. There is only one major loss of energy – the loss to space of radiation emitted by the surface, by clouds, and by atmospheric gases. This will be at IR wavelengths. Let W_{abs} denote the power absorbed from solar radiation, W_{int} the rate of heat flow from the interior, and L_{out} the radiant power emitted to space. A steady state requires

$$L_{out} = W_{abs} + W_{int} \tag{9.3}$$

Because of short-term variations in these three quantities, the steady state of interest is in the longer term. Each of the three quantities is then an average over a period typically much greater than the orbital period. Table 9.2 gives the ratio L_{out}/W_{abs} for the planetary bodies with substantial atmospheres, and also for Mercury and the Moon, which are two of the many bodies that have tenuous or negligible atmospheres.

❑　What is the value of L_{out}/W_{abs} if there is negligible heat flow from the interior?

L_{out} then is equal to W_{abs}, and $L_{out}/W_{abs} = 1$. You can see from Table 9.2 that only for Jupiter, Saturn, and Neptune is this ratio substantially greater than one. Their IR excesses were discussed in Section 5.3. For all other planetary bodies listed, W_{int} is much less than W_{abs}. Among those not listed, Io is the only exception, because of its huge tidal heating. Its value of L_{out}/W_{abs} is uncertain but is substantially smaller than the Jovian value. We shall not consider Io further here.

Table 9.2　L_{out}/W_{abs}, a_B, and T_{eff} for some planetary bodies

Planetary body	L_{out}/W_{abs}^a	a_B	$T_{eff}/$ K
Venus	1.000	0.76	229
Earth	1.0002	0.30	See Q9.2
Mars	1.000	0.25	210
Titan	1.000	0.21	85
Triton	1.000	~ 0.85	~ 32
Pluto^b	1.000	~ 0.5	~ 37
Jupiter	1.67	0.34	124
Saturn	1.77	0.34	95
Uranus	1.06	0.30	59
Neptune	2.62	0.29	59
Mercury^c	1.000	0.10	436
Moon^c	1.000	0.11	271

^{a} Values shown as 1.000 depart from being exactly 1 in the fourth or lower decimal places. This is also the case for the bodies not included here, except for Io. Digits in brackets are uncertain.

^{b} Pluto's distance from the Sun varies far more than those of the other planets, and it has a long orbital period. The values are for the late 1990s.

^{c} Mercury and the Moon differ from the other bodies here in that they only have extremely tenuous atmospheres.

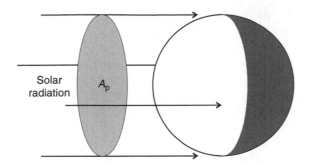

Figure 9.4 The projected area of a planetary body.

W_{abs} is related to the flux density F_{solar} of solar radiation at the distance r of the planetary body from the Sun, via

$$W_{abs} = F_{solar} A_p (1 - a_B) \tag{9.4}$$

where A_p is the projected area of the surface (Figure 9.4), which for a spherical body of radius R is πR^2, and a_B is the **Bond (or planetary) albedo**, i.e. the fraction of the intercepted solar radiation that is reflected back to space by the surface and atmosphere (William Cranch Bond, American astronomer, 1789–1859). Table 9.2 lists values of a_B. F_{solar} is given by the solar luminosity $L_\odot$ divided by the area of the sphere over which the energy is spread at the distance r:

$$F_{solar} = \frac{L_\odot}{4\pi r^2} \tag{9.5}$$

Equation (9.5) assumes that solar radiation is uniformly distributed over all directions.

It is usual to express L_{out} as the **effective temperature** of the planetary body. This is defined in terms of the radiation from an ideal thermal source (Section 1.1.1). The power radiated by unit area of such a source depends only on its absolute temperature T, and is given by σT^4 where σ is a universal constant called Stefan's constant (Table 1.6). For an ideal thermal source of surface area A the power radiated is therefore given by

$$L = A(\sigma T^4) \tag{9.6}$$

If A is the total area of a planetary body, and if L_{out} is the power radiated, then equation (9.6) *defines* T to be the effective temperature T_{eff} of the planetary body. Thus, *by definition* $L_{out} = A(\sigma T_{eff}^4)$, which can be rearranged as

$$T_{eff} = \left(\frac{L_{out}}{A\sigma}\right)^{1/4} \tag{9.7}$$

If W_{int} is negligible, then, for a spherical body (see Question 9.2),

$$T_{eff} = \left(\frac{F_{solar}(1 - a_B)}{4\sigma}\right)^{1/4} \tag{9.8}$$

This is useful in that T_{eff} depends only on the solar flux density at the orbit of the planetary body and on its Bond albedo – the radius of the body has been eliminated. Table 9.2 gives values of T_{eff}.

However, a planetary body does not radiate in the manner of an ideal thermal source, which raises the question of what T_{eff} means in terms of actual temperatures. For a body without an atmosphere the global mean surface temperature T_s will be approximately equal to T_{eff} if the surface is a very good absorber of radiation at the wavelengths of its emission. The less good it is, the greater will be T_s for a given T_{eff} (for a given L). For a body with a substantial atmosphere, the radiation emitted to space comes from a range of altitudes in the atmosphere, as well as from the surface. In this case, T_{eff} is the temperature at some altitude above the surface, but this is not from where all the emission comes.

As well as total quantities, it is also important to consider the spectral distribution of the radiation. Figure 9.5(a) shows a typical spectrum of the radiation emitted to space by the Earth, and Figure 9.5(b) the spectrum of solar radiation. Note that whereas planetary emission is mainly at middle IR wavelengths, solar radiation is mainly at visible and near-IR wavelengths (see also Figure 4.13). This is a consequence of the different source temperatures – the solar photosphere has a temperature around 5800 K, whereas the temperatures of planetary atmospheres and surfaces are much lower, typically a few hundred kelvin. Because these two wavelength ranges

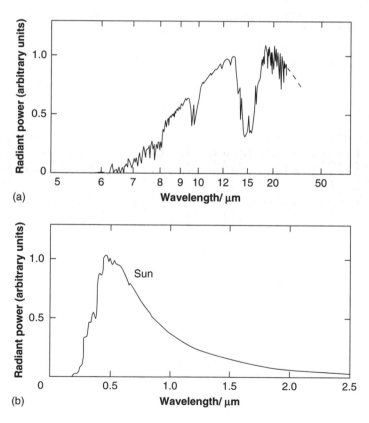

Figure 9.5 (a) The Earth's emission spectrum (per unit frequency interval). (b) The solar spectrum (per unit wavelength interval).

are so separate, a surface or atmosphere can absorb a very different proportion of radiation in one range than in the other.

Question 9.2

(a) Derive equation (9.8).
(b) Use it to calculate the missing value of T_{eff} in Table 9.2.

9.2.2 Pressure, Density, and Temperature Versus Altitude

In this section we shall be concerned with global averages across the surface of a planetary atmosphere.

The temperatures and pressures in a planetary atmosphere vary from point to point. Here we shall concentrate on altitude variations, globally averaged, deferring the horizontal variations to Section 9.2.6.

❏ If we know the temperature and pressure versus altitude, how can the density versus altitude be obtained?

The density versus altitude can be obtained from the perfect gas equation of state (equation (9.2)), provided we know at each altitude the mean mass of the molecules that constitute the atmosphere there.

Pressure and density versus altitude

Figure 9.6 shows pressure versus altitude in the Earth's atmosphere, which will serve to introduce the general ideas. How do we account for the rapid drop in pressure? On average over the globe, the atmosphere is neither expanding nor contracting, and so we can apply the hydrostatic equation introduced by equation (4.11). This gives the change in pressure δp for an increase δr in the distance from the centre of a spherically symmetrical body. A planetary atmosphere is a relatively thin layer on the surface and so in place of δr it is more sensible to use the increase δz in altitude z above the surface. If we again ignore planetary rotation, which has a relatively small effect on pressure, then

$$\delta p = -\frac{GM}{r^2}\rho\delta z \tag{9.9}$$

where r is the distance from the centre of the body to a point in the atmosphere, M is the total mass within r, and ρ is the density at altitude z. This equation shows that as altitude increases the pressure *must* decrease.

In equation (9.9) GM/r^2 is the magnitude g of the gravitational field at r (equation (4.6)), and with the atmosphere being a thin, low-mass layer we can use the surface gravitational field g_s everywhere within it. Thus, to a good approximation,

$$\delta p = -g_s\rho\delta z \tag{9.10}$$

Substituting for ρ from equation (9.2)

$$\delta p = -p[g_s m_{av}/(kT)]\delta z \tag{9.11}$$

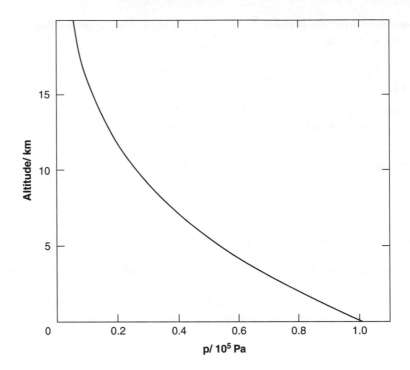

Figure 9.6 Pressure versus altitude in the Earth's atmosphere.

where T is the temperature at altitude z. If the temperature is the same at all altitudes (isothermal atmosphere), and if m_{av} is also independent of altitude, then it can be shown that equation (9.11) leads to

$$p = p_s e^{-z/h} \tag{9.12}$$

where p_s is the pressure at zero altitude and h is $kT/(g_s m_{av})$. The quantity h is called the isothermal scale height. For every increase in z of h the pressure falls by a factor of e, which is 2.718 (to four significant figures). For the Earth, h is about 8 km in the lower atmosphere – very much less than the Earth's radius.

Though atmospheres are not isothermal, and though m_{av} can change with altitude, equation (9.12) possesses the observed feature that the pressure falls rapidly with altitude. It follows from equation (9.2) that, for any realistic temperature variation with altitude, the density must also fall rapidly with altitude.

Temperature versus altitude

The temperature versus altitude depends on the various ways that energy is gained and lost at each level, and at the surface. These processes are summarised in Figure 9.7.

Consider first the surface of a body. It receives the solar radiation that has survived passage through the atmosphere. A fraction of this is absorbed by the surface, and the rest is reflected. The surface also receives radiation emitted by the atmosphere. Recall from Figure 9.5 the very different wavelength ranges of these radiation inputs, with solar radiation mainly at visible and

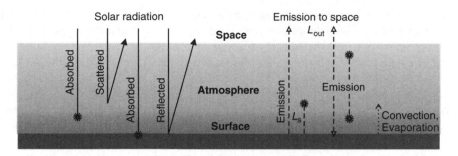

Figure 9.7 Energy gains and losses in a planetary atmosphere.

near-IR wavelengths (around 1μm) and atmospheric radiation mainly at middle IR wavelengths (around 10μm). The surface also receives heat from the interior, but (except for the four giant planets and Io) this is a sufficiently small fraction of the absorbed solar radiation that it can be ignored.

The surface loses energy by emitting its own radiation, again in the middle IR. It will also lose energy through the conduction of heat into the atmosphere in contact with the surface, usually followed by convection in the atmosphere. Convection was described in general terms in Section 4.5.2.

❑ What condition on the variation of temperature with altitude must be met for convection to occur?

For convection to occur, the temperature decrease with altitude must be at least as great as in the adiabatic gradient. In the atmosphere convective plumes then rise upwards, surrounded by denser, descending atmosphere that replaces the plume material. The convective plumes break up, and mix with the surrounding atmosphere, thereby transferring heat from the surface to the atmosphere. In planetary atmospheres the rate of temperature decrease with altitude is called the **lapse rate**, and so the adiabatic gradient is called the adiabatic lapse rate. Note that a positive lapse rate corresponds to a decrease in temperature as altitude increases (Figure 4.15). If the actual lapse rate is less than the adiabatic value, there is no convection. If it is greater than the adiabatic value the rate of convection is so large that the adiabatic value, or something close to it, is quickly established.

For an atmosphere in which there is no condensation of any component in the rising plumes, the adiabatic lapse rate is given by

$$\Gamma_d = \frac{g_s}{c_p} \tag{9.13}$$

where c_p is the atmospheric specific heat at constant pressure, i.e. the quantity of heat that must be transferred (at constant pressure) to unit mass of the atmosphere to cause unit rise in temperature. The value of c_p depends on atmospheric composition. The only significant condensate in the Earth's atmosphere is water, and if water condensation is *not* occurring then the value of Γ_d is about $10 \, \text{K km}^{-1}$. The derivation of equation (9.13) is in standard texts on planetary atmospheres (see Further Reading), though you can perhaps convince yourself that the equation should contain g_s and c_p, because g_s determines the rate of decrease of pressure with altitude (equation (9.10)), and c_p determines the temperature change of a parcel of atmosphere when it is in a new pressure environment.

The surface can also lose energy through the evaporation of liquids or solids (we shall use the term 'evaporation' to include sublimation). It takes energy to convert a liquid or a solid into a gas, energy properly called the enthalpy of evaporation, but more colloquially known as the latent heat of evaporation (Section 4.5). For this to be a significant energy loss, the gas has to be removed so that further evaporation can take place. Convection provides one means of removal, by carrying the evaporated gas upwards.

In a convective column the temperature declines and might reach a value where the evaporant condenses in the atmosphere, to form liquid or solid droplets, which might constitute a cloud. The atmosphere is then said to be saturated with the condensable gas. When condensation occurs, energy is released, and so the energy lost by the surface is gained by the atmosphere. The energy released in condensation is called the latent heat of condensation, and it has the effect of introducing additional thermodynamic properties of the atmosphere to the right hand side of equation (9.13). For the Earth, if a region of the atmosphere is saturated with water vapour then Γ_d can be as low as $3\,\mathrm{K\,km}^{-1}$. The global average is about $6.5\mathrm{K\,km}^{-1}$.

Each level in the atmosphere where convection is occurring is thus heated not only directly by convection, but also indirectly through the effect that convection has on promoting latent heat deposition. A level is also heated by absorbing a proportion of the solar radiation passing through it, and by absorbing some of the IR radiation emitted by the other levels in the atmosphere and by the surface. Energy is lost by each level through its emission of (IR) radiation. A further process that is effective in some circumstances is conduction. All these processes are shown in Figure 9.7. Together they determine the temperature at each atmospheric level.

A steady state

Though surface and atmospheric temperatures exhibit daily and seasonal changes, the average of the overall rate of gain at the surface is very nearly equal to the overall rate of loss if the average is taken over several orbits of the Sun. The same is true at each level in a planetary atmosphere. Therefore, the gains and losses in Figure 9.7 are nearly in balance, and so we are close to a **steady state**. It follows that the average temperatures are very nearly constant. Note that this also applies to the giant planets, where substantial heat flow from the interior occurs at a quasi-steady rate.

The actual value of the mean temperature at the surface and at each atmospheric level depends on the way the various rates of energy gain and loss vary with temperature. Any change in any of the gains and losses, and the temperature is likely to change. For example, if the flux density of solar radiation F_{solar} were suddenly reduced, then the immediate effect at the surface would be a rate of gain of energy that was smaller than the rate of loss. This would result in surface cooling and a reduction in the rate of energy loss from the surface until a steady state was re-established at a new, lower surface temperature. Correct though this conclusion is, the details are *very* complicated, involving changes in energy transfer rates throughout the atmosphere, and between the atmosphere and surface.

Atmospheric domains

The lapse rate is one basis on which an atmosphere can be divided into different domains, and Figure 9.8 shows these domains in an idealised way – the Earth's profile in particular is more complicated, as you will see in Section 10.1.1.

The **troposphere** is where the temperature decreases with increasing altitude, the lapse rate having a large positive value, in many cases equal or close to the adiabatic value. Indeed, the

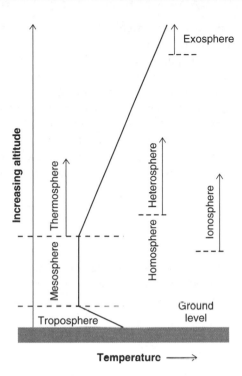

Figure 9.8 A schematic profile of temperature versus altitude, and the major atmospheric domains.

term 'troposphere' means 'turning sphere', which implies convection. The troposphere is where most clouds occur and is where weather changes can be rapid and large. In addition to the ubiquitous heat exchanges through the emission and absorption of IR and solar radiation, a convecting troposphere also transports heat by convection, the result of the heating of the surface (and perhaps the lower troposphere) by the Sun, and in the case of the giant planets, also by heat from the interior. The troposphere extends up to the tropopause. Above the tropopause the atmosphere is sufficiently thin that nearly all of the IR radiation emitted upwards there escapes to space, and this is the essential feature that determines the altitude of the tropopause. The altitude varies with latitude, season, and weather.

Above the tropopause there is the **mesosphere**, where the temperature hardly varies with altitude. The mesosphere gains energy by the absorption of IR radiation from below, by solar radiation, particularly at the UV wavelengths that lie just beyond visible wavelengths, and loses it by the emission of IR radiation downwards and to space. There is no convection. The mesosphere ends at the mesopause. The **thermosphere** extends above the mesopause and is characterised by a *negative* lapse rate – the temperature *increases* with increasing altitude. It is heated by the absorption of solar UV radiation at much shorter wavelengths. Energy is lost through IR emission and by the downward conduction of heat.

☐ Is there any convection in the thermosphere?

With a negative lapse rate there is no convection. The lapse rate is a complex outcome of the way the energy gains and losses vary with altitude.

You can see from Figure 9.8 that the atmosphere is also divided on bases *other* than the lapse rate. In the **homosphere**, atmospheric motions and frequent atomic and molecular collisions keep the atmosphere well mixed and therefore it has the same composition at all altitudes, except for any substances that condense (such as water vapour in the case of the Earth) – condensation depletes the local gas phase in the condensable substance, and the locality will be depleted as a whole if the condensed solid or liquid particles fall downwards (precipitate). In the **heterosphere** the mixing is weaker, with the result that molecules with greater mass are more concentrated by gravity into the lower heterosphere. In the **exosphere** the density is so low that collisions between molecules are sufficiently rare to allow the escape to space of a large fraction of any particles moving upwards with sufficient speed. Below the exosphere the density is higher, and so a particle is likely to collide before it escapes. In the **ionosphere** solar UV radiation has ionised a sufficient fraction of the molecules for the medium to display special properties, e.g. high electrical conductivity.

Question 9.3

(a) Rewrite equation (9.12) in terms of density instead of pressure.
(b) In what domain in a planetary atmosphere should the actual decrease of pressure with altitude follow most closely the form of equation (9.12)? Give reasons.

Question 9.4

Describe the possible effects on a planetary atmosphere if the UV radiation from the Sun were to become much weaker.

9.2.3 Cloud Formation and Precipitation

A cloud consists of a thin dispersion of solid or liquid particles in a gas. Though a cloud is very effective at scattering light, the particles are separated by distances that are much greater than their diameter, and so the density of a cloud is not much greater than the density of the cloud-free atmosphere adjacent to the cloud.

Crucial to our understanding of cloud formation is the phase diagram of the substance that can condense to form the cloud particles. Let us take a region of the Earth's atmosphere as an example, where the condensate is water, and start with the region in a state where clouds cannot form. We thus have a mixture of water vapour and the other atmospheric gases, mainly N_2 and O_2. The atmospheric pressure is the sum of the pressures of the individual gases – O_2 exerts a pressure, so does N_2, so does water vapour, and so does each of the other constituents. Each gas in the atmosphere behaves much like a perfect gas, and it is useful to note that $\rho = nm_{av}$, and to rewrite equation (9.2) as

$$p = nkT \tag{9.14}$$

where n is the total number of molecules per unit volume – the number density (not the number fraction, which is global). For a mixture,

$$p = (n_1 + n_2 + n_3 + \ldots)kT \tag{9.15}$$

where T is the temperature of the mixture, and n_i is the number density of the molecules of type i. The pressure contributed by a particular gas is called its partial pressure. For the ith constituent the partial pressure is

$$p_i = n_i kT \qquad (9.16)$$

In considering cloud formation we can usually use the phase diagram of the condensable substance, provided that we interpret the pressure as its partial pressure.

Figure 9.9 shows the phase diagram of water. Suppose that at some point in the atmosphere the partial pressure of water vapour (gas) is at point A, and that the temperature of the vapour is that of the atmosphere. Starting at A, consider an increase in partial pressure at constant temperature. This could be achieved by putting more water vapour into the atmosphere – equation (9.16) shows that the partial pressure is proportional to the number density. When the pressure reaches the 'liquid + gas' line, liquid water droplets form. The number density of water molecules in the vapour is then fixed at the value corresponding to the partial pressure on this line. Any attempt to increase the number density will be negated by further condensation.

Returning to A, consider a decrease in temperature at constant partial pressure.

☐ What sort of cloud particles form if the temperature is reduced sufficiently?

The 'solid + gas' line is reached, so solid particles of water (ice) form. The formation of ice particles fixes the number density of water molecules in the vapour at the value corresponding to the partial pressure on this line. The 'liquid + gas' and 'solid + gas' lines together constitute the saturation line. The partial pressure on this line is called the **saturation vapour pressure** and the atmosphere there is said to be saturated with water vapour.

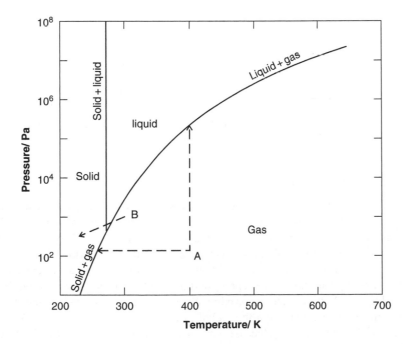

Figure 9.9 Phase diagram of water. Note that water is unusual in that the solid + liquid line slopes to the left, though this is barely apparent on the scale here. The triple point is at 273.16 K, 610 Pa.

Droplet or ice particle formation is delayed without the presence of other particles, such as dust, to act as condensation nuclei. Without such nuclei, and particularly if the decrease in temperature is rapid, the partial pressure will overshoot the saturation vapour pressure, and the atmosphere becomes supersaturated. This is not an equilibrium condition so ultimately the gas will condense, thus reducing the partial pressure to the saturation value – remember that a phase diagram, like an equation of state, is valid only for equilibrium conditions. It is also common for liquid droplets to appear initially, even if the temperature is low enough for ice to form. Such supercooled droplets are also in non-equilibrium, and will freeze.

Cloud formation is common in planetary atmospheres. Gas borne aloft by convection cools, and clouds form at a base altitude at which the saturation line of the condensate is reached. The curve in Figure 9.9 starting at B is one particular possibility in the Earth's troposphere. Clouds can also form above the cloud base if the conditions at higher altitudes lie on or to the left of the saturation line. Also, some cloud particles are carried upwards by vertical currents, even to the top of the convective column. Until condensation occurs, the partial pressure of the condensate at all altitudes is a fixed proportion of the total pressure. Above the cloud base the gas phase is depleted in the condensable substance, and so p_i/p is less.

If cloud particles become sufficiently large, or if the upward currents are too slow, then the particles fall out of the cloud. If they do not wholly evaporate before they reach the surface the residue will reach the surface as rain, snow, or hail – different forms of precipitation. Water vapour can also condense directly from the atmosphere onto the surface if the surface is sufficiently colder than the atmosphere. We then get dew, or frost. This return of water to the surface is the opposite physical process to the loss of water from the surface to the atmosphere through evaporation or sublimation.

Question 9.5

Suppose that in some region of the Earth, the surface temperature is 293 K, and the lapse rate in the troposphere up to the cloud base is $7 \, \mathrm{K \, km^{-1}}$. If the partial pressure of water in the lower troposphere in this region is 100 Pa, use Figure 9.9 to estimate the altitude of the cloud base. Will the clouds consist of water droplets or ice crystals?

9.2.4 The Greenhouse Effect

In Section 9.2.1 you saw that a surface–atmosphere system loses energy to space in the form of radiation, largely at IR wavelengths, at a rate given by the radiant power L_{out}. The radiant power L_s emitted by the surface alone is generally greater than L_{out}. The name given to this phenomenon is the **greenhouse effect** and it can be quantified by the ratio L_s/L_{out}. You saw that L_{out} can be represented by an effective temperature T_{eff}, which is an equivalent ideal thermal source temperature. L_s can likewise be represented by an effective temperature, but because the surface of a planetary body radiates approximately like an ideal thermal source, the effective temperature is not very different from the global mean surface temperature T_s. For this reason the magnitude of the greenhouse effect is more often quantified as $T_s - T_{\mathrm{eff}}$. The greenhouse effect raises surface temperature above the value it would otherwise have. (The effect gets its name from the passive heating of greenhouses, though detailed calculations show that only a small fraction of the temperature rise in a greenhouse is associated with IR radiation. Most of the rise is due to the trapping within the greenhouse of warmed air that cannot be convected away.)

It might come as a surprise to you that L_s can exceed L_{out} – at first sight it seems to imply a continuous build-up of energy in the atmosphere. However, this is not so. L_s and L_{out} are only two of the many energy flows involving the surface–atmosphere system (Figure 9.7). As long as the total rate of energy gain by the surface equals its total rate of loss then it will be in a steady state, and the same applies to the atmosphere and to the system as a whole. Any single rate of flow need not equal any other single rate – energy can flow at a much greater rate within a system than the rate at which it enters or leaves the system. In the case of L_s and L_{out}, the former originates from the surface, but the latter originates not only from the surface but from throughout the atmosphere, notably in and just above the troposphere. The positive lapse rate in the troposphere then ensures that T_{eff} is less than T_s.

The greenhouse effect depends on the absorption by atmospheric constituents of the radiation that is emitted by the atmosphere and by the surface. If a constituent absorbs radiation then it must also emit radiation. Some of this emission will be absorbed by the surface, thus raising its temperature above what it would have been had the atmosphere been transparent to the radiation (and if all else was the same). If the atmosphere is non-absorbing then the surface would emit all of its radiation direct to space. Furthermore, there would then be no emission from the atmosphere. Therefore, L_s would equal L_{out} and the greenhouse effect would be zero. Emission from planetary surfaces and atmospheres is at mid-IR wavelengths, and so it is the atmospheric absorptivity at such wavelengths that is important. Molecules consisting of a single atom do not absorb significantly at such wavelengths, and molecules with two atoms – diatomic molecules – are weak absorbers unless the atoms are of different elements. Molecules with more than two atoms are relatively strong absorbers.

If a strong absorber is introduced into an atmosphere then, if all other factors remain the same, T_s will rise, and the difference $T_s - T_{eff}$ will become greater. There is, however, a limit. For example, consider the CO_2 absorption line in Figure 9.1 centred on 15μm. As more CO_2 is introduced, the atmosphere will become completely opaque around 15μm. Ultimately this will be the case for all the absorption lines of CO_2, and adding more of it will not appreciably increase the greenhouse effect.

The effects of cloud particles

As well as the gaseous absorption–emission greenhouse effect, there is also a greenhouse effect from the scattering of mid-IR radiation by cloud particles. Some of the radiation from the surface and lower troposphere will be scattered back to the surface by this means, resulting in further warming. Clouds can also cool, by increasing the albedo of a planet. Which effect 'wins' depends on the size of the cloud particles/drops, their number density, their distribution with altitude, and, in atmospheres with more than one condensate, on their composition. Overall, the effects of clouds on temperature are complex and poorly understood.

Question 9.6

(a) Suppose that the atmosphere of a planet consisted solely of single atoms of argon. Discuss whether the greenhouse effect would be significant.

(b) Explain why T_s would rise if a dry atmosphere had water vapour added to it.

9.2.5 Atmospheric Reservoirs, Gains, and Losses

The specific molecules in a planetary atmosphere are transient. They are removed to the surface by precipitation and also by chemical combination with surface materials. They are lost to space, and they can undergo chemical reactions within the atmosphere. The average time that a molecule of a particular constituent spends in a planetary atmosphere is called its **residence time**, and it is rather short, typically less that a few tens of days. Therefore, unless the constituent is replaced, it will quickly disappear from the atmosphere. The almost constant mass of the various constituents of planetary atmospheres shows that molecules are being replaced today almost exactly as rapidly as they are being removed.

Figure 9.10 illustrates this balance for a typical constituent. The constituent resides in a set of interconnected reservoirs, and the chemical form of the constituent need not be the same in all of these. The reservoirs are connected by the processes through which the constituent moves from one reservoir to another, and each process transfers the constituent at a particular rate. As long as the sum of the rates of loss from a reservoir is (very nearly) equal to the sum of the rates of gain, then the quantity of the constituent in that particular reservoir will be (very nearly) in a steady state. Note that Figure 9.10 can be applied with care to the giant planets, with surface and subsurface reservoirs being regarded as at considerable depths, at which the atmosphere has become far denser than the atmospheres at the surfaces of the other bodies in the Solar System. We shall now examine the various processes show in Figure 9.10.

Physical processes between the atmosphere and surface

You have already met precipitation as a physical process that removes a condensable constituent from the atmosphere, such as water from the Earth's atmosphere. Atmospheric gases are also

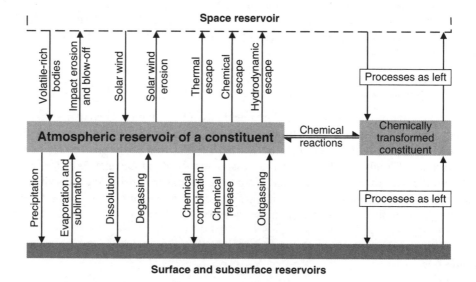

Figure 9.10 The reservoirs of some atmospheric constituent, and the various types of process by which it can be transferred from one reservoir to another.

removed through other physical processes, notably dissolution in liquids (such as the Earth's oceans) and, much more slowly, dissolution in solids. The reverse processes are evaporation (liquid to gas), sublimation (solid to gas), and degassing (the reverse of dissolution).

A constituent can also pass from surface to atmosphere through volcanism. This is called **outgassing**, and it can be accompanied by chemical changes not far below the surface so that the gases emitted are not the same as those at the source. The opposite process to outgassing is when an atmospheric constituent that has been physically trapped or chemically combined at the surface is subsequently buried by geological activity. Later, it might be outgassed again, though some proportion of outgassed material will never before have been part of the atmosphere – such gases are called juvenile.

Note that for an atmospheric constituent to be in a steady state, there is no requirement for these individual gains and losses to balance.

❏ Why not?

For a steady state, it is the sum of the rates of loss from a reservoir by *all* the various loss processes that must equal the sum of the rates of gain by *all* the various gain processes.

Thermal escape

You saw in Section 9.2.2 that in the exosphere collisions between molecules are sufficiently rare to allow the escape to space of a large fraction of any molecules moving upwards with sufficient speed. The sufficient speed is called the **escape speed**. It is given by

$$v_{esc} = \sqrt{\frac{2\,GM}{r}} \qquad (9.17)$$

where G is the gravitational constant, M is the mass of the planetary body, and r is the distance from the centre of the body to the base of the exosphere. As you might expect, equation (9.17) shows that the escape speed increases as the mass of the planetary body increases, and decreases as the height of the exosphere increases.

The fraction of any constituent with sufficient speed to escape depends on the distribution of molecular speeds at the base of the exosphere. Here, collisions are just about frequent enough for the atmosphere to be treated as if it were in thermal equilibrium. In this case, the atmosphere has a well-defined temperature, and it determines the speed distribution for molecules of a given mass. The distribution is called the **Maxwell distribution** after the British physicist James Clerk Maxwell (1831–1879), who did much theoretical work in this area.

Figure 9.11 shows the Maxwell distribution for CO_2 at 400 K and 600 K, and for H_2O at 400 K.

❏ How does the distribution vary with molecular mass and with temperature?

At a given temperature, the greater the mass of the molecule, the lower the speeds. For a given molecular mass, the higher the temperature, the greater the speeds. Note that a particular molecule changes its speed at every collision, and so a particular range of speeds in Figure 9.11 does *not* apply to a particular group of molecules, but to the fraction of the whole population that at any instant has speeds in that range. The speed at the peak of the distribution is called the most probable speed, and is given by

$$v_{mps} = \sqrt{\frac{2\,kT}{m}} \qquad (9.18)$$

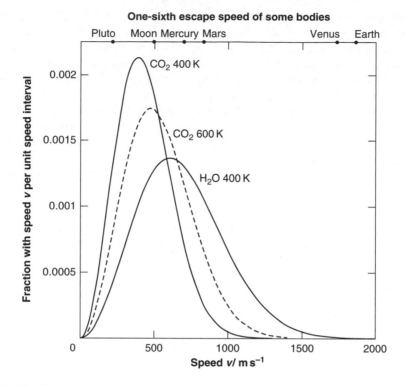

Figure 9.11 The Maxwell distribution of speeds of CO_2 and H_2O, compared with one-sixth of the escape speed of some planetary bodies.

where k is Boltzmann's constant, T is the absolute temperature, and m is the mass of the molecule.

In considering escape, T can be taken to be that at the base of the exosphere. The greater the value of T, the greater the value of v_{mps}, and the larger the fraction of a constituent that will have speeds in excess of the escape speed v_{esc}. The prominence of temperature in this escape process leads to its name, **thermal escape**. For a given exospheric temperature, the greater the mass of a molecule, the slower the rate of thermal escape, and this leads to discrimination between one type of molecule and another, between CO_2 and H_2O in the example in Figure 9.11. Thermal escape is thus mass selective. Mass selection extends to isotopes. The graphs in Figure 9.11 correspond to the isotopes ^1H, ^{12}C, ^{16}O, but had, for example, some water molecules included an atom of deuterium ^2H, to form $^2H^1HO$ instead of ^1H_2O, then these molecules would have had a mass 19/18 times that of ^1H_2O, and a correspondingly slower rate of thermal escape.

Figure 9.11 includes $v_{esc}/6$ for some planetary bodies. If, throughout the 4600 Ma history of the Solar System, it has been the case that for some constituent $v_{mps} < v_{esc}/6$ then the accumulated fraction of the constituent lost by thermal escape will be small. Figure 9.12 shows more extensive data. Each planetary body is plotted at a point corresponding to $v_{esc}/6$ and the temperature at the base of its exosphere (which is at its surface if it has little or no atmosphere). This gives a broad indication of which constituents are more likely to have suffered extensive thermal escape.

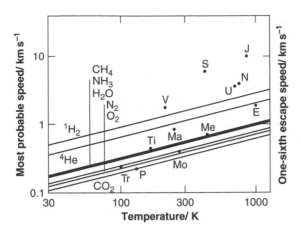

Figure 9.12 The most probable speed v_{mps} versus temperature T for some major atmospheric gases. Planetary bodies are at the temperatures at the base of their exospheres.

Other physical processes between the atmosphere and space

Molecules in the exosphere can also be removed by the impact of charged particles coming from the solar wind or from any substantial magnetosphere of the planetary body. Ionised molecules can also be removed by entrainment in the solar wind's magnetic field, particularly if there is no substantial planetary magnetosphere. Early in Solar System history, when the Sun was going through its T Tauri phase (Section 2.1.3), the greater UV flux must have generated ions in exospheres in great abundance, and the stronger wind must have been more effective in ejecting them. Therefore, the losses could have been considerable.

The opposite processes to losses to space are gains from volatile-rich bodies and the solar wind (Figure 9.10). These processes (as you will see) are thought to have been of great significance early in Solar System history, but the fluxes today are so low that they are negligible for all major atmospheric constituents in all substantial planetary atmospheres. Today, significant replacement of the major constituents can come only from the surface (possibly en route from the interior).

Early in Solar System history, during the heavy bombardment, giant impacts must have been common, and instead of *adding* to an atmosphere they must have caused huge net losses. If the impactor had a radius greater than about the isothermal scale height of the atmosphere (Section 9.2.2) then pretty well the whole atmosphere would have been ejected, largely as a result of the shock-induced vibrations of the planet. This is called **blow-off**. Smaller impactors, less than the scale height of the atmosphere, would have caused more modest losses – partial ejection largely due to the explosion of hot vapour generated at the impact site. Once this **impact erosion** has started, the scale height falls, making the atmosphere subject to erosion by smaller impactors, and these are more numerous. Considerable losses can consequently occur. Blow-off and impact erosion are *not* mass selective.

Chemical processes

Chemical reactions can increase the kinetic energy of the products of the reaction, thus increasing their speed, particularly for products of lower molecular mass than the reactants. For example, a nitrogen molecule can be dissociated by a photon of solar UV radiation to give fast-moving nitrogen atoms:

$$N_2(+UV \text{ photon}) \rightarrow N + N \tag{9.19}$$

This can lead to escape to space, adding **chemical escape** to thermal escape.

☐ Where does the reaction need to take place for escape to be possible?

The reaction needs to take place in the exosphere.

For some constituents, chemical escape can be far more rapid than thermal escape, and like thermal escape it is mass selective: the greater the mass, the slower the rate of escape. Hydrogen H or H_2 are the least massive molecules, and can therefore escape particularly rapidly through chemical and thermal means. If there is a large quantity of hydrogen escaping, then the high-volume flow rate can entrain other molecules and carry them off in a process called **hydrodynamic escape**. It is thought to have been important early in Solar System history. Hydrodynamic escape *is* mass selective: the greater the mass of the molecule, the less likely it is to be entrained, and the slower its rate of escape.

The involvement of the UV photon in reaction (9.19) makes this an example of a **photochemical reaction**. Such reactions are not only important in increasing loss rates. Thermospheres are in part warmed by the energy acquired through photochemical reactions, again involving solar UV photons. In the case of the Earth, the stratosphere is also warmed in this way, through the absorption of solar UV photons by ozone (O_3). O_3 is itself a product of photochemical reactions (more on this later). Indeed, there is a great variety of important photochemical and other types of chemical reactions within planetary atmospheres that modify their compositions.

Chemical processes also lead to gains and losses at planetary surfaces. For example, if an oxygen-poor mineral is exposed on the surface, then atmospheric oxygen can combine with it in a process called **oxidation**. Gases can also be removed through adsorption onto grains, and through the formation of clathrates (Section 3.2.3), in which water ice entrains small molecules in its open, cage-like structure – molecules such as CO_2. Chemical processes can also *release* gases from the surface, such as in the weathering that occurs in chemical reactions between surface materials and water.

Question 9.7

State, with justification, which gain and loss rates are affected by each of the following factors:

(a) the mass of a planet (the radius of the exosphere being fixed);
(b) the temperature at the base of the exosphere;
(c) the mass of the molecule of an atmospheric constituent;
(d) the flux density of solar UV radiation.

9.2.6 Atmospheric Circulation

Atmospheric circulation is the combination of a variety of atmospheric motions on a global scale. The circulation distributes energy and materials throughout a planetary atmosphere, and is a crucial factor in regional climates. Some types of motions are important on only a single planetary body, and those are excluded from consideration in this general section.

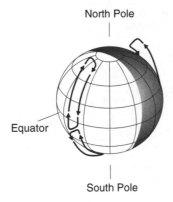

Figure 9.13 A single Hadley cell per hemisphere on a slowly rotating terrestrial planet with a small axial inclination.

Hadley circulation

Consider the case of a planet with an axial inclination considerably less than the 23.4° of the Earth. The solar radiation per unit area of its surface is then always greatest at the equator and least at the poles. This is because, as latitude increases, the surface tilts more and more away from the Sun, and also because a greater thickness of atmosphere has to be traversed. The surface at the equator thus becomes hotter than at the poles, and consequently the atmosphere becomes warmer and more buoyant, rises, moves polewards, and radiatively cools sufficiently to sink to the surface. The cycle is completed by the equatorward motion of the atmosphere at the surface, where it constitutes an equatorward wind. In the lower atmosphere of each hemisphere a huge convection cell is thus established, stretching from equator to pole, as in Figure 9.13. These convection cells transport energy away from the equator, and thus reduce the temperature change with latitude.

This kind of convection cell is called a **Hadley cell** after the British scientist George Hadley (1685–1768) who proposed the existence of such cells in the atmosphere of the Earth. They must not be confused with the smaller scale convective plumes described in Section 9.2.2, by which an adiabatic lapse rate can be established locally in a troposphere.

You will see shortly that for the circulation in Figure 9.13 to occur it is necessary for the planet to rotate no more than slowly. Among the planetary bodies with significant atmospheres, Venus rotates the most slowly, and it also has a small axial inclination. It is therefore reassuring to find that in its troposphere there is indeed a prominent single Hadley cell per hemisphere. The other bodies with significant atmospheres rotate much more rapidly, and so we need to consider the effect of this on Hadley circulation. This brings us to the Coriolis effect.

The Coriolis effect

Consider a parcel of atmosphere of mass m at the equator of a planet, and suppose that the rotation of the planet is carrying the parcel west to east at a speed v_e, the same as the equatorial surface speed. These speeds are with respect to an external observer fixed in space. The parcel starts to move polewards. The motion could be at the top of a Hadley cell, though any other poleward motion will do. It moves to a latitude l, and gravity keeps it close to the surface,

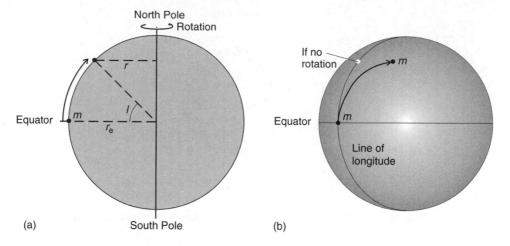

Figure 9.14 The Coriolis effect at a planetary surface.

as in Figure 9.14(a). The parcel will increase its speed in the direction of rotation (not in the northerly direction) in accord with the principle of conservation of angular momentum. This principle emerges from Newton's laws of motion, and as applied to the parcel it states that the angular momentum mvr (equation (2.1)) is constant, where v is its west-to-east speed, and r is its perpendicular distance from the axis of rotation, which has a value r_e at the equator (Figure 9.14(a)). The parcel mass m is constant, and thus as r decreases then, if the atmosphere is free to slip over the surface, v increases in proportion. At a latitude l the value of r is reduced to $r_e \cos(l)$, and therefore the west-to-east speed has increased to $v_e / \cos(l)$.

We can now obtain an expression for the speed of the parcel *with respect to the surface* at a latitude l. With respect to the external observer, the surface at l is moving west to east at a speed $v_e \cos(l)$, and the parcel is moving west to east at a speed $v_e / \cos(l)$. Therefore, the speed of the parcel relative to the surface is given by

$$v_{rel} = \frac{v_e}{\cos(l)} - v_e \cos(l) \tag{9.20}$$

With $0 < l < 90°$ it follows that $0 < \cos(l) < 1$, and therefore v_{rel} is greater than zero. The speed relative to the surface is west to east. The parcel started with zero west-to-east relative speed, so in moving northwards it has picked up relative speed in the direction of rotation. This would also be the case for a poleward-moving parcel in the other hemisphere. A notional path of the parcel across a planetary surface is shown in Figure 9.14(b). This path indicates the change in wind direction as the atmosphere heads northwards.

☐   What happens to the parcel when it travels to the equator at the bottom of a Hadley cell, if it starts there with v_{rel} as given by equation (9.20)?

The parcel will lose the relative speed it has acquired – as the distance from the rotation axis increases, the west-to-east speed of the surface increases, and the parcel loses speed to keep its angular momentum constant. In equation (9.20) we need to put $l = 0$ (the equator), and we see that then $v_{rel} = 0$. In reality the motion of the parcel is moderated in various ways, e.g. by friction between the atmosphere and the surface.

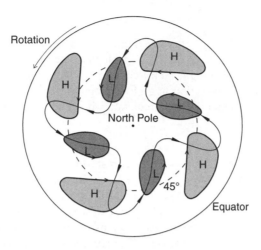

Figure 9.15 Mid-latitude atmospheric waves.

The west-to-east speeds of the atmosphere in Figure 9.14 are an example of the general case of a body (here a parcel of atmosphere) *viewed from a rotating frame of reference* (here the surface of a planetary body) having a tendency to accelerate in a direction that is perpendicular both to the motion of the body (here polewards along the planet's surface) and to the axis of rotation. It will actually accelerate in this direction if it is free to do so. The size of the acceleration is proportional to the sine of the angle between the rotation axis and the body's direction of motion. This tendency is called the **Coriolis effect** after the French physicist Gaspard Gustave de Coriolis (1792–1843).

The Coriolis effect has a profound influence on atmospheric circulation. From equation (9.20) you can see that it is proportional to the equatorial surface speed v_e, and is therefore greater, the more rapid the rotation. One effect on a rapidly rotating planet, such as the Earth, is that the Hadley cells in Figure 9.13 are disrupted and do not reach as far as the poles.

Other circulation patterns

Rotation can have a further important effect. Figure 9.15 shows mid-latitude circulation on a rapidly rotating planet. The rotation has resulted in a wave-like pattern of motion in the middle troposphere, which at the surface is correlated with the wind circulations labelled L for 'cyclones' and H for 'anticyclones' that are familiar from terrestrial weather maps. From such weather maps you will know that cyclones (low-pressure areas) and anticyclones (high-pressure areas) come and go, and so the details of the pattern in Figure 9.15 are not fixed. When atmospheric waves such as those in Figure 9.15 are present at mid latitudes, they are the dominant mechanism of the poleward transport of energy. Many other sorts of atmospheric wave exist.

As well as winds associated with Hadley circulation, cyclones, and anticyclones, there are winds at a more local scale that arise, for example, from local temperature differences, such as occur between land and sea on the Earth. Winds, however caused, can be deflected by large-scale topography into stable patterns called stationary eddies.

On the Earth, and some other planets, a rich variety of cloud forms is associated with the winds. You will see that the motions of clouds can help discern the pattern of the winds.

Question 9.8

(a) On the Earth there is a Hadley cell extending from the equator to about 30°N and another extending from the equator to about 30°S. Calculate the east-west speed picked up by a parcel of atmosphere in travelling from the equator to a latitude of 30°, and state the direction of motion in the east–west direction in each hemisphere.

(b) Speculate on the effect on this Hadley circulation were the Earth to rotate much more slowly.

9.2.7 Climate

We all have some idea of what is meant by climate on the Earth. Climate can be defined regionally – the climate of the coast of Kenya is unlike the climate of the Loire Valley in France – and it can be defined globally. Regional or global, the main elements that specify climate are temperature, precipitation, cloud cover, and wind speed. These are the same as those that specify weather, the difference being that whereas weather can change on a daily or hourly basis, with climate we take a longer term view. The **climate** is the average of each of the various elements over a sequence of years, usually 30 for the Earth. These averages do not change much from one 30 year sequence to the next, whereas larger changes are common on the shorter time scales of weather.

In addition to the 30 year average, we can also specify climate variability. For example, as well as noting that the 30 year mean temperature for some region is 8°C, we could add that in three years out of four the annual mean temperature is between 6°C and 10°C. We can also specify climate and its variability month by month, or season by season.

All planetary bodies have climates in some sense, those without appreciable atmospheres having comparatively simple ones. Ignoring the latter, climate, global or regional, depends on a huge number of factors. For example, global climate depends on the energy gains and losses for the surface and atmosphere as a whole, on energy exchanges within the atmosphere, and on the exchanges between the atmosphere and the surface. The greenhouse effect depends on atmospheric composition. Further factors include surface topography, and, at least in the case of the Earth, the behaviour of the oceans. Among many further significant factors is the rate of rotation of the planet, which determines the influence of the Coriolis effect on atmospheric circulation. So complex are climate systems that many aspects of planetary climates are poorly understood. Climate change is just as poorly understood, though there is ample evidence for large climate changes on several planets throughout their long histories, including the Earth, and climates will continue to change. Particular cases are outlined in the next chapter.

9.3 Summary of Chapter 9

A few atmospheres have been studied by means of instruments within them. Any atmosphere can be studied remotely, and spectral absorption lines produced within an atmosphere have been used to establish the composition, pressure, and temperature at various altitudes. Refraction of spacecraft radio waves and of starlight during occultation are additional sources of information.

For the surface–atmosphere system as a whole, solar radiation and heat flow from the interior are the two major gains of energy, with the latter being significant only for the giant planets (and Io). Energy loss is through the emission of radiation to space. This loss is often expressed as an effective temperature.

Atmospheric pressure decreases rapidly as altitude increases. The variation of temperature with altitude is more complicated, and depends on the energy gains and losses at each altitude, as summarised in Figure 9.7. On the basis of the way the temperature changes with altitude, an atmosphere is divided into a troposphere, mesosphere, and thermosphere, though for the Earth there are more divisions. Subdivisions are also made on other bases. These subdivisions are summarised in Figure 9.8.

Clouds can form if the partial pressure of the condensable substance exceeds its saturation vapour pressure, though cloud formation is delayed without the presence of condensation nuclei. Ultimately the liquid droplets or ice crystals that form a cloud will precipitate out of it.

The greenhouse effect depends on the absorption by atmospheric constituents of (some of) the IR radiation that is emitted by the atmosphere and by the surface. If a constituent absorbs radiation then it must also emit radiation. Some of this emission will be absorbed by the surface, thus raising its temperature above what it would have been had the atmosphere been transparent to the radiation (and if all else was the same).

Atmospheric constituents are gained and lost by a variety of physical and chemical processes operating between a number of reservoirs, as summarised in Figure 9.10.

Atmospheric circulation is the combination of a variety of atmospheric motions. Motions that occur in several planetary atmospheres include Hadley circulation, Coriolis winds, and atmospheric waves with their associated cyclones and anticyclones.

Global or regional climates on a planetary body are specified in the main by the medium-term averages of temperature, precipitation, cloud cover, and wind speed, along with their variability. Climate and climate change depend on a huge number of factors, and are not well understood.

Table 9.1 reveals the rich diversity of the substantial planetary atmospheres.

10 Atmospheres of Rocky and Icy–Rocky Bodies

In this chapter, attention is focused on the substantial planetary atmospheres. This means that the majority of planetary satellites are excluded from detailed consideration, and also the planet Mercury. From the rocky bodies only the Earth, Venus, and Mars are included, and from the icy–rocky bodies only Titan, Triton, and Pluto. The atmospheres of the four giant planets are the subject of Chapter 11. Tables 9.1 and 9.2 list the main properties of the atmospheres to be considered in detail.

10.1 The Atmosphere of the Earth

The Earth's atmosphere is, of course, the most familiar of all. From Table 9.1 you can see that the column mass today is modest, $10\,300\,kg\,m^{-2}$, but that it is the only substantial atmosphere in which O_2 is more than a minor constituent, with a number fraction of 0.21 (most of the rest being N_2). This is not the only way in which our atmosphere is atypical.

10.1.1 Vertical Structure; Heating and Cooling

Figure 10.1 shows the present-day globally averaged vertical structure of the Earth's atmosphere up to the top of the homosphere (Section 9.2.2). Only the lowest part of the thermosphere is included – it extends beyond the base of the exosphere, which is at about 500 km, the temperature rising as altitude increases. The thermosphere is heated mainly through the absorption of solar photons with wavelengths shorter than 91 nm – the extreme ultraviolet (EUV). The main absorbers are oxygen atoms that are themselves largely created with the aid of solar UV radiation with wavelengths in the range 100–200 nm. This radiation photodissociates oxygen molecules

$$O_2(+UV\ photon) \rightarrow O + O \tag{10.1}$$

At times of high solar activity the EUV flux is greater, and consequently the thermospheric temperatures are higher, particularly above 120 km, and can reach 2000 K or more at 500 km. The heating at each level in the thermosphere is balanced by cooling, mainly through the downward conduction of heat, but also by radiation from O and NO.

Discovering the Solar System, Second Edition   Barrie W. Jones
© 2007 John Wiley & Sons, Ltd

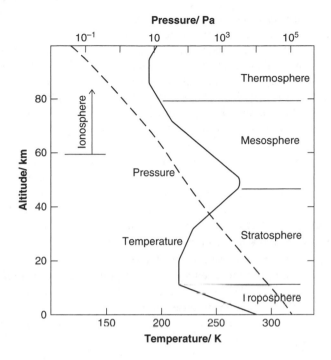

Figure 10.1 The globally averaged vertical structure of the Earth's atmosphere, up to the top of the homosphere. The angularity of the profile is an averaging artefact.

If you compare Figure 10.1 with the generic Figure 9.8 you will see that between the thermosphere and the troposphere there is a temperature bulge, a feature unique to the Earth. The upper part of the bulge is the mesosphere, and though the temperature decreases with altitude this is insufficient to promote much convection, heat losses and gains at any level being mainly through the absorption and emission of IR radiation by CO_2. The lower part of the bulge, in which the temperature increases with altitude, is called the **stratosphere**. There is negligible convection here. The temperature bulge results largely from the absorption by ozone (O_3) of solar UV radiation in the wavelength range 200–300 nm. Heating is balanced by cooling, mainly through IR radiation emitted by CO_2 and H_2O. The ozone is produced from molecular oxygen O_2 by a sequence of two chemical reactions that starts with the photodissociation of O_2 (reaction (10.1)) and finishes with

$$O + O_2 + M \rightarrow O_3 + M \tag{10.2}$$

where M is any atom or molecule involved simultaneously in this collision of O and O_2. O_3 production is significant in the altitude range 10–80 km.

As well as creating O_3, solar UV photons also destroy it. It is also destroyed by reactions with a variety of atoms and molecules. The quantity of O_3 at any altitude depends on the local rates of production and destruction. At high altitudes the atmosphere is thin, and three-body collisions as in reaction (10.2) are rare, so there is little O_3. At low altitudes there is little O_3 because much of the UV radiation that creates it has been absorbed at higher altitudes. The optimum is at about 25 km, so this is where the quantity of O_3 is greatest.

O_3 is important to life on Earth because it screens the surface from most of the solar UV radiation that is damaging to life. It is therefore a matter of concern that substances are being released by human activity that are reducing the quantity of O_3. The reduction is particularly severe over the poles and mid latitudes, where so-called holes in the ozone layer have appeared. The main substances that destroy O_3 are nitrogen oxides released by high-flying aircraft, and chlorine from the dissociation of chlorofluorocarbons (CFCs). In contrast, O_3 in the troposphere is increasing, owing to human activities, and direct contact with O_3 damages life. It would be fortunate indeed if the benefit to life of the increased UV screening offset the deleterious effects of direct contact.

In and above the stratosphere the atmosphere is sufficiently thin for significant cooling to take place by IR radiation to space. This ceases to be significant at an altitude marked by the tropopause, which is the rather sharp upper boundary of the troposphere, at an average altitude of about 11 km. The troposphere contains about 75% of the mass of the Earth's atmosphere. It is convective throughout most of its volume most of the time.

☐ What does this imply about the lapse rate in the troposphere?

It thus normally has an adiabatic lapse rate, for the reasons discussed in Section 9.2.2, where you also saw that the value depended on the water vapour content. In Figure 10.1 a global average of about $6.5\,\mathrm{K\,km^{-1}}$ is shown. The heating at any level in the troposphere is mainly through the absorption of IR radiation from the ground and from other atmospheric levels, plus the latent heat of condensation of water. This water is made available through the convection in the troposphere, the convection itself also contributing to the heating. In the lower troposphere absorption of solar IR radiation by water vapour makes a further heating contribution. The troposphere cools mainly through the emission of IR radiation.

Most clouds are in the troposphere. These consist of liquid water droplets, or of flakes or particles of water ice. Some icy clouds occur in the lower stratosphere. At any one time approximately 50% of the Earth's surface is covered in clouds, and this raises the Earth's albedo considerably (Table 9.2). The contribution made by a particular cloud depends on its altitude, thickness, particle/drop size, and composition, in poorly understood ways.

The global mean surface temperature T_s of the Earth is 288 K, whereas the effective temperature T_{eff} is 255 K.

☐ What is the reason for the 33 K difference?

This is caused by the greenhouse effect (Section 9.2.4). About 21 K is due to water vapour, a minor constituent, and most of the rest is due to CO_2, which is an even more minor constituent with a number fraction of only 0.000 37. Figure 10.2 shows the contributions of these two gases to the mid-IR absorptivity of the Earth's atmosphere. There are small contributions by other gases, such as methane. High-altitude cloud particles also contribute, and this includes condensation trails from aircraft. We shall return to the greenhouse effect in the next section.

Question 10.1

Explain why convection and condensation play no role in heat transfer in the Earth's stratosphere.

10.1.2 Atmospheric Reservoirs, Gains, and Losses

Figure 9.10 in the previous chapter summarises the various gains and losses that can affect the quantity of an atmospheric constituent. For the Earth today a major source is outgassing through volcanic activity. Most of the gases are recycled atmospheric constituents that have been buried

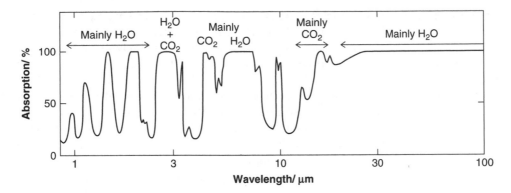

Figure 10.2 The absorptivity of the Earth's atmosphere versus wavelength.

by geological activity, though a small proportion of volcanic emissions might be juvenile (never outgassed before). A major loss of atmospheric constituents is through the formation of minerals that then constitute rocks. Thermal escape and chemical escape occur, but at very low rates.

☐ Why are the thermal escape rates of the major atmospheric constituents oxygen (O_2) and nitrogen (N_2) so low?

Figure 9.12 shows that for both of these gases, at the present temperature at the base of the Earth's exosphere, v_{mps} is much less than $v_{esc}/6$, and so these gases must be escaping at very low rates.

The generic nature of Figure 9.10 obscures two components of the Earth's surface that have played a major role in determining the quantities of various constituents of the Earth's atmosphere. These components seem to be unique in the Solar System today. One of them is the oceans, and the other is the biosphere. The **biosphere** is the assemblage of all living things and their remains. It is a thin veneer covering most of the land (down to a depth of a few kilometres), and it also occupies the oceans, particularly the surface layers. One of several important effects of the oceans and the biosphere is the dominant role they play in determining the atmospheric quantities of O_2 and CO_2, as we can see if we examine the carbon cycle.

The carbon cycle

The main features of the present-day carbon cycle are shown in Figure 10.3. The most important reservoirs of carbon are shown as a set of rectangles. Though the predominant repository of carbon in the Earth's atmosphere is CO_2 (with a number fraction of 0.000 37), other chemical forms predominate in other reservoirs, and therefore it is the mass of *carbon* in each reservoir that is shown. The arrows represent transfers between reservoirs, only the main ones being included. Each arrow is labelled with the rate of transfer of carbon between the two reservoirs connected by the arrow. The greatest rates involve the biosphere.

Consider first the arrow labelled 'Photosynth' connecting the atmosphere to the land surface. This is **photosynthesis**, the process of building body tissue from CO_2 and water, and it is performed by green plants and cyanobacteria with the aid of photons of solar radiation at visible and near-UV wavelengths. A long sequence of chemical and photochemical reactions can be represented as

$$nCO_2 + nH_2O(+\text{photons}) \rightarrow (CH_2O)_n + nO_2 \tag{10.3}$$

where n can have a variety of values. $(CH_2O)_n$ are carbohydrates, and are the basis of further tissue building. The CO_2 is taken from the atmosphere, the water is taken from the soil, and the O_2 is released into the atmosphere. Most other organisms build body tissue by consuming other organisms. The carbon in tissue derived directly or indirectly from photosynthesis is called organic carbon.

Respiration and decomposition ('Rsp + decomp' in Figure 10.3) have the reverse effect of that of photosynthesis. Respiration is performed by most organisms, and in higher animals it is called breathing. In respiration, oxygen is taken in, and converts some tissue to CO_2 and water. The chemical energy released by this process enables the organism to live and breed. After death, the dead tissue might be consumed, but if not it can decompose, and this too takes up atmospheric O_2 and releases water and CO_2. Not all death results in the reversal of photosynthesis. Most dead tissue on land is in the form of leaf fall. A small proportion of this becomes buried, and over many millions of years it is transformed into the organic carbon components of rocks. Particularly rich examples are the fossil fuel deposits, of coal, oil, and natural gas. This transformation is included in 'Formation of rocks' in Figure 10.3

Photosynthesis also occurs in the oceans – *inside* the ocean reservoir in Figure 10.3 – where it is performed by cyanobacteria, particularly in the upper 100 metres or so, where solar radiation can penetrate. As on land, the carbon involved in photosynthesis is in the form CO_2, in this case dissolved in the oceans. However, about 99% of the carbon in the oceans is in the form of the bicarbonate ion HCO_3^-, and the carbonate ion CO_3^{2-}, both of which result from reactions in solution between water and CO_2. Ocean organisms make use of this to form shells of calcium carbonate, $CaCO_3$. The carbon in these shells is called inorganic carbon to distinguish it from the organic carbon that is derived from photosynthesis.

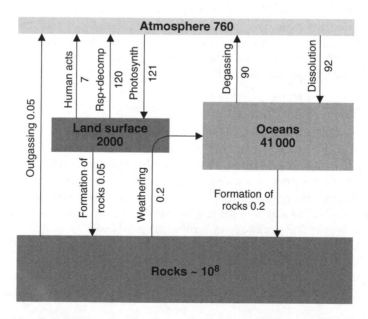

Figure 10.3 The main features of the Earth's present-day carbon cycle. The size of each reservoir is given in 10^{12} kg of carbon, and the rates of transfer in 10^{12} kg of carbon per year.

Respiration and decomposition also occur in the oceans, where, as on land, a small proportion of the organic carbon escapes, in this case by settling to the ocean floors as a component of the sediments that can later form rocks. In addition, inorganic carbon also settles into the sediments, and this has led to the formation of huge deposits of sedimentary carbonate rocks, such as limestone and chalk, and their metamorphic products such as marble. Consequently, rocks are by far the largest reservoir of carbon on Earth (Figure 10.3). Carbonate rocks are exposed by geological processes and are weathered, particularly by water. The resulting dissolved (bi)carbonate ions return to the oceans. Volcanic activity returns CO_2 to the atmosphere.

The atmospheric quantity of CO_2 is determined by the accumulated effect of all the various rates of transfer in Figure 10.3. The quantity will be in a steady state if the sum of the rates of transfer into the atmosphere equals the sum of the rates out.

☐ Are these two sums equal in Figure 10.3?

The rate of gain is slightly greater than the rate of loss. Consequently the atmospheric quantity must be rising, and Figure 10.4 shows the observational evidence that it is. In fact, it was this sort of evidence that showed that there must be an inequality. The slight imbalance in recent decades has probably been the result of human activities, mainly the burning of fossil fuels and forest clearance ('Human acts' in Figure 10.3). The effect of these activities is to transfer some of the carbon in the rocks and on the land into the atmosphere, as CO_2. The rate of rise in Figure 10.4 is only about half of the rate of release by human activities. This is because the rate of dissolution in the oceans and the rate of photosynthesis have increased in partial compensation.

Global warming today

The main concern about the increase in atmospheric CO_2 is that it is an important greenhouse gas (Figure 10.2) and therefore the Earth's surface temperature will rise if nothing else changes in compensation. Global data since about 1870 show that the Earth's global mean surface temperature (GMST), though exhibiting short-term fluctuations, has risen by about $0.7 \pm 0.2\,°C$, the rate of rise being greatest in the last few decades. This is not entirely due to CO_2. Good data from orbiting spacecraft since 1978 have shown that the flux density of solar radiation F_{solar} has risen, and though the rise is slight, it can account for 20–30% of the rise in the GMST from 1980

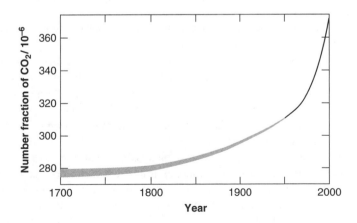

Figure 10.4 The recent increase in the atmospheric quantity of carbon dioxide.

to 2002. For the remaining rise, at least some of it is thought to be the result of the increase in atmospheric CO_2, a conclusion supported by the similarity of the curve in Figure 10.4 to the broad trend in the GMST over the same period. A rise in the atmospheric abundance of water vapour and decreases in cloud cover would also contribute to a rise in the GMST, but such changes, and their effects, are not well known.

As well as substances that promote global warming, global cooling can result from increases in albedo, resulting from increased cloud cover, increased dust and aerosol content of the atmosphere, and increased snow and ice cover. Clearly, the factors that promote warming are currently winning, and indeed one consequence seems to be a *decrease* in snow and ice cover.

The rate of rise of the GMST is slowed by the thermal mass of the oceans, which can absorb large quantities of heat with little temperature rise. But as oceanic temperatures rise, their moderating effect on the rise in the GMST is diminished. Thus, if the excess of heating over cooling were held constant from now, the GMST would continue to rise for about 100 years. In fact, human activities will increase the greenhouse effect, so the rise in the GMST might well be greater. So, what does the future hold?

Climate modelling is fraught with difficulties – lack of data, lack of understanding of physical, chemical, and biological processes, lack of computing power, and uncertainty about future human activities. Different modellers come up with different assumptions and different compromises. As a consequence there is a range of predictions of the rise in the GMST over the next 100 years, from 1.4 K to 5.8 K or more. Even the lower limit, which is fairly firm, will give rise to significant climate change, with loss of surface snow and ice, and more extreme weather. The changes will be more severe in some regions than others. Ocean levels are predicted to rise somewhere in the range 0.1–0.8 m, mainly through the thermal expansion of water, which will result in the inundation of many heavily populated coastal regions.

The rise in the GMST could be greatly increased if methane became more abundant. It is a greenhouse gas that fills some of the absorption gaps in Figure 10.2, and, molecule for molecule, it is much more absorbing of IR radiation than CO_2. What are its sources and sinks?

Methane

Currently methane (CH_4) accounts for only 1.7 ppm of the molecules in the atmosphere. It is rapidly oxidised by oxygen, so to sustain a constant amount requires some fairly copious sources.

☐   If atmospheric CH_4 is destroyed at a certain high rate, at what rate must it be released into the atmosphere to keep the amount constant?

To sustain a constant amount, it must be released at the same high rate. Less than 1% comes from non-biogenic sources, such as volcanoes. The rest comes from the biosphere. About 21% comes from wetlands, 20% from the guts of ruminants, 15% from bacteria in termites and other creatures, and about 12% from rice paddies. Nearly all of the remainder comes from natural gas and the burning of biomass, and it is the increase in these rates through human activities that could, in principle, increase the greenhouse effect.

But a much larger increase of CH_4 is possible. Large quantities are locked in methane clathrates in permafrost ice crystals, even more is trapped beneath continental shelves, and it is observed to be seeping from a few ocean floor sites. Rises in temperature could release this CH_4, rather suddenly, producing a huge, if temporary, increase in atmospheric CH_4. The rise in the GMST could far exceed the few degrees currently predicted. This would present humanity with huge challenges.

Oxygen

The biosphere also plays the major role in determining the amounts of the two main constituents of the Earth's atmosphere, O_2 and N_2. Photosynthesis is the main source of O_2 in the Earth's atmosphere. A much smaller contribution comes from the photodissociation of water vapour and from the conversion of sulphates to sulphides (with the aid of organic matter) and their subsequent burial. Therefore, if respiration and decay completely reversed the effect of photosynthesis, then the number of molecules of O_2 in the atmosphere and oceans would be only a little greater than the number of carbon atoms in the biosphere. In fact there is *far* more O_2 than this in the atmosphere, and an even greater quantity is dissolved in the oceans (mainly combined with sulphur, in the sulphate ion SO_4^{2-}). This is a consequence of the formation of the organic carbon component of rocks. From carbon stored in this way a huge amount of oxygen has been released from CO_2.

And yet most of the oxygen corresponding to the organic carbon in the rocks is not in the atmosphere and oceans. Over the ages most of the O_2 liberated from CO_2 has combined with oxygen-poor volcanic gases, and with freshly exposed oxygen-poor minerals in crustal rocks. This includes crustal rocks that have become (more) oxygen poor as a result of oxygen ions supplied by the rocks to the ocean in the formation of carbonates. As a result, the total mass of oxygen released through the formation of the organic carbon component of rocks is split today between the atmosphere, oceans, and surface rocks in the approximate ratios 1:2:40.

Atmospheric O_2, and oxygen dissolved in the oceans, are currently removed mainly by the oxidation of organic matter and sulphides freshly exposed by geological processes, and by volcanic activity that generates under-oxidised magma, and gases such as H_2, SO_2, H_2S, and CO.

☐ Why not H_2O and CO_2?

These are already fully oxidized.

Nitrogen

The biosphere and the oceans also influence the N_2 content of the atmosphere. N_2 is removed mainly by lightning and by combustion. Nitrogen oxides are thus created that are washed out of the atmosphere, to form the nitrate ion NO_3^- in the oceans, and in the form of nitrates (NO_3 group(s)) and nitrites (NO_2 group(s)) in minerals. Various bacteria also remove nitrogen, at an overall rate roughly double that of lightning and combustion. N_2 is returned very slowly to the atmosphere by the weathering of crustal rocks, but much faster by certain bacteria. The outcome is that nitrogen is now roughly equally split between the atmosphere, where it is present as N_2, and sedimentary rocks (and their metamorphic products), where it is stored mainly as nitrates and nitrites.

There is only a small amount of nitrogen in the biosphere. But without the action of the biosphere and the oceans, the atmospheric mass of N_2 would be nearly double its present value. The quantity stored in nitrates and nitrites would be correspondingly small. For example, without the biosphere the O_2 content of the atmosphere would be much less, and thus the rate of formation of nitrogen oxides by lightning would be greatly reduced. Without the biosphere but with the oceans nearly all the nitrogen would be in the oceans, in the form of the nitrate ion NO_3^-.

Question 10.2

Suppose that the Earth's biosphere is entirely destroyed, and that sufficient time passes for the Earth's atmosphere to reach a new steady-state composition.

(a) Explain what this new composition might be in terms of the major overall constituents.
(b) Why is it not possible for you to estimate the change in the Earth's GMST?
(c) Why would the solar UV radiation levels at the Earth's surface probably be greater?

10.1.3 Atmospheric Circulation

The atmospheric circulation in the Earth's troposphere, averaged over time, is shown in Figure 10.5. There are three cells per hemisphere. The ones nearest the equator are Hadley cells (Section 9.2.6). They extend to about latitudes 30° north and south, and thus cover about half of the Earth's surface. They are stabilised in latitude against seasonal changes by the slow thermal response of the oceans. Disruption by wave-like disturbances at mid latitudes prevents them reaching higher latitudes. The polar cells are less prominent and less permanent. The mid-latitude cells are also impermanent, and are artefacts of averaging around the Earth various complex patterns, including waves. They are *not* Hadley cells – for example, the circulation is in the *opposite* direction to that conforming to the decrease of temperature with latitude. At mid latitudes there is the system of cyclones and anticyclones shown in Figure 9.15. The Coriolis effect modifies the circulation within each cell by deflecting the north–south flow in the east–west direction, to give a diagonal flow. The resulting winds at the surface are shown schematically in Figure 10.5. In the tropics the winds are called the trade winds, and were utilised by sailing ships.

Wind directions are modified by friction between the atmosphere and the surface, and by other disturbances such as mountain ranges and continents. They are considerably modified by the oceans. The oceans also have circulation patterns, strongly influenced by the configuration of the ocean basins. Oceans and the atmosphere interact, and so the circulation of the one influences the other. This is a unique feature of the Earth. As mentioned earlier, it makes climate modelling particularly difficult.

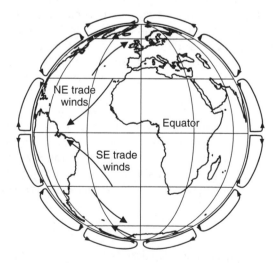

Figure 10.5 Circulation in the Earth's troposphere. The vertical scale is greatly exaggerated.

10.1.4 Climate Change

In this section we shall examine climate change extending back from today to about 3000 Ma, leaving even earlier times to Section 10.6.2. At 3000 Ma we are in the middle of an eon called the Archaean, defined as starting at the end of the heavy bombardment, 3900 Ma ago, and ending 2500 Ma ago, when atmospheric O_2 started to rise. The time before the Archaean is called the Hadean.

Ice ages – the record

Perhaps the most dramatic aspect of climate change in the past is the occurrence of ice ages. An **ice age** is characterised by a cooling of mid-latitude and polar regions to the point where polar ice caps, and any deposits of ice elsewhere, spread to give glacial conditions down to mid latitudes. Within an ice age there can be short-lived warmings to give interglacial conditions in which the ice sheets retreat.

Figure 10.6 shows the Earth's over the past 3000 Ma, inferred from geological and biospheric evidence. There was an ice age around 2300 Ma ago (the earliest dip in Figure 10.6) and another at 750–600 Ma. There have been others since. The one around 2300 Ma was probably the most severe the Earth has ever experienced – the dip in temperature shown is probably an underestimate. At this time all oceans seem to have been covered in thick ice, leading to the appellation 'snowball Earth'. The ice age at 750–600 Ma might have been less severe, with large areas of ocean covered in thin ice, or no ice at all, as indicated by marine fossils from this time that photosynthesised. If so 'slushball Earth' is an appropriate name!

We are presently in an ice age that started about 2.4 Ma ago, though in an interglacial period that has so far lasted about 10 000 years. The present extent of the permanent ice is shown in Figure 10.7(a), and the extent at the height of the last glacial period, 18 000 years ago, is shown in Figure 10.7(b).

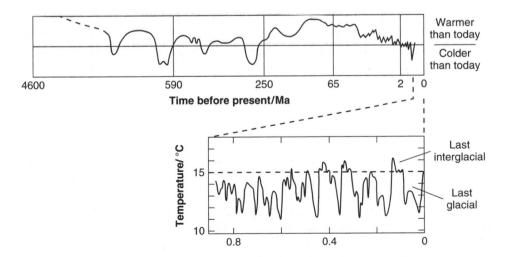

Figure 10.6 The Earth's GMST. The time axis has changes of scale at the vertical lines. Between adjacent lines the scale is linear. The age ranges marked off, from right to left, denote (approximately): the Quaternary period, the Tertiary period, the Mesozoic era, the Palaeozoic era (see Further Reading). (Adapted from L A Frakes, *Climates Throughout Geological Time*, Elsevier, 1979)

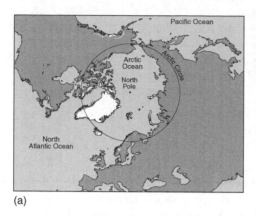
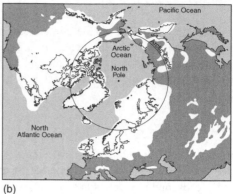

(a)   (b)

Figure 10.7   (a) The present extent in the northern hemisphere of permanent ice on land. (b) The maximum extent at the height of the last glacial period in the current ice age, about 18 000 years ago. (Adapted from B S John, *The Ice Age*, Collins, 1977)

The transition from glacial to interglacial conditions can be rapid, occupying as little as a century, or so. It is thus possible that the current interglacial period could be over in 100 years time, with permanent ice spreading again to mid latitudes. However, any incipient glacial conditions could be delayed by human activities.

☐   How could human activities delay the onset of a glacial period?

A delay could result from the burning of fossil fuels and forest clearance. These activities are contributing to an increase in atmospheric CO_2 (Figure 10.4) and consequently to an increase in the greenhouse effect.

The causes of ice ages

The underlying causes of ice ages and of interglacials within ice ages have been much disputed, and the matter is not yet clear. There are several factors that have certainly had some influence on climate. The primary source of energy for the Earth's climate system is the Sun, and it is known that the luminosity of the Sun varies by up to about 0.5% on a time scale of centuries. On a longer time scale, stretching back to the origin of the Solar System 4600 Ma ago, there has been a gradual increase in luminosity from about 70% of its present value.

A further factor is changes in the Earth's orbital eccentricity and in the precession of its perihelion (Section 1.4.6), largely due to the gravitational field of Venus, which is nearby, and of Jupiter, which is massive. Also, there is precession of the rotation axis (Section 1.5.1) and changes in the axial inclination, largely due to the gravitational field of the Sun and Moon acting on the slightly non-spherical mass distribution of the Earth. These changes alter the seasonal and latitudinal distribution of solar radiation. Another factor that influences climate is changes in the latitude distribution of the continents, and in the shape of the ocean basins.

☐   What is the cause of these changes?

The cause is plate tectonics (Section 8.1.2). The time scale is 1–10^3 Ma. Yet another factor is atmospheric composition. This has changed on time scales ranging from a year (a major volcanic eruption) to billions of years. The operation of one or more of these factors might not in itself be sufficient to cause the recorded climate changes. They can, however, act as triggers.

Let us turn from generalities to a few specific examples, namely the ice age around 2300 years ago, the one from 750 to 600 Ma, and the current ice age.

A currently popular view is that the ice age that occurred around 2300 Ma was triggered by the increase at that time of O_2 in the atmosphere. (The history of atmospheric O_2 is presented in Section 10.6.2.) At that time the luminosity of the Sun was only 85% of its present value, and it is proposed that the Earth was being protected from a deep freeze by the powerful greenhouse gas, CH_4. With the increased production of O_2 the CH_4 was destroyed by oxidation, with a consequent fall in the GMST, perhaps by as much as 30 K. One possible cause of the increase in O_2 was the flourishing of oxygenic cyanobacteria, perhaps due to an increase in organic nutrients in the oceans, though the reasons for any such increase are unclear. It is possible that the prime cause of the rise in O_2 at 2300 Ma was the change in the nature of plate tectonics at about 2500 Ma (Section 8.1).

So, with such a dramatic fall in the GMST, how did the Earth escape from its snowball state? The **carbonate–silicate cycle** was surely responsible. With the Earth covered in ice, the rate of loss of CO_2 from the atmosphere through the weathering of silicate rocks, with the aid of liquid water, would have declined sharply. Volcanic emission of CO_2 would have carried on regardless, and so the amount of atmospheric CO_2 would have increased, increasing the greenhouse effect.

The 750–600 Ma ice age might also have had biospheric changes as an essential contribution. It has been suggested that a population spurt of marine plants removed so much CO_2 from the atmosphere that the greenhouse was diminished.

The current ice age came at the end of a gradual cooling over the past few tens of million years (Figure 10.6). This coincides with a reduction in atmospheric CO_2, but the reason for this decline remains obscure. One suggestion arises from the growth of the huge Tibetan Plateau during this period, caused by the collision of the Indian craton with the Asian Plate. This growth could have increased the rate of removal of CO_2 through weathering. Also, plate motion was placing more continental area at high latitudes, which would accumulate ice more readily than the oceans, and thus increase the Earth's albedo.

One feature that might be unique to the current ice age is the relatively short-term oscillations of the GMST, apparent in Figure 10.6, that give rise to interglacial periods. It is thought that these oscillations are largely due to the orbital and rotational changes outlined above. These are quasi-periodic, and much work has been done to try to match these periods to the temperature record. One example will suffice. The interval between interglacial periods over the past 1 Ma is of order 0.1 Ma. One suggestion is that this is due to the variation in the inclination of the Earth's rotation axis. This varies from 22.5° to 24.0° (the current value is 23.43°, decreasing). Glacial periods end near times when it is 24°, at which times the poles are warmer. However, the period of this variation is 41 000 years, about three times too short. A model shows that the missed deglaciations are because it takes two–three cycles for ice sheets to build to the size required for them to become sensitive to inclination variations. That either two or three cycles are required explains the variability in interglacial interval. The precession of the equinoxes might also modulate the timing.

Other climate changes

As well as ice ages, there have been other dramatic changes in global climate, though with considerably shorter durations. The most famous of these is around the time of the extinction of the dinosaurs, 65 Ma ago. Indeed, about 70% of marine species and a large proportion of land species became extinct. There was a major asteroid or comet impact at that time, which is partly to blame, mainly by injecting dust into the upper atmosphere that would have enhanced the global cooling, which (cause unknown) was in any case occurring. This episode of cooling was too brief to be shown in Figure 10.6. Other mass extinctions are also thought to have resulted from relatively brief cool periods, with a variety of possible (non-impact) causes.

Throughout Earth history there have been relatively small climate changes, with relatively short durations. Recorded history includes the Little Ice Age that lasted from roughly 1550 to 1850, when average temperatures in Europe fell by somewhat less than 1 °C. A contributory factor might be indicated by the Maunder minimum – the virtual absence of sunspots from 1645 to 1715. There is evidence that low solar activity, of which a symptom is few sunspots, is associated with reduced solar luminosity. It has been estimated that F_{solar} was 2–4 W m^{-2} lower during this period.

Question 10.3

Present a qualitative argument to show that the solar flux density *at the Earth's surface* at the poles, averaged over a year, is less when the Earth's axial inclination is zero than when it has its present value. (Assume no change in cloud cover.)

10.2 The Atmosphere of Mars

From Table 9.1 you can see that the atmosphere of Mars is much less substantial than that of the Earth.

☐ On a square metre of Martian surface, by what factor is the mass of the Martian atmosphere less than that of the Earth?

This factor is the ratio of the column masses, $(1.03 \times 10^4)/(0.015 \times 10^4) = 69$. As well as a much smaller column mass, you can also see that the composition is very different. Whereas the Earth's atmosphere is dominated by N_2 and O_2, the Martian atmosphere is dominated by CO_2.

10.2.1 Vertical structure; heating and cooling

Figure 10.8 shows the globally averaged vertical structure of the Martian atmosphere when its dust content is low. In the kilometre or so near the ground there are huge diurnal variations, with surface temperatures at night plunging typically to roughly 100 K below that shown in Figure 10.8. A major difference from the Earth (Figure 10.1) is that there is no mesospheric bulge of temperature. This is a direct consequence of the near absence of O_3, itself a consequence of the very small quantity of O_2. The lapse rate in most of the troposphere is close to the dry adiabatic value.

☐ What does the 'dry' mean?

This is the lapse rate when there is no condensation. In fact, water vapour does condense, and thin clouds of water ice can form. But the number fraction of water vapour is only about 0.0002, and so the latent heat release is slight. A more serious disturbance of the adiabatic lapse rate would be the condensation of CO_2, but its low condensation temperature restricts its condensation to the coldest regions of the atmosphere.

Some water-ice clouds are just visible in Figure 7.11 edge on in the upper left. The water-ice clouds are rather like cirrus, and not the billowy sort. They are seen from the Earth, from Martian orbit, and from the surface, e.g. from *Spirit* and *Opportunity*. They form from air lifted

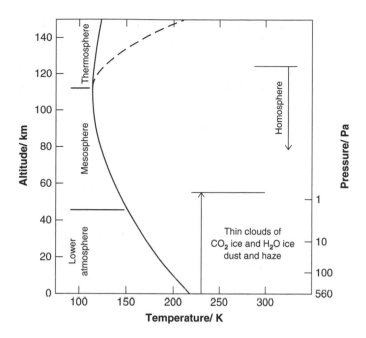

Figure 10.8 The globally averaged vertical structure of the Martian atmosphere when its dust content is low. The two curves in the thermosphere correspond to extremes of low and high solar activity.

by Hadley circulation, mainly at low latitudes when Mars is near aphelion and thus colder. They also form in the winter polar region, where the atmosphere is so cold that clouds of CO_2 ice crystals also form. Elsewhere, at ground level, water-ice fogs have been observed, and water frost has condensed on landers and on the ground in their vicinity.

The atmosphere is generally transparent to solar radiation, but there is so much fine dust at the Martian surface that surface winds (up to the order of $10\,m\,s^{-1}$) often raise enough of it into the lower atmosphere to make it the major absorber there. This makes the lapse rate in the lower troposphere smaller. There is then little convection and it is more appropriate to refer to the 'lower atmosphere' rather than to the 'lower troposphere'. The dust consists of particles typically $1–3\,\mu m$ across, composed of basalt and clay minerals. There is also about 1% of an iron oxide that gives the dust its reddish tint, similar to that often seen in the sky from the surface of Mars (Plate 9). Carbonates might also have been detected in atmospheric dust.

The exploration rovers have seen spectacular evidence of wind-raised dust, in the form of 'dust devils', earlier seen by Pathfinder. These are caused by thermal whirlwinds, which raise dust columns, some of which are more than $100\,m$ across. They rotate with surface speeds of the order of a metre per second, and migrate across the landscape at a few metres per second.

On a much larger scale, seen from Martian orbit or from Earth, there are the dust storms noted in Section 7.3.1. They typically appear in the southern summer, when Mars is around perihelion. They vary from less than about $100\,km$ across, lasting a few days, to global storms, lasting many weeks. Only 10 global or near global storms have occurred in the past 130 years. Many of the largest storms originate in the Hellas Basin, probably because of cold air moving northwards from the warming south polar region. This air sinks into this huge deep basin, thus creating winds.

The Martian GMST is considerably lower than that of the Earth, 218 K instead of 288 K. Though Mars has a lower planetary albedo than the Earth, it is at a greater distance from the Sun, and has a greenhouse effect of only about 5 K, compared with about 33 K for the Earth. The smaller greenhouse effect is largely a consequence of the much lower water vapour content on Mars – the column mass of CO_2 is actually *greater* than it is in the Earth's atmosphere.

10.2.2   Atmospheric Reservoirs, Gains, and Losses

The dryness of the Martian atmosphere is the result of the low temperatures – the atmosphere is in fact close to saturation. It is thus to be expected that atmospheric water vapour is in equilibrium with surface deposits. As well as frosts, there is at least one substantial deposit of water – the permanent cap at the North Pole. There is ample evidence for other surface deposits too. This evidence includes the water-related surface features described in Section 7.3.6, and the observations at the surface outlined in Section 7.3.7. The Martian meteorites contain rather modest proportions of water, though it is not possible to infer from this anything very useful about the water content of the regions from where they came. There is a wide range of estimates of the present quantity of water at and near the surface of Mars. These are usually given in terms of the uniform depth to which the water would cover the entire planet if the water were all liquid. For the Earth the total in the (near-)surface and atmospheric reservoirs corresponds to a depth of 2700 m, whereas for Mars the estimates range from several tens of metres to a few kilometres. Even the smallest estimates are far larger than the equivalent of around 10^{-5} m presently in the Martian atmosphere.

The major constituent of the Martian atmosphere, CO_2, also condenses onto the surface, notably at the coldest parts. The winter pole can be as cold as 130 K, which is well below the condensation temperatures of CO_2 at the surface of around 150 K. You saw in Section 7.3.5 that the seasonal caps at both poles on Mars consist largely of CO_2 ice, and that this is also the case for the permanent cap at the South Pole (perhaps underlain by water ice). Frosts of CO_2 also occur. This is unlike the major constituents of the Earth's atmosphere, where it never gets cold enough for O_2 and N_2 to condense. Over the seasonal cycle about a third of the CO_2 content of the atmosphere is exchanged with the surface, causing a noticeable oscillation in atmospheric pressure.

Additionally, CO_2 can be locked up in water ice as a clathrate, and adsorbed onto clay minerals. It can also be present in carbonates, which have been detected in small amounts in the Martian surface dust by spacecraft and landers, probably also in the airborne dust, and in Martian meteorites. Is there a lot more to be found? The evidence that Mars was warmer and wetter in the distant past (Section 7.3.6) includes possible evidence for bodies of water of various size at the surface, in which case extensive carbonate deposits are likely to have formed. However, as pointed out in Section 7.3.6, even if there were such bodies, there could have been chemical inhibition to the deposition of carbonates. All in all, it is not surprising that estimates of the present quantity of CO_2 in some form in Martian (near-)surface reservoirs range from the order of 10^2 times the amount in the atmosphere to far greater quantities.

Evidence that there are significant reservoirs of some sort is provided by carbon isotopes. Atmospheric CO_2 is photodissociated by solar UV radiation. This results in some loss of carbon atoms to space. A greater proportion of the lighter, common isotope ^{12}C is lost than ^{13}C. Over 4600 Ma, this would lead to a substantial enrichment of ^{13}C in the atmosphere, compared with, for example, the Earth, where the higher gravity has retained nearly all the carbon. This enrichment is not seen, either in the Martian atmosphere, or in the Martian meteorites. This indicates a substantial subsurface reservoir that replenishes the atmospheric CO_2. This need not be in the form of carbonates – it could, for example, be a water–CO_2 clathrate.

☐ Atmospheric O_2 is similarly unenriched. Why?

There must be a (sub)surface reservoir for oxygen too. This is water. This enters the atmosphere as vapour where it is photodissociated by solar UV radiation. The hydrogen readily escapes to space, oxygen far more slowly, but enough over 4600 Ma to produce enrichment of the heavier isotopes if there were no reservoir. No such enrichment is seen. By contrast, the $^{15}N/^{14}N$ ratio in the atmosphere of Mars is 1.6 times that on the Earth (Section 10.4). It is not surprising that there is no significant nitrogen reservoir on Mars, without persistent rain and oceans (Section 10.1.2).

Atmospheric CO_2 is photodissociated at a high rate by solar UV radiation – the atmosphere is thin and almost devoid of ozone, so UV radiation can do its damage through the whole depth of the atmosphere. If there were no regeneration of CO_2 then it would be almost entirely photodissociated in 3000 years and at any instant there would be almost no CO_2 in the atmosphere! That this is not the case is the result of various chemical reactions in the lower atmosphere that put the molecule back together again. Of crucial importance in these reactions is the trace of atmospheric water.

10.2.3 Atmospheric Circulation

Figure 10.9 shows three important modes of atmospheric circulation in the lower atmosphere of Mars. The contrast with the Earth is striking (Figure 10.5). With a rotation rate and axial inclination similar to the Earth, a similar type of circulation might be expected. However, the absence of oceans on Mars leads to a rapid change of surface temperature in response to seasonal changes in insolation. Therefore, in the summer the hottest surface is not near the equator but at the subsolar point. As a consequence there is a single Hadley cell stretching from the subsolar latitude across the equator to the less heated hemisphere (Figure 10.9(a)). In this hemisphere at mid latitudes there are frequent cyclones and anticyclones. Around the equinoxes the picture is more like that on the Earth, with a Hadley cell in each hemisphere, and perhaps a smaller one at 60°–70°, circulating the opposite way. Also as on the Earth, the Coriolis effect causes Hadley cells to break down far from the equator, to be replaced by cyclones and anticyclones.

In addition to Hadley cells there is **condensation flow** (Figure 10.9(b)). This arises from the condensation of CO_2 at high latitudes in the winter hemisphere, notably at the seasonal polar cap. The associated reduction in pressure draws the atmosphere towards this region.

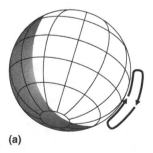

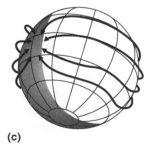

(a) (b) (c)

Figure 10.9 Atmospheric circulation in the lower atmosphere of Mars. (a) Hadley cell. (b) Condensation flow. (c) Thermal flow. (Adapted by permission from Figure 5 (p97) of *The New Solar System*, J K Beatty and A Chaikin (eds.), Sky Publishing, 1990)

☐ Why is condensation flow unimportant in the Earth's atmosphere?

Condensation of a major atmospheric constituent does not occur on the Earth. Another component of the circulation is a **thermal tide** (Figure 10.9(c)). This is the result of the rapid cooling of the Martian atmosphere at night, leading to low pressure near dawn and a consequent flow from the warmer and consequently higher pressure afternoon hemisphere. Thermal tides occur on the Earth, but the thin atmosphere of Mars, dominated by the efficient IR radiator CO_2, leads to far greater night-time falls in tropospheric temperatures, typically by 100 K. This leads to far stronger flow on Mars. Among other thermal effects there is a correlation with topography, with downhill drainage of cold air at night, and uphill ascent during the day.

10.2.4 Climate Change

In Section 7.3.6 it was noted that many water-related features on Mars indicate a warmer, wetter climate in the distant past. This important issue will be considered in Section 10.6.3. You also saw (Section 7.3.5) that the polar deposits suggest an intricate series of more recent climate changes. It is thought that changes in the axial inclination of Mars have been particularly important in the climatic changes reflected in the polar deposits, as follows.

Calculations of these changes show that the axial inclination, currently 25.2°, has ranged in the past millions of years from 0° to 60° – far greater than the 22.5° to 24.0° range for the Earth. This is because the Earth's axis is stabilised by the Moon. The Martian changes are quasi-periodic, with an underlying period of about 0.12 Ma, superimposed on a longer period of about 2 Ma. At low axial inclinations the total annual solar radiation in equatorial regions is much greater than in polar regions, but the totals become more equal as the inclination increases (Question 10.3). The greater high-latitude temperature at large inclinations has the effect of reducing the quantities of CO_2 and water in the polar reservoirs, and increasing those in the equatorial and atmospheric reservoirs. The net effect is an increase in atmospheric mass, and this is expected to increase the frequency of dust storms, which increases the dust content of the atmosphere and promotes deposition at the poles. A competing effect is the depletion of the polar reservoirs and a possible increase in wind scouring there. Thus, as the axial inclination varies with its two characteristic periods, there is a complicated interplay of erosion, deposition, and the dust/water ratio in the atmosphere.

Additionally, there is a 51 000 year cycle in the season at which perihelion occurs – the combined effect of axial and perihelion precession. Also, it is likely that the dust/water ratio in the atmosphere has also been changed by volcanic eruptions and by impacts. These effects would have been particularly significant on shorter time scales. Combined with the changes in axial inclination, it is plausible that they have led to the patterns of layering, deposition, and erosion seen in the polar deposits. Swings in climate across the globe are a necessary accompaniment to the changes reflected in the polar deposits.

Changes in axial inclination can also account for other surface features, such as the gulleys (Section 7.3.6), and the evidence for past glaciation at mid latitudes (debris in valleys, arcs of debris on slopes, and so on) – at high inclination the polar regions get more radiation, and so water sublimes and condenses closer to the equator.

More recently, there is evidence of global warming on Mars. Data from the past 150 years or so show that the frequency of (near) global dust storms has increased. Also, the springtime retreat of the south polar cap in 1999–2000, observed by Mars Global Surveyor, was earlier that the one observed in 1977 by the Viking orbiters, and the north polar cap receded faster in the 1980s and 1990s than in the 1960s, with its residual (water-ice) cap diminishing. The cause of this warming is uncertain. One cause, or contributory factor, is an increase in solar luminosity. Accurate measurements extend back to only 1978 (Section 10.1.2). For earlier times, back to about 1750, data on the Earth's GMST has been shown to anticorrelate with the length of the solar cycle, which has varied from 18 to 26 years. This suggests an anticorrelation between cycle length and solar luminosity. Since 1950 the cycle length has been shortening, so the presumption is that the solar luminosity was increasing between 1950 and 1978, as well as later.

Question 10.4

(a) Starting at midsummer in the southern hemisphere on Mars, describe the changes in the location of the Hadley cell(s) as Mars goes around its orbit.
(b) Describe how the location of the Hadley cell at midsummer in the southern hemisphere changes as the axial inclination of Mars varies.

10.3 The Atmosphere of Venus

For a planet that has a mass, size, and internal structure not very different from those of the Earth, the atmosphere of Venus is astonishingly different from ours. Like the Martian atmosphere it is dominated by CO_2 (Table 9.1), but the column mass is $102 \times 10^4 \, kg \, m^{-2}$ compared with $0.015 \times 10^4 \, kg \, m^{-2}$ for Mars and $1.03 \times 10^4 \, kg \, m^{-2}$ for the Earth. Additionally, Venus is 100% cloud covered, giving it a planetary albedo of about 0.76. In spite of this high albedo its GMST is an astonishing 740 K, which exceeds the melting points of lead, tin, and zinc. The mean surface pressure and density are $93 \times 10^5 \, Pa$ and $67 \, kg \, m^{-3}$.

10.3.1 Vertical structure; heating and cooling

Figure 10.10 shows the globally averaged vertical structure of the Venusian atmosphere. Like the Martian atmosphere it has no mesospheric bulge of temperature.

☐ Why is this?

Again, this is because there is no strong absorber of solar radiation concentrated at these levels. The lapse rate is invariably adiabatic in the lower 35 km or so of the troposphere, but from 35 km up to the cloud base at about 45 km it can be less than adiabatic. This is because at these higher altitudes the atmosphere is more transparent to IR wavelengths, so can more readily exchange heat radiatively.

The main cloud deck extends from about 45 km to 65–70 km, completely shrouding the planet all the time. The upper 10 km or so consists primarily of droplets of sulphuric acid, H_2SO_4, 1–10 µm in diameter. The clouds have a pale-yellow tint, possibly due to a trace of sulphur or $FeCl_3$. Below this upper deck the particles might be crystalline, which is incompatible with H_2SO_4. X-ray fluorescence analysis has shown the presence of sulphur and chlorine, and crystals involving chlorine are a possibility. The non-volatile content of the clouds is thought to be sustained by large impacts raising crustal materials high into the atmosphere. A sufficiently energetic impact is estimated to occur roughly every million years.

Above the cloud deck there is a haze of 1–3 µm particles that on the night side has been shown by Venus Express to extend to an altitude of about 105 km, and is fairly thick up to about 90 km. Below the deck a haze of 1–2 µm droplets extends down to about 30 km. The haze particles are dominated by H_2SO_4. It seems that they form at an altitude around 80 km, the H_2SO_4 being

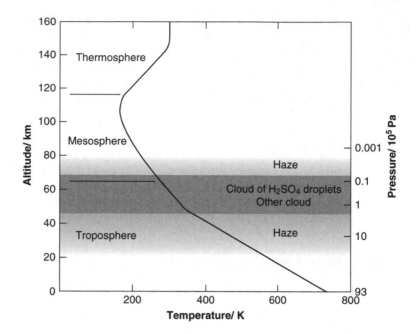

Figure 10.10   The vertical structure of the Venusian atmosphere.

produced by photochemical reactions involving minor atmospheric constituents including SO_2, the sulphur initially entering the atmosphere in the form of sulphur vapour as a minor component of volcanic gases. Once formed, the particles descend to the upper tropopause where they are kept aloft by convection to form the main cloud deck. The droplets leak downwards, particularly those that grow large, and as they descend below the cloud deck they gradually evaporate in the ever-increasing temperatures, with none surviving below about 25 or 30 km.

Though the cloud deck is thick, its density is rather less than that in typical Earth clouds. One consequence is that the daytime surface is not that dark. We would perceive an illumination equivalent to that on a surface a few metres from a reading lamp. Daylight on Venus is tinted red by the atmosphere.

Below the haze the atmosphere is much clearer, and the temperature continues to increase to reach an astonishing GMST of 740 K. The effective temperature is 229 K, and so the size of the greenhouse effect is a huge 510 K or so. The greenhouse effect on Venus is largely sustained by the combined effects of CO_2, water vapour, SO_2, and H_2SO_4 droplets. These substances, plus the huge atmospheric pressure that broadens the spectral lines, make the atmosphere strongly absorbing over a wide range of mid-IR wavelengths. However, these substances cannot account for all of the IR opacity. Some unknown constituents are plugging some of the IR gaps.

10.3.2   Atmospheric Reservoirs, Gains, and Losses

The atmosphere is very probably the main reservoir of carbon on Venus today. This is unlike the Earth where most of the carbon is in organic carbon deposits and in sedimentary carbonate rocks. The absence of oceans on Venus means that the carbonate generation rate is extremely low, whereas volcanism and high surface temperatures will gradually have destroyed any ancient sedimentary carbonate deposits and liberated CO_2 into the atmosphere. Support for this conclusion is the approximately similar mass of carbon in the Venusian atmosphere to that in all of the Earth's reservoirs. This is to be expected from their proximity in the Solar System.

Volcanic activity presumably still releases CO_2, but with so much already in the atmosphere, volcanic emission at the present estimated rate would cause only a very slow increase.

Similar considerations apply to nitrogen. The quantity of N_2 in Venus's atmosphere is somewhat greater than that in the Earth's atmosphere, but the Earth also has nitrate and nitrite deposits, and when these are included, the two quantities are not so very different.

☐ Why are extensive deposits of nitrates and nitrites on Venus unlikely?

Nitrate and nitrite formation is promoted on Earth by the oceans and biosphere. It is possible that there were short-lived oceans early in Venusian history, but it is very unlikely that there was ever a biosphere. The widespread view is that there are no extensive deposits of nitrates and nitrites on Venus, and that consequently most of Venus's nitrogen is in the atmosphere.

With the global total quantities of carbon dioxide and nitrogen on Venus being similar to those on the Earth, it might be expected that this would also be the case for water. This is not so. The number fraction in the Venusian atmosphere is about 4×10^{-5}, and the surface is far too hot for much water to be present, even if it were bound to minerals. The quantity of water deep in the crust and in the mantle is unknown, but is unlikely to exceed terrestrial quantities. Excluding potential deep reservoirs in both planets, if all the Venusian water is in the atmosphere, then Venus has about 10^{-5} times the water known on Earth! However, as for Mars, the trace of atmospheric water plays a role in reversing the effects of the photodissociation of CO_2.

The present low water content in the Venusian atmosphere could represent a balance between the gains from slow outgassing by volcanic activity, plus the occasional impact of comets and asteroids, and a slow rate of loss by photodissociation. Quite why there is so little water is a topic for Section 10.5.

10.3.3 Atmospheric Circulation

On such a slowly rotating planet with a small axial inclination, a single Hadley cell extending from equator to pole in each hemisphere is to be expected (Figure 9.13). There is evidence that such cells exist in the lower atmosphere. However, above these cells there are others, and the reasons for this are not well understood. There is also a thermal tide (Figure 10.9(c)), and though this is not as significant a component of the circulation as it is on Mars, the mass of the Venusian atmosphere is so great that the thermal tide could have influenced the rotation of the planet. Also, the associated mass redistribution leads to a *gravitational* tidal force exerted by the Sun, and this too influences planetary rotation. The net accumulated effect of the atmosphere on the rotation rate of Venus is unclear, though some astronomers have tentatively concluded from computer models that the atmosphere has played a prominent role in causing the present slow (243 day), retrograde rotation of Venus.

The Hadley circulation is apparent in the $5-10\,m\,s^{-1}$ winds from equator to pole at the cloud tops. At the surface, owing to friction between the atmosphere and the surface, the pole-to-equator flow is less than $1\,m\,s^{-1}$, but with such a massive atmosphere this transfers heat sufficiently rapidly to reduce equator-to-pole temperature differences to only about $1-2\,K$. Surface friction also causes a low-altitude east–west wind. At the Venera and Vega sites these were in the range $0.3-1\,m\,s^{-1}$, and at altitudes around $10\,km$ were still only a few metres per second. But this suffices to reduce day–night temperature differences to about $1-2\,K$ also. This east–west drift continues to increase with altitude, becoming about $100\,m\,s^{-1}$ at the cloud tops,

Figure 10.11 Atmospheric circulation at cloud-top level on Venus, imaged in UV by the Pioneer Venus Orbiter. (NASA P790226)

particularly at high latitudes. These high speeds can be explained as a result of the upward transfer of angular momentum from the lower atmosphere, though the details are complicated and not at all fully worked out. Figure 10.11 shows the combined effect of the north–south and east–west drifts at the cloud tops. This image is at UV wavelengths to enhance cloud features – the dark places are *not* breaks in the cloud. Their cause is unknown.

Since April 2006, Venus Express has been in orbit around the planet and is returning much atmospheric data. As well as investigating the haze (see above), it is revealing atmospheric motions by tracking cloud features. For example, it has clarified the structure of a double vortex over the South Pole, and has revealed the existence of a single vortex over the North Pole. Why the poles differ in this respect is not known. Over the next year or so, this and many other aspects of the atmospheric circulation of Venus should be elucidated.

Question 10.5

If *no* solar radiation penetrated the clouds of Venus, why would the lapse rate below the clouds be close to zero?

10.4 Volatile Inventories for Venus, the Earth, and Mars

You have seen that the volatile substances that constitute a planet's atmosphere might also be present at or under the surface of the planet, perhaps in a different chemical form. In considering the origin of atmospheres we must consider the origin of volatiles totalled over all these possible reservoirs, and an essential first step is to attempt a volatile inventory for the present day. This inventory is then a basis for estimating the inventory at earlier times. In this section we will confine ourselves to the main constituents of the volatile inventories of the three planets described in the preceding sections – Venus, the Earth, and Mars.

Figure 10.12 shows the *present* volatile inventories of water, carbon, and nitrogen for Venus, the Earth, and Mars. The quantities are the **global mass fractions** of H_2O, CO_2, and N_2, i.e. the total mass of each substance divided by the total mass of the planet. The quantities in other forms, such as water as hydroxyl (OH), carbon in carbonates, nitrogen in nitrates, are included by adjusting them to the masses of H_2O, CO_2, and N_2 that these other forms would yield.

☐ Why do the major volatiles consist of hydrogen, oxygen, carbon, and nitrogen?

These are cosmically abundant elements (Table 1.5), and they form relatively volatile compounds. (Note that volatiles are *not* the major repositories of oxygen. Most of a planet's oxygen is present in silicates and metal oxides. By contrast, only a small fraction of the carbon and nitrogen is *not* in the volatile reservoirs.)

The atmospheric quantities in Figure 10.12 are well known, but you have seen that the quantities at and beneath the surface are far less certain. Consequently, the global mass fractions shown in Figure 10.12 are lower limits, and in many cases the actual quantities could be very much greater. This is true even for the Earth – the quantities in Figure 10.12 are for the surface and crust. For example, there is also water in the upper mantle, but estimates vary from 60 ppm to over 200 ppm by mass, and even a trace in the greater volume of the lower mantle could add hugely to the inventory. To illustrate the uncertainty further, note that Figure 10.12 places lower limits on the Martian global mass fractions that are less than those for the Earth. This is at

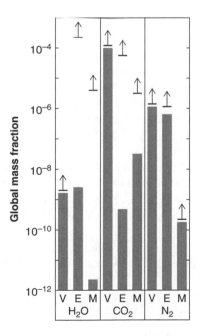

Figure 10.12 Present volatile inventories of Venus, the Earth, and Mars. The bars are for the atmospheres, and the lines with arrows are the lower limits for the global mass fractions.

variance with studies that indicate that Mars is richer than the Earth in volatiles. If this is so, it is either because Mars was made from materials richer in volatiles, or because the Moon-forming impact depleted the Earth in volatiles. If Mars is indeed volatile rich, it does not seem to be much outgassed – this is discussed in Section 10.5.5.

This is an unpropitious start! Nevertheless, Figure 10.12 indicates that the present volatile endowments of Venus, the Earth, and Mars are very different. But what about the past? Were the inventories more similar then? For the global mass fractions, only loss of different proportions to space could have made a difference.

Losses to space

There is strong evidence that Mars has lost to space a considerable proportion of its nitrogen. The evidence is in the atmospheric isotope ratio $n(^{14}N)/n(^{15}N)$, where $n(^{14}N)$ is the number density of atoms of the common isotope ^{14}N in the atmosphere, and $n(^{15}N)$ is the number density of ^{15}N atoms. Each number density includes the isotope present as single atoms and as contained in the molecule N_2. In the Earth's atmosphere today $n(^{14}N)/n(^{15}N) = 272$, whereas in the Martian atmosphere it is about 170. The accepted explanation is the greater rate at which the less massive isotope has escaped from Mars, thus leading to its relative depletion there.

Nitrogen has been lost from Mars mainly by chemical escape, e.g. by the photochemical reaction

$$N_2(+UV\ photon) \rightarrow N + N \tag{10.4}$$

where the UV photon is provided by solar radiation. This reaction is only one of several, but in all of them the resulting nitrogen atoms are boosted to sufficient speeds to escape at a far greater rate than they do by thermal escape. The lighter isotope is lost at a faster rate because above the homosphere the atmosphere is no longer well mixed, and the number of lighter N_2 molecules decreases less rapidly with altitude than does the number of heavier molecules. This leads to enrichment of the lighter isotope in the exosphere, from where chemical escape occurs. If the Earth and Mars started out with the same $n(^{14}N)/n(^{15}N)$ ratio, then it can be shown that the present difference in the ratios indicates that Mars has lost to space the order of ten to a few hundred times its present atmospheric content of nitrogen.

☐ Why has the Earth lost less nitrogen than Mars through chemical escape?

Chemical escape of nitrogen from the Earth has been less because of the Earth's higher escape speed (Table 9.1).

Some escape to space of Martian oxygen and carbon must occur, but, as noted in Section 10.2.2, there must be sufficient reservoirs of water and CO_2 at or near the surface to have prevented enrichment of the lighter isotopes. For nitrogen there seem to be no such reservoirs, presumably because any oceans and rainfall have, at best, been confined to the Noachian (Section 7.3.9).

Escape to space can also go some way towards explaining the dryness of Venus. The explanation starts with photodissociation of water in the upper atmosphere. The net effect of a series of reactions is

$$2H_2O(+UV\ photon) \rightarrow 2H_2 + O_2 \tag{10.5}$$

yielding hydrogen and oxygen molecules as gaseous components of the upper atmosphere. H_2 thermally escapes at only a low rate from Venus today because the base of the exosphere is cold (Figure 9.12), but in the past the upper atmosphere could have been warmer. Also, when the Sun was young, though its luminosity was lower than today, there could have been a greater solar UV flux and a more copious solar wind, thus enhancing the rate of escape. Furthermore, though the exosphere is too cool today for thermal escape, loss of hydrogen might still be occurring through the action of electric and magnetic fields in the solar wind. These fields could accelerate hydrogen ions in the exosphere to escape speeds, the ions having been formed by photodissociation. It is thus likely that large quantities of hydrogen have been lost to space over the lifetime of Venus.

We must also get rid of Venus's oxygen – there would have been far more produced from reaction (10.5) than the very small upper limit that we have for the atmosphere today.

☐ Why is thermal escape inadequate?

The mass of the O_2 molecule is much greater than that of H_2, so the molecular speeds of O_2 would have been much lower than those of H_2, and the rate of thermal escape correspondingly small (Figure 9.12). Presumably the oxygen has combined with the crust, and some evidence for this is provided by the lander Venera 13, which found highly oxidised surface rocks.

The series of reactions summarised by equation (10.5) also occurs on Earth, but most terrestrial water is in the oceans and in other surface reservoirs, where it is protected from dissociation. Consequently, the Earth has lost little of its initial endowment of water. It was the hot troposphere of Venus and the consequent lack of precipitation that allowed water to reach high altitudes, thus exposing it to rapid photodissociation.

Support for this water loss mechanism from Venus is provided by the present-day isotope ratios $n(^2H)/n(^1H)$ for Venus and the Earth. The lower mass of ^1H enables its molecules and ions to escape to space at a greater rate than molecules and ions that include ^2H, and therefore on both planets $n(^2H)/n(^1H)$ tends to increase with the passage of time. Atmospheric and surface reservoirs exchange water, and with most terrestrial water residing in the surface reservoirs, protected from dissociation, this tends to maintain the terrestrial $n(^2H)/n(^1H)$ ratio at its initial value. On Venus there have been no oceans for a long time, if ever. Therefore, we would expect the $n(^2H)/n(^1H)$ ratio on Venus to be greater than on the Earth, and this is exactly what we find – about 150 times the terrestrial value of 1.6×10^{-4}.

With detailed models it is possible to estimate the amount of water that must have been lost by Venus to give such an enhancement. These models include the greater UV flux in the past. If the initial ratio of $n(^2H)/n(^1H)$ was much the same on both planets, and if the atmosphere of Venus has not been resupplied with water, then this loss process has got rid of the order of 10^2–10^3 times the mass of water presently in the atmosphere. This loss is equivalent to a global layer of liquid water of the order of 3–30 m. This is evidence that the crust of Venus is dry, though mantle water, as in the case of the Earth, cannot be ruled out.

The hydrogen isotope ratio $n(^2H)/n(^1H)$ on Mars also indicates loss to space – the ratio is five times greater than that of the Earth, indicating a far less severe fraction lost than in the case of Venus. This is consistent with the existence of (sub)surface reservoirs of water.

The models that predict the losses to space make various assumptions, and therefore in tracing volatile inventories into the past there are further uncertainties to add to those about the present inventories. Moreover, processes that were concentrated early in Solar System history, such as impact erosion, blow off, and hydrodynamic escape (Section 9.2.5) probably removed huge quantities of volatiles. Therefore, we must now start at the earliest times and try to work forward, under the constraint that there has to be an acceptable match with what we can establish by working back.

Question 10.6

If Venus has lost to space 10^2–10^3 times the mass of water presently in its atmosphere, what would its global mass fraction of water have been before this loss, and how does this compare with the global mass fraction of water on Earth?

10.5 The Origin of Terrestrial Atmospheres

Important evidence about the earliest times is provided by the inert gases.

10.5.1 Inert Gas Evidence

The complete suite of inert gases is helium (He), neon (Ne), argon (Ar), krypton (Kr), xenon (Xe), and radon (Ra). Recall that they are called inert because they are chemically unreactive, and they are called gases because at all but very high pressures or very low temperatures they exist in the gaseous phase – they are extremely volatile.

☐   Would you expect argon gas to be Ar or Ar_2?

In accord with their chemical unreactivity, the inert gases are present in atomic form, rather than combined into molecules.

Of particular importance here are the inert gas isotopes that are stable (not radioactive), and that have not been produced by radioactive decay – such radiogenic isotopes will have had their atmospheric quantities increased over time in some uncertain manner. All of radon's isotopes are unstable, and the non-radiogenic isotopes of helium are excluded because the atoms are light enough for helium to escape from the terrestrial planets. By contrast a high and similar proportion of the other isotopes will have been retained. Moreover, the unreactivity and extreme volatility of the inert gases make it likely that nearly all of the initial endowment of each of these isotopes is in the atmosphere. The global mass fractions of the atmospheric isotopes should therefore bear a primordial imprint.

The non-radiogenic isotopes used most often in studies of planetary volatiles are ^{20}Ne, ^{36}Ar, ^{84}Kr, and ^{132}Xe. Figure 10.13 shows the present-day global mass fractions of these isotopes in the atmospheres of Venus, the Earth, and Mars. Also included are the values for the Sun, which must surely have had a composition similar to the nebula from which the planets formed. A striking feature of Figure 10.13 is that the graphs for the terrestrial planets have very *different* shapes from that for the Sun. Greater similarity would be expected if a high proportion of today's planetary volatiles had been captured from the solar nebula as a whole. This expectation remains even allowing for subsequent mass-selective loss mechanisms such as thermal escape and hydrodynamic escape. Therefore, if the terrestrial planets ever did capture a large quantity of volatiles from nebular gas and dust (mainly from the gas) then it has since been lost or overwhelmed by other sources. Section 10.5.3 discusses loss mechanisms.

The fifth graph in Figure 10.13 is for the C1 chondrites (the dashed line). These are a particularly primitive subclass of the carbonaceous chondrite meteorites (CCs, Section 3.3.2), and are thought to have a broadly similar chemical composition to the planetesimals from which the terrestrial planets formed (Section 10.5.2). Moreover, the heavier inert gases would *not* have been lost to space when the planetesimals impacted the growing planet. In contrast to the Sun, the shape of the graph for the C1 chondrites is not very different from the shapes for the terrestrial planets. It is therefore an easy step to conclude that a large fraction of the mass of the terrestrial volatiles must have been delivered within planetesimals of broadly chondritic

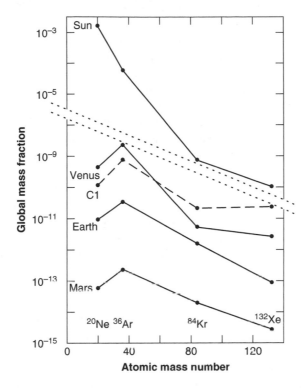

Figure 10.13 The atmospheric global mass fractions of the important non-radiogenic inert gas isotopes, for Venus, the Earth, and Mars. The mass fractions for the Sun and for the C1 chondrites are also shown.

composition, or embryos built of such planetesimals. The C1 chondrites certainly contain the right sort of substances – they have water, mostly in hydrated minerals such as serpentine ($3MgO.2SiO_2.2H_2O$), and they have carbon and nitrogen, mostly within organic compounds. Furthermore, the non-volatile composition of comets is not very different from that of the C1 chondrites, and so if this similarity extends to the volatiles then one cannot rule out comets as important sources of terrestrial volatiles too.

But though the shapes of the graphs for the planets and the C1 chondrites are similar, they are not identical. Also, the graphs are separated by large differences in the values of the global mass fractions. We shall account for these features in Section 10.5.3. First we need to consider more closely the acquisition of volatiles from planetesimals and other bodies, during planet formation.

10.5.2 Volatile Acquisition During Planet Formation

The initial volatile endowment of the terrestrial planets is inextricably linked with the formation of the terrestrial planets as a whole. The solar nebular theory of their formation was outlined in Chapter 2, and we can pick up the story at the point where the inner Solar System was full of a swarm of planetesimals, plus embryos. The volatile content of these bodies increased somewhat from the region where Venus will form, to the region now occupied by the asteroid belt.

☐　Why only 'somewhat'?

An increase in volatile content with heliocentric distance is to be expected from the associated decline in nebular temperatures. However, with the growth of planetary embryos, close encounters between embryos, and between an embryo and a planetesimal, made many orbits highly elliptical. Therefore, embryos and planetesimals began to criss-cross the terrestrial and asteroid zones, blurring any compositional differences.

By the time that Venus, the Earth, and Mars had grown to about $0.1 m_E$ (m_E is the present mass of the Earth), the impacts of planetesimals were sufficiently violent for the planetesimals to be fully devolatilised. This resulted in massive atmospheres, mainly of H_2O, which is the most abundant volatile, but with CO_2, plus some NH_3, SO_2, and other volatiles. Impact devolatilisation is expected to have been less overall for Mars because $0.1 m_E$ is about the same as the planet's present mass. Nevertheless, the later arriving material would certainly have been devolatilised, and this would have created a considerable atmosphere on Mars too. Note that devolatilisation, which would be followed by substantial losses to space (Section 10.5.3), would result in a trend in global composition away from the C1 chondrites and towards the ordinary chondrites.

For the Earth, a computer model of its formation has shown that many planetesimals came from regions well beyond its orbit, where a broadly C1 composition is expected. The Earth was still forming after the gaseous component of the nebula had been dissipated. Enough time had then elapsed so that embryos had formed in what is now the outer asteroid belt. These would be rich in volatiles. A few of them collided with the Earth. For an embryo mass a few per cent of that of the Earth, the model indicates that they brought to our planet the bulk of its water and other volatiles. Consistent with this model is the ratio of deuterium (^2H) to ^1H in the Earth's oceans. It is close to the mean value in the water-rich components of CCs. A late veneer of volatile-rich bodies from the Uranus–Neptune region and the E–K belt could distort this picture. But the model shows that the amount of material impacting the Earth from these outer zones would deliver less than 10% of the earlier endowment. Late veneers are the subject of Section 10.5.4.

So, we have arrived at the point where Venus, the Earth, and, to a lesser extent, Mars have massive water-rich atmospheres. This led to large rises in surface temperature, partly because of the associated large greenhouse effect, and partly because of aerosols that not only added to the greenhouse effect but also absorbed solar radiation. With most of the kinetic energy of the impactors converted to heat, a magma ocean at about 1500 K was formed on Venus and the Earth (Sections 8.1.5 and 8.2.9). Mars might have lacked a magma ocean partly because the greater heliocentric distance would have reduced the solar heating of the surface, and partly because of lower impact speeds – the combined result of lower orbital speeds at the greater distance and the smaller gravitational field of Mars. On Venus and the Earth, water dissolved in the magma oceans, stabilising the atmospheric mass at the order of 10^{21} kg. With a general decline in impacts the magma oceans solidified, giving up some of the water and other dissolved volatiles as they did so. In the case of the Earth most of the atmospheric water condensed to form oceans, but this might not have happened on Venus because of the greater solar heating closer to the Sun. If no magma ocean appeared on Mars, and if there was little subsequent geological activity that folded the surface into the mantle, then the mantle might be dry, with any water that condensed confined to no more than shallow depths.

The earliest atmospheres were subject to enormous modifications by various processes, notably those outlined in the next section.

10.5.3 Early Massive Losses

Whereas the smaller planetesimals brought a net increase of the volatile endowment, larger planetesimals caused impact erosion, and the very largest and any left-over embryos would have caused blow-off (Section 9.2.5). Not all volatiles were equally affected. Water can condense or combine with surface materials more readily than the other common volatiles, and can thus obtain some protection. This can help to explain the dryness of Venus, where incorporation in the surface was limited by the high surface temperatures resulting from the proximity to the Sun. The extent to which CO_2 was protected depends largely on whether any oceans of water appeared quickly enough for dissolution and carbonate formation to occur before extensive impact erosion. Note that blow-off is consistent with the delivery of volatiles by embryo impact. Not all the volatile endowment of embryos is necessarily lost. (The Moon formed from largely *vaporised* material from a grazing impact by a massive embryo, which explains its low volatile content – Section 5.2.1.)

Impact erosion can also explain the low global mass fractions of the inert gases in the Martian atmosphere (Figure 10.13) – the low mass of Mars and its proximity to the asteroid belt resulted in the loss of a great proportion of these gases. This was also the case for other volatiles, and could explain any scarcity of carbonates on Mars, at least in part. It is also possible that impact erosion accounts for any shortfall between the large quantities of water that seem to be needed to explain the Noachian fluvial features (always supposing that they were not caused by the cycling of a small quantity of water), and the possible inadequacy of subsurface reservoirs to have supplied this water.

Another major early process is hydrodynamic escape (Section 9.2.5), resulting from a huge pulse of H_2 production. Any iron in a magma ocean would react with water to generate such a pulse. The simplest reaction is

$$Fe + H_2O \rightarrow FeO + H_2 \qquad (10.6)$$

where the water might be present as H_2O or chemically combined with minerals. Water in hydroxyl form undergoes an equivalent reaction. Hydrogen would have been generated in a similar way during core formation as liquid iron trickled through the mantle. The hydrogen built up in the atmosphere, and little water was left in the crust and mantle. Additional hydrogen in a water-rich atmosphere would come from the photodissociation of water. But regardless of how it was generated, the presence of large quantities of hydrogen in the atmosphere led at once to a large outflux of hydrogen to space.

☐ Why is this?

Hydrogen as H or H_2 has a low mass, so readily suffers thermal escape. Other molecules are entrained and so are lost too. Recall that the process is mass selective and would have led to an increase in the $n(^2H)/n(^1H)$ ratio.

One entrained molecule was O_2 from the photodissociation of water, though models indicate that on all three terrestrial planets most of the O_2 was retained, and was removed later by the oxidation of rocks. If there was sufficient CO, it too would have taken up a significant mass of O_2 in being oxidised to CO_2.

There is evidence of hydrodynamic escape. For example, it explains differences in xenon isotope ratios between the Martian atmosphere and interior. Other isotope data suggest the escape happened within 160 Ma of the birth of Mars.

Less dramatic, but very effective at removing atmospheres, is the solar wind, which was particularly copious when the Sun was young. This could have stripped atmospheres in a few thousand years, unless protected at that time by a powerful magnetic field, as in the case of the Earth, and perhaps Mars too.

It seems likely that all the terrestrial planets suffered huge losses through hydrodynamic escape, blow-off, and erosion by impacts and perhaps the solar wind. Blow-off would have been most likely about 4600 Ma ago, and hydrodynamic escape before about 4200 Ma ago. Impact erosion would have been in step with the heavy bombardment, so would have been in steep decline about 3900 Ma ago. Almost entire atmospheres could have been removed, in which case the present atmospheres must largely be the result of subsequent outgassing (including from the incorporated fragments of bodies that caused blow-off) and, additionally, late veneers.

10.5.4 Late Veneers

With the growth of the giant planets, it is expected that icy planetesimals (comets) and other volatile-rich bodies were thrown across the terrestrial zone, and some were captured by the then (nearly) fully formed terrestrial planets, thus providing them with a late veneer. This veneer could probably have provided them with considerable quantities of volatiles, supplementing significantly, or even dominating, any residue from earlier times. Subsequent geological activity would have buried much of this veneer into the mantle, except perhaps on Mars because of its lower level of activity. Today, the rate of acquisition of volatiles by this means is very low.

☐   Why did these bodies not cause impact erosion?

They were small bodies.

Evidence that at least some of the water on Mars came as a late veneer is provided by the Martian meteorites. Oxygen isotope ratios in some of the water are different from those in the oxygen in the silicate components. The water presumably came largely from water deposits at or near the surface, whereas the silicates were derived from deeper in the lithosphere. It thus seems that the lithosphere and surface have had rather separate histories.

Further evidence that there was a late veneer, and that it was due at least in part to icy planetesimals, is provided by Figure 10.13. The C1 chondrite graph at its right hand end is much flatter than the graphs for the terrestrial planets. This can be explained by volatile-rich bodies additional to bodies with a C1 chondrite composition, and icy–rocky planetesimals is a popular choice. In the Jupiter region the volatile complement would have been dominated by water ice, plus small quantities of ices rich in carbon and nitrogen. Further out there were increasing quantities of ices of the more volatile substances, in accord with the decreasing temperatures. The gas trapping efficiency also increases with decreasing temperature. The planetesimals from the furthest reaches, from beyond Neptune, would have had, as well as water, ices and trapped gases rich in nitrogen and carbon so that, overall, the elements C and N are in solar proportions. Of relevance to Figure 10.13 is that the inert gases would also have been present in solar proportions, except for an underabundance of He and, to a lesser extent, of Ne. A proportionate input from these icy–rocky planetesimals can give a match to the terrestrial graphs in Figure 10.13.

A late veneer on the Earth is indicated by a handful of crystals of zircon, $ZrSiO_4$, from Jack's Hills in Western Australia. The oldest of these hardy crystals have survived from 4400 Ma ago, not long after the Moon-forming impact. They formed in crustal rocks, long gone, and have oxygen isotope ratios that can be shown to indicate that in their vicinity there were large lakes, even oceans. Outgassing might not have had enough time to meet the need, in which case icy planetesimal impacts are indicated.

In order to account for many of the details of the volatile endowments of the terrestrial planets it is necessary to assume that each terrestrial planet received a particular mixture of the various bodies, the mix depending on the heliocentric distance of the planet. It is then possible to explain why Venus, the Earth, and Mars have global mass fractions of the inert gases that, on the whole, are smaller than their global mass fractions of H_2O, CO_2, and N_2. Furthermore, these three fractions can be turned into two ratios by dividing by the global mass fraction of H_2O. This is useful because we can be far more certain about the values of these ratios than we can about the individual global mass fractions. It is the different values of these ratios from planet to planet that can be explained. The details are beyond our scope.

Finally, note that icy–rocky planetesimals are hydrogen-rich bodies, and therefore they can give rise to chemical reactions that produce NH_3 and CH_4 in terrestrial atmospheres. These are powerful greenhouse gases and if they were abundant they would have been important in the evolution of the atmospheres, as you will see.

10.5.5 Outgassing

A late veneer does not preclude the possibility of continued outgassing, long after any magma ocean solidified and any core formation was complete. Indeed, outgassing must have been important on even the least geologically active of the three planets, Mars. Outgassing continues to this day on the Earth, probably on Venus, and possibly on Mars.

The mix of gases emerging from the interior depends on the mix of materials there and on the temperature. Any metallic iron present in the mantle, such as there would be before core formation, would remove oxygen from molecules, and thus convert much of the H_2O in the mantle to H_2 and much of the CO_2 to CO. The decomposition of any carbonates present would yield CO rather than CO_2. With iron concentrated in the core, leaving little or none in the mantle, then at the sort of temperatures occurring in terrestrial planet mantles, the volatile-rich materials would outgas H_2O, CO_2, and N_2. Recent experiments in which CCs are heated, experiments supplemented by extensive chemical calculations, indicate that if the terrestrial planets have a significant CC content, then the early atmospheres might have consisted not only of water vapour but of CH_4 and NH_3 too, plus several tens of per cent of H_2. This seems to be consistent with outgassing *before* the formation of an iron-rich core.

Whatever the main constituents, there were traces of other volatiles. Among these were ^{40}Ar from the radioactive decay of the unstable isotope of potassium, ^{40}K. Potassium is confined to rocky materials, and so if we know the atmospheric quantity of ^{40}Ar, and if we can estimate the amount of potassium in the interior of a planet, we can estimate the degree to which the planet has outgassed. We know the atmospheric quantities of ^{40}Ar for Venus, the Earth, and Mars, and a rough estimate of the potassium content has been made for the Earth, and also for Mars from studies of Martian meteorites. It seems that Mars is less outgassed than the Earth. This is in accord with its lower level of volcanic activity.

If indeed Mars *is* less outgassed than the Earth, then this is a factor in accounting for its lower atmospheric global mass fractions of inert gases in Figure 10.13. It is also a factor in accounting for any deficiencies in the near-surface quantities of other Martian volatiles. However, we do not know the initial volatile inventory of the interior, and so we cannot use the degree of outgassing as a means of calculating the quantities of water and other volatiles that have appeared on the surface.

In spite of the many uncertainties, we nevertheless have rough estimates of the early volatile inventories of the terrestrial planets. What happened next?

Question 10.7

Describe the CO_2 data in Figure 10.12, and outline the processes common to the three planets that have determined their global CO_2 endowments. State to what extent the differences between the CO_2 endowments have been explained so far in this chapter.

Question 10.8

Present a plausible argument for the absence of significant atmospheres on

(a) the Moon
(b) Mercury.

10.6 Evolution of Terrestrial Atmospheres, and Climate Change

In considering the origin of the terrestrial atmospheres we have been concerned largely with events that occurred up to no later than the end of the heavy bombardment at about 3900 Ma ago (perhaps a bit later on Mars). We shall now be more concerned with events from about 4200 Ma ago to the present. By 4200 Ma any magma oceans had solidified, any blow-off had occurred, and core formation, rapid outgassing, and hydrodynamic escape were over. Impact erosion was continuing, but at a declining rate. Since then the atmospheres of Venus, the Earth, and Mars have taken dramatically different evolutionary paths. This has been against the backdrop of a gradual increase in solar luminosity as shown in Figure 10.14.

Some plausible scenarios are now presented. In considering each planet in turn some familiar material will be encountered as we blend the past into the present. You will see that distance from the Sun has been a crucial factor.

☐　What are the ratios of the solar flux densities at the orbits of Venus, the Earth, and Mars? These are the ratios of $1/r^2$ (equation (9.5)), where we can use the semimajor axes for r. Thus, the ratios compared with the Earth are $(1/0.723)^2 : 1 : (1/1.524)^2$, i.e. 1.91:1:0.43. These are substantially different from one. For example, in the beginning, when the Sun was only 0.7

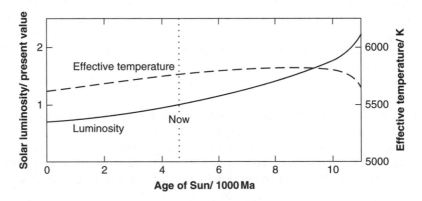

Figure 10.14　The increase in solar luminosity from its birth 4600 Ma ago until the end of its main sequence lifetime, about 6000 Ma into the future, according to recent models.

times its present luminosity, the flux density at Venus was still 1.3 times the present flux density at the Earth.

10.6.1 Venus

About 4200 Ma ago Venus probably had an atmosphere of CO_2, water vapour, and N_2, perhaps overlying oceans of liquid water. The atmosphere was not as massive as it is today, but it contained sufficient water and CO_2 for a modest greenhouse effect. This, coupled with the proximity of Venus to the Sun, made the lower atmosphere warm. The warm temperatures sustained a high partial pressure of water, and the release of latent heat on condensation reduced the lapse rate to the point that atmospheric temperatures decreased rather slowly with height. Therefore, the number fraction of water was sustained to high altitudes, where it was subject to photodissociation and the consequent loss of hydrogen to space. The oxygen liberated has probably been lost through oxidation of the surface rocks, particularly as the surface temperature rose (Section 10.4).

Though water was being lost by photodissociation it was replenished by evaporation from surface reservoirs, thus maintaining its partial pressure. As the solar luminosity rose, the partial pressure of water in the lower atmosphere crept up the saturation curve. This increased the greenhouse effect which caused a further increase in partial pressure. This is an example of positive feedback, an inherently unstable feature that is thought by many to have resulted in the complete evaporation of the oceans within the order of 100 Ma, by which time the temperatures were high, though not as high as today, and the atmosphere contained a huge mass of water vapour. Models indicate that this **runaway greenhouse effect** would have occurred very early in Venus's history, probably well before 3000 Ma ago

A variant on this is the moist greenhouse effect. Depending mainly on the atmospheric properties, and particularly by lowering the solar luminosity, models can produce an outcome in which Venus lost its oceans much more slowly than in a runaway. Consequently the temperatures rose more slowly. The dissolution of CO_2 in the oceans and the formation of carbonates would have further moderated the rate of temperature rise. Though the oceans were hot, they did not boil. In this model there is still a moist stratosphere, and thus there was again a steady loss of water by photodissociation and the escape of hydrogen. Gradually the water was lost, and as temperatures rose the carbonates were dissociated. The end point is much the same, but reached less dramatically, and perhaps as late as 3000 Ma ago.

Though most of Venus's water was lost through photodissociation, the process slowed as the atmosphere dried out and the upper atmospheric temperatures consequently declined. Therefore, the last 10^6 Pa or so of water vapour must have been lost some other way. One possibility is incorporation into the crust and subsequent burial in the mantle through geological processes. Volcanic activity can subsequently release such water, but in a steady state it is feasible that only a small proportion is ever in the atmosphere, which has therefore remained dry.

You have seen that photodissociation also explains the high ratio of $n(^2H)/n(^1H)$ in the atmosphere of Venus today.

☐ What is the explanation?

Photodissociation liberates hydrogen, and the lower mass of ^1H enables its molecules and ions to escape to space at a greater rate than molecules and ions that include ^2H. In addition, the earlier hydrodynamic escape of hydrogen also favoured the lighter isotope.

With the loss of oceans, sedimentary carbonate formation became negligible, and so the CO_2 content of the atmosphere increased as volcanic activity released CO_2 from the crust and mantle, including from any remaining carbonates that might have formed as ocean sediments, to form the massive CO_2 atmosphere that we have today. This sustains the present temperatures through the greenhouse effect of CO_2, enhanced by pressure broadening, aided by the greenhouse effect of the clouds, and of small quantities of other constituents, some of which remain unknown.

There is an alternative scenario in which oceans *never* formed on Venus, but the end point is again the same. This alternative is unlikely, if, as is widely believed, the devolatilisation of impactors and the late veneer produced large quantities of water over a relatively short time. If Venus *did* once have oceans of liquid water for hundreds of millions of years then it is just possible that life began a short Venusian career. But there is little chance that any evidence has survived to today.

Global resurfacing at 500–800 Ma

The one major climatic event that we are confident has occurred on Venus was due to the near global resurfacing that occurred over a short interval between 500 Ma and 800 Ma ago (Section 8.2.3). The copious volcanic emissions would have included CO_2, but there was so much in the atmosphere already that this barely made a difference. Of more significance was the H_2O and SO_2, which it is estimated would have increased the atmospheric quantities by 10 times and 100 times respectively. This increased the greenhouse effect by (partially) closing windows at IR wavelengths, but also increased cloud thickness and hence increased the albedo. One model indicates that initially the increase in albedo 'won', and the GMST fell, perhaps by 100 K. The clouds thinned. The excess H_2O and SO_2 were disposed of in the usual way (loss to space, reactions with rocks), and in a few hundred million years, via a trickle of volcanic gases, including H_2O and SO_2, the clouds were restored and the earlier conditions re-established.

During this considerable climatic disturbance, there could have been a dramatic increase in temperature if the H_2O content of the atmosphere rose sufficiently for the 'leaky' IR window at 2.1–2.7 μm to be much less transparent. A 20 times increase to around 0.5% would have caused a rapid rise in the GMST. This rise would have been limited to about 920 K. At that temperature the peak in the IR spectrum of the radiation welling upwards from the surface would have been right in this window, and the GMST stabilized. In the usual way H_2O was lost, the atmosphere cooled, and the earlier conditions re-established as before.

We have no evidence of substantial climatic change since then.

10.6.2 The Earth

In Section 10.5.4 mineralogical evidence was outlined for extensive bodies of liquid water on the Earth 4400 Ma ago. Metamorphosed sedimentary rocks 3800 Ma old must have been derived from ocean sediments. Biological evidence for water is the likely continuous presence of life since about 3850 Ma ago – only about 50 Ma after the end of the heavy bombardment. Though the details of the origin of life on Earth are still largely obscure, we do know that liquid water is essential for life in all its forms.

This stark contrast with Venus arises from the greater solar distance of the Earth. The atmosphere never became warm enough for water vapour to be abundant in the upper atmosphere – the upper troposphere has always been a cold trap where water has condensed and returned to the surface. Indeed, we have the opposite problem. The low luminosity of the youthful Sun (Figure 10.14) means that if the Earth's albedo and greenhouse effect were initially as they are at present, the GMST would have been about 265 K, and the oceans in that far-off time would have completely frozen over. The albedo would then have become so high that it would not have been until 1000–2000 Ma ago that the solar luminosity would have risen to the point where the ice melted. And yet there was no extensive glaciation throughout this time. This is called the **faint Sun paradox**.

❑ How could different atmospheric composition have helped prevent global freezing?

If the Earth's greenhouse effect was sufficiently larger, the effect of reduced solar luminosity could have been offset. Something like 300–3000 times the present atmospheric mass of CO_2 is needed around 3800 Ma ago, with a slow decline rather well matched to the increasing solar luminosity, otherwise the temperatures would have become higher than we know they were. The total volatile inventory of CO_2 on the Earth is at least 10^5 times that in the atmosphere, so the problem is one of controlling the atmospheric fraction. Could a carbon cycle accomplish this?

As soon as oceans appeared, they began to dissolve the CO_2, and they reached a steady state with the atmosphere on the short time scale of the order of centuries. Carbonate formation in the oceans was slower, operating on a time scale of order 10^2 Ma. On this longer time scale a carbon cycle became established, but it was not in a steady state – the atmospheric CO_2 content must have declined as the outgassing rate decreased with the reduction in volcanic activity. It is therefore conceivable that there was always sufficient atmospheric CO_2 to prevent a global freeze but not so much as to cause temperatures to be too high. This rather fine tuning is aided by the carbonate–silicate cycle (Section 10.1.4).

A higher CO_2 content early on could also have helped warm the Earth through the scattering greenhouse effect by CO_2 cloud particles (Section 9.2.4). The low temperatures high in the atmosphere when the Sun's luminosity was well below its present value could have led to extensive CO_2 cloud cover. This would have increased the planetary albedo – just what we *do not* want. However, if the cloud particles were larger than a few micrometres they would have scattered IR radiation emitted from below them and enhanced the greenhouse effect to an extent that more than compensated for the increased albedo.

It is quite possible that the CO_2 content of the atmosphere was never sufficient to compensate for the Sun's faintness. The shortfall could have been bridged by greenhouse gases that filled the windows in the CO_2 absorption spectrum, such as NH_3 or CH_4. A further compensation would have come from the lower albedo of the Earth's surface, because of smaller land area, and the shorter day (14 hours) that reduced the night-time during which ice could form – these two factors together could have raised the GMST by about 5 K.

NH_3 or CH_4 are rapidly broken up by solar UV radiation so we need either a steady resupply, or protection. The icy planetesimals that contributed to the late veneer could have been the early resupplier of both gases. Subsequently, if methanogens became the dominant form of life, they could have taken over as the main resupplier of CH_4. For protection from UV we also look to CH_4. It would be photodissociated in the upper atmosphere to yield a fine dust of solid hydrocarbons that could shield NH_3. The amounts of NH_3 and CH_4 in the atmosphere would have declined as the atmosphere became more oxidising. This could have been due to the descent of metals and sulphides into the Earth's core, because of their high density. Models show that NH_3 and CH_4 could have warmed the Earth significantly until about 3800 Ma ago, and perhaps for a few hundred million years later, by which time the solar luminosity had risen. The early presence of NH_3 and CH_4 would have facilitated the synthesis of the huge organic molecules that life is based on, notably proteins, RNA, and DNA.

The occurrence of ice ages (Section 10.1.4) shows that the fine tuning that is needed to keep the Earth's surface temperature fairly constant as the solar luminosity rose has not been perfect. It is nevertheless impressive, and this has led to the suggestion by the British chemist James Ephraim

Lovelock that the biosphere (unconsciously) took an active part in the control. The cumulative mass of all organisms that have ever lived is at least an appreciable fraction of the Earth's total mass! Therefore, a significant effect is not out of the question, though this does not of itself indicate that the effect has been stabilising. The general notion of active biospheric control that tends to preserve optimum conditions for the biosphere is called the **Gaia hypothesis**. One of many ways in which it could operate is through the biogenic release of CH_4, and the release and uptake of CO_2.

The biosphere has certainly had a profound effect on atmospheric composition. You saw in Section 10.1.2 that it sustains almost the whole O_2 and N_2 content of the atmosphere, and that is has promoted the removal of CO_2. Let us look more closely at oxygen, and not just at the biosphere's role, though that is where we start.

The history of oxygen in the Earth's atmosphere

When did the biosphere become active? We know that by about 3850 Ma ago simple single cells probably existed on Earth and that at about the same time an early form of photosynthesis was probably operating. By no later than 3400 Ma ago photosynthesis that generated O_2 existed, and this soon overtook the photodissociation of water as the main source of O_2.

☐   Why has photodissociation always been a rather weak source of O_2?

Most of the water has been in the oceans, beyond the reach of solar UV radiation.

At first the O_2 from photosynthesis was almost entirely consumed by the oxidation of less than fully oxidised substances – rocks, volcanic gases, and biological materials. Rocks older than about 2300 Ma have sufficient pyrite (FeS_2), uraninite (UO_2), and siderite ($FeCO_3$) to indicate that the O_2 partial pressure must have been less than about 20 Pa, otherwise oxidation would have largely destroyed these minerals. A lower limit of about 10^{-6} Pa has been suggested by some scientists, on the basis of the abundance of Fe_3O_4, one of the more oxygen-rich iron oxides. Photodissociation of water alone might have yielded a partial pressure of O_2 of around 5×10^{-4} Pa. The present value is 21 000 Pa.

Further evidence for a low partial pressure comes from the decline in the formation of banded iron formations (BIFs) after 2400 Ma ago. BIFs are widely distributed, finely layered sedimentary rocks that consist of dark layers with up to 30% iron alternating with lighter layers of silica. The iron-rich layers require the iron to have been dissolved in oceans, from which they precipitated. This is not possible unless the oxygen content of those ancient oceans was far smaller than it is today. At about 2000 Ma ago redbeds appeared. These are formed when iron is weathered out of rocks in the presence of oxygen, not necessarily as much as in the atmosphere today, but significantly more than before 2000 Ma. BIFs enjoyed a resurgence from around 2000 Ma to about 1800 Ma, indicating a temporary and modest decrease in atmospheric oxygen, presumably due to a decrease in the organic carbon burial rate, for reasons unknown.

About 2000 Ma ago a type of cell began to spread upon which all of the higher forms of life is based. This is the eukaryotic cell, which has a nucleus and other internal structures, in contrast to the less structured prokaryotic cells that had earlier been the only kind. Even though all life was still in the sea (and almost entirely single celled) we can estimate the atmospheric O_2 that would have been needed to sustain the minimum O_2 content of water for eukaryotic cells to exist, and this corresponds to about 10^2 Pa of atmospheric O_2.

From this and other evidence it is clear that after about 2400 Ma, O_2 levels rose hugely. In fact, they have never since been less than about 100 Pa, and in the past 1600 Ma never less than about 2000 Pa. What caused the increase from 2400 Ma?

One possibility is that the decline in tectonic activity as the Earth aged reduced the rate of appearance of under-oxidised substances at the surface, or (perhaps additionally) there was a change in composition, with a decline in the proportion coming from the mantle, which is a source of under-oxidised substances. It could be that a change in the composition of volcanic gases sufficed, with a reduction in oxygen devouring gases such as H_2.

An alternative view is that oxygenic cyanobacteria, though present much earlier, flourished, perhaps due to an increase in organic nutrients in the oceans, as mentioned in Section 10.1.4. Another possible biogenic cause is that methanogens captured volcanic H_2 to make CH_4, thus reducing the removal of O_2 by H_2 to make water. The CH_4 was dissociated in the upper atmosphere, and the hydrogen escaped.

For about 1000 Ma after this rise in O_2, the eukaryotes did not evolve much. Evidence from sulphur isotopes is consistent with anoxic oceans at all but shallow depths, which would have restricted the living space available, and thus retarded evolution.

From 800–600 Ma ago, sulphur isotopes (and other evidence) indicate a rise in O_2 from about 2000 Pa to about 18 000 Pa, not far short of the present value of 21000 Pa. This could have been caused by an increased rate of burial of organic carbon, itself perhaps a result of higher rates of deposition of sediments due to the continental break-up known to be occurring at that time. Another factor could have been the growth in biomass, e.g. with the increase in marine plants during this period. This, in turn, could have been the result of the ending of the 750–600 Ma ice age (Section 10.1.4) or the increased availability of nutrients, or both factors. At present the cause(s) of the rise in atmospheric O_2 at this time is very uncertain.

There is yet other geological and biological evidence. The totality of evidence leads to the inferred record of the partial pressure of atmospheric oxygen in Figure 10.15. There are large uncertainties, particularly in the pre-Cambrian, i.e. before about 570 Ma ago, where the data are more qualitative than quantitative. The onset of the Cambrian is marked by the emergence of

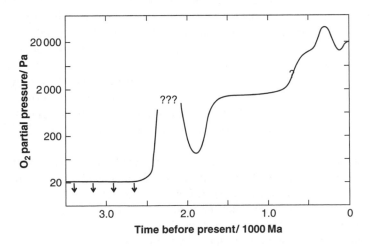

Figure 10.15 The build-up of atmospheric oxygen on Earth.

life with hard shells (e.g. trilobites) that have well-preserved fossils, and consequently provide good evidence for sea temperatures.

10.6.3 Mars

The Noachian

The big question concerning atmospheric evolution and climate change on Mars is posed by the abundant evidence outlined in Section 7.3.6 that liquid water was stable at the surface during the Noachian epoch, timed to end at the end of the heavy bombardment, about 3700 Ma ago at Mars. This requires temperatures over much of the surface to exceed 273 K, and pressures to exceed 610 Pa. The pressure requirement is not difficult to meet. The temperature requirement is a much bigger problem. At a time of about 75% the present solar luminosity this requires an enhanced greenhouse effect just as in the case of the Earth, but the requirements for Mars are more severe because of its greater distance from the Sun.

Various atmospheric models have attempted to produce sufficient warming with a CO_2–H_2O atmosphere, with different amounts of cloud, with or without a contribution to the greenhouse effect from scattering by CO_2 cloud particles. This scattering contribution is now thought to have been slight. Also, the most optimistic models of outgassing in these early times, consistent with the composition of Martian meteorites, give column masses no more than about $1.5 \times 10^4 \, kg \, m^{-2}$, which, though about 100 times the present column mass, no longer seems enough. But even if huge quantities of CO_2 and water vapour were present, there would still be gaps in the IR absorption spectrum, to the extent that a CO_2–H_2O atmosphere would never suffice.

☐ So, how could sufficient warming have occurred?

One way is if the powerful greenhouse gases NH_3 and CH_4 had been present. Each of these gases fills some of the IR absorption gaps. Much of them could have been derived from icy planetesimals. Sufficient impacts could have been sustained to the end of the Noachian. Also, enhanced volcanic activity could have provided significant amounts. What have we found?

NH_3 has not yet been detected, and the upper limit is a number fraction of only about 5 parts per billion. It would be rapidly destroyed on the surface today, so would need to be emitted at a prodigious rate to be discovered. We thus have no evidence that NH_3 was ever a significant component of the Martian atmosphere. The case of CH_4 is more encouraging.

CH_4 was first definitely detected on Mars in 1999, by Earth-based telescopes. Since then there have been other detections, notably by the Fourier IR spectrometer on Mars Express, which obtained a number fraction of 10 ± 5 parts per billion (and water vapour) above three equatorial regions that have subsurface ice. All this CH_4 would be destroyed by solar radiation in only about 600 years, and so there must be a source. Comets and meteorites are insufficient, so a subsurface source is needed, which could be geological, even biological! Whether there was enough atmospheric CH_4 in the Noachian for the greenhouse effect to raise the temperature above 273 K is unknown.

If not, there are other ways to achieve a warm, wet Noachian. One is through major impacts that vaporised subsurface ice, and delivered their own water content, to form clouds that rain out in about 10 years at a rate of 1–2 m per year. Longer lasting is surface warmth from hot ejecta that could have released subsurface water for hundreds of years. There are about 30 large impact craters with ages exceeding 3500 Ma – the 'smoking guns'. Another way is via major volcanic episodes that release water vapour from the magma. All in all, there are plausible models for a warm Noachian.

After the Noachian

But gradually the atmosphere became less massive. CO_2, a greenhouse gas, was removed through impact erosion, adsorption into the regolith with no plate tectonics to restore it, and the formation of sedimentary carbonates. Removal by carbonates would have been particularly rapid if there were large open bodies of liquid water and if there was no chemical inhibition. It should be noted, however, that very little carbonate has been detected on Mars (Section 7.3.6), perhaps because of high acidity in the water, though 'dry' formation of carbonates can occur if water vapour is present. Volcanic activity returned CO_2 to the atmosphere, but because of Mars's small size and consequent rapid interior cooling, volcanic activity declined steeply early in Mars's history. With CO_2 removal still occurring, its atmospheric mass declined. Any NH_3 and CH_4 content was eliminated as these gases were destroyed by solar UV radiation faster than the declining rate of supply by volcanic activity, The atmosphere cooled and so became dry, the water joining the surface and subsurface deposits. The clement conditions came to an end.

Since the end of the Noachian, impact erosion has been slight, but CO_2 has continued to be removed from the atmosphere by regolith adsorption and by dry carbonate formation.

What of the other volatiles? For water, one view is that since early in the Noachian the equivalent of a global depth of at most about 100 m of liquid water has been outgassed, and perhaps a lot less. A fraction of this water has been lost by photodissociation and the escape of hydrogen, though much of it has been protected in some surface or regolith reservoir. The fraction lost has been estimated from the Martian atmospheric isotope ratio $^2H/^1H$ (or $D/^1H$). This is about five times greater than that on the Earth, indicating that Mars might have lost more than half of its initial endowment of water, though interpretation of the Martian data is fraught with uncertainty.

Chemical escape has deprived Mars of far more N_2 than is presently in its atmosphere, and even if there ever were open bodies of water on Mars they were too short lived to trap much nitrogen (as nitrates), otherwise the observed atmospheric enrichment of ^{15}N over ^{14}N would not be observed today, as discussed in Section 10.4.

10.6.4   Life on Mars?

It seems probable that during the first 1000 Ma or so after the planet's formation, liquid water and the associated clement conditions existed on its surface even if only in discrete episodes. Life probably got going on Earth by about 50 Ma after the end of the heavy bombardment, so the question arises of whether life arose on Mars, and whether it survives today.

This latter question was put to the test in 1976 when Vikings 1 and 2 landed on Mars and made surface observations for several years. Though three of the experiments were designed to detect life, none of them gave unequivocally positive results. Nor was there any evidence of organisms in the surface images. At the other extreme, organic compounds (which in any case can have a non-biological origin) were found to be present in only very small quantities. Nevertheless it is just possible that in some warmer and perhaps wetter regions life clings on today. As to whether life once existed but is now extinct, many astrobiologists are optimistic, and therefore a search for Martian fossils is a high priority for future missions. One Martian meteorite, ALH84001, was once thought to display evidence of an early Martian biosphere, but most scientists now believe that the evidence is not there.

The discovery of atmospheric CH_4 in 1999 has raised the possibility that it has a biological source, analogous to methanogens of Earth. The subsequent detection of CH_4 in association with water above three regions known to by icy raises hopes still further. Moreover, traces of atmospheric H_2 have also been detected. Though this could come from the photodissociation of water, some at least could have an underground origin where it could provide an energy source for methanogens in their production of CH_4. Methanogens exist in the Earth's crust. Could they exist today on Mars, at depths where water is liquid?

Question 10.9

Suppose that the solar luminosity increased considerably. Describe the possible consequences for the atmospheres of the Earth and Mars.

10.7 Mercury and the Moon

Mercury and the Moon have extremely tenuous atmospheres, with column masses of the order of $10^{-10}\,kg\,m^{-2}$. On Mercury, O, H, and He were detected by Mariner 10, and Na, K, and Ca by observations from the Earth. Other substances are expected to be there, presumably below present instrumental limits. On the Moon H, He, and Ne dominate. The atmospheres are entirely exospheric. Figure 9.12 shows that thermal escape is not entirely to blame. As well as thermal escape, substantial proportions of such tenuous atmospheres are also lost through the UV ionisation of atoms and molecules, the ions then being carried off by the magnetic field in the solar wind. The solar wind is also a *source* of atmosphere, partly by supplying ions and atoms for capture, and partly by ejecting particles from the surface. UV radiation also ejects surface particles. Other sources are feeble outgassing, and the capture of volatile-rich bodies such as comets and CCs. In each case the atmospheric turnover is rapid.

The main source of the H and He on Mercury is almost certainly directly from the solar wind. For Na and K the surface must also be a source, the material being ejected by solar UV radiation and by micrometeorite impact vaporisation. If Mercury was initially well endowed with an atmosphere, and if this survived impact erosion, then Mercury's proximity to the Sun would have ensured that it developed a large greenhouse effect. Its low escape speed and the high UV radiation levels and intense solar winds so close to the Sun would soon have stripped it of all atmosphere and surface volatiles, leaving it with the tenuous atmosphere we see today. The surface of Mercury shows it to have been volcanically rather inactive, and this is because of its small size and rapid cooling. It might therefore be only partially outgassed, in which case juvenile volatiles might be appearing on its surface today – perhaps contributing to the ice deposits near the poles (Section 7.2.3). On the other hand, if much of Mercury's mantle was stripped by a giant impact (Section 5.1.3), the remaining mantle might be almost free of volatiles, in which case Mercury has never had an appreciable atmosphere. There might, however, still be a contribution from the interior to Mercury's atmosphere.

It is likely that the lunar interior has no atmosphere in waiting, and never has had.

☐   Why is this?

It is thought to have been created almost volatile free by the manner of its birth (Section 5.2.1). Any late veneer was not retained in the face of the various loss processes, because of the Moon's low gravitational field. The present tenuous atmosphere, from the mass spectrometer on Apollo 17, consists of H, He, and Ne, in roughly equal amounts, plus traces of other substances, including ^{40}Ar, which is the daughter product of ^{40}K, and thus probably comes from the lunar interior. Otherwise, the solar wind is the major source, supplemented by micrometeorite impacts that bring their own volatiles.

10.8 Icy–Rocky Body Atmospheres

The only icy–rocky bodies with more than negligible atmospheres are Titan, Triton, and Pluto. Among the icy–rocky bodies these have comparatively high escape speeds (Table 9.1). However, this cannot be the only factor. The Galilean satellites have comparable escape speeds yet they all have negligible or extremely tenuous atmospheres.

☐ What is another factor in determining atmospheric retention?

Temperature must also be considered. Titan, Triton, and Pluto have low surface temperatures (Table 9.1), a result of the great distance of these bodies from the Sun. Escape is therefore at a low rate. However, a surface can be *too* cold, resulting in very low vapour pressures. For example, in the cases of Europa, Ganymede, and Callisto, the water ice that is so abundant at their surfaces is too involatile to have a significant vapour pressure even at their comparatively high surface temperatures. Therefore a further factor is the types of volatile material that have been available.

10.8.1 Titan

Saturn's Titan, at 2575 km radius slightly larger than Mercury, has a remarkable atmosphere with a column mass 11 times that of the Earth. In the lower troposphere it consists of 95% N_2, with nearly 5% CH_4, and traces of many other substances. Photochemical hazes screen the surface from view at visible wavelengths and give an orange tint (Plate 17). The vertical structure of the atmosphere is shown in Figure 10.16 up to 350 km.

Vertical structure; heating, cooling, and circulation

At the surface of Titan the atmospheric pressure is 1.5×10^5 Pa. GMST is 94 K. Above the surface there is a troposphere in which the temperature declines with altitude, to the tropopause at about 42 km. Above this there is a stratosphere, where the temperature increases

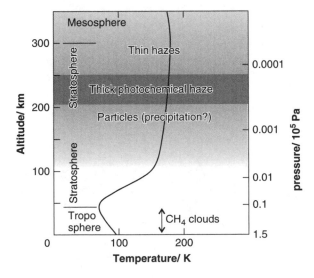

Figure 10.16 The vertical structure of Titan's atmosphere.

as altitude increases. The stratosphere continues through the thickest part of the haze layers. There is then a mesosphere extending well beyond the top of Figure 10.16, and then a thermosphere. A layered ionosphere peaks lies around 1180 km, and is produced, at least in its upper half, by galactic cosmic rays. There are layers of haze above the stratosphere (Plate 17). Changes in these layers since the Voyagers in 1980–1981 could be due to seasonal changes affecting either photochemistry or atmospheric circulation (the solar 'year' on Titan is 30 Earth years).

There is a greenhouse effect due mainly to CH_4, with a magnitude of about 21 K. This is partly offset by the hazes, because they are transparent at mid-IR wavelengths but absorb 90% of the incident solar radiation. Without hazes the surface temperature would be several degrees higher. Atmospheric circulation and the massive atmosphere ensure very little temperature variation across the globe. In spite of the solar 'day' on Titan being rather long, 16 days, the diurnal variation is only about 0.5 K. The main haze layer is very deep and it is this layer that completely screens the surface from us at visible wavelengths.

The surface temperature is close to methane's triple point temperature (90.4 K) and so at the surface CH_4 could exist as a solid, liquid, or gas, as shown by the methane phase diagram in Figure 10.17. If the CH_4 number density at the surface is sufficiently high, then the partial pressure of CH_4 will be on the saturation line in Figure 10.17. Any tendency for it to be higher causes condensation, to form CH_4 clouds, and precipitation. This would maintain the partial pressure close to the saturation value in regions of precipitation. It is the surface temperature that determines the mass of CH_4 in the atmosphere, just as in the case of water in the Earth's atmosphere.

The altitude range of the expected CH_4 clouds is shown in Figure 10.16, in the troposphere. Such clouds have been seen from the Cassini spacecraft. In 2005, when it was summer in the southern hemisphere, there were white, fluffy/billowy clouds in the south polar region,

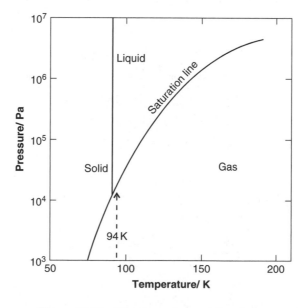

Figure 10.17   The phase diagram of methane.

which disappeared a few months later as autumn came on. There were more persistent clouds, wispy, cirrus-like, at 40°S. If these are related to atmospheric circulation they should move to about 10°N by about 2015, following the Sun. Alternatively, these clouds might be due to cryovolcanism at 40°S. In this case, the rate of release of CH_4 is comparable with that required to sustain the CH_4 content of the atmosphere against photochemical disruption. Elsewhere, small CH_4 clouds are often seen, of order 100 km across. Longer cloud streaks, with lengths of order 1000 km, are present at a few locations. They seem to originate from particular locations on the surface, either from where CH_4 is released, or from where the wind blows over local topography.

As noted above, CH_4 will precipitate from the clouds. This will fall as rain at the surface. A variety of other hydrocarbons presumably also precipitate onto the surface, conjuring a vision of a somewhat murky landscape of water ice and hydrocarbon ices, and hydrocarbon lakes, even oceans, though observations indicate that CH_4 lakes or oceans are not widespread and many are transitory (Section 8.4.2).

The clouds have also enabled winds to be measured, from which the circulation of the atmosphere can be deduced. Observations of cloud motions by Cassini have revealed that winds in the middle and lower troposphere flow west to east, at speeds up to $34\,m\,s^{-1}$, as predicted by circulation models. In the northern polar regions, when it was winter, the winds circulated around the pole, isolating it from the rest of the atmosphere. This happens on Earth at the South Pole, when it is winter there. Winds were also measured from the motion of the Huygens Lander as it descended through the atmosphere. The on-board wind sensor failed, so the motion was detected from the Earth by radiotelescopes that measured the Doppler shift of the radio signals from Huygens. There was a decrease in wind speed from about $120\,m\,s^{-1}$ at 60 km altitude, to about $30\,m\,s^{-1}$ at 50 km altitude, to zero at about 7 km. There was a reversal in direction in the lowest kilometre or so, to east to west. This reversal is consistent with the motion of the longer cloud streaks, also east to west, and with comparable speeds of a few metres per second. The speeds throughout the troposphere are not very different from those obtained from cloud motions.

Composition, sources, and sinks

It had long been known that above the thick photochemical haze the atmosphere consists largely of N_2 and CH_4 with mere traces of other substances. The gas chromatograph–mass spectrometer (GCMS) and aerosol collector and pyrolyser (ACP) on board the Huygens Lander showed that this is also the case below the haze. Below an altitude of about 5 km the GCMS gave 4.7% CH_4 by number fraction, which corresponds to a CH_4 relative humidity of about 50%. The isotope ratio $^{15}N/^{14}N$ shows enrichment in ^{15}N.

☐ What does this indicate?

It indicates that much of the N_2 has escaped. It is estimated that Titan once had 1.6–100 times the present mass in its atmosphere. The main cause of escape is through bombardment by energetic particles from Saturn's magnetosphere.

A rich variety of carbon compounds exists as traces in Titan's atmosphere, particularly hydrocarbons (additional to CH_4), HCN, and more complex molecules. These come from a series of chemical reactions that start with the dissociation of N_2 and CH_4 high in the atmosphere, caused by solar UV radiation and by energetic electrons from Saturn's magnetosphere. At the pressures and temperatures of the atmosphere many of these traces will condense to form solid or liquid particles, and it is these that constitute the haze layers in Figure 10.16. Typical particle sizes are $0.1-0.5\,\mu$m. In the particles the ACP found NH_3, HCN, and many other species. Not all of these are condensed directly from the atmosphere. For example, atmospheric NH_3 has not been detected.

Different complex chemistry occurs at the winter pole, where the various atmospheric constituents are in the dark, in a zone isolated by circumpolar winds. The atmosphere is sinking at the winter pole so a great variety of organic compounds is carried down to the surface, where layers must build up. The surface of Titan is thus greatly enriched by its atmosphere.

10.8.2　Triton and Pluto

Triton, the large, 1353 km radius satellite of Neptune, has a tenuous atmosphere consisting largely of N_2, with a few per cent of CH_4 and a trace of CO. Other common volatiles would have extremely low vapour pressures at the 38 K surface temperature measured by Voyager 2 in 1989. It also measured the surface pressure, 1.4 ± 0.1 Pa, corresponding to a column mass of $1.8\,\mathrm{kg\,m^{-2}}$. Voyager 2 imaged a haze layer about 3 km thick centred at an altitude of about 3 km, through which the surface was readily seen (Plate 21). N_2 ice covers much of the surface, and there are also ices of CH_4, CO, and CO_2. With N_2 dominating the surface and atmosphere, and with the expected equilibrium between these two phases, the atmospheric pressure is determined by the mean surface temperature. At 38 K the pressure of N_2 would indeed be around 1.4 Pa. Because of this equilibrium the surface temperature is the same everywhere to within about 1 K, through the sublimation of surface frosts on the day side, and their condensation on the night side, releasing latent heat. The 'day' on Triton is 5.88 days – if it were much longer then heat transport would not be able to equilibrate the temperature, and a really long 'day' would result in complete condensation of the atmosphere on the night side.

In November 1997 Triton occulted a star bright enough to allow further atmospheric investigations, this time from Earth. The atmospheric pressure was determined with low precision, but lay somewhere in the range 2.0–4.5 Pa, distinctly higher than in 1989. This increase might be due to the shift in the latitude of the subsolar point, which moved from 45.4 °S in 1989 to 49.6 °S in 1997, thus increasing the insolation on the south polar region. This could have increased the sublimation rate of surface frosts if they are non-uniformly distributed, or there could have been an increase in nitrogen geyser activity (Section 8.4.4). The subsolar point oscillates between about 50 °N and 50 °S with a period of about 688 years, the outcome of the configuration of the Neptune–Triton system and the orbital period of Neptune.

The temperature increase corresponding to this rise in pressure, assuming equilibrium, is a modest 1–2 K. This is because pressure is very sensitive to temperature (as can be seen for methane in Figure 10.17).

Evidence of winds on Triton was obtained by Voyager 2, which observed the plumes from the active geysers being swept into streaks. Note that the geysers, predominantly emitting N_2, could sustain the atmosphere against losses to space.

The high orbital inclination of Triton gives rise to huge seasonal changes over the 164 year orbital period of Neptune and so there is presumably a component of atmospheric circulation akin to the condensation flow on Mars.

Pluto also has a surface dominated by N_2 ice, and in addition there are ices of CH_4 and a trace of CO ice (Section 7.4.1). A stellar occultation in 1988 revealed a very thin, slightly hazy atmosphere consisting mainly of N_2 with some CH_4, and presumably some CO. The surface pressure was about 5 Pa. A surface temperature somewhere near 40 K indicates equilibrium between surface and atmospheric nitrogen, as on Triton. For the same reason as on Triton, the temperature might well be fairly uniform across the surface. Another stellar occultation in 1994 showed a near tripling of atmospheric pressure, though the corresponding rise in temperature is only about 2 K, as for Triton (again assuming equilibrium). There are several possible causes of this increase: first, the slight decrease in albedo of Pluto observed in recent years; second, a frost layer near the North Pole, where it is now spring, might be subliming more rapidly; third, thermal inertia in the surface ices, 1994 being not long after Pluto's 1989 perihelion.

Pluto has an eccentric orbit that carries it within the orbit of Neptune for 20 years. Since 1999 it has been further from us than Neptune. At aphelion, in 2113, it will be about 1.7 times further from the Sun than at perihelion, with a correspondingly far colder surface and an atmosphere far more tenuous than it is now. As Pluto goes around its orbit the column mass is estimated to vary over a considerable range.

10.8.3 The Origin and Evolution of the Atmospheres of Icy–Rocky Bodies

Titan

When Voyager 1 in 1980 provided the first direct measurement of the composition of Titan's atmosphere, the predominance of nitrogen caused some surprise, though there were theoretical indications that nitrogen would be present.

There are two possible sources of the copious quantity of N_2 in Titan's atmosphere. First, when Titan formed in the outer solar nebula, the temperature was so low that N_2 molecules were trapped within the water ice that constituted a large fraction of the icy–rocky planetesimals that formed Titan.

☐ What kind of substance is this water-ice–N_2 combination?

This is a clathrate. Much of this N_2 was released during accretion. The subsequent slight heating of Titan, tidal or radiogenic, has since released more.

Second, in the cold outer solar nebula, NH_3 is expected to be a significant constituent of the icy planetesimals. At the low surface temperature of Titan, NH_3 has a very low vapour pressure, which is why, if it is still present in the ices, it has not been detected in the atmosphere. However, via a series of chemical and photochemical reactions, atmospheric NH_3 can be converted into N_2 and H_2. The H_2 will suffer thermal escape, which explains the small quantity in the atmosphere, but the nitrogen is retained. Operating over the 4600 Ma history of Titan, this can account for some of the nitrogen. It can only account for most or all of it if the surface of Titan was 50 K or so warmer in the past. This condition is necessary to prevent the condensation of an intermediate product in the reaction sequence, namely hydrazine (HNNH). If hydrazine condenses then it cannot undergo a crucial photochemical reaction and the sequence halts. It is plausible that the surface was sufficiently warm to prevent condensation, through greater tidal or radiogenic heating in the past. Isotope data favour this second source as the major one, but the matter is far from settled.

The origin of CH_4 is better understood. It is readily photodissociated in the atmosphere, with escape of hydrogen and some carbon. However, there is no enrichment of ^{13}C over ^{12}C so there needs to be an active source of CH_4. Argon and carbon isotopes indicate that it comes from geological activity, and not, for example, from biospheric activity, which would enrich ^{12}C over ^{13}C in the CH_4, which is not the case. There is evidence from certain CH_4 clouds (see below), and from surface features that might be cryovolcanoes, that cryovolcanism could sustain the atmospheric CH_4. This could be driven by tidal heating. As well as gaseous CH_4, there would be 'lava' of liquid water, liquid NH_3, and other icy materials.

The whole family of icy–rocky body atmospheres

Among Pluto and the large satellites of the outer Solar System you have seen that it is only Titan that has a massive atmosphere. Triton and Pluto have tenuous atmospheres, and the remaining large satellites have negligible atmospheres. The explanation starts with the sort of ices that could have condensed in the outer Solar System during its formation, and in the sort of gases that the ices could have trapped. At Jupiter's distance from the Sun the nebular temperature was so high that the icy planetesimals contained water ice as the only significant volatile, with only small quantities of carbon-rich and nitrogen-rich ices. Little gas would have been trapped. The present surface temperatures of Europa, Ganymede, and Callisto are not sufficiently high to generate a significant atmosphere from the relatively involatile water ice. Io probably formed too close to Jupiter to have ever had much water ice, but if it did, tidal heating would have driven it off.

From Saturn outwards the more volatile ices of CH_4 and NH_3 could condense in the planetesimals, and at the lower temperatures, trapping of significant quantities of gases like N_2 and CO was possible. Titan is sufficiently warm for a massive atmosphere to have been derived from these ices. Further out, Triton and Pluto are too cold for such a massive atmosphere – the N_2 and CH_4 are largely condensed on the surface, the less volatile substances even more so. Only small quantities of inert gases would have been trapped in the planetesimals, so it is unsurprising that these have not yet been detected.

☐ At the low nebular pressures around where Pluto formed, in what gases would most of the nitrogen and carbon occur?

In Section 7.4.1 you learned that most nitrogen would be in N_2 rather than NH_3, and most carbon would be in CO rather than CH_4 or CO_2. This is consistent with the atmospheric and surface compositions of Pluto (and also of Triton, which might have been captured from the Pluto region (Section 2.3.1)).

In Section 10.5.2 you were reminded that planetesimals were not confined to the zone in which they formed, which raises the question of why the Galilean satellites did not acquire atmospheres similar to that of Titan, as a late veneer. A possible explanation is the high speeds acquired by planetesimals as they fall towards the Sun, made even higher as they subsequently fall towards Jupiter. The impacts would then be too violent for volatiles to be retained. Moreover, if the Galileans already had atmospheres, this is one way in which they could have been removed – by impact erosion.

Question 10.10

Suppose that Titan and Triton swapped places today! Discuss whether the atmosphere of Titan would become like that of Triton, and vice versa.

10.9 Summary of Chapter 10

Tables 9.1 and 9.2 and Figure 10.12 present the basic properties of the atmospheres discussed in this chapter.

The Earth's atmosphere today differs from that of Mars and Venus in that it consists largely of O_2 and N_2 rather than CO_2. This is largely a consequence of the action of the oceans and the biosphere. The abundance of O_2 has led to the unique feature of a stratospheric 'bulge' in temperature, caused by the absorption of solar UV radiation by ozone (O_3), derived from O_2.

The greenhouse effect is small on Mars, because of its thin, dry atmosphere. It is larger on Earth, because of the greater quantity of atmospheric water vapour, supplemented by CO_2 (and traces of other greenhouse gases). On Venus the greenhouse effect is very large, sustained by a huge mass of CO_2, with the assistance of far smaller quantities of water vapour, sulphur dioxide, and the sulphuric acid droplets that constitute the planet-wide cloud.

The circulation of the Earth's atmosphere is dominated in the tropics by one Hadley cell per hemisphere, by planetary waves at mid latitudes, and by a cell in each polar region. The Coriolis effect limits the extension of the Hadley cells, and also deflects the winds. On slower rotating Venus there is just one Hadley cell per hemisphere in the lower troposphere, stretching from equator to pole. Mars, except near the equinoxes, when its circulation resembles Earth's, has a single Hadley cell stretching from the subsolar latitude to the winter hemisphere. There is also condensation flow, and a thermal tide stronger than that on the Earth and Venus.

The present volatile inventories of the Earth, Mars, and Venus show substantial differences. It is not possible to deduce their initial volatile inventories by adjusting the present inventories using loss processes that are still operating. This is because the effects of blow-off, impact erosion, and hydrodynamic escape, all of which are thought to have occurred early in Solar System history, are very uncertain and yet caused huge losses. A consistent picture is one in which there is rapid loss of volatiles from embryos and planetesimals during planet formation, followed by outgassing due to core formation. Much of these early atmospheres were then lost, and volatiles subsequently reached the surfaces through further outgassing and through collisions with a variety of volatile-rich planetesimals.

The three terrestrial atmospheres have evolved quite differently from each other largely because of their different distances from the Sun. On Venus the volatiles that have been retained are largely in the correspondingly massive atmosphere, except for water, most of which has been photodissociated, with loss of hydrogen to space and incorporation of oxygen into rocks. On the Earth, the oceans and the biosphere have 'locked' most of the carbon into carbonates, and about half the nitrogen into nitrates. Photosynthesis has resulted in a build-up of atmospheric oxygen over the last 2000 Ma. Mars probably has much of its volatiles in a variety of near-surface deposits, because of its greater distance from the Sun. Martian volatiles have also been lost to space, through thermal and chemical escape.

The increase in solar luminosity, and the greenhouse effect as determined by evolving atmospheric compositions, have been of enormous importance in the evolution of the volatile inventories on Venus, the Earth, and Mars, and consequently in their long-term climate changes too. In the case of the Earth there is much concern over climate change that seems to be a result of human activities, particularly those that increase the greenhouse effect.

The smaller terrestrial bodies – Mercury and the Moon – are devoid of significant atmospheres because of their low escape speeds and proximity to the Sun, and in the case of the Moon, the low volatile content of the material from which it formed.

Titan, Triton, and Pluto have atmospheres determined by the types of icy materials available to them when they formed, and their surface temperatures. At Titan, ices of CH_4 and NH_3 are expected in addition to the water ice that can also condense closer to the Sun. The water ice could also contain N_2, as a clathrate. Titan's atmosphere is dominated by N_2 with a few per cent of CH_4, but there is uncertainty about whether the nitrogen was delivered predominantly as N_2 or as NH_3. At Triton and Pluto, N_2 and CO would have been the dominant repositories of nitrogen and carbon, and this is reflected in their atmospheric and surface compositions. The atmosphere of Titan is more massive than that of Triton and Pluto because it is closer to the Sun and hence warmer.

Other large icy–rocky bodies, and Io, lack significant atmospheres because of the lack of volatile materials that would be gaseous at their surface temperatures. Small bodies have insufficient gravity to retain atmospheres.

11 Atmospheres of the Giant Planets

With Jupiter, Saturn, Uranus, and Neptune we come to four planets where, as Table 9.1 shows, the atmospheres are dominated by molecular hydrogen (H_2) and atomic helium (He), and not by water and substances rich in carbon and nitrogen. Moreover, for these giant planets the distinction between the atmosphere and the interior is blurred. This is apparent from the interior models discussed in Section 5.3. In Jupiter and Saturn the atmospheres blend seamlessly into the molecular hydrogen envelope, with no surface separating them. In Uranus and Neptune the atmospheres acquire high densities, though not terrestrial liquid densities, by the time we reach a fairly sharp transition to a liquid mantle consisting of icy and rocky materials.

Because of cloud formation, but additionally because of the huge depths of the atmospheres, variations of composition with depth are to be expected. In Table 9.1 the compositions of the giant planet atmospheres are given for the atmosphere above that at which the pressure is a few times 10^5 Pa, which is well into the troposphere. These are number fractions. In the case of the hydrogen-dominated giants it is also common to specify the composition as the number of molecules of each constituent in a volume divided by the number of hydrogen molecules (H_2) in the same volume. This is the **mixing ratio** of the constituent, in this case with respect to H_2. Table 11.1 gives the mixing ratios corresponding to the values in Table 9.1 for He and CH_4, plus values for NH_3 and H_2O. Several trace constituents have not been listed.

These values have been obtained by a variety of techniques, including visible reflection, and emission at IR and micro-wavelengths. The emission includes radiation from beneath cloud and haze that obscures visible radiation. After early visits by Pioneer spacecraft in the 1970s (Table 4.1), Jupiter and Saturn were visited by the Voyagers, and Uranus and Neptune by Voyager 2. Since then, Jupiter has been orbited by Galileo, which also delivered a probe into the atmosphere in December 1995. Cassini flew past Jupiter around New Year 2001, and since July 2004 it has been orbiting Saturn. From Earth, large optical and radiotelescopes have been used.

Though there are similarities between the four atmospheres, there are also differences, and consequently the atmospheres can be grouped into two pairs, Jupiter and Saturn forming one pair, Uranus and Neptune the other. The atmospheres within each pair resemble each other more closely than they resemble the atmospheres in the other pair. Even so, Saturn is not Jupiter's twin, and Neptune is not Uranus's twin, as you will see.

Discovering the Solar System, Second Edition   Barrie W. Jones
© 2007 John Wiley & Sons, Ltd

Table 11.1 The atmospheric composition of the giant planets, given as mixing ratios with respect to H_2

Species[a]	Jupiter	Saturn	Uranus	Neptune
H_2	1	1	1	1
He	0.156	0.13	0.18	0.18
CH_4	2.1×10^{-3}	5.1×10^{-3}	0.019	0.027
NH_3[b]	7.1×10^{-4}	$< 7.7 \times 10^{-5}$	—	—
H_2O[b]	$< 6 \times 10^{-4}$	—	—	—

[a] Only abundant species are shown, as measured in the upper molecular hydrogen envelopes at depths where the pressures are a few times 10^5Pa.

[b] A dash indicates that no useful upper limit is available.

11.1 The Atmospheres of Jupiter and Saturn Today

Plates 11 and 16 show the highly coloured, richly structured cloud tops of Jupiter and Saturn, at once strikingly different from the planetary bodies that we have so far considered. The patterns are more distinct on Jupiter partly because the overlying haze there is thinner (because of the higher temperatures), and partly because on Saturn the lower gravity has caused the upper cloud deck to be spread over a greater range of altitudes, and it is consequently less sharply defined.

11.1.1 Vertical Structure

Figure 11.1 shows the vertical structure of the atmospheres of Jupiter and Saturn, obtained from many observations. For both planets the lapse rate up to the tropopause is close to the adiabatic value for the mix of gases present, and so the atmosphere is probably convective. The energy source is mainly heat welling up from the interior, supplemented among the clouds by the absorption of solar radiation. In the mesospheres and thermospheres there is no convection, and the lapse rate is determined by radiative transfer between the different altitudes, by radiation to space, and by the absorption of solar radiation. The generally lower temperatures for Saturn (at a given pressure) are due to the greater distance of Saturn from the Sun and the lower heat flux from its interior. The Jovian exosphere is particularly hot, perhaps because of atmospheric waves travelling upwards, or bombardment by magnetospheric electrons, or a reduction in methane content – methane promotes radiative cooling.

For each planet, three layers of cloud are shown in Figure 11.1, plus some layers of haze. In each case the uppermost cloud layer consists of NH_3 ice particles that form cirrus-like sheets. This layer is readily observable, but it is not unbroken. On Jupiter it has bands where it is very thin, called belts, which appear dark at visible wavelengths, in contrast to the brighter bands where it is thick, forming zones – see Plate 11. In polar regions the band structure is absent, and the NH_3 clouds are patchy. Plate 16 shows a comparable belt structure on Saturn. At lower altitudes on both planets there is thought to be a layer of ammonium hydrosulphide (NH_4SH) cloud, perhaps mixed with ammonium hydroxide (NH_4OH). The NH_4SH is formed from NH_3 and H_2S – the latter has been detected as a trace in the atmosphere. Lower still a cloud layer of water is thought to exist. In all three layers the cloud particles would be solid.

Cloud formation was outlined in Section 9.2.3.

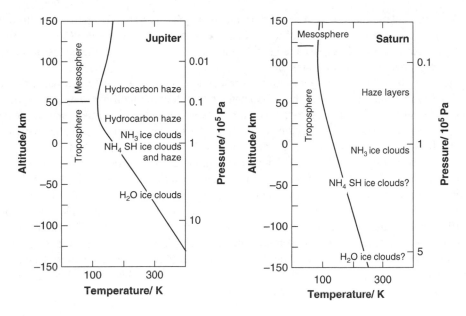

Figure 11.1 The vertical structure of the atmospheres of Jupiter and Saturn.

☐ Try to recall the essential features of the process.

As altitude increases the partial pressure of each atmospheric gas decreases, but in the troposphere the temperature decreases too. If, for any constituent, the partial pressure and temperature meet the saturation line in the phase diagram of the constituent then the constituent can condense, either as ice crystals or as liquid droplets, depending on the temperature. The lowest altitude at which this occurs will be at the base of the clouds. The higher the partial pressure of a constituent in the atmosphere below the cloud base, the lower the altitude of the cloud base will be.

Jupiter

Much of the general picture in Figure 11.1 was for a long time based more on modelling than on measurement. Though the earlier spacecraft provided valuable data, it was IR measurements from the Galileo Orbiter that played a large part in confirming the general picture, and in adding detail, such as the altitudes of the various deeper cloud layers. It was the orbiter that first confirmed that Jupiter's upper cloud layer consisted of NH_3 ice particles.

The Galileo probe returned data from an altitude range of 150 km, over which the atmospheric pressure increased from 0.4×10^5 Pa to 22×10^5 Pa. It thus held the promise of the direct sampling of cloud particles at accurately measured altitudes. But it went through an almost cloud-free hole, called a hot spot. At hot spots the Jovian atmosphere is rather clear, allowing IR radiation to escape from warm lower levels. In spite of this bad luck, the probe detected a patchy cloud layer with a base at about 0.6×10^5 Pa and a very tenuous layer with a base at about 1.5×10^5 Pa. An increase in opacity below 9×10^5 Pa might have been the thin upper reaches of a third cloud layer. The composition of the particles in these layers was not measured by the probe, but the highest level cloud is presumed to be the NH_3 cloud. The very tenuous layer could be the NH_4SH cloud – the probe's mass spectrometer detected sufficient NH_3 and sulphur compounds in the gas in this region to make NH_4SH particles. If there was a third layer then it might have been H_2O. The Galileo Orbiter, with its solid state imager (SSI) and near-IR mapping spectrometer (NIMS), detected patchy clouds at 4×10^5 Pa near the edges of hot spots. These were probably clouds of water ice.

Figure 11.1 also shows haze layers. In the mesosphere there is a thin brown haze, the result of solar UV radiation and aurorae that dissociates CH_4. Recombination produces hydrocarbon molecules larger than CH_4, giving particles with an overall brown tint. In the upper troposphere there is a thicker brown haze, possibly consisting of particles settling from the mesosphere, mixed with particles from below. The haze reaches the NH_3 clouds, so NH_3 particles could readily be delivered upwards by convection. There might also be phosphine (PH_3) particles in this lower haze. The cloud is so thin in the belts and in dark streaks that the brown haze has been observed in near-IR and red wavelengths down to a thin cloud layer at about 0.9×10^5 Pa, presumably NH_4SH, below which there is a visually dark, and therefore fairly clear, atmosphere.

In 1994, before the Galileo spacecraft reached Jupiter, the fragmented Comet Shoemaker–Levy 9 plunged into the Jovian atmosphere. Unfortunately, less has been learned than had been hoped, in large part because it is difficult to distinguish between comet material and Jovian material. However, gases were observed leaving Jupiter at greater than the escape speed, so the impact did teach us something about impact erosion.

Saturn

Until Cassini arrived in 2004, Saturn was not nearly as well observed as Jupiter. Therefore, the compositions of the clouds and the altitudes of the cloud bases had been inferred from the measured temperatures and pressures at each altitude, and from atmospheric composition. With Cassini's arrival the situation improved, but there are still large gaps in our data.

As for Jupiter, the uppermost cloud consists of NH_3 ice particles. It is readily observable, though the overlying haze is deeper than on Jupiter and so features at the cloud surface are less distinct. Cassini's visual and IR mapping spectrometer (VIMS), observing at about $5\,\mu$m, where the obscuration by higher level hazes and clouds is largely overcome, also saw clouds deeper down, about 30 km beneath the NH_3 clouds. At this wavelength they were seen in silhouette against the IR radiation from Saturn's hot interior. Many of the clouds were isolated, and took a variety of forms. The composition of these clouds was not measurable, but it is presumed to be either NH_4SH or H_2O.

Cassini also saw dozens of planet-girdling lanes of clouds at comparable depths. These are not the bands seen at higher level in the NH_3 clouds (Plate 16), but deeper, narrower, and more numerous. Many of them connect with what appear to be convective cells.

11.1.2 Composition

Hydrogen and helium

Table 11.1 gives the mixing ratios at depths down to a few times 10^5 Pa, which is well into the homospheres. H_2 is readily detected, but even though He is a substantial constituent it has not been easy to establish its quantity. This is because its strong spectral signatures are at UV wavelengths. The Earth's atmosphere is opaque at such wavelengths, and in any case the UV signatures of He are obscured by the UV signatures of H_2. Spacecraft have detected He spectrally in the atmospheres of Jupiter and Saturn, but only above the homopause, where the mixing ratio is not the same as in the homosphere. Then, in 1995, the He mixing ratio for

Jupiter was obtained by direct sampling of the homosphere by the Galileo probe, and it is this value that is given in Table 11.1. Up until then, the quantities of He in the homosphere had been obtained for Jupiter only by indirect techniques. This is still the case for Saturn and the other giants.

In one indirect technique the total pressure of the atmosphere is obtained from the collisional broadening (Section 9.1) of the IR spectral lines of H_2. It is assumed that, except for H_2, the atmosphere contains little else except atomic helium (He). Thus, the partial pressure of He is assumed to be the difference between the measured total pressure and the measured partial pressure contributed by H_2 (and any other measured substances, notably CH_4). In another indirect technique, data from radio occultation have been combined with IR measurements to obtain the mean molecular mass of the atmosphere. On the assumption that the atmosphere is dominated by H_2 and He, the mixing ratio of He is then calculated. Because both techniques use IR and radio wavelengths at which the giant planet atmospheres are relatively transparent, the He content has been determined to some way below the uppermost cloud layer, to where the pressure is several times 10^5 Pa.

H_2 and He do not condense in the giant atmospheres, nor are they significantly depleted by chemical reactions. Therefore the values in Table 11.1 apply to great depths. In the case of Jupiter and Saturn the values probably apply throughout the molecular hydrogen (H_2) envelope down to the metallic hydrogen interface, deep in the interior. This is because the envelope is certainly fully convective. Models indicate, however, that the He mixing ratios in Table 11.1 are lower now than in the past because of the downward settling of He in the *metallic* hydrogen mantle – this depletes the upper mantle in He which in turn depletes the envelope. Models of the evolution of Jupiter and Saturn outlined in Section 5.3.1 suggest that the depletion should be greater in Saturn.

☐ Do the values in Table 11.1 bear this out?

If Saturn and Jupiter started out with similar mixing ratios of helium, then today we would expect the mixing ratio in the atmosphere of Saturn to be less, and this is exactly what we find, as Table 11.1 shows. Table 11.2 gives the present-day He content of the molecular envelopes and metallic hydrogen mantles, as mass fractions. (These values were quoted in Section 5.3.1). The envelope values are those for the observable troposphere, and the mantle fractions are from models.

Other substances

With the other atmospheric constituents we have to be aware of the possibility of strong compositional variations with altitude. That the values in Table 11.1 apply at a level well into the homosphere is no guarantee against this.

Table 11.2 Mass fractions of helium in Jupiter and Saturn

Mass fraction	Jupiter envelope	Jupiter mantle	Saturn envelope	Saturn mantle
Helium (Y)	0.238	~ 0.27	~ 0.20	~ 0.30

The division between the envelope and mantle are for the present time.

☐   Why not?

In a homosphere, by definition, there is sufficient mixing to prevent *gases* from separating. This does not rule out separation through condensation. Condensation will lead to the formation of cloud and haze. In the troposphere convection will carry the cloud particles above the cloud base, but this will only partially offset the depletion. Moreover the *gas* phase will still be heavily depleted, and it is the gas phase that many compositional detection methods sense. Another cause of altitude variation is chemical reactions within the atmosphere. Above the clouds it is photochemical reactions driven by solar UV radiation that are important, giving rise to layers of photochemical smogs consisting largely of hydrocarbons that are much scarcer lower down. The ozone layer in the Earth's atmosphere is another example of a photochemical product.

Table 11.1 shows that in the upper molecular hydrogen envelopes the next most abundant atmospheric constituents after H_2 and He are (in order) CH_4, NH_3, H_2O. To what extent do the mixing ratios of these compounds represent the whole molecular hydrogen envelope?

CH_4, the main repository of carbon, does not condense anywhere in the atmospheres of Jupiter and Saturn, nor is it significantly depleted by photochemical reactions. There is, however, a significant chemical conversion very deep in the molecular hydrogen envelope, where the temperatures exceed about 1200 K. Models show that a proportion of the CH_4 is converted to CO via the reaction

$$CH_4 + H_2O \rightarrow CO + 3H_2 \tag{11.1}$$

Evidence that this occurs is the detected trace of CO, presumably brought up by convection. But even if a significant fraction of CH_4 has been converted to CO at these depths, the C/H_2 ratio for the whole envelope must be almost exactly equal to the measured CH_4 fraction in its upper reaches. This is partly because it makes no difference whether the C is in a CH_4 molecule or in a CO molecule, and partly because the quantity of H_2 produced by reaction (11.1) is very small. What *would* make a difference would be the production deep in the envelope of significant quantities of a *condensable* carbon compound that was then confined there. There is no evidence, theoretical or observational, that this has happened.

The dominant repository of nitrogen in a hydrogen-rich atmosphere at the pressures and temperatures of the observed homospheres is NH_3, and the altitude variation in its mixing ratio has been followed down to several times 10^5 Pa. In Figure 11.1 you can see that in Jupiter such pressures occur beneath the NH_3 and NH_4SH clouds, and it is this deep measurement that is given in Table 11.1. Therefore, it is thought that the NH_3/H_2 fraction in Table 11.1 is typical of the molecular hydrogen envelope of Jupiter. The position with Saturn is more marginal – the measurements might not reach the supposed NH_4SH cloud level.

For each planet the variation of the NH_3 mixing ratio with depth provides additional information on the composition of the cloud layers. The ratio increases by several orders of magnitude as we move down through the uppermost cloud deck, lending support to the view that this deck consists of NH_3 crystals. For Jupiter there is also a clear increase in the NH_3 fraction as we traverse lower altitudes, and this is in accord with the Galileo Orbiter findings that a second cloud deck exists, and that it consists of particles of NH_4SH, presumably with some NH_4OH. Chemical models predict the formation of such substances in the giant atmospheres by the reactions

$$NH_3 + H_2S \rightarrow NH_4SH \tag{11.2}$$

$$NH_3 + H_2O \rightarrow NH_4OH \tag{11.3}$$

the amounts of the products depending on the partial pressures of the reactants and on the temperature. The models predict that NH_4SH is more abundant than NH_4OH and that at altitudes above about 1.5×10^5 Pa it is an important repository of sulphur. However, H_2S is the main repository of sulphur in the *gas phase* in a hydrogen-rich atmosphere – NH_4SH condenses. If NH_3 was initially much more abundant than H_2S then reaction (11.2) is a plausible explanation of the very small quantities of H_2S found in the Jovian atmosphere above this cloud level.

Before the Galileo probe measured the composition of the Jovian atmosphere, H_2O (the oxygen repository) had been detected only above the presumed water cloud deck. Unsurprisingly the mixing ratios are very small. The value in Table 11.1 is from the probe at a depth of 19×10^5 Pa, at which depth the value was still increasing.

Neither H_2O nor H_2S has been detected in Saturn's atmosphere. Both are presumed to be present, but would be condensed at levels well below those so far observed.

In Section 11.3 we will compare these compositions with the values thought to have been present in the solar nebula, to see what more we can learn about the origin of the giant planets.

Question 11.1

(a) From Table 9.1, use the number fractions for Jupiter to calculate the average mixing ratios in the atmosphere. Compare your values with those in Table 11.1. Comment briefly on the outcome.

(b) If, in Jupiter, measurements were to be made at much greater depths than a few times 10^5 Pa, would the mixing ratios differ from those in Table 11.1? Justify your conclusions (in a few sentences).

11.1.3 Circulation

The atmospheric circulation of the two planets is revealed by the cloud patterns (Plates 11 and 16) and their motions. Though the details of the patterns change on a time scale of days, the largest scale features have changed little in their broad appearance over the three centuries or so of observations. This indicates a stable system of atmospheric circulation.

Belts and zones

The most prominent, most widespread large-scale feature is the banding parallel to the equator. The bright bands are called zones and the dark ones are called belts. Figure 11.2 illustrates a widely accepted model of what distinguishes belts from zones, based on temperature measurements and on vertical cloud motions.

The zones are marked by high, cool clouds freshly formed near the tops of convective columns, and therefore consisting of clean crystals (presumably of NH_3). The belts are where the atmosphere is sinking to complete the convective cycle. This sinking causes warming. The belts are freer of cloud because the condensates have been largely frozen out as cloud particles in the zones, from where a substantial fraction of the particles precipitate (presumably), rather than get carried into the belts. The sparseness of cloud in the belts exposes deeper lying, warmer regions that are more richly coloured, and are darker at visible wavelengths because they are poorer reflectors of sunlight. This convective model of the banding also explains, for example, the presence at cloud-top altitudes of phosphine (PH_3). Though this is present only as a trace

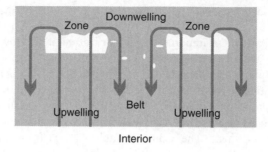

Figure 11.2 An explanation of the belts and zones of Jupiter and Saturn.

gas, its abundance is far above the quantities that should exist in chemical equilibrium at the low temperatures at these altitudes.

❒ What does this indicate?

It indicates that PH_3 has been borne aloft from deep below where the higher temperatures have created it.

On Saturn the correlation between zones and ascending atmosphere and belts and descending atmosphere is weak, and so the convective model in Figure 11.2 might not apply. Cassini has provided evidence that supports this conclusion. Images with a superb 58 km resolution at Saturn show fast-rising, towering plumes of white clouds in the *belts*. This indicates that the atmosphere is rising there, not sinking. The plumes are thought to originate at about the 5×10^5 Pa level, where water condensation (to form clouds) releases latent heat that drives convection. It might be that such upwelling is also occurring in the zones, but is obscured by the clouds. Of course, the atmosphere cannot be rising everywhere! In the belts it could be a local phenomenon, with other regions in the belts sinking. This problem is unresolved.

Belts, zones, winds, and deep circulation

For both planets we still have to account for the zones and belts forming bands rather than some other geometrical form, and for the orientation of these bands parallel to the equator. Data are provided by the winds, as revealed by the motion of the observable clouds, mainly the NH_3 clouds. The motions are very predominantly parallel to the equator with speeds typically as in Figure 11.3, where positive values are west to east (the direction of planetary rotation), and negative are east to west. The flow can be thought of as a set of parallel jets. The speeds are with respect to the rotation of the deep interior, which is deduced from the periodicity of kilometric radio emissions from charged particles orbiting under the influence of the planets' magnetic fields. The rotation period of the deep interior of Jupiter is secure at close to 0.4135 days. For Saturn the value from the Voyagers is 0.4440 days. Cassini's magnetometer, however, in May 2006 detected a small periodic variation of 0.4484 days. This difference from the earlier Voyagers' value is unexplained.

The wind speeds for Jupiter in Figure 11.3 are from Cassini during its flyby in 2000. The wind profiles have hardly changed in the 21 years since Voyager 2 in 1979. You can see that the same is *not* true for Saturn at equatorial latitudes. Two results from Cassini show substantially lower speeds. The three sets of measurements were obtained at three different wavelengths, which sense different depths. Detailed analysis has led to two conclusions about Saturn. First, there is an increase of wind speed as depth into the atmosphere increases. Second, the altitude

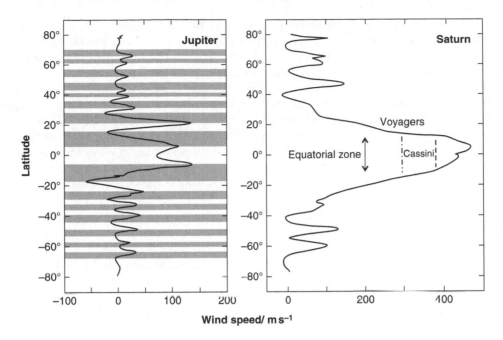

Figure 11.3 Winds on Jupiter and Saturn. Positive values are winds from west to east.

of the visible clouds has increased since 1979, to where the wind speeds are lower. It is also possible that seasonal changes are a factor, including the motion of the shadow of the rings from one hemisphere to another.

In Jupiter the wind speeds also increase with depth. As well as observations of clouds, this has been observed directly by the Galileo probe, which entered the Jovian atmosphere at $7.4\,°N$, near the northern edge of the equatorial zone. It obtained $80–100\,m\,s^{-1}$ at the top of the NH_3 cloud, increasing to about $170\,m\,s^{-1}$ some $35\,km$ deeper, and then remaining at about this value for as far as the probe reached, about $100\,km$ below the cloud top.

Figure 11.3 shows that on Jupiter there is a strong correlation between the wind profiles and the (idealised) bands. In the zones, the wind speed increases as latitude increases, and in the belts it decreases, with speed maxima and minima generally coinciding with the boundaries between belts and zones. In the case of Saturn the strong equatorial jet correlates with the equatorial zone, but elsewhere the bands correlate rather better with the mean wind speed in the band.

Presumably, the convection circulation between belts and zones, plus the strong Coriolis effect on these rapidly rotating bodies, plays a role in establishing the predominantly west–east flows in the observable atmosphere. Nevertheless, the winds are not fully understood, and in particular whether they are a shallow phenomenon, where solar radiation plays a major role, or deep seated where internal energy sources dominate. One clue is provided by the Galileo probe measurements of wind speeds. It was noted above that these remained at about $170\,m\,s^{-1}$ from $35\,km$ below the top of the NH_3 clouds to at least about $100\,km$. At such a depth, over most of Jupiter, little solar radiation penetrates. If the winds are powered only by the absorption of solar radiation, the speeds might be expected to have decreased considerably as the probe sank. That they did not indicates that the outward flow of internal energy might be the main power source of these winds. Only in the upper troposphere could solar radiation play a significant role.

Further insight into the nature of the circulation in Jupiter and Saturn is provided by the temperatures in the upper tropospheres – these vary only by a few kelvin with latitude. The decrease of solar irradiation with latitude on Jupiter tends to produce a temperature decrease from equator to pole. On Saturn, with its greater axial inclination (26.7° versus 3.1°) and ring shadowing, the summer pole can be *warmer* than the equator, but again only by a few kelvin. That the measured differences are so slight must be the result of equator-to-pole circulation, plus heat flow from the interior which is expected to be independent of latitude. Table 9.2 shows that about 1.7 times as much energy is radiated by Jupiter and Saturn as is absorbed from the Sun. Therefore, the internal source is comparable with the solar source, and is surely important, even essential, in reducing latitude variations of temperature. Loss of internal heat sustains convection throughout the molecular envelopes, which must have adiabatic temperature gradients.

These considerations support the circulation model illustrated schematically in Figure 11.4. There are concentric cylinders of fluid, extending throughout the molecular hydrogen envelope, and with axes coincident with the rotation axis of the planet. Models of rapidly rotating fluid spheres with adiabatic temperature gradients develop such concentric cylinders of fluid. If this kind of circulation exists in Jupiter and Saturn then the observable wind jets that form the bands would be at the tops of the cylinders. The cylinders terminate at the metallic hydrogen mantle. At equatorial latitudes the cylinders extend from one hemisphere to the other, but at higher latitudes the metallic hydrogen mantle gets in the way. The curvature of this boundary helps

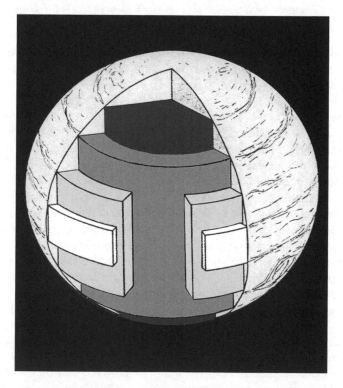

Figure 11.4 A possible circulation pattern in the molecular hydrogen envelopes of Jupiter and Saturn. This is a schematic illustration. (Adapted from *The New Solar System*, J. K. Beatty and A. Chaikin (Eds), Figure 14, p 148, Sky Publishing 1990)

organise the flow into narrower jets, as observed. The difference between the thicknesses of the molecular hydrogen envelopes in Jupiter and Saturn could then help account for the differences in the wind patterns between these two giant planets.

To test the model in Figure 11.4, we await more data, including more detailed measurements on the planets' gravitational fields.

Ovals and spots (vortices)

As well as belts and zones, Jupiter and Saturn display ovals of various sizes, from the limit of instrument resolution (a few tens of kilometres), up to thousands of kilometres across. Cloud motions within and around the ovals show them to be vortices, also called eddies – regions of rotating atmosphere rather like cyclones and anticyclones on Earth. In the northern hemisphere of the giant planets (and the Earth) cyclones rotate anticlockwise, and anticyclones clockwise. In the southern hemisphere these directions are reversed. The vortices come and go, with lifetimes from the order of weeks for the smaller ones to many years for the largest. Vortices are seen to merge.

The atmosphere is generally rising at the core of an anticyclone and descending in a spiral outside the core. In a cyclone the atmosphere generally descends in the centre, though there are small regions where upwelling occurs. Vortices are concentrated where the horizontal wind shear is greatest, which on Jupiter is at the boundaries between the bands. The vortices rotate in accord with this shear (rather like cylinders between two plates moving at different speeds), which often transforms vortices into wavy streaks. The vortices drift westwards, but do not drift in latitude. They do not exchange material laterally, though they might dredge up material from below.

On Jupiter, several hundred vortices can be seen at any one time, though not within about $10°$ of the equator, where wavy streaks develop instead. Most vortices have sizes in the range 1000–10 000 km across. The smaller vortices are usually dark. The larger ones are usually white, and extend to several kilometres above the top of the surrounding cloud layer, presumably raised by local convection. At high latitudes, beyond the visible bands, a mottled appearance has long been observed. At these latitudes Cassini has seen numerous spots that might be small vortices. A weak narrow band structure is indicated by the motion of these spots. If, over a narrow range of latitudes, the motion is east to west, over adjacent narrow ranges it is west to east.

The largest vortex, the Great Red Spot (GRS) on Jupiter (Plate 11) measures about 22 000 km east–west and about 11 000 km north–south. For comparison, the Earth's (equatorial) diameter is 12 756 km. It has existed for at least 100 years, and probably for over 300 years. Its longevity is a result of its size. From when it became sufficiently large, its rotational energy has been so huge that it has survived encounters with smaller features. Longevity also owes something to the lack of a solid surface to dissipate rotational energy. The GRS is in the southern hemisphere and rotates anticyclonically, once every 6 days. It extends to no more than about 100 km above the surrounding clouds, and no more than about 500 km below them, so it is rather shallow for its huge lateral size. However, it extends well below any water-ice clouds, from where it could dredge up material. If this includes traces of phosphorus compounds, their dissociation would yield phosphorus that could account for the red colour of the GRS. Three white vortices south of the GRS were each about the diameter of Mars, and lasted for over 50 years.

Other patchy or wispy features are not vortices. In some cases they are glimpses through the upper cloud layer. These include irregular dark-brown spots and irregular dark-blue spots, about 10^3 km across. The dark-brown spots are thought to be glimpses of a lower cloud layer, whereas the blue-grey/purple spots are the hot spots mentioned briefly in Section 11.1.1. In these almost cloud-free areas, our view is limited only by the depth to which sunlight can penetrate, a limit imposed by the scattering of sunlight from atoms and molecules. This is **Rayleigh scattering**, named after the British physicist, John William Strutt Rayleigh (1842–1919), and it accounts for the blue tint of the hot spots (and also for the blue skies on Earth).

❑ What does the existence of hot spots indicate about the spatial distribution of the deeper cloud layers?

Clearly, these cloud layers are patchy. This is in accord with the findings of the Galileo probe.

Saturn's major bands are fewer in number than those of Jupiter (Plate 16). The bands have wavy edges that often break into vortices. Cassini has penetrated below the upper cloud layer by observations of deep clouds at $5\,\mu$m wavelength. About 30 km below the upper cloud layer, there are planet-girdling lanes, many of them connected with clouds that indicate convective cells. As noted above, there is also an increase in jet speed with depth, at least in the equatorial region. Cassini has seen more than 10 oval spots greater than 500 km across, mostly dark but many with bright haloes. As on Jupiter, they come and go, and merge, with lifetimes exceeding a week. The spots are anticyclonic, and, again as on Jupiter, rotate in accord with the shear between adjacent wind jets. They migrate along the jet boundary at speeds up to about $25\,\text{km s}^{-1}$, which, compared with the jet speeds (Figure 11.3), is modest. Nothing on the scale of the GRS has ever been seen on Saturn.

The vortices on Jupiter and Saturn probably originate from instabilities in the wind jets, and derive their rotational energy from the motion of the rising currents of atmosphere. However, there is much still to be understood, not only about the vortices on each planet separately, but also about the differences between the circulation and its manifestations on these two giant planets.

11.1.4 Coloration

It remains to account for the colours of Jupiter and Saturn. The substances that are thought to dominate each cloud layer – NH_3, NH_4SH, NH_4OH, and H_2O – are all colourless. Clouds consisting of these substances would therefore appear white. Whitish features are confined to high altitudes and are presumably freshly condensed NH_3 particles almost free of colouring agents. Other features have presumably had time to become coloured. It needs only a trace of a coloured substance to cause the intensity of the observed colours.

One group of possible colouring agents, as on Io, is various forms of solid particles of sulphur, derived photochemically from sulphur compounds below the upper cloud deck. Some forms of solid sulphur are yellow, others are brown. Another possibility is phosphorus, solid particles of which can be yellow or red. Though phosphine (PH_3) has been detected as a trace in both Jupiter and Saturn, it is not known if the appropriate chemical reactions occur to form phosphorus particles. If they do form, then, as noted earlier, they could account for the red colour in the GRS. Other possible colouring agents are various compounds that can be formed by chemical reactions involving some or all of the compounds CH_4, NH_3, and H_2S. The hazes, as noted earlier, are likely to be dominated by hydrocarbon particles above the clouds, but possibly enriched by phosphorus and its compounds at and below the clouds. But in spite of all these possibilities the actual colouring agents and the reasons for their spatial distribution are unknown.

Question 11.2

Suppose that the atmospheres of Jupiter and Saturn consisted *only* of hydrogen and helium.

(a) Discuss plausible differences this would make to: the clouds; the temperatures in the upper molecular envelope; the facility with which the atmospheric circulation could be investigated.
(b) State why the solar driven component of the circulation would be changed.

11.2 The Atmospheres of Uranus and Neptune Today

Table 11.1 shows that, like Jupiter and Saturn, the atmospheres of Uranus and Neptune consist mainly of H_2 and He. These massive atmospheres are each bounded by a liquid icy–rocky mantle. Plates 19 and 20 show that Uranus and Neptune are blander in appearance than Jupiter and Saturn, and that Uranus is somewhat blander than Neptune. The predominant visual impression in Plates 19 and 20 is the result of the Rayleigh scattering of sunlight from a deep layer of gas, the bluish-green tint arising from absorption of red wavelengths by methane. Uranus has a slightly greener tint, perhaps due to a deep-lying layer of methane cloud.

11.2.1 Vertical Structure

Figure 11.5 shows the vertical structures of the atmospheres of Uranus and Neptune, deduced from much the same variety of measurements used for Saturn before Cassini. The observable atmospheres are generally colder at comparable pressures than those of Jupiter and Saturn, a consequence of the greater solar distance. Heating is by solar radiation down to about the 10×10^5 Pa level, which includes the upper troposphere. It is not known whether the lapse rates are as large as the adiabatic values. Below this level the increase of temperature with depth depends on internal heat sources. These greater depths have been explored by microwave observations, which have revealed that whereas in Neptune the increase is probably at the adiabatic rate, in Uranus the lapse rate is generally a bit smaller. Therefore, the deep troposphere of Neptune is probably convecting, whereas that of Uranus is probably not, or only weakly so. One would therefore predict that there must be a far lower rate of escape of heat from the

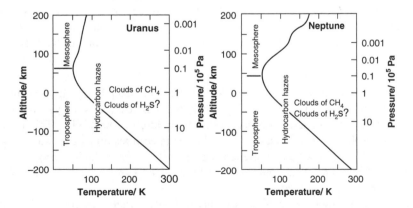

Figure 11.5 The vertical structures of the atmospheres of Uranus and Neptune.

interior of Uranus than from the interior of Neptune. This is consistent with the barely detectable IR excess from Uranus and the greater excess from Neptune (Table 9.2). Models indicate that deeper into the interior the lapse rate is adiabatic in both planets.

In the upper troposphere there are flecks and streaks of cirrus-like clouds. These are white because they are at too shallow a depth to be seen though a tinted atmosphere. Higher still there is a thin haze of photochemical compounds, mainly hydrocarbons. The low temperatures result in extensive freezing out at low altitudes of many condensable atmospheric components, giving rise to deep-lying cloud decks. The composition of the visible and deep-lying clouds can only be inferred, but evidence from atmospheric composition, atmospheric temperatures, and the CH_4 phase diagram strongly suggests that the high-altitude white clouds are CH_4 ice crystals, whereas any deep-lying cloud deck is probably CH_4 droplets, or, less likely, particles of H_2S. If there is a deep-lying CH_4 cloud deck, then models suggest it might be underlain by H_2S clouds, perhaps underlain in turn by H_2O clouds.

11.2.2 Composition

Table 11.1 shows the main constituents of the atmospheres of Uranus and Neptune as mixing ratios with respect to H_2. Table 9.1 shows H_2, He, and CH_4 as number fractions. The atmospheres consist mainly of H_2 and He (the He having been measured indirectly, as for Saturn). Compared with Jupiter, and particularly Saturn, He is rather more abundant in these two subgiants. These differences have a ready explanation based on our models of the interiors (Section 5.3). The interiors of Uranus and Neptune do not attain pressures sufficiently high for metallic hydrogen mantles to form. Therefore, there is no possibility of the downward settling of Hc arising from its insolubility in metallic hydrogen. Consequently the He mixing ratio measured in the observable atmospheres of Uranus and Neptune is presumed to be the same as that in the whole planet. The interiors of Jupiter and Saturn *do* attain sufficient pressures for metallic hydrogen mantles to form, but whereas Jupiter is so hot that only slight segregation has occurred, the cooler interior of Saturn has led to greater segregation. It is thus possible that the whole-planet He mixing ratios are similar in all four giant planets. We shall return to this important point in Section 11.3.

Unlike He, CH_4 has had its abundance measured directly. It was important to measure it at depths where the temperature is sufficiently high to prevent condensation. The measurements in Table 11.1 are at such depths, and you can see that CH_4 is considerably more abundant in Uranus and Neptune than in Jupiter and Saturn. This is another important point for Section 11.3. High in the atmospheres, Uranus and Neptune are greatly depleted in CH_4. This lends strong support to the view that the observed clouds and any deep-lying cloud deck consist of CH_4.

H_2O has not been detected in either atmosphere. This is not surprising in view of the low temperatures.

❏   Why is it not surprising?

H_2O is much less volatile than CH_4, so it would condense very deep in the atmospheres. Whether there is a very deep-lying H_2O cloud deck is unknown, but if there is, and if the particles are liquid, then we can explain another observation – the failure to detect NH_3. This is no surprise in the upper troposphere, again because of the low temperatures, but microwave observations indicate that NH_3 depletion persists to greater depths. NH_3 is very soluble in liquid H_2O, so this scarcity could be the result of its solution in H_2O droplets. However, there are other possible explanations. Here are two of them.

The nitrogen/sulphur ratio could be so low that NH_3 has largely disappeared into the formation of NH_4SH, which would form deep-lying clouds, beyond present detectability. Another possibility relates to the chemical reaction that produces NH_3 from N_2

$$N_2 + 3H_2 \rightarrow 2NH_3 \tag{11.4}$$

This reaction is favoured in the low-temperature conditions in the atmospheres of Uranus and Neptune, but it requires a catalyst if it is to proceed at a high rate. If there are no suitable catalysts, and if the original form of nitrogen were N_2 rather than NH_3, then this could also help to explain the scarcity of ammonia.

In the upper troposphere of Neptune CO and HCN have been detected, but not in the case of Uranus. In hydrogen-rich atmospheres at the low temperatures in the upper tropospheres, chemical reactions would quickly reduce CO and HCN to very small quantities unless they were being replenished at a sufficient rate from the deep interior. Convection can fulfil this role. In the case of CO, it is CO itself that would be brought up, whereas for HCN it is N_2, which then participates in reactions that yield HCN. The direct detection of N_2 is beyond present capabilities because of its weak (IR) spectral lines. The absence of CO and HCN from the upper troposphere of Uranus might be another indication, in addition to the barely detectable IR excess, that convection is absent in some layers of the atmosphere of Uranus.

11.2.3 Circulation

The circulation of the atmospheres, as for Jupiter and Saturn, is revealed largely by bands, and by the motion of the rather sparse cloud features. The speeds of motion are again with respect to the rotation period of the interior, and this period is again determined from the periodicity of radio emissions. It has been more difficult to establish the circulation of Uranus. This is because until 2000 it displayed fewer atmospheric features than Neptune. The rare, elusive white flecks of cloud seen in the upper troposphere of Neptune were even rarer and more elusive on Uranus. That changed in 2003–2004, when near-IR images from the Keck II telescope revealed dozens of clouds on Uranus, as in Plate 19. They varied in size, brightness, and longevity, from a few hours to possibly the two decades since Voyager 2 flew by in 1986. This change is probably the result of the large climate swings on this tipped-over planet as it orbits the Sun every 84 years. In 1986 the southern hemisphere was in early summer, with the South Pole pointing nearly towards the Sun. Now, the southern summer has ended with the equinox in 2007. The larger, longer-lived clouds might be underpinned by vortices, but these are unlikely to be anywhere near as energetic as a terrestrial hurricane – solar radiation is far weaker at Uranus, and the heat from the interior is slight (Table 9.2). Such as it is, vortex formation is aided by the absence of a surface at shallow depths – this reduces frictional dissipation that would otherwise hamper vortex formation.

Neptune too has displayed an increase in cloudiness since Voyager 2 flew past in 1989 (Plate 20). Neptune's clouds are probably particles of CH_4 ice in the stratosphere. That new, bright bands of cloud have appeared in the sunny south suggests some kind of seasonal change. The HST has shown a clear increase in cloudiness on Neptune since 1996. With solar radiation 900 times less than on Earth, and each season lasting over 40 years, this increase might be due partly to an internal change, such as one that has promoted convection. Certainly, the considerable IR excess of Neptune (Table 9.2) could make cloud formation sensitive to small internal changes.

Though no main cloud decks are visible for either planet, a deep-lying haze layer on Neptune exhibits dark belts and dark spots. A large dark spot was discovered in the southern hemisphere by Voyager 2 in 1989, and named the Great Dark Spot. It covered about the same fraction of Neptune's surface as the GRS covers of Jupiter's surface, and also seems to have been a vortex. It had vanished when Neptune was examined by the HST a few years later, but another one had appeared in the northern hemisphere. The dark spots can be interpreted as holes in the haze layer, giving a view into deeper regions that appear darker because they scatter less sunlight.

The deep haze layer of Uranus has weaker banding, but one gigantic storm has been seen, in the southern hemisphere. Images from the HST and the Keck telescopes show that the storm had lasted since at least the arrival of the new millennium. It extends to higher altitudes than observed for any cloud features before, presumably because of vigorous convection raising material from below. Strangely, and unlike other features on Uranus and Neptune, it drifts appreciably north–south, across about 5° of latitude. The reason is unknown.

The wind speed data for both planets are sparser than for Jupiter and Saturn. Figure 11.6 shows a rather idealised picture for both planets. The equatorial wind speed for Uranus is from radio occultation data. Though there are considerable differences in the details between Figure 11.6 and the corresponding Figure 11.3 for Jupiter and Saturn, all four planets exhibit one prominent feature common to their circulations.

☐ What is this?

The observed circulation is predominantly parallel to the equator. As in the case of Jupiter and Saturn, the Coriolis effect promotes such an outcome, but any further contribution from any deep circulation is largely unknown. One striking feature on Neptune, possibly due to deep-seated circulation, is the large negative wind speeds – the atmosphere is predominantly rotating more slowly than the interior. At present the cause of this slow rotation is unknown. There is, however, a plausible explanation for the small temperature difference between poles and equator

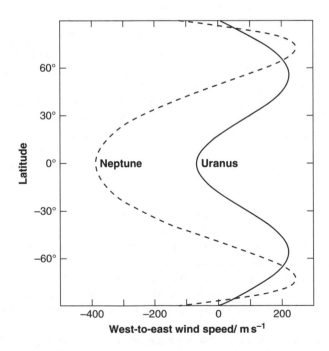

Figure 11.6 Winds on Uranus and Neptune.

on Neptune. As in the cases of Jupiter and Saturn, this could well be the combined effect of a high rate of heat flow from the interior, which is independent of latitude, and equator-to-pole circulation efficient enough to offset the greater equatorial insolation.

In the case of Uranus, the solar radiation incident at each latitude, averaged around the orbit, is greater at the poles than at the equator. Moreover, the thermal response time of the troposphere is of the same order as the orbital period. Consequently, a Hadley circulation transferring heat from the poles to the equator could be a fairly persistent feature in each hemisphere. It can be shown that such a flow is consistent with the variation of wind speed with latitude in Figure 11.6. However, measurements have show that the South Pole is not much hotter than the equator, but is at much the same temperature. Therefore, there is an efficient redistribution of absorbed solar radiation across the whole planet. The mechanism is unknown.

Question 11.3

In Table 11.1, the mixing ratios of He in the atmospheres of Uranus and Neptune are indirect estimates, based on the amounts needed to top up the atmospheres to their total masses. These estimates have assumed that the proportion of N_2 is negligible. In which of these two planets might this assumption fail, and what is the evidence? What (qualitatively) would be the effect on the estimate of the He mixing ratio (in one sentence)?

11.3 The Origin of the Giant Planets – A Second Look

The atmospheres of the giant planets are so massive that their origin is inseparable from the origin of the giant planets as a whole.

In the solar nebula theory (Chapter 2), the planets form from a disc of gas and dust around the Sun. This disc has much the same composition as the young Sun, which is preserved today outside the core in which fusion reactions are occurring. The composition by mass that we have adopted is 70.9% hydrogen, 27.5% helium, and 1.6% 'heavy' elements (Section 2.2). The *gas* in the disc has much the same He/H mass ratio as the young Sun, but it is somewhat depleted in those heavy elements that form the icy and rocky compounds that make up the dust grains in the disc. Conversely, the dust grains are enriched in heavy elements, and depleted in hydrogen and helium. Helium is almost entirely absent from the grains, and only a small fraction of the hydrogen is present. At the distances that the giant planets formed from the Sun, hydrogen in the dust grains is in compounds such as H_2O, NH_3, and CH_4. In the core-accretion theory of the formation of a giant planet (Section 2.2.5), the dust accretes to form icy–rocky planetesimals and a number of these become assembled into a kernel of several Earth masses of rocky and icy materials, massive enough to capture nebular gas and other planetesimals. We thus have two distinct sources of giant material – nebular gas (with a trace of dust) and icy–rocky planetesimals. The planetesimal material tends to concentrate into a central core and the nebular gas into a surrounding envelope.

Can we reconcile the data in Table 11.1 with this theory? To see whether we can it is convenient to convert the data into elemental mass ratios with respect to hydrogen. For example, the He/H mass ratio is the ratio of the mass of helium in any region to the mass of hydrogen in the same region. For elements such as N, C, and O that form compounds, the mass of the element is the sum over all the forms in which it has been measured. Table 11.3 gives the elemental mass ratios for C, N, O, and He for the four giant planets. Values for the young Sun have been added.

Table 11.3 Elemental mass ratios with respect to hydrogen in the giant planet molecular envelopes and in the young Sun

Ratio	Jupiter	Saturn	Uranus	Neptune	Young Sun
He/H	0.31	0.26	0.36	0.36	0.39
C/H	1.3×10^{-2}	3.1×10^{-2}	0.11	0.15	3.9×10^{-3}
N/H^a	5.0×10^{-3}	$< 7.7 \times 10^{-4}$	—	—	1.2×10^{-3}
O/H^a	$< 4.8 \times 10^{-3}$	$< 1.6 \times 10^{-6}$	—	—	1.1×10^{-2}

^{a} A dash indicates that no useful upper limit is available.

The He/H ratio

The He/H mass ratio in the captured gas is expected to be much the same as in the young Sun.

☐　Why is this?

This is because He has remained almost entirely in the gas, and only a small fraction of the hydrogen has been removed from the gas in the form of compounds in the planetesimals. This expectation is borne out in Uranus and Neptune within the uncertainties of the data for these planets. Jupiter has a slightly but significantly smaller ratio, and Saturn a yet smaller ratio. Remembering that the data are for the outer envelopes of the giants, this is a further indication that there has been a slight settling of He in the metallic hydrogen mantle of Jupiter, and rather more in that of Saturn. This is in accord with the expectation that the mantle of Jupiter is at a considerably higher temperature than that of Saturn (Section 5.3.1).

There seems to have been very little settling in Uranus and Neptune, and this is in accord with the lower interior pressures and the consequent absence of metallic hydrogen.

The C/H ratio

The C/H ratios in Table 11.3 are thought to apply to the whole envelope – CH_4 does not form clouds in Jupiter and Saturn, and the values for Uranus and Neptune are probably below the methane cloud base. You can see that for all four giant planets the C/H ratio exceeds the solar value. Moreover, the ratio increases from Jupiter, to Saturn, to Uranus and Neptune. How can these excesses be reconciled with the supposed formation of the envelopes from captured nebular gas that was at least mildly depleted in heavy elements?

The answer is probably that the planetesimal material becomes hot enough during and after planet formation to lose some of its more volatile constituents. These constituents thus enrich the surrounding envelope in certain heavy elements, including carbon. If all four giant planets have acquired about the same mass of planetesimals (of order 10 Earth masses), and if their planetesimal materials have all outgassed to roughly the same extent, then the C/H ratio in the envelope will be inversely related to the mass of the envelope – the greater the mass of the envelope, the greater the dilution of the outgassed carbon. It follows that there would then be an inverse relationship between the C/H ratio and the total mass of the planet.

☐　Is this what we find?

This is exactly the feature in Table 11.3. Further carbon enrichment in all four cases would come from the subsequent capture of volatile-rich bodies, notably icy–rocky planetesimals and comets.

In the alternative, gravitational instability model of giant planet formation (Section 2.2.5), all four giants are predicted to have an overall initial composition similar to the Sun. It is therefore very difficult to account for an excess of carbon in the atmospheres. Indeed, core separation of heavy elements would cause a depletion. Capture of planetesimals and comets could offset this depletion, though perhaps not completely, and moreover this might well not produce the strong differences between the giants. Further difficulties facing the gravitational instability model have been outlined in Section 2.2.5.

Other mass ratios

The atmospheric $^2H/^1H$ mass ratio (not shown in Table 11.3) is around 4×10^{-5} for Jupiter and Saturn, which is about the same as the Sun, but increases to about 12×10^{-5} for Uranus and Neptune. In spite of large uncertainties, the increase from Jupiter and Saturn (and the Sun) to Uranus and Neptune is significant. A good explanation of this increase starts in the interstellar medium (ISM), where the dust was formed that subsequently became a component of the solar nebula. The formation of most compounds in the ISM depends on reactions between ions and molecules. In ion–molecule reactions involving hydrogen, there is a tendency for the heavy isotope to become concentrated in condensable substances. For example, in HCN the ratio $^2HCN/^1HCN$ is greater than the overall $^2H/^1H$ ratio. In the solar nebula the dust is thereby enriched in ^2H (deuterium), and consequently the planetesimal material is likewise enriched. The remainder of the argument accounts for the differences between the giants, and parallels the argument just given for the C/H ratio. The protoplanet theory has difficulty explaining the $^2H/^1H$ ratios too, and for similar reasons to its difficulty with the C/H ratios.

The N/H ratio for Jupiter is a few times that of the Sun. We have only an upper limit for Saturn, comparable with the solar value, and no values at all for Uranus and Neptune. Jupiter has presumably benefited from outgassing from the kernel and from the icy component of captured planetesimals. Saturn could have more nitrogen beneath any NH_4SH clouds, so might also have N/H in excess of the solar ratio, and for the same reasons. As noted in Section 11.2.2, the failure to detect NH_3 in the atmospheres of Uranus and Neptune might be because NH_3 has been taken up in the formation of NH_4SH, or that it has dissolved in deep-seated water clouds. It is also possible that the planetesimals that formed in the cold outer reaches of the solar nebula had much of the carbon and nitrogen in CO and N_2, rather than as CH_4 and NH_3, in which case the planetesimals would preferentially acquire carbon, CO being less volatile than N_2. Moreover, you have seen that if the nitrogen is largely in the form N_2 in the atmospheres it would not yet be detectable.

The only measurements we have of O/H are from measurements of H_2O in Jupiter and Saturn. The atmospheres above any H_2O clouds are expected to be very dry, but you have seen (Section 11.1.2) that though the Galileo probe penetrated to depths below where water clouds in Jupiter lie, it entered a dry hole. Therefore, we only have an upper limit on O/H of rather less than a half of the solar ratio. If this is typical of the whole of Jupiter outside the core, it is possible that water, being less volatile than the repositories of carbon and nitrogen, outgassed from planetesimal material to a far smaller extent. It is also possible that different compounds have different solubilities in metallic hydrogen. About 85% of Jupiter's hydrogen is in the metallic phase, and so a modest excess solubility of water could lead to a large depletion in the molecular hydrogen envelope. The position with Saturn is much the same.

The data in Table 11.3 cover only a few elements, though they are among the most abundant. Taking the data as a whole, and in spite of some remaining puzzles, we can conclude with some confidence that the atmospheres of the giant planets are consistent with an origin of the giant in which a kernel of icy–rocky materials captured nebular gas (with a trace of dust) and other icy–rocky planetesimals. The composition of the envelopes was then modified by outgassing of the kernels, and to some extent by the subsequent capture of volatile-rich bodies.

Question 11.4

Neon (Ne), like helium, is a volatile inert gas. Discuss whether you would expect the Ne/H ratio to be the same in the giant planets as in the young Sun. Given that Ne is soluble in liquid helium, would you expect the Ne/H ratio measured in the outer envelope of each giant to be the same as for the planet as a whole?

11.4 Summary of Chapter 11

Tables 9.1 and 9.2 list basic properties of planetary atmospheres, and Tables 11.1–11.3 give further data for the giant planet atmospheres.

All four giants have atmospheres dominated by molecular hydrogen (H_2) and helium (He), with small quantities of less volatile substances that form clouds. There are traces of other (unknown) substances that colour the clouds. The clouds are banded parallel to the equator. The circulation revealed by the clouds is predominantly west to east, and a factor in this is the large Coriolis effect in these rapidly rotating planets that acts on convection patterns. For Jupiter and Saturn there is probably a major contribution from deep-seated circulation patterns in the molecular envelope, in the form of concentric cylinders.

Jupiter, Saturn, and Neptune have adiabatic lapse rates in their troposphere, the result of heat welling up from the interior, supplemented by the absorption of solar radiation in the upper troposphere. Uranus has a subadiabatic tropospheric lapse rate in the lower troposphere, consistent with the low rate of upwelling heat.

The composition of the giant planets' atmospheres can be explained by the formation of icy–rocky kernels and subsequent capture of nebular gas during the formation of the Solar System, with subsequent modifications due to the downwards segregation of helium in the metallic hydrogen mantles of Jupiter and Saturn, the outgassing of planetesimal material, and the capture of volatile-rich bodies.

As you will now see, the Solar System will undergo huge changes when the Sun evolves into a red giant and then into a white dwarf.

11.5 The End

We have now completed our study of the Solar System as it is today, of how it might have originated, and of how it might have evolved to its present condition. We hope you have found this exploration of our own planet and of our neighbours in space as exciting, fascinating, and intriguing as it deserves to be. By way of valediction let us consider briefly what awaits the Solar System in the future.

The interiors of all planetary bodies will continue to lose energy, and for those bodies whose geological activity is dominated by interior sources of heat, the level of geological activity will decline. Atmospheric loss mechanisms will continue, and with the decline in outgassing that accompanies a decline in geological activity, a decrease in atmospheric mass will occur until the growing luminosity of the Sun (Figure 10.14) begins to drive volatiles from surface repositories.

In the case of the Earth, for the next 1000 Ma or so, the increasing solar luminosity will make the Earth warmer, though negative feedback should stabilise the temperature to some extent. The feedback operates through a decrease in the size of the greenhouse effect, due to the carbonate–silicate cycle outlined in a different context in Section 10.1.4. But the feedback will not be fully effective and so the surface temperature will rise, and this will put more water into the atmosphere. The consequent increase in the greenhouse effect now introduces a positive feedback. One scenario for 1400 Ma into the future has GMST of about 325 K, the result of a moist greenhouse runaway (Section 10.6.1). A bacterial biosphere could survive to at least 400 K – some of Earth's present bacteria flourish at such temperatures – but at 400 K there would be no liquid water at the surface. The Earth is now firmly on the evolutionary trail followed by Venus early in Solar System history. As the GMST rises, the crust softens, causing a rise in volcanism, which increases the CO_2 content of the atmosphere. In spite of water loss through UV photodissociation, the GMST rises further, which promotes a further rise in volcanism. This positive feedback causes a dry greenhouse effect. As a result, perhaps by 3500 Ma from now, perhaps earlier, the Earth will have a Venus-like atmosphere and a correspondingly high surface temperature.

At some time during the demise of the Earth as we know it, Mars might (again) become clement, as CO_2 and water vapour are released from surface and subsurface reservoirs.

The increasing solar luminosity in Figure 10.14 is the result of the continuing conversion of hydrogen to helium in the solar core (Section 1.1.3). An inert inner core of helium grows in size, surrounded by a shell in which hydrogen is still undergoing fusion. About 6000 Ma from now, with the Sun about 11 000 Ma old, this shell will be so thin that the rate of energy release by fusion within it will be insufficient to support the interior. The Sun has then reached the end of its main sequence lifetime. The core will contract and, in a complicated process, the rest of the Sun will expand and cool, the surface taking on an orange tint. The huge increase in the surface area of the Sun will far outweigh the decrease in surface temperature, and so the solar luminosity will increase greatly. The Sun will reach beyond the present orbit of Venus, nearly as far as the Earth. However, the Sun loses mass in the form of an intense solar wind, so that the planetary orbits increase in size, in the case of the Earth to about 1.7 AU. This might spare the Earth from consumption, and perhaps Venus too, though not Mercury. On the other hand, the tidal interaction between the bloated Sun and the Earth will tend to reduce the size of the Earth's orbit, so it will be touch and go! In any case, the atmospheres of Venus, the Earth, and Mars, and all surface volatiles, will have escaped to space long before. The peak surface temperature of the Earth will be about 1600 K.

The Sun will have then made the transition to a red giant, where the gravitational energy released by the contraction of the He core will have raised core temperatures to the point where helium undergoes fusion to form carbon, releasing energy at a sufficient rate to maintain a temperature gradient that stabilises the star. The red giant phase will last about 1000 Ma, during which time Pluto and some of the satellites of Uranus and Neptune might become suitable for life.

But this is a turbulent time, with luminosity variations and a copious solar wind. It comes to an end in a runaway instability in which the Sun will cast off a significant fraction of its mass, a supersonic blast of hot gas tearing though the Solar System on its way to forming a huge shell of gas called a planetary nebula (nothing to do with planets). The solar remnant will contract to become an Earth-sized body called a white dwarf. Though its surface will be far hotter than that of the Sun today, it will be so small that its luminosity will be feeble. Its gravitational field will also be much reduced, and it is likely that some planets will escape from the Solar System. These, and the planets that remain in solar orbit, having been cooked and blasted, will then freeze. Any intelligent beings that might have survived will probably have to journey to other stars to find somewhere to live. The existence of extrasolar planets will give them great expectations.

Question Answers and Comments

Note that comments are in curly brackets { }. These are not expected as part of your answer.

Question 1.1

At the lower photospheric temperature in the past the spectrum was at longer wavelengths. Also, the lower luminosity in the past gave a smaller area under the spectrum. Since then, the spectrum has shifter to shorter wavelengths and the area under the spectrum has increased. {See Figure 1.1.}

Question 1.2

Figures 1.4 and 1.5 together show that the medium-sized planets – the four terrestrial planets – are closest to the Sun, the largest, the Earth, being the third one out. Then comes the largest planet, Jupiter, then the somewhat smaller planet Saturn, and then the two smaller giant planets, Uranus and Neptune. Finally we come to Pluto, the smallest planet of all. The broad correlation of size with distance is therefore that the largest planets lie at intermediate distances, with smaller planets closer to, and further from, the Sun. {The reasons for this are discussed in Chapter 2.}

Question 1.3

The semimajor axis of Fortuna is obtained from the equation (Section 1.4.1)

$$\frac{a_F}{a_E} = \left(\frac{P_F}{P_E}\right)^{2/3}$$

where the subscript E denotes the Earth, and F denotes Fortuna. For the Earth, a_E is 1 AU and P_E is 1 year. The orbital period P_F of Fortuna is given in the question as 3.81 years. Therefore

$$a_F = (1 \text{ AU}) \times \left(\frac{3.81 \text{ years}}{1 \text{ year}}\right)^{2/3} = 2.44 \text{ AU}$$

Discovering the Solar System, Second Edition   Barrie W. Jones
© 2007 John Wiley & Sons, Ltd

Question 1.4

The perihelion distance of a planet from the Sun is $a(1-e)$ (Figure 1.8). Taking a and e from Table 1.1, the perihelion distances of the Earth and Venus are, respectively, 0.983 AU and 0.718 AU. Thus the Earth–Venus distance is 0.265 AU.

From the radar data, this distance is also $(3.00 \times 10^5 \, \text{km s}^{-1}) \times (264 \, \text{s}/2)$, which is 3.96×10^7 km. Therefore, $1 \, \text{AU} = 3.96 \times 10^7 \, \text{km}/0.265 = 1.5 \times 10^8$ km to two significant figures. {The accurate value in metres is given in Table 1.6.}

Question 1.5

(a) {One possible sketch is shown. You might have adopted a different viewpoint, but the relationship between the orbit of Comet Kopff, the ecliptic plane, and ♈ should be the same. Also, the shape of the orbit should (roughly) reflect its eccentricity, with the Sun distinctly off centre, and with some allowance for the oblique viewpoint.}

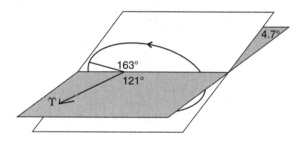

(b) Denoting the perihelion distance by q, Figure 1.8 shows that

$$q = a(1-e)$$

and so

$$a = \frac{q}{1-e}$$

Thus, putting in the values of q and e given in the question,

$$a = \frac{1.58 \, \text{AU}}{1-0.54} = 3.43 \, \text{AU}$$

The aphelion distance is (Figure 1.8) $a(1+e)$, which is 5.28 AU. The orbital period P_K is obtained from Kepler's third law (equation (1.3)), which gives

$$P_K = P_E \left(\frac{a_K}{a_E}\right)^{3/2} = (1 \text{ year}) \times \left(\frac{3.43 \, \text{AU}}{1 \, \text{AU}}\right)^{3/2} = 6.35 \text{ years}$$

This is 6 years and 0.35×365.24 days, which is 128 days. Thus, the first perihelion date in the twenty-first century is 6 years 128 days after 2 July 1996. This puts it in November 2002.

(c) If the orbits are in the same plane they will intersect, but they are not – the orbits have different inclinations. In this case they will intersect only if the two orbits have just the right relative sizes for the difference between the longitudes of the ascending node. Given that the orbits do not intersect we must conclude that this condition is not met.

Question 1.6

Applying equation (1.6) to the Earth,

$$M_\odot + m_E = \frac{4\pi^2 \, a_E^3}{G \, P_E^2}$$

$M_\odot \gg m_E$, so

$$M_\odot \approx \frac{4\pi^2 a_E^3}{G \, P_E^2} = \frac{4\pi^2 (1.496 \times 10^{11} \, \text{m})^3}{(6.672 \times 10^{-11} \, \text{N} \, \text{m}^2 \, \text{kg}^{-2})(3.156 \times 10^7 \, \text{s})^2}$$

Therefore

$$M_\odot \approx 1.989 \times 10^{30} \, \text{kg}$$

Applying the same procedure to Jupiter,

$$M_\odot \approx \frac{4\pi^2 (7.782 \times 10^{11} \, \text{m})^3}{(6.672 \times 10^{-11} \, \text{N} \, \text{m}^2 \, \text{kg}^{-2})(3.743 \times 10^8 \, \text{s})^2}$$

Therefore

$$M_\odot \approx 1.990 \times 10^{30} \, \text{kg}$$

The values of a and P in Table 1.1 are only given to four significant figures, so these two values of the solar mass agree within this precision. {We have actually calculated $M_\odot + m_{\text{planet}}$, and m_J is about $M_\odot/1000$, whereas M_E is only about $M_\odot/300\,000$. Therefore, the second value needs to be reduced by a factor of about 1000, giving $M_\odot \approx 1.988 \times 10^{30}$kg. The accurate value for $M_\odot$ is 1.9891×10^{30} kg.}

Question 1.7

(a) If the Sun rotated much more rapidly then its departure from spherical symmetry would be greater. Therefore, the orbital elements of Venus would become more variable. {This would not apply to a planet in an orbit in the equatorial plane of the Sun. No planets orbit in this plane.}

(b) If the mass of Jupiter were doubled it would exert a greater gravitational force on the other planets. Again, the orbital elements of Venus would become more variable. {Also, because Jupiter would more greatly distort the other planets tidally, they would have a greater influence on Jupiter itself.}

(c) If the Sun entered a dense interstellar cloud the gravitational and non-gravitational forces on Venus would increase. Once more, the orbital elements of Venus would become more variable.

Question 1.8

From Table 1.1, the orbital period of Neptune is 165.07 years and that of Pluto is 250.88 years. The ratio of these periods is 1.52. Bearing in mind the (slight) variations in orbital periods, and that a resonance has a 'width', it is entirely plausible that Neptune and Pluto are in a 3:2 mean motion resonance. {This is the case. A secular resonance might fortuitously occur here as well, but it is beyond our scope to see if this is the case. In fact there is no secular resonance here.}

Question 1.9

Table 1.1 shows that the axial inclination of Venus is 177.4°, and so makes an angle of only 2.6° with the plane of its orbit – nearly 10 times less than the case of the Earth. Therefore, seasonal variations would be small, unless its orbit were very eccentric. However, Table 1.1 shows that the orbital eccentricity is only 0.0068, less than half that of the Earth. {The seasonal variations on Venus are indeed small.}

Question 1.10

If our calendar were based on the sidereal year, then, because of the precession of the equinoxes, the seasons would gradually move through the year. For example, 12 900 years from now, in the northern hemisphere, spring would start in September, and the summer solstice would be in December.

Question 1.11

The orbit of Mars is moderately eccentric, whereas the Earth's orbit is only slightly eccentric (Table 1.1). Consequently, the opposition distance is more sensitive to the position of Mars in its orbit than to the position of the Earth in its orbit. It is thus particularly important to have Mars near perihelion when it is at opposition. From Figure 1.24 it is clear that when Mars is at perihelion, then for it to be in opposition the Earth has to be about two-thirds of the way around its orbit from its own perihelion in early January. Therefore, the date of such an opposition is late August/early September {in fact, late August}. Because the Earth is at aphelion in July, the closest oppositions will be slightly earlier {in fact, in mid August}.

Question 1.12

Eclipses can only occur when the Moon is at or near a node of its orbit when at the same time the nodes are on or near the line from the Earth to the Sun. If these nodes are *not* fixed with respect to the Earth's orbit around the Sun then the line-up will move through the months of the year, and consequently eclipses are not confined to particular months.

Question 2.1

(a) The greater the mass of a planet, the larger the displacement of the star from the centre of mass of the system (equation (1.7)), and therefore the larger the orbit of the star around the centre of mass. This makes it easier to detect and measure the stellar orbital motion.

(b) For a stellar orbit of given size, the angular size of the orbit is greater, the nearer the star, and therefore the easier the orbital motion is to detect and measure.

Question 2.2

With the giant planet spiralling inwards (Figure 2.2), there is no reason to expect high eccentricities in hot Jupiters unless the process that stops migration, and the subsequent disc dispersal, causes this. {In fact they do not. Hot Jupiters have low eccentricities. High eccentricities do occur in some of the other giant exoplanets.}

Question 2.3

The feature already apparent in circumstellar discs is feature 2 in Table 2.1 – that the orbits of the planets lie in almost the same plane {the plane of the circumstellar disc}, and that the Sun lies near the centre of this plane.

Question 2.4

(a) The magnitude of the average orbital angular momentum of a planet is given to a good approximation by

$$l_{orb} = mva \tag{2.1}$$

where m is the mass of the planet, v is its average orbital speed, and a is the semimajor axis of the orbit. The average speed is given by $2\pi a/P$ where P is the orbital period. Therefore

$$l_{orb} = m\left(\frac{2\pi a}{P}\right)a = m\left(\frac{2\pi a^2}{P}\right)$$

Using Kepler's third law

$$P = ka^{3/2} \tag{1.3}$$

we obtain

$$l_{orb} = m\left(\frac{2\pi a^2}{ka^{3/2}}\right)$$

Therefore

$$l_{orb} = \frac{2\pi}{k}ma^{1/2} \tag{2.2}$$

(b) We use equation (2.1) in the form in part (a)

$$l_{orb} = m\left(\frac{2\pi a^2}{P}\right)$$

with values for m, a, and P from Table 1.1. Thus, for the Earth

$$l_{orb} = 5.974 \times 10^{24}\,\text{kg} \left(\frac{2\pi(1.496 \times 10^{11}\,\text{m})^2}{3.156 \times 10^7\,\text{s}} \right)$$

$$= 2.662 \times 10^{40}\,\text{kg}\,\text{m}^2\,\text{s}^{-1}$$

From equation (2.2)

$$l_{orb} = l_{orb}(\text{Earth}) \frac{ma^{1/2}}{m_E a_E^{1/2}}$$

So, from Table 1.1, for Jupiter

$$l_{orb} = (2.662 \times 10^{40}\,\text{kg}\,\text{m}^2\,\text{s}^{-1}) \left(\frac{1.8988 \times 10^{27}\,\text{kg}}{5.974 \times 10^{24}\,\text{kg}} \right) \sqrt{\frac{7.782 \times 10^{11}\,\text{m}}{1.496 \times 10^{11}\,\text{m}}}$$

$$= 1.930 \times 10^{43}\,\text{kg}\,\text{m}^2\,\text{s}^{-1}$$

Similarly

$$l_{orb} = 2.504 \times 10^{42}\,\text{kg}\,\text{m}^2\,\text{s}^{-1}$$

for Neptune.

Question 2.5

The ice line divides the dust into two zones, with water ice dominating in the outer zone and rocky materials in the inner zone. The ice line during the formation of the Solar System was located roughly where the present division between the terrestrial planets and the giant planets lies (Figure 2.7).

Question 2.6

(a) Assume that all the dust in the sheet in the distance range 0.8–1.2 AU forms planetesimals in that range, and that no planetesimals come from elsewhere. In this case the total mass M of the planetesimals is the area between these distances times the average column mass. To sufficient accuracy, from Figure 2.7

$$M = \pi[(1.2\,\text{AU})^2 - (0.8\,\text{AU})^2] \times 10^2\,\text{kg}\,\text{m}^{-2}$$

Converting AU to metres (Table 1.6), we get

$$M = 6 \times 10^{24}\,\text{kg}$$

This is similar to the Earth's mass of $5.974 \times 10^{24}\,\text{kg}$, and so this suggests that much of the dust sheet in the range 0.8–1.2 AU went to form the Earth. This is consistent with 0.8 AU being beyond the orbit of Venus, and 1.2 AU being within the orbit of Mars.

(b) The size range of planetesimals is given in Section 2.2.3 as 0.1–10 km across, so use a typical size of 1 km across. {A typical size of 0.5 km is acceptable, but 5 km is not – there will be only a small proportion this big.} Taking the typical size to be the diameter of a sphere, the typical volume V is

$$V = \frac{4}{3}\pi(500\,\text{m})^3 = 5 \times 10^8\,\text{m}^3$$

With a mean density given in the question as $2500\,\text{kg}\,\text{m}^{-3}$, the typical mass m is $1 \times 10^{12}\,\text{kg}$. {It is sufficient to work to one significant figure.} Thus the number of planetesimals in the range 0.8–1.2 AU is of order

$$\frac{M}{m} = \frac{6 \times 10^{24}\,\text{kg}}{1 \times 10^{12}\,\text{kg}} = 6 \times 10^{12}\{\text{A lot!}\}$$

Question 2.7

The features in Table 2.1 that apply to the terrestrial planets are (in the numbering in the table) 2, 3, 6, 7, and 9. All of these features can be explained by solar nebular theories. {The rotation of Venus is not discussed explicitly in Chapter 2. Its retrograde rotation is discussed briefly in Section 10.3.3.}

Question 2.8

If the proto-Sun went through its T Tauri phase much earlier than in Figure 2.13, then the growth of the giant planets would have been stunted. Jupiter and Saturn would have acquired less massive envelopes of hydrogen and helium, and Uranus and Neptune might have got no further than the kernel stage.

Question 2.9

The inner boundary of the Oort cloud is thought to be spherical, centred on the Sun, with a radius of order 1000 AU. EKOs beyond about 40 AU from the Sun are thought to be a mixture of those formed *in situ* and those scattered by the giant planets. It is not known whether there is a significant population of EKOs at 1000 AU, but this cannot be ruled out, in which case the E–K belt would blend into the Oort cloud.

Question 2.10

The satellites of the giant planets in Table 1.2 (additional to Triton), that are likely candidates for a capture origin, are as follows:

Jupiter: Ananke, Carme, Pasiphae, Sinope

Saturn: Phoebe

Uranus: Caliban, Sicorax

Neptune: 2002 N1

These all have orbital inclinations greater than 90°, so are in retrograde orbits. They are also in large orbits, well away from the giant planet.

Question 2.11

(a) The present ring system of Saturn is the outcome of the accumulated effects of various processes that replenish ring particles, and various processes that remove them. The time scale on which these processes operate is much shorter than the age of the Solar System. Therefore, if the replenishment rate were to fall substantially below the removal rate, then at some time in the future the ring system would be much less extensive.

(b) By the same arguments as in part (a), it is possible that the ring system of Jupiter will be much more extensive in the future: this would require the replenishment rate to rise substantially above the removal rate, such as could follow the disruption of a small satellite.

Question 3.1

(a) The orbital inclinations of Pallas and Euphrosyne are unusually large for belt asteroids, the inclinations of belt asteroids being predominantly less than 20°. The eccentricity of Bamberga is considerably larger than the typical range of 0.1–0.2 for belt asteroids.

(b) To be at or near L_4 or L_5, the semimajor axis of the orbit must be similar to that of the planet (Figure 3.3) {and the eccentricity has to be similar}. Comparison of the semimajor axes in Table 1.3 with those in Table 1.1 shows that none of the asteroids in Table 1.3 is at or near any L_4 or L_5 points.

Question 3.2

It depends on how rapidly the number of asteroids increases with decreasing size. Consider the total mass in a certain range of size ΔR centred on R_1. The total mass is the number of asteroids in the range times the average mass of an asteroid in the range. If ΔR is centred on a smaller size R_2, the average mass decreases, and this can offset the greater number of asteroids in this new range, so that the total mass is considerably less than in the first case. {In Figure 3.5, unless the graph turns up unrealistically sharply at sizes below those shown, there is only a small fraction of the total mass in the asteroid belt awaiting discovery. You might have answered the question in a different way, perhaps without using symbols. Does your answer contain the essential idea, and does it express it succinctly?}

Question 3.3

(a) All of the asteroids in Figure 3.6 are smaller than the approximate 300 km radius above which gravitational forces make a body roughly spherical. {Vesta comes close.}

(b) A small spherical asteroid could have been born spherical, in which case it is not a collisional fragment, which would be non-spherical. Alternatively, regardless of its initial shape, it could have since suffered extensive erosion by dust impact, which tends to make asteroids spherical, and little by way of volatile loss, which tends to make them non-spherical.

Question 3.4

Among the five most populous asteroid classes – S, M, C, D, and P – the reflectance spectrum of Eros (Figure 3.9) has a similar shape (up to a wavelength of $1.1\,\mu$m) to that of class S, with class D as the only other plausible possibility (Figure 3.7). But the albedo of Eros is medium, like class S, and not low, like class D. Consequently, it is placed in the S class. Its surface is thus likely to be made of iron–nickel alloy mixed with appreciable proportions of silicates.

Though Eros is an NEA, its composition indicates that it probably originated in the inner region of the asteroid belt.

Question 3.5

The semimajor axis is given by (Figure 1.8)

$$a = \frac{q}{1-e}$$
$$= \frac{0.976\,586\,\text{AU}}{1-0.905\,502} = 10.334\,\text{AU}$$

The orbital period, from Kepler's third law (equation (1.3)), is

$$P = 10.334^{3/2}\ \text{years} = 33.22\ \text{years}$$

It is thus a short-period comet, with rather a long period for the Jupiter family. It thus seems to be a member of the Halley family. This is confirmed by its large orbital inclination. {Its inclination is over 90°, so it is in a retrograde orbit.}

Question 3.6

At 30 AU from the Sun there might be a coma consisting of the most volatile icy substances evaporated from the nucleus by the (feeble) heat of the Sun. Any such coma could well be too faint to be seen from the Earth. As the comet approaches the Sun the coma grows as less volatile materials, notably water, begin to evaporate, and the coma becomes visible. Evaporation is from vents in a residue of ice-depleted dust that covers the surface of the nucleus, and the venting gas carries dust into the coma. A huge hydrogen cloud forms, and also ion tails, dust tails, and perhaps other types of tail. The ion tail points away from the Sun; the dust tail only approximately so, and it is curved. As it recedes from the Sun the gases recondense onto the nucleus so the tails fade away and the coma also. There might be sporadic outbursts from reactivated vents that break through the dusty surface.

Question 3.7

The other property is the time spent near the Sun. For example, a comet in a low-eccentricity orbit will spend most of its orbital period not far from its perihelion distance, whereas a comet in a high-eccentricity orbit will dash through the inner Solar System, and therefore spend only a very short time near perihelion.

Question 3.8

The active lifetime of a JFC is of the order of a few thousand years. The order of 100 perihelion passages is thus consistent with an orbital period of the order of 10 years, which is typical of JFCs. {These are very much order-of-magnitude correspondences.}

Question 3.9

The feature of the orbits of HFCs that indicates an inner Oort cloud origin is their generally larger orbital inclinations than the other short-period comets. The orbits of EKOs generally have low-inclinations, and would tend to give rise to low-inclination short-period comets.

Question 3.10

Most meteorites are never found because

- they fall in the oceans (which cover most of the Earth's surface);
- they fall in remote parts of the continents;
- they get buried or otherwise obscured before anyone passes by;
- they are mistaken for terrestrial rocks.

Question 3.11

For the ratio of carbon to iron to be much the same as in the Sun, the meteorite would be of the most primitive kind, i.e. a carbonaceous chondrite (CC), particularly a C1 chondrite. Such meteorites have relative abundances of all the elements, much like those in the Sun, except for those elements that remained as gases, and so were not retained by the meteorites. Helium is one of these, and this is why in CCs the ratio of helium to carbon is far smaller than in the Sun.

Question 3.12

Suppose a mineral becomes isolated from its environment at $t=0$ with a number $N_0(^{87}\text{Rb})$ of ^{87}Rb nuclei. Equation (3.3) gives the number of radioactively unstable ^{87}Rb nuclei at time t as

$$N(^{87}\text{Rb}) = N_0(^{87}\text{Rb})e^{-t/\tau}$$

Therefore, the *radiogenically* produced ^{87}Sr grows as

$$N(^{87}\text{Sr}) = N_0(^{87}\text{Rb}) - N(^{87}\text{Rb}) = N_0(^{87}\text{Rb})(1 - e^{-t/\tau}) = \frac{N(^{87}\text{Rb})}{e^{-t/\tau}}(1 - e^{-t/\tau})$$

Therefore

$$N(^{87}\text{Sr}) = N(^{87}\text{Rb})(e^{t/\tau} - 1)$$

We must now add the ^{87}Sr present in the mineral at $t = 0$. This gives us

$$N(^{87}\text{Sr}) = N(^{87}\text{Rb})(e^{t/\tau} - 1) + N_0(^{87}\text{Sr})$$

We now divide by $N_0(^{86}\text{Sr})$, where ^{86}Sr is a stable isotope of strontium. Thus

$$\frac{N(^{87}\text{Sr})}{N_0(^{86}\text{Sr})} = \frac{N(^{87}\text{Rb})}{N_0(^{86}\text{Sr})}(e^{t/\tau} - 1) + \frac{N_0(^{87}\text{Sr})}{N_0(^{86}\text{Sr})}$$

At a given time t, $(e^{t/\tau} - 1)$ is a constant, so a graph of $N(^{87}\text{Sr})/N_0(^{86}\text{Sr})$ versus $N(^{87}\text{Rb})/N_0(^{86}\text{Sr})$ is a straight line with a slope $(e^{t/\tau} - 1)$.

Question 3.13

If the meteorite came from the outer metre or so of a larger body then it will have been exposed before it was liberated, and so its exposure age would be greater than the time from when it was liberated. However, its exposure age could not exceed its solidification age because cosmic ray tracks cannot survive in a liquid.

Question 3.14

A possible origin of the stony-iron meteorites is the interface between the iron core and the silicate mantle of differentiated asteroids. Stony-irons, as their name indicates, consist of mixtures of silicates and iron (alloyed with nickel), clearly in accord with their suggested origin. The OCs are thought to come from S class asteroids – these are undifferentiated. The OCs consist of silicates, with an elemental composition (excluding volatile substances) similar to the Sun. The S class asteroids also consist of silicates, and, as undifferentiated bodies formed at a few AU from the Sun, are thought also to have an elemental composition (excluding volatile substances) similar to the Sun.

Question 3.15

One reason is that the comet is in such a long-period orbit that it has not passed through the inner Solar System in recorded history. An alternative reason is that the comet has devolatilised, or become disrupted, so is no longer visible.

Question 4.1

In Figure 4.1 there is a clear gap in size between the pair of giant planets Jupiter and Saturn and the other pair Uranus and Neptune. Moreover, because of their greater mass, the mean densities of Jupiter and Saturn have been more increased by compression than have those of Uranus and Neptune, with Jupiter more compressed than Saturn. Thus, in equal states of low compression, Jupiter and Saturn could be distinctly less dense than Uranus and Neptune. This points to different compositions. Therefore, on the basis of size and composition, Jupiter and Saturn could form one subgroup, and Uranus and Neptune another. {This is the case, as you will see later. Note that no reference was made to the *uncompressed* mean densities of the giant planets. This is because much of their mass would be gaseous in such a state, and an uncompressed gas will expand indefinitely.}

Question 4.2

At double the distance the $1/r^2$ term in equation (4.7) is reduced by a factor of 4, but the J_2 term, which is proportional to $1/r^4$, is reduced by a factor of 16. Therefore, the ratio of these terms is reduced by a factor of 4. Any further extra terms fall off no less rapidly, and so, at double the distance, the field is more closely that of a spherically symmetrical body.

Question 4.3

For a hollow cylindrical shell, with a radius R, every mass element δM is a distance R from the longitudinal axis. Thus the sum of $\delta M R^2$, which is C, is MR^2 where M is the total mass of the shell. Therefore $C/MR^2 = 1$.

Question 4.4

A self-exciting dynamo generates a large magnetic dipole moment if there is an electrically conducting fluid in the interior of the planetary body, an energy source to sustain convection in the fluid, and rapid rotation of the planet to coordinate the motions. The Earth and the four giant planets rotate relatively rapidly (Table 1.1). In Section 1.2.1 it was stated that the Earth has an iron-rich core, largely liquid, and in Section 1.2.2 it was stated that the giants are fluid throughout. Therefore, a reasonable hypothesis is that these five planets not only rotate sufficiently rapidly, but also contain electrically conducting fluids in convective motion. {This is the case.}

For the other planets one or more of these conditions must not be met. Venus, which is the Earth's twin in size and mass, and probably has an iron-rich, largely liquid core (Section 1.2.1), differs in that it rotates slowly, so this might explain the absence of a large magnetic dipole moment. {Another reasonable possibility is that internal energy sources are insufficient to sustain convection.} Mercury also rotates slowly, so that might be the explanation for its lack of a large magnetic dipole moment too. {As well as a lack of sufficient internal energy sources, another reasonable possibility is that there are no conducting fluids.} Mars rotates about as rapidly as the Earth, so reasonable hypotheses are that either it contains no electrically conducting fluids, or, if it does, the internal energy sources are insufficient to sustain convection. Pluto rotates slowly, and it is a small body a long way from the Sun, so could well be solid throughout. {Chapter 5 will explore these hypotheses in more detail.}

Question 4.5

{With an entirely fluid interior, only P waves exist. The P wave speed will change with depth. The likely case of an increase with depth is shown here. Half way to the centre, the P wave speed will increase in accord with equation (4.9) by the factor $\sqrt{2/1.2}$, which is 1.29. The speed is then assumed to increase further to the centre, also likely.}

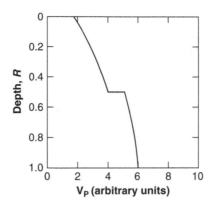

Question 4.6

We use equation (4.12) with ρ_m and R for Jupiter taken from Table 1.1, and G taken from Table 1.6. Thus

$$p_c \approx \frac{2\pi \times 6.672 \times 10^{-11}\,\text{N}\,\text{m}^2\,\text{kg}^{-2}}{3}(1330\,\text{kg}\,\text{m}^{-3})^2(7.149 \times 10^7\,\text{m})^2$$

$$\approx 1.3 \times 10^{12}\,\text{N}\,\text{m}^{-2} = 1.3 \times 10^{12}\,\text{Pa} = 1300\,\text{GPa}$$

{This is about 3.4 times the pressure at the Earth's centre. Models of Jupiter yield the more accurate value of about 4000 GPa, so equation (4.12) underestimates the pressure, as expected.}
  At shallower depths the pressure must be less – pressure increases with depth.

Question 4.7

As the water is compressed at constant temperature the solid + gas phase boundary is encountered {just to the left of the triple point}. The water then solidifies. On further compression the solid + liquid phase boundary is encountered {this is because this boundary slopes to the left from the triple point}. The water then melts to become a liquid. Further compression increases the density of the liquid.

Question 4.8

{Much of the detail in the figure is arbitrary. The important features are

- the rates of energy release from accretion and initial differentiation rise to an early maximum and then decline to zero, with differentiation later than accretion;
- the rate of energy release from short-lived radioactive isotopes is now negligible, whereas that from long-lived isotopes, though declining, is still significant {many short-lived isotopes decay far more quickly than the example shown};
- ongoing differentiation, tidal energy (where it is significant), and solar energy are more sustained, with solar energy (in this example) making a relatively minor contribution (and increasing as the solar luminosity increases).

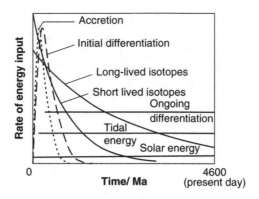

The details will differ greatly from one planetary body to another.}

Question 4.9

(a) From equation (4.6), the difference Δg in the magnitude of the gravitational field across a body with a radius R, due to another body with a mass M a distance r away, is

$$\Delta g = GM\left(\frac{1}{(r-R)^2} - \frac{1}{(r+R)^2}\right)$$

This can be rearranged to give

$$\Delta g = GM\left(\frac{(r+R)^2 - (r-R)^2}{(r-R)^2(r+R)^2}\right)$$

Expanding the brackets on the top line and cancelling terms gives

$$\Delta g = GM\left(\frac{4rR}{(r-R)^2(r+R)^2}\right)$$

If r is much greater than R, then $(r-R)$ and $(r+R)$ are each equal to r to a good approximation. Thus

$$\Delta g \approx GM\left(\frac{4rR}{r^4}\right) = 4GM\left(\frac{R}{r^3}\right)$$

(b) For the Earth, the ratio of the solar tidal field to the lunar tidal field is $4GM_\odot(R_E/a_E^3)$ divided by $4GM_M(R_E/a_M^3)$, where R_E is the radius of the Earth, $M_\odot$ is the mass of the Sun, a_E is the semimajor axis of the Earth's orbit, M_M is the mass of the Moon, and a_M is the semimajor axis of the Moon's orbit. The ratio is thus

$$\frac{M_\odot a_M^3}{M_M a_\odot^3}$$

Putting in the values from Tables 1.1 and 1.2, this ratio is 0.46, or 46%. {As well as Δg, the magnitude of the tidal 'squeeze' (Section 1.4.5) is also proportional to $GM(R/r^3)$.}

Question 4.10

(a) Radiative transfer is important only at high temperatures, so it is unlikely to be significant. Local regions of molten material are also unlikely, so advection is probably not occurring. With low temperature gradients beneath the surface layer, there is no convection below the surface layer, which itself is probably at too low a pressure for (solid state) convection. Therefore, conduction is very probably the only means by which energy is being transferred to the surface layer.

(b) The thin surface layer must have a lower heat transfer coefficient than the rest of the body.

Question 4.11

(a) Accretional energy was endowed in a short period a long time ago, so the temperature will long since have become uniform throughout the interior.

(b) Long-lived radioactive isotopes will still be releasing energy in the core, so there will be a temperature increase towards the core.

(c) Because A/M is smaller in the larger body, it will have cooled less, so the temperature increase towards the centre will be greater. {This assumes that the heat transfer coefficients in the interior are similar in both bodies.}

Question 5.1

Were the Earth's outer core to solidify

- there would be no electrically conducting fluid and so (according to the self-exciting dynamo theory) no magnetic dipole moment;
- seismic S waves would be able to travel throughout the interior, unless the latent heat released on solidification melted the lower mantle.

{The contraction of the core on solidification would wreak havoc with the crust. Further crustal havoc might result from the latent heat released, which could make the solid state convection in the mantle more vigorous.}

Question 5.2

Venus might have an asthenosphere as extensive as that of the Earth, i.e. extending throughout most of the mantle. Support for this view is that Venus probably has a similar composition to the Earth, with comparable pressures and temperatures at a given depth. In this case the 'plasticity profile' could be similar in the two planets. Moreover, the surface displays evidence of extensive geological activity a few hundred million year ago, consistent with a convective interior presently overlain by a lithosphere that acts as a stagnant lid.

If Mercury has an asthenosphere at all, it will extend over a small range of radii, and be located deep down, where pressures and temperatures are sufficiently high. Mercury is a small world, and in spite of its high mean density its internal pressures are not as high as in the Earth (Table 5.2). Its surface indicates that the lithosphere is thick.

Question 5.3

If Mars rotated very slowly it would be less flattened by rotation, and so the gravitational coefficient J_2 would be smaller and consequently more difficult to measure. Also, the more spherical form of Mars would result in a reduction in the torques that cause precession of the rotation axis, and a consequent increase in the precession period. At present the precession period is 0.1711 Ma, so would be more difficult to measure if it were even longer.

We would thus have a far less precise value of C/MR_e^2 and consequently a much poorer constraint on the variation of density with depth.

Question 5.4

Planetary body A will have a cooler interior than planetary body B for either one or both of the following reasons:

(1) body A has had less copious energy sources;
(2) body A has lost energy at a greater rate.

Accretional energy per unit mass is roughly proportional to $\rho_m R^2$ (Section 4.5.1), and so the Earth would have received more accretional energy per unit mass than the considerably smaller Mars and Moon. The Earth's core is also a greater proportion of the mass of the Earth than the cores of Mars and the Moon in proportion to those bodies, so any energy of differentiation would have been greatest in the Earth. Insufficient information is given about radioactive sources and any other energy sources, to decide whether, overall, the Earth has had more copious energy sources than Mars and the Moon.

Equation (4.14) shows that, other things being equal, the energy loss rate per unit mass has been greater from Mars and the Moon than from the Earth because of their greater area-to-mass ratios. However, the generally higher temperatures in the Earth have promoted solid state convection, and advection could also have been more effective (Table 4.5), thus increasing the Earth's heat transfer coefficient, and hence obscuring the issue.

{In fact, the smaller sizes of the Moon and Mars are indeed thought to be the main reason for their cooler interiors.}

Question 5.5

Pluto is an icy–rocky body because it formed well beyond the ice line in the solar nebula (Section 2.2.2). It is small because beyond Neptune there was a low spatial density of planetes-imals and embryos, and slow orbital motion, resulting in a low collision rate. There was also a lack of nebular gas sufficient to reduce the eccentricities of these bodies. Consequently, their collision speeds were so high that fragmentation was more likely than accretion (Section 2.2.6).

Question 5.6

Adapting the answer to Question 4.9, the tidal field produced by the Earth across the Moon is given by $4GM_E(R_M/a_M^3)$, where M_E is the mass of the Earth, R_M is the radius of the Moon, and a_M is the semimajor axis of the Moon's orbit. In the case of tides produced *in the Earth*, the Sun is only 46% as effective as the Moon (Question 4.9(b)), and it will be even less effective for the Moon because the Earth/Sun mass ratio is much greater than the Moon/Sun mass ratio. So we can neglect the tide produced in the Moon by the Sun.

The tidal field produced across Io by Jupiter is given by $4GM_J(R_I/a_I^3)$, where M_J is the mass of Jupiter, R_I is the radius of Io, and a_I is the semimajor axis of Io's orbit. The tide produced in Io by the Sun and other bodies is negligible by comparison.

The ratio of the tidal field produced across the Moon by the Earth, to that produced across Io by Jupiter, is thus

$$\frac{M_E R_M a_I^3}{M_J R_I a_M^3}$$

Putting in the values from Tables 1.1 and 1.2, this ratio is 0.004 i.e. 250 times greater in Io.

Equation (4.13) shows that tidal heating is proportional to orbital eccentricity. The eccentricity of the lunar orbit (0.0554) is 13.5 times that of Io (0.0041). The Moon being much less geologically active than Io, it must be the case that the far greater tidal field across Io more than offsets the greater eccentricity of the lunar orbit in determining tidal heating. {This is indeed the case.}

Question 5.7

As Jupiter's interior cools the rate of settling of helium will increase, releasing energy of differentiation. It is triggered by cooling because the lower the temperature of the metallic hydrogen mantle, the lower the miscibility in it of helium. The heat from this differentiation will tend to sustain the internal temperatures, and so their decline will be retarded.

Question 5.8

The important radioactive isotopes are elements found in rocky materials (Table 4.4). Rocky materials are a minor ingredient of the giant planets, and therefore radioactive isotopes are only a minor energy source.

Question 5.9

(a) If the Earth had no atmosphere then the atmospheric source of electrons and ions for the Earth's magnetosphere would be replaced by the Earth's surface. It is thus possible that the nature of the plasma belts would change {they would}. There would be no aurorae.

(b) If the solar wind had a lower speed v and a lower number density n of charged particles, then the factor $\mu^{1/3}/n^{1/6}v^{1/3}$ (Section 5.4.1) would be increased, and the magnetosphere would be larger. The number density of solar wind particles in the magnetosphere would be reduced.

Question 6.1

Because of its slow rotation, Titan will not be rotationally flattened, and in view of its large size it is likely to be approximately spherical. Therefore, a sphere centred on Titan's centre of mass, and lying within or near the range of radii at the surface, could serve as a zero of altitude. {This is the approach adopted, zero altitude being the sphere with a radius of 2575.0 km.}

Alternatively, a specific value of mean atmospheric pressure could be used, somewhere near the mean surface pressure. {A body does not need to be rotationally flattened for atmospheric pressure to be used in this way.}

Question 6.2

If the atmosphere of Venus were transparent to radar, then radar could be used to establish surface morphology and composition. If circularly polarised radar were used then extra information on surface composition would be obtained. {The atmosphere of Venus is indeed transparent to radar, and much of what we know about the surface comes from radar.}

Gamma and X-ray fluorescence can provide further information on composition. {Neutron spectrometry is not feasible, because the neutrons would be unable to traverse the atmosphere.}

Question 6.3

Pure silica is not subject to partial melting or fractional crystallisation because it is a single mineral, not a mixture of minerals.

Question 6.4

For the summit crater in Figure 6.5(b) to be of impact origin the impact would have had to be in the centre of a pre-existing mountain – rather unlikely. If the composition were typical of extrusive volcanism, then this would strongly indicate that the crater was a volcanic caldera. If around the crater the layers of rock were not turned over {as they are in Figure 6.9(c)} then this would be strong evidence against an impact origin. {You might have spotted other indicators, such as the profile of the shield volcano in Figure 6.5(b) being unlike the impact crater profiles in Figure 6.10.}

Question 6.5

(a) With a crater density on Chryse Planitia of 2.2×10^{-4} per km^2 for craters greater than 4 km in diameter, there are $(10^5 km^2)(2.2 \times 10^{-4}$ per $km^2)$ such craters on a typical area of $10^5 km^2$. This is 22 craters greater than 4 km diameter.

(b) If the Earth–Moon data in Figure 6.14 apply to Mars, then Chryse Planitia was resurfaced about 3500 Ma ago. However, the impact rate on Mars throughout much of Solar System history has been greater than in the Earth–Moon system. Therefore Chryse Planitia must have been resurfaced more recently than 3500 Ma ago. {Quite when is uncertain. One estimate is 3400 Ma.}

Question 6.6

If a planet has an atmosphere but no surface liquids then gradation will be via

- disintegration by impact cratering, dry thermal cycling, seismic waves, chemical reactions, UV irradiation, removal of adjacent material;
- erosion by micrometeorites and wind-borne dust;
- mass wasting and aeolian transport processes;
- deposition from the atmosphere only (not from rivers, lakes, or oceans).

Question 7.1

The anorthositic composition of the lunar highlands is thought to have been derived by differentiation from a magma ocean of peridotite composition. If the highlands had a peridotite composition this would imply that no such magma ocean formed.

Question 7.2

(a) Mare infill derived through gradation from the highlands would have the mineralogical composition of the highlands, i.e. anorthosite rather than basalt.
(b) Because of lithospheric thickening and interior cooling, lava has been unavailable since about 3100 Ma ago, and so the only infill of an impact basin formed today would be through gradation.

Question 7.3

{This question is similar to Question 5.4.}

A thicker lithosphere is a consequence of a cooler interior, so we have to look at energy gains and losses.

Accretional energy per unit mass is roughly proportional to $\rho_m R^2$ (Section 4.5.1), and so Mercury would have received less accretional energy per unit mass than the considerably larger Earth and Venus. This might have been offset by energy of differentiation – Mercury's core is a greater proportion of the mass of Mercury than the cores of the Earth and Venus in proportion to those bodies (Figure 5.1). We have no surface analyses of Mercury so can say little about radiogenic heating, and so it is not possible to say, overall, if Mercury has had less copious energy sources than the Earth and Venus.

Equation (4.14) shows that, if the heat transfer coefficients have been comparable, then the energy loss rate has been greater from Mercury than from the Earth and Venus, because Mercury has a higher area-to-mass ratio. However, the generally higher temperatures in the Earth and Venus have promoted solid state convection, thus increasing these two planets' heat transfer coefficients, hence obscuring the issue.

{Detailed thermal models show that, in essence, the higher area-to-mass ratio of Mercury is the underlying reason for its cooler interior, and consequently for its greater lithospheric thickness.}

Question 7.4

Syrtis Major, the triangular dark feature to the right, is an example of a dark region. These are where the light dust is streaky, with a higher proportion of dark dust, and with exposure of dark underlying terrain. The dark areas are at locations determined by small-scale topography, wind speed, and direction.

To the left of Syrtis Major is an example of a bright region. These are where the light-red dust provides the dominant covering. The red tint is the result of iron-rich minerals. It is thought that the light dust is derived from the dark dust (also iron rich) by a variety of physical and chemical processes. The dark dust is basaltic, whereas the light dust includes various clay minerals. The dark dust itself is derived from the basaltic crust.

To the north are the smooth plains of the northerly hemisphere. These are the result of basaltic lava flows, resulting in a comparatively young surface, almost free from impact craters. The volcanic activity is thought to be the result of a thinner northerly crust, the result of crustal thinning early in Martian history, perhaps caused by convection.

To the south of Syrtis Major is the Hellas Basin, the result of a giant impact when Mars was young. To the left of Hellas there are examples of smaller impact craters.

The generally cratered southerly hemisphere is apparent in the lower half of the image. Here the crust is thicker, and therefore volcanic activity has for long been rare. Consequently, many impact craters are present.

At the bottom of the disk the south polar cap is visible. Its rather small extent is consistent with it being summer/near summer in the southern hemisphere {in fact, summer}. This (receding) cap is formed from deposits of CO_2 ice particles (plus dust). The residual cap will have the same composition.

The atmospheric haze at the top of the disc is cloud over the winter/near-winter north polar cap. It must consist of the icy particles of the condensibles in the atmosphere, i.e. water and/or CO_2 {in fact, mainly CO_2 – see Section 10.2.1}.

Question 7.5

Among the water-related features on Mars, ejecta blankets such as that in Figure 7.13 are the most unlikely to be found on the Earth. Though such blankets might be produced in areas of deep permafrost {rare on the Earth}, the production rate would be very low and the removal rate, by weathering and geological processes, high. {No such blankets are known on the Earth.}

Question 7.6

The northerly hemisphere of Mars has many volcanic and tectonic features, and the Martian meteorites indicate that magma was solidifying as recently as 165 Ma ago. There is evidence that volcanism was occurring in the Tharsis and Elysium regions as recently as 3 Ma. Therefore, though there is no direct evidence for volcanic and tectonic activity today, it is possible that it is still ongoing. If this were not very rare, the material on Mars in this chapter would fit better in the chapter on active surfaces. {In fact volcanic and tectonic activity is so sparse today that Mars fits comfortably in this present chapter.}

Question 7.7

Had the Martian crust in the southerly hemisphere been as thin as that in the northerly hemisphere then it is possible that the southerly hemisphere would resemble the northerly hemisphere. In this case there would be no heavily impact-cratered surface, and the evidence from the Martian channels for an early warmer, wetter epoch would be absent. Instead, there would be more volcanic and tectonic features.

Question 7.8

To cause cryovolcanism, the outer shell of each body needs to have its temperature raised so that (partial) melting can occur. If Pluto were moved closer to the Sun then solar radiation could accomplish this. In the case of Ganymede, if it were moved closer to Jupiter, or if its orbital eccentricity were increased, then tidal heating could cause cryovolcanism.

Question 8.1

(a) Plate tectonics promotes rapid recycling of oceanic crust, and promotes volcanism that, along with erosion, resurfaces continental crust. Therefore, most of the Earth's surface is young and impact craters consequently rare.

(b) The rift valley is explained as a new constructive margin. {Iceland is oceanic crust, so this is not a rift valley within continental crust.}

(c) Mountain ranges in the midst of continents are often sites where two slabs of continental crust have met, and crumpled. {As is the case here.}

(d) At shallow depths near mid-oceanic ridges and ocean trenches there is a comparatively large amount of upward and sideways motion, and partial melting. This general activity is expected to generate stresses and fractures in the crust and in the solid mantle beneath, which will be observed as seismic activity.

(e) Figure 8.2 shows that the Aleutian Islands border a destructive margin, and so they could be an island arc. {This is the case.} Island arcs are caused by partial melting, fractional crystallisation, and volcanism at destructive margins where there are no continents at the margin.

Question 8.2

The Venusian surface dates from no more than about 800 Ma ago. Large impacts have been so rare since the end of the heavy bombardment 3900 Ma ago that it is no surprise that Venus has not suffered an impact large enough to produce an impact basin in the last few hundred million year.

Question 8.3

On the Earth granites are intrusive rocks formed at destructive margins by fractional crystallisation and the partial melting of continental crust (Section 8.1.2). An essential feature of these margins is subduction. Subduction probably occurs on Venus only in a few locations (at certain chasmata), involving small quantities of crust, so perhaps granites are formed at such sites.

Question 8.4

If Io were devoid of sulphur then the magma would be devoid of what seems to be major volatile components, such as sulphur dioxide and sulphur itself. Therefore, the volcanic plumes would not reach as high, and perhaps the volcanism would be almost entirely effusive. The surface would be less colourful, because much of the present colour comes from a veneer of sulphur and its oxides.

Question 8.5

The decreasing tidal heating with increasing distance from Jupiter (equation (4.13)) leads one to expect that the surface of Io will show signs of greater activity than Europa, with Callisto showing the least signs of activity. This is as observed. Io has extensive silicate volcanism, and a young surface devoid of impact craters. Europa has cryovolcanism, liquid or slushy oceans of water, and perhaps some volcanic activity on the rocky floor of the oceans. Ganymede displays evidence of viscous relaxation and of episodic partial resurfacing by cryovolcanism, perhaps at times when tidal heating was greater than today. Callisto displays less viscous relaxation, and little sign of cryovolcanic resurfacing since soon after its formation – it is heavily cratered.

Question 8.6

At the low atmospheric and surface temperatures on Titan, water is well below its triple point and is thus very hard, behaving rather like rock on the Earth. Therefore, there is no flow of liquid water at the surface, and nothing that behaves like glaciers on Earth. There is very little water in the atmosphere, and it cannot be present as a liquid. Consequently there is a negligible water–frost cycle, negligible snow, and no rain of water.

Question 8.7

It is only in the outer Solar System that temperatures can always have been low enough for the very volatile N_2 to have been retained as a solid (Section 5.2.2). N_2 has only been detected on satellites beyond Saturn and on Pluto itself. If there is subsequent modest internal heating, this can vaporise some of the N_2, giving nitrogen cryovolcanism, though sufficient internal heating would be restricted to the larger of this group of bodies. Triton is the largest of all, but still only has weak N_2 volcanism in spite of powerful tidal heating early in its history. Pluto is next down in size, and is thought to be inactive, partly because of its smaller size, and partly because it probably never experienced as much tidal heating as Triton (Section 7.4.1).

Question 9.1

(a) If the number of CO_2 molecules in the Martian atmosphere were to decrease, the *area* of the spectral absorption line in Figure 9.1 would decrease.
(b) If the temperature of the Martian atmosphere were to increase, the Doppler width of the line would increase. If Doppler broadening was not initially determining the line shape it would increasingly do so as the temperature rose. {Such a rise could also reduce the line area, if a proportion of the CO_2 was thereby in a higher energy state to start with. The absorption lines of such molecules would be at longer wavelengths.}
(c) If the pressure of the Martian atmosphere were to increase the collisional broadening of the line would increase. If collisional broadening was not initially determining the line shape it would increasingly do so as the pressure rose.

Question 9.2

(a) If W_{int} is negligible, then, from equations (9.3) and (9.4),

$$L_{out} = F_{solar} A_p (1 - a_B)$$

Substituting this expression for L_{out} into equation (9.7) gives

$$T_{eff} = \left(\frac{F_{solar} A_p (1 - a_B)}{A\sigma} \right)^{1/4}$$

For a spherical body the projected area $A_p = \pi R^2$ and the total surface area $A = 4\pi R^2$. Thus

$$T_{eff} = \left(\frac{F_{solar}(1 - a_B)}{4\sigma} \right)^{1/4}$$

(b) The missing value is for the Earth. The Earth's average distance from the Sun is 1.496×10^{11} m (Table 1.1), and so, from equation (9.5) and with $L_\odot$ from Table 1.6,

$$F_{solar} = \frac{3.85 \times 10^{26} \, W}{4\pi (1.496 \times 10^{11} \, m)^2}$$

$$= 1370 \, W \, m^{-2}$$

Thus, using σ from Table 1.6 and a_B from Table 9.2,

$$T_{eff} = \left(\frac{1370 \, W \, m^{-2}(1 - 0.30)}{4 \times 5.67 \times 10^{-8} \, W \, m^{-2} \, K^{-4}} \right)^{1/4} = 255 \, K$$

Question 9.3

(a) To rewrite equation (9.12) in terms of density, the perfect gas equation of state (equation (9.2)) needs to be used in the rearranged form

$$\rho = p \left(\frac{m_{av}}{kT} \right)$$

Because m_{av} and T are assumed to be the same at all z in equation (9.12), they are the same at the surface as elsewhere, and so

$$\rho = \rho_s \, e^{-z/h}$$

where ρ_s is the density at zero altitude. {Thus, because density is proportional to pressure, the form of the equation is the same for density as for pressure.}

(b) A planetary atmosphere is approximately isothermal in the mesosphere {the Earth is an exception – Section 10.1.1}. If (as in Figure 9.8) the mesosphere is in the homosphere then (except for any condensable substances) it will also have uniform composition. It is thus in such a mesosphere that the pressure is expected to vary with altitude in accord with equation (9.12).

Question 9.4

Though solar UV radiation is a small fraction of the total luminosity (Figure 9.5(b)), it is the main energy source for the mesosphere and thermosphere, and so these domains would be cooler, and the increase of temperature with increasing altitude in the thermosphere would be less steep. Also, without solar UV radiation there would be only a feeble ionosphere {from cosmic ray impacts}.

Question 9.5

At a partial pressure of 100 Pa, the solid + gas line in Figure 9.9 is at about 253 K. Therefore, if the partial pressure did not change with altitude the cloud base would be at an altitude of $(293\,\text{K} - 253\,\text{K})/7\,\text{K}\,\text{km}^{-1}$, i.e. 5.7 km. This is a large change in altitude, and so the pressure will be less, and consequently the partial pressure of water too. A better estimate is to follow a path parallel to that shown from point B in Figure 9.9, starting at 293 K and 100 Pa. This shifted path intercepts the solid + gas line at a temperature of about 247 K. In this case the altitude of the cloud base would be about 6.6 km.

In either case, the cloud will consist of water-ice crystals, though if supercooling persists, then it will consist of liquid water droplets.

Question 9.6

(a) An atmosphere consisting of single atoms would be a weak absorber of mid-IR radiation, and therefore the greenhouse effect would be weak.

(b) Water molecules consist of three atoms, and so water vapour is a strong absorber of mid-IR radiation. Therefore, if a dry atmosphere were to have water vapour added to it the greenhouse effect would be larger, and T_s would rise. {However, if clouds became a lot more extensive, this rise could be offset by the increase in albedo and the consequent decrease in temperature.}

Question 9.7

(a) With a given exosphere radius, the escape speed is proportional to $\sqrt{M}$ (equation(9.17)), where M is the mass of a planetary body. Therefore, the thermal escape rate is dependent on M. Chemical escape and hydrodynamic escape must also depend on M, because the greater the escape speed, the less likely it is that molecules attain it, even if their speeds are non-thermal. {Impact erosion is a more difficult case. Greater planetary mass makes it less likely that atmosphere will be ejected, but greater mass also results it greater impact energy, and for a population of small potential impactors, greater planetary mass increases the collection rate.}

(b) The temperature at the base of the exosphere affects the thermal escape rate: the greater the temperature, the greater the rate.

(c) The greater the mass of a molecule, the less likely it is to have the escape speed. The mass thus affects the rates of thermal escape, chemical escape, and impact erosion. The greater the mass of a molecule, the less likely it is to be entrained in outflow, and so the rate of hydrodynamic escape is reduced too.

(d) The solar UV flux is essential for at least some forms of chemical escape, and any reduction in this flux will reduce the rate of chemical escape. {A reduction in solar UV will also affect photochemical reactions deep in the atmosphere, and at the surface.}

Question 9.8

(a) Equation (9.20) gives the required speed. We need first to obtain the equatorial surface speed v_e. This is obtained from the Earth's equatorial radius and sidereal rotation period in Table 1.1. Thus

$$v_e = \frac{2\pi \times 6.378 \times 10^6 \, m}{0.9973 \times 24 \times 3600 \, s}$$
$$= 465 \, m \, s^{-1}$$

Thus from equation (9.20)

$$v_{rel} = \frac{465 \, m \, s^{-1}}{\cos(30°)} - 465 \, m \, s^{-1} \, \cos(30°)$$
$$= 134 \, m \, s^{-1}$$

{Quite a wind! The wind would be reduced for example by friction at the surface.}
In the northern hemisphere the parcel moves from west to east, and in the southern hemisphere from east to west.

(b) If the Earth were to rotate much more slowly, then as well as lower east–west speeds, the Coriolis effect would be less disruptive on the Hadley circulation, and so the equatorial Hadley cells might extend to higher latitudes. Also, the cyclonic–anticyclonic activity at mid latitudes might be less.

Question 10.1

Figure 10.1 shows that in the Earth's stratosphere the lapse rate is zero or negative (temperature *increases* with increasing altitude), and so there will be no convection, and consequently no condensation.

Question 10.2

(a) The Earth's biosphere sustains both of the major present constituents of the atmosphere – O_2 and N_2. Without the biosphere the O_2 would largely be lost in oxidising surface substances and certain volcanic gases, and most of the N_2 would end up as the nitrate ion in the oceans. The Earth's atmosphere would then consist mainly of water vapour, argon (Table 9.1), and CO_2. {The carbon cycle will become altered so that the atmospheric mass of CO_2 is increased, and it will become the main constituent.}

(b) Without knowing the new masses of the greenhouse gases H_2O and CO_2 in the atmosphere, and any changes in albedo {due to changes in cloud cover and in ice and snow at the surface}, it is impossible to estimate the change in the Earth's GMST.

(c) Solar UV radiation at the surface would probably be greater because of the greatly reduced O_2 content of the atmosphere, and the consequent reduction in oxygen atoms in the thermosphere and ozone in the stratosphere. {This could be offset by greater cloud cover.}

Question 10.3

When the Earth's axial inclination is zero, the flux density on the Earth's surface at the poles is always extremely small because the Sun is always on the horizon. With its present axial inclination, though at the poles the Sun is below the horizon for about half the year, it is well above the horizon in midsummer. Consequently, the solar flux density at the poles, averaged over a year, is less when the Earth's axial inclination is zero than it is at present.

Question 10.4

(a) At midsummer in the southern hemisphere the Hadley cell on Mars extends from the subsolar latitude (in the southern hemisphere) to the northern (winter) hemisphere. After midsummer the subsolar latitude moves northwards, and with it the Hadley cell. Around the equinox the Hadley cell transforms into one each side of the equator, perhaps with smaller cells at higher latitudes, but with the approach of midsummer in the northern hemisphere the single Hadley cell becomes re-established, now extending from a northerly subsolar latitude to the southern hemisphere. Thereafter, the cell migrates southwards, transforms around the equinox as before, and becomes re-established as at the start of this answer.

(b) As the axial inclination of Mars varies, the midsummer subsolar latitude swings towards and away from the equator, being nearer the equator at low inclinations. For example, the location of the Hadley cell at midsummer in the southern hemisphere lies further south at high inclinations, and further north at low inclinations, possibly splitting into the configuration seen at the equinoxes at the present time.

Question 10.5

For a positive lapse rate it is necessary that there is some heating from below. With no solar radiation penetrating the clouds, heating would be from above (ignoring the small outflow from the interior). Consequently the temperature would vary little with altitude, i.e. the lapse rate below the clouds of Venus would be close to zero. {This point was made in relation to planetary interiors, in Question 4.11(a).}

Question 10.6

From Figure 10.12, the global mass fraction of water on Venus today has a lower limit of about 2×10^{-9}, so $10^2 - 10^3$ times this value is a lower limit of between 2×10^{-7} and 2×10^{-6}. On the Earth the lower limit is much higher, about 2×10^{-4}, so before the loss Venus would probably still have been drier than the Earth.

Question 10.7

Figure 10.12 shows that the lower limits of the global mass fractions of CO_2 for Venus, the Earth, and Mars are, respectively, 1×10^{-4}, 6×10^{-5}, and 3×10^{-6}. The atmospheric quantity on Venus is only slightly less than the lower limit on its global value, whereas for the Earth it is less by a factor of about 10^5, and by a factor of about 100 for Mars.

The very first atmospheres seem to have been lost, presumably when the Sun went through its T Tauri phase. Subsequently, the late stages of accretion would have brought CO_2 to these three planets. Even if all of this endowment was lost through hydrodynamic escape, impact erosion, and blow-off, a late veneer would have resupplied the planets. It is more likely that some of the earlier endowment was retained inside the planet, and that a proportion of this was outgassed, to add to the late veneer. Since the end of the heavy bombardment there has been little by way of loss to space, and so we can account for the quantities of CO_2 for Venus, the Earth, and Mars being substantial.

The differences in the global mass fractions of CO_2 have only been addressed in the broadest of terms, in relation to the composition of the materials from which these three planets formed, and the subsequent acquisition of volatile materials and losses to space. There are great uncertainties.

Question 10.8

(a) The Moon has a very low abundance of volatiles. It also has low gravity (see Figure 9.12), so that retention even of CO_2 would be marginal, particularly as the surface temperatures can reach 400 K at equatorial noon (Section 7.1), rather than the 275 K or so surface average in Figure 9.12. Furthermore a tenuous atmosphere would be stripped by the Sun. Solar UV disrupts molecules, reducing the particle masses, making escape easier, and the solar wind assists by blowing the fragments away.

(b) Mercury might have suffered a massive impact that stripped away most of the silicate mantle and most of the volatiles. Though Figure 9.12 indicates that Mercury could retain an atmosphere of N_2, O_2, and CO_2, the temperature in Figure 9.12 (about 440 K) is some average over the surface: as mentioned in Section 7.2, the equatorial temperature at noon is about 725 K, which would make the retention even of CO_2 marginal. Moreover, Mercury is so close to the Sun, that its stripping power is much greater than at the Moon.

{Both of these bodies might have had thin atmospheres today, if they had ever had atmospheres sufficient to moderate the peak surface temperatures. Given the likely paucity of available volatiles, this is unlikely.}

Question 10.9

If the solar luminosity increased considerably, the Earth's atmosphere might well come to resemble that of Venus. The mass of water vapour in the atmosphere would increase, and this would increase the greenhouse effect, leading to positive feedback, until most or all of the oceans had evaporated to give a massive atmosphere initially dominated by water vapour. The mass of CO_2 would then increase as sedimentary carbonates decomposed in the high surface temperatures. The proportion of water would gradually fall through photodissociation of H_2O and the escape to space of the liberated hydrogen.

In the case of Mars, a significant proportion of the surface and near-surface reservoirs of water and CO_2 would be liberated, raising atmospheric pressure. The enhanced greenhouse effect would lead to higher surface temperatures, and there would be large open bodies of liquid water. For Mars to end up something like Venus, the luminosity of the Sun would have to rise more than that required for the Earth to become like Venus. This is because of the greater solar distance of Mars. In any case a Venus-like phase might be short lived because of the lower escape speed of Mars.

Question 10.10

The differences between the atmospheres of Titan and Triton are believed to be due largely to their different distances from the Sun. Therefore, if they were to swap places they would also swap atmospheric characteristics. Much of the N_2 and CH_4 in Titan's atmosphere would condense onto its surface, and N_2 and CH_4 ices on the surface of Triton would be evaporated into the atmosphere.

Question 11.1

(a) To obtain the Jovian mixing ratios, the number fractions in Table 9.1 have to be divided by the number fraction of H_2. This gives the following table, with the values from Table 11.1 added for comparison. The values for He and CH_4 are the same because in both tables the data for these substances are for the atmosphere above a few times 10^5 Pa.

Species	Mixing ratio from Table 9.1	Mixing ratio in Table 11.1
H_2	(1)	(1)
He	0.156	0.156
CH_4	2.1×10^{-3}	2.1×10^{-3}

(b) Helium is thought to be well mixed in the molecular hydrogen envelope, and therefore the mixing ratio in Table 11.1 probably applies down to the metallic hydrogen envelope. CH_4 does not condense anywhere in Jupiter's atmosphere, is also well mixed, and so the same conclusion applies. The NH_3 mixing ratio in Table 11.1 is at a pressure of a few times 10^5 Pa, where we are below the level at which NH_3 clouds and any NH_4SH clouds lie (Figure 11.1). Therefore, the Table 11.1 value also typifies the envelope.

Question 11.2

(a) If the minor constituents of the atmospheres of Jupiter and Saturn were absent there would be no cloud formation. The absence of clouds would decrease the albedo, and this would lead to greater solar heating of the upper molecular envelope and (presumably) to greater temperatures there, except perhaps where the latent heat of condensation of cloud particles had been important.
Without clouds it would be less easy to discern the circulation.
(b) The solar driven component of the circulation would be changed because of the greater solar heating of the upper envelope, and the absence of condensation. {Plausible details are beyond the scope of this question.}

Question 11.3

The assumption that the proportion of N_2 is negligible in the atmospheres of Uranus and Neptune might fail for Neptune. The evidence is the detection of HCN in the upper troposphere of Neptune. HCN can be formed from N_2, and the mixing ratio of HCN for Neptune is large enough to require a source of N_2 from the interior. The mixing ratio of HCN for Uranus is certainly much less, and this difference between the two planets is consistent with other evidence for the lack of convection at some depths in Uranus.

If indeed there is N_2 in Neptune's atmosphere, then the mixing ratio of helium must be reduced correspondingly. {Likely quantities would reduce it only slightly.}

Question 11.4

Neon, because it is volatile and unreactive, would have largely resided in the nebular gas, with only a small proportion trapped in the planetesimals. Because of its great abundance, most of the hydrogen was also in the gas. Similar proportions of the hydrogen and the neon in the nebula would have been captured, and therefore it is to be expected that the Ne/H ratio is the same in the giant planets as in the young Sun.

Because Ne is soluble in liquid helium, it will become depleted in the outer envelopes of the giant planets if helium settling has occurred. Therefore, the measured Ne/H ratios in the outer envelopes of Uranus and Neptune are expected to be about the same as in the young Sun, whereas the ratio for Jupiter should be lower, and that for Saturn lower still. {The Galileo probe found that in the outer Jovian envelope the Ne/H ratio is only about 10% of the value for the young Sun. There are no measurements for the other three giant planets.}

Glossary

Cross-references to other glossary entries are *italicised*.

accretion   The acquisition by a larger body of smaller bodies. It is an essential stage in planet building in *solar nebular* theories.

achondrite   A type of *stony meteorite* that lacks *chondrules*.

adiabatic process   A process during which no *heat* is transferred to or from a substance. If a vertical temperature gradient exceeds the adiabatic gradient, then *convection* occurs.

advection   A process of energy transfer by local bulk motion of warm liquids, such as molten rock or water, or of solids.

angular momentum   The angular momentum of a body of mass m moving at a speed v a perpendicular distance r from an axis has a magnitude mvr. If no angular momentum is transferred to or from a system, its angular momentum is constant – this is the principle of conservation of angular momentum.

aphelion   The point in the orbit of a body at which it is furthest from the Sun.

argument of perihelion   The orbital element of a body in the Solar System that is the angle around an orbit between the direction from the Sun to the *ascending node* and the direction from the Sun to *perihelion*, measured in the direction of orbital motion.

ascending node   The point where the orbit of a body in the Solar System intersects the *ecliptic plane* when the body crosses from south to north of this plane.

asteroid belt   A belt that includes the great majority of *asteroids*. It lies between the orbits of Mars and Jupiter.

asteroids   Small rocky bodies, largely confined to the *asteroid belt*.

asthenosphere   A plastic region in the interior of a body, usually located in the *mantle*, and perhaps covering a great range of depths.

astrometric technique   A means of detecting the presence of a celestial body through the effect it has on the position of a visible body.

astronomical unit (AU)   The *semimajor axis* of the Earth's orbit around the Sun.

aurora   A dynamic display of light in the upper atmosphere of a body, mainly in the polar regions. Aurorae are caused by energetic charged particles that enter the upper atmosphere and collide with the atoms there.

axial inclination (obliquity)   The angle between the rotation axis of a body and the perpendicular to its orbital plane.

basaltic–gabbroic rocks   *Extrusive* (basalt) and *intrusive* (gabbro) *rocks*, consisting largely of the *silicates* feldspar and pyroxene.

biosphere   The assemblage of all living things and their remains. The Earth is the only planet known to have a biosphere.

black body   See *ideal thermal source*.

blow-off   The loss of nearly all of a planetary atmosphere through a large impact from space.

Bond (or planetary) albedo   The fraction of the intercepted solar radiation that a body reflects back to space.

C1 chondrites   A particularly primitive subgroup of the *carbonaceous chondrites*.

caldera   A crater created by *volcanism* and not by impact.

carbonaceous chondrite   A type of *chondrite* (meteorite) that contains *carbonaceous material* and *hydrated minerals*. Carbonaceous chondrites have never been subjected to much heating or compression, and overall they have been little altered since they formed. Therefore they are primitive bodies.

carbonaceous materials   Carbon and *organic compounds*.

carbonate   A chemical compound containing the chemical unit CO_3.

carbonate–silicate cycle   A feedback mechanism that stabilises the Earth's *climate* by regulating the amount of carbon dioxide in the atmosphere, and hence modifying the *greenhouse effect*.

Centaurs   A group of *asteroids* with orbits that lie among the giant planets. They are *icy–rocky bodies* from the *Edgeworth-Kuiper Belt*.

centre of mass   The point in a body (or in a system of bodies) that accelerates under the action of an external force as if all the mass were concentrated at that point.

chalcophiles   Chemical elements that tend to form compounds with sulphur, e.g. zinc.

chaos   That property of a system whereby its configuration in the future/past is so sensitive to its present configuration that aspects of its future or past configuration cannot be predicted or 'retrodicted' at all, or only within certain limits.

chemical escape   The escape to space of a component of an atmosphere, promoted by chemical reactions that give the reaction products high speeds.

chondrite   A type of *stony meteorite* that contains small globules of silicates – *chondrules* – that have solidified from molten droplets. Most chondrites are classified as ordinary, but a small, important proportion are *carbonaceous chondrites*.

chondrule   A component of a *chondrite*, being a small silicate globule, solidified from a molten droplet.

clathrate   An icy solid plus another substance enclosed within its crystal structure, such as carbon dioxide enclosed in water ice.

clay mineral   A *silicate mineral* that has been chemically modified by water.

climate   The medium-term (e.g. 30 year) average of temperature, precipitation, etc., plus the variability of these factors.

collisional broadening   The broadening of spectral lines due to collisions between atoms and molecules in the source region. Also called pressure broadening.

column mass   (1) The total mass in a column of unit cross-sectional area running perpendicular to the disc of the *solar nebula*. (2) The total mass of atmosphere in a column of unit cross-sectional area stretching vertically upwards from the surface of a body.

comets   Small, *icy–rocky bodies* that develop huge fuzzy heads and tails when they are in the inner Solar System.

condensation flow   A component of atmospheric circulation whereby an atmospheric constituent flows to where it is condensing.

convection   The transport of energy by bulk flows arising from vertical temperature gradients.

core-accretion model   The model of the formation of a *giant planet*, in which a kernel of *icy* and *rocky materials* forms, that captures gas from the *solar nebula*.

Coriolis effect   As viewed in a rotating frame of reference, the tendency of a body to accelerate in a direction that is perpendicular both to the rotational motion of the body and to the axis of rotation.

cosmic rays   Atomic particles that pervade interstellar space, moving at speeds close to that of light. They are primarily nuclei of the lighter elements, notably hydrogen.

Cowling theorem   This states that the *magnetic field* of a planet cannot have the same symmetry as the rotation. This means that the rotation and magnetic axes must have an angle between them.

crust   The relatively thin outermost solid layer of a body, chemically distinct from the underlying layer.

cryovolcanism   *Volcanism* involving *icy materials*.

day   By definition, exactly $24 \times 60 \times 60$ seconds in length. It is very nearly equal to the *mean solar day*, which varies slightly over a period of years.

delamination   The loss of the base of a lithosphere through any compressional thickening that makes the base sufficiently dense to break away and sink.

differentiation   A process whereby denser substances settle downwards, and less dense substances float upwards.

Doppler effect   The dependence of the observed wavelength (or frequency) of radiation on the speed of the source relative to the observer along the line of sight. The Doppler shift is the change in wavelength. The Doppler effect can lead to the broadening of a spectral line – Doppler broadening.

dust   In space, particles less than about 0.01 mm across.

dwarf planet   A controversial relabelling of Pluto, Eris, Gres, and comparable bodies.

eccentricity   A measure of the degree of departure of an *ellipse* from circular form. It is defined as $\sqrt{1-b^2/a^2}$ where a is the *semimajor axis* and b is the semiminor axis.

eclipse   The passage of a body into the shadow of another body. As seen from the Earth, in a total solar eclipse the Moon completely obscures the Sun, whereas in a total lunar eclipse the whole Moon enters the umbral shadow of the Earth. There are also partial eclipses.

ecliptic plane   The plane of the Earth's orbit.

Edgeworth–Kuiper belt   A belt of *icy–rocky bodies* extending from just beyond the orbit of Neptune perhaps as far as the inner *Oort cloud*. Only its innermost, largest members have been observed. The belt supplies some of the *comets*.

effective temperature   A temperature defined by $(L_{out}/A\sigma)^{1/4}$, where L_{out} is the radiant power that a body emits to space, A is its total area, and σ is Stefan's constant.

ellipse   A closed curve that has the shape of a circle when viewed obliquely. Half its long dimension is its *semimajor axis*, half of its short dimension is its semiminor axis.

embryos   Bodies of the order of the mass of the Moon or Mars that appear at a fairly late stage in the growth of planets in solar nebular theories (*solar nebula*).

enthalpy   See *latent heat*.

equation of state   For a substance in equilibrium, the relationship between the pressure on the substance and its density and temperature.

equinox   A point in an orbit where the rotation axis of a body is perpendicular to the line from the body to the Sun. At any latitude, day and night have equal length at this point.

escape speed   The minimum speed a particle requires to escape from a body.

exoplanetary system   A system of planets around a star other than the Sun.

exosphere   A layer in an atmosphere from where atoms and molecules can readily escape directly to space if they are travelling upwards at greater than the *escape speed*.

extrusive rocks   *Rocks* that form from *magma* that solidifies at the surface of a body (via *volcanism*).

faint Sun paradox   The lack of extensive glaciation early in Earth history in spite of the lower solar luminosity. A larger *greenhouse effect* is common to many resolutions of the paradox.

fault   A fracture in a *lithosphere*, along which relative motion can occur.

first point of Aries   The direction from the Earth to the Sun when the Earth is at the *vernal equinox*.

flux density   The *power* in the electromagnetic radiation that is incident on a unit area facing the source.

fractional crystallisation   The crystallisation of a *mineral* from a *magma* so that the composition of the remaining melt is modified.

Gaia hypothesis   The hypothesis that active biospheric control tends to preserve optimum conditions for the Earth's biosphere.

Galactic tide   The *tidal force* exerted on bodies in the Solar System by the stars and interstellar matter in the Galaxy. It is thought to have been important in modifying the orbits of *comets*.

Galilean satellites   The four large satellites of Jupiter: Io, Europa, Ganymede, Callisto. They are named after their discoverer, Galileo Galilei.

gamma ray fluorescence spectrometry   The investigation of the chemical composition of a substance using the gamma rays emitted when its atomic nuclei return to the state they were in before they were excited into higher energy states.

general relativity   A theory by Albert Einstein that is superior to *Newton's laws* (of motion, and gravity), though Newton's (simpler) laws can be used for most purposes in the Solar System.

geometrical albedo   A particular measure of the reflectance of a surface with respect to solar radiation.

giant planet   In the Solar System, one of the four planets Jupiter, Saturn, Uranus, Neptune. These massive planets consist largely of hydrogen, helium, and *icy materials*.

global mass fraction   The total mass of a substance divided by the total mass of the body of which it is a component.

graben   A valley produced by slumping between two parallel normal *faults*.

gradation   Any process by which material is eroded from a surface, and then transported and deposited elsewhere.

granitic–rhyolitic rocks   *Extrusive* (rhyolite) and *intrusive* (granite) *rocks* consisting largely of feldspar (a *silicate*) and quartz.

gravitational coefficient   A factor that defines the size of each extra term that has to be added to GM/r^2 to represent the *gravitational field* of a body.

gravitational field   The gravitational force per unit mass at any point in space; equivalently, the acceleration of any unrestrained mass placed at the point.

gravitational instability model   The model of formation of a *giant planet* in which the whole planet forms through gravitational contraction of a fragment of the *solar nebula*.

gravitational potential The energy required to move unit mass from a point in space outside a body to infinity, assuming that no other mass is present.

greenhouse effect The phenomenon in which the radiant power emitted by the surface of a body exceeds that emitted by the body to space. This excess arises from the absorption and scattering of infrared radiation by atmospheric constituents. The greenhouse effect is often quantified as T_s–T_{eff}, where T_s is the global mean surface temperature, and T_{eff} is the *effective temperature*.

Hadley cell A large convection cell in an atmosphere that arises from the decrease in solar heating with increasing distance from the subsolar latitude.

half-life The time it takes half of the nuclei of a radioactive isotope to decay into another isotope.

Halley family comets A small subgroup of *short-period comets* with orbital periods in the range 15–200 years, and with orbital inclinations typically larger than the *Jupiter family comets*.

heat Energy transferred in which the transfer is random at a microscopic level.

heavy bombardment The heavy bombardment of bodies in the Solar System that persisted until about 3900 Ma ago. In solar nebular theories (*solar nebula*) it is due to the mopping up of remnant *planetesimals*. A possible peak in the bombardment before 3900 Ma ago is called the late heavy bombardment.

heavy elements All chemical elements other than hydrogen and helium.

heterosphere The layer in a planetary atmosphere where, aside from any substances that condense, the composition depends on altitude. It lies above the *homosphere*.

Hirayama family A group of *asteroids* with orbits sufficiently similar to suggest that they have originated from the break-up of a larger asteroid,

homosphere The layer in a planetary atmosphere where, apart from any substances that condense, the composition is the same at all altitudes. It lies below the *heterosphere*.

hydrated mineral A *mineral* that includes water, either as attached molecules of H_2O, or as attached hydroxyl, OH.

hydrocarbon A compound consisting of carbon and hydrogen. One of the simplest is methane, CH_4.

hydrodynamic escape The escape of an atmospheric constituent through its entrainment in the high-volume flow of a rapidly escaping lighter constituent, notably hydrogen.

hydrostatic equilibrium The equilibrium state of a body when it has responded to forces in the manner of a fluid, i.e. in the manner of material with zero shear strength.

hyperbolic orbit An open orbit in which the two arms at infinity are diverging. A parabolic orbit is a marginal case, where the two arms are parallel at infinity. Elliptical and circular orbits are closed.

hypsometric distribution A histogram showing the frequency distribution of different altitudes of the surface of a body.

ice age A long period on Earth of cooler climate, particularly outside the tropics, where glacial conditions reach to mid latitudes. An ice age is punctuated by warmer intervals called interglacial periods.

ice line The distance from the proto-Sun beyond which water condenses in the *solar nebula*.

icy materials A group of *volatile* substances, such as water, ammonia, methane, carbon dioxide, nitrogen, etc.

icy–rocky bodies Bodies consisting of *icy* and *rocky materials*, with icy materials accounting for a significant proportion of the mass.

ideal gas See *perfect gas*.

ideal thermal source A body that absorbs all of the electromagnetic radiation that falls on it. One consequence is that the spectrum of the radiation emitted by the body is uniquely determined by its temperature.

igneous rocks *Rocks* produced from *magma*.

impact crater A crater formed by the impact of a projectile. They are the commonest feature of surfaces in the Solar System.

impact erosion The loss of a proportion of a planetary atmosphere through the impacts of numerous small bodies from space.

inert gases (noble gases) A group of chemically unreactive elements: helium, neon, argon, krypton, xenon, radon.

internal energy The energy within a body, as opposed to the energy of the body as a whole. It consists of the random kinetic energy of its atoms and molecules, plus their (chemical) potential energy of interaction.

intrusive rocks *Rocks* that form from *magma* that solidifies beneath the surface of a body.

ionosphere A layer in a planetary atmosphere sufficiently ionised for the layer to display special properties. Solar ultraviolet radiation is an important cause of ionisation.

iron meteorite A *meteorite* consisting almost entirely of iron alloyed with a few per cent of nickel.

isostatic equilibrium A state in which, above some depth in a body, there are equal masses in each vertical column. The minimum depth in any region for which this condition holds is the level of compensation.

J_2 The *gravitational coefficient* of a particular term additional to GM/r^2 in the *gravitational field* of a body. For *planetary bodies* the J_2 term is the largest extra term.

Jupiter family comets These comprise the majority of the *short-period comets*, and are characterised by orbital periods less than 20 years and low-to-modest orbital inclinations. Their aphelia (see *aphelion*) lie in the region of Jupiter's orbit.

Kepler's laws of planetary motion Three empirical rules that describe fairly accurately the orbital motions of the planets. They are explained by *Newton's laws* (of motion and of gravity).

Kirkwood gaps Ranges of *semimajor axes* in the *asteroid belt* where there are few *asteroids*.

Lagrangian points Five points at locations fixed with respect to two bodies in (a low-eccentricity) orbit around each other. In principle, a third body with a small mass can be placed at each of these points, and remain there, though in practice some of the points are unstable.

lapse rate The rate of decrease of temperature with increasing altitude in a planetary atmosphere. The *adiabatic* value is the adiabatic lapse rate.

latent heat *Heat* that produces no temperature change. Instead, the energy goes to increase the (chemical) potential energy contribution to the *internal energy*.

lava *Magma* that reaches a surface owing to *volcanism*, or icy liquids that reach a surface owing to *cryovolcanism*.

light year   The distance travelled by light in a vacuum in a year of 365.2425 *days*. It equals 9.460536×10^{15} metres.

lithophiles   Chemical elements that tend to be found in silicates and oxides, e.g. magnesium and aluminium.

lithosphere   The rigid outer region of a body, usually comprising the *crust* and upper *mantle*.

longitude of perihelion   The orbital element of a body in the Solar System that is the sum of the *longitude of the ascending node* and the *argument of perihelion*.

longitude of the ascending node   The orbital element of a body in the Solar System that is the angle, in the direction of the Earth's motion, from the direction to the *first point of Aries* to the line from the Sun to the body's *ascending node*.

long-period comet   A *comet* with an orbital period in excess of 200 years.

Love numbers   Parameters that give us a measure of how much the interior of a *planetary body* deforms under *tidal forces*.

lunar eclipse   See *eclipse*.

magma   Molten *rocky* or *icy materials*.

magnetic dipole field   The *magnetic field* some distance away from a loop or loops of electric current.

magnetic dipole moment   The strength of the source of a *magnetic dipole field*.

magnetic field   The force field sustained by electric current. A current loop generates a field that has the form of a dipole field far from the loop, and a strength proportional to the *magnetic dipole moment* of the loop.

magnetosphere   The magnetic 'sphere of influence' of a body, bounded by the magnetopause. The Earth and the giant planets have extensive magnetospheres.

main belt   A zone within the *asteroid belt* with a particularly high concentration of *asteroids*.

main sequence star   A star in that phase of its lifetime when it is sustained by hydrogen *nuclear fusion* in its core.

mantle   The layer of a body that underlies any *crust*, and is chemically distinct from the crust.

mascon   A region on the Moon with an excess of mass.

Maxwell distribution   The distribution of molecular speeds in a gas in thermal equilibrium.

mean density   The total mass of a body divided by its total volume.

mean motion resonance   The situation in which the orbital periods of two bodies have the ratio $(p+q)/p$, where p and q are integers. The gravitational interaction between the bodies in such cases can enhance orbital stability, or lead to instability.

mesosphere   A layer in a planetary atmosphere where the *lapse rate* is less than the adiabatic value, and so *convection* does not occur. It lies above any *troposphere* but below any *thermosphere*.

metallic hydrogen   A high-pressure form of hydrogen in which the electrons detach from the nuclei. The substance then has the properties of a metal.

metamorphic rock   An *igneous* or *sedimentary rock* that has been modified in any way short of complete melting.

meteor   A small body that enters the Earth's atmosphere at high speed. They are commonly seen as bright, transient streaks of light across the night sky.

meteor shower   A period of a few days during which the rate at which *meteors* are seen is enhanced. The members of a shower have similar orbits in space.

meteorite   A body, or its fragments, that has come from beyond the Earth and has survived passage to the Earth's surface. A micrometeorite is a meteorite less than a few millimetres across; some of these have recondensed after being vaporised or melted in the atmosphere.

meteoroids   *Asteroids* smaller than about a metre across.

micrometeorite   A *meteorite* less than a few millimetres across.

micrometeoroid   A smaller meteoroid, in the approximate size range of 0. 01 millimetres to a few millimetres across.

mineral   A naturally occurring (solid) substance with a basic unit that has a particular chemical composition and crystal structure. A *rock* is an assemblage of one or more minerals,.

minimum mass solar nebula   A *solar nebula* with the least mass that could have produced the Solar System.

minor planets   See *asteroids*.

mixing ratio   The number of molecules of a constituent in a local volume divided by the number of molecules of some other constituent in the same volume. In the giant planet atmospheres, mixing ratios are normally given with respect to H_2.

near-Earth asteroids   *Asteroids* that can make close approaches to the Earth, with the possibility of collisions with the Earth.

neutron spectrometry   The investigation of the chemical composition of a substance using the energy spectrum of neutrons ejected from the substance.

Newton's law of gravity   Enunciated by Isaac Newton. If two point masses M and m are separated by a distance r then there is a gravitational force of attraction between them with a magnitude GMm/r^2, where G is the universal gravitational constant.

Newton's laws of motion   Enunciated by Isaac Newton. The three laws are as follows:

(1) An object remains at rest or moves at constant speed in a straight line unless it is acted on by an unbalanced force.

(2) If an unbalanced force of magnitude F acts on a body of mass m, then the acceleration of the body has a magnitude F/m, and the direction of the acceleration is in the direction of the unbalanced force.

(3) If body A exerts a force of magnitude F on body B, then body B will exert a force of the same magnitude on body A but in the opposite direction.

nuclear fusion   A nuclear reaction in which a more massive nucleus is formed from less massive nuclei. The pp chains that power the Sun have the overall effect of converting four protons (^{1}H) into a nucleus of the helium isotope ^{4}He.

number fraction   The number of molecules of an atmospheric constituent divided by the number of molecules in the whole atmosphere.

occultation   A body is occulted when it passes from view behind another body.

Oort cloud   A thick spherical shell of *icy–rocky bodies* inferred to extend from about 10^3 to 10^5 AU from the Sun. It supplies some of the *comets*.

opposition   As viewed from the Earth, a body is in opposition when it is at that point in its orbit that places it nearest to being in the opposite direction to the Sun. At inferior and superior conjunctions, the body is nearest to being in the same direction as the Sun. These terms can be applied to vantage points other than the Earth.

orbital elements Five quantities that specify the shape, size, and orientation of an orbit. For an elliptical orbit they are: the *semimajor axis*, the *eccentricity*, the *inclination*, the *longitude of the ascending node*, and *the longitude of perihelion*.

orbital inclination The angle between an orbital plane and a reference plane. The reference plane for planetary orbits is the *ecliptic plane*.

ordinary chondrite The most common sort of chondrite. There is little *carbonaceous material*.

organic compound A compound of carbon and hydrogen, often with other elements. *Hydrocarbons* are a subclass of organic compounds.

outgassing The volcanic emission of gases from the interior of a body.

oxidation A variety of chemical processes including those in which oxygen is incorporated into the reaction products. The opposite process – the extraction of oxygen – is one form of a chemical process called reduction.

partial melting The process in which a mixture of *minerals* starts to melt, such that the composition of the molten material is different from that of the original mixture.

partial pressure The pressure exerted by the molecules of any one component in a mixture of gases.

perfect gas (ideal gas) A gas that obeys the relatively simple *equation of state* $p = \rho kT/m_{av}$, where p, ρ, and T are respectively the gas pressure, density, and temperature. k is Boltzmann's constant, and m_{av} is the mean mass of the molecules in the gas. Real gases approach perfect gas behaviour at low densities and/or high temperatures.

peridotite A *rock* consisting largely of the *silicates* olivine and pyroxene. The *minerals* in the Earth's lower mantle have different crystal structures but the chemical composition is still that of peridotite.

perihelion The point in its orbit at which a body is nearest to the Sun.

phase diagram A diagram that shows the equilibrium phase of a substance at each pressure and temperature. Solid, liquid, and gas are three phases. The solid phase region is subdivided in accord with different crystal structures (different minerals).

photochemical reaction A chemical reaction involving electromagnetic radiation, e.g. as in *photodissociation* and *photoionisation*. Such radiation can also promote chemical reactions by exciting molecules/atoms, short of dissociation/ionisation.

photodissociation The disruption of molecules by electromagnetic radiation. Ultraviolet radiation is particularly effective.

photoionisation The ionisation of atoms and molecules by electromagnetic radiation. Ultraviolet radiation is particularly effective.

photometry The measurement of the radiation received from a source in up to a few, broad ranges of wavelengths. Compare *spectrometry*.

photon The particle of electromagnetic radiation in those circumstances where the radiation can be regarded as a stream of particles.

photosphere The bright surface of a star (including the Sun).

photosynthesis The process in which certain life forms, notably green plants and cyanobacteria bacteria, synthesise carbohydrates from smaller molecules with the aid of solar radiation. Through photosynthesis, green plants and blue-green bacteria release molecular oxygen.

planetary body A body large enough for its own gravity to ensure that it is in, or close to, *hydrostatic equilibrium*. In this case an isolated body will be spherical.

planetesimals 'Little planets', small bodies 0.1–10 km across, formed from dust in the *solar nebula*. In solar nebular theories, many are incorporated in *embryos* en route to forming planets and the cores of *giant planets*.

plasma A highly ionised medium. In the Solar System, examples include the solar interior, the *solar wind*, and the plasma in *magnetospheres*.

plate tectonics The sculpting of the Earth's surface through the relative motion of lithospheric plates (see *lithosphere*). The plates are thought to be driven by *convection* in the mantle.

polar moment of inertia (C) The moment of inertia of a body with respect to its rotation axis.

power The rate at which energy is transferred, by any mechanism.

Poynting–Robertson effect The deceleration of a small particle in space due to the extra *photon* bombardment on its leading surface. It is more important the smaller the particle, and is insignificant for particles greater than about 0.1 m across.

precession of the equinoxes The coning of the rotation axis at fixed *axial inclination*. This causes the location of the *equinoxes* to move around the orbit.

precession of the perihelion The motion of the perihelion direction around the plane of an orbit.

precession of the rotation axis The coning of the rotation axis at fixed *axial inclination*. It leads to *precession of the equinoxes*.

prograde direction The predominant direction of orbital motion and rotation in the Solar System – anticlockwise as viewed from above the Earth's North Pole.

radial velocity technique The measurement of the orbital motion of a body (usually a star) from the oscillating Doppler shifts of its spectral lines (see *Doppler effect*).

radiation pressure The pressure exerted by *photon* bombardment.

radiative transfer The transfer of energy by means of electromagnetic radiation.

radiogenic heating Heating resulting from the decay of radioactive isotopes.

radiometric dating The dating of a variety of events from the relative abundances of radioactive isotopes and the isotopes they create when they decay. The measured *half-lives* of the radioactive isotopes enable absolute ages to be obtained.

Rayleigh scattering The scattering of electromagnetic radiation by objects that are much smaller than the wavelengths in the radiation.

refractory substance A substance with low *volatility*.

regolith The mixture of dust and small pieces of rock that is abundant on the surface of the Moon (and other bodies).

residence time The average time that a molecule spends in a reservoir, e.g. the atmosphere.

retrograde motion The opposite of *prograde motion*.

Roche limit The distance between a smaller body and a larger body within which the smaller body is disrupted by the *tidal force* exerted by the larger body. It applies only if the smaller body is held together by gravitational forces.

rock An assemblage of one or more *minerals*, in solid form.

rocky materials A group of relatively *refractory substances*, including silicates, iron–nickel, etc.

rotational flattening (oblateness) The flattening of a body along its axis of rotation, caused by the rotation.

runaway greenhouse effect A *greenhouse effect* enhanced greatly by positive feedback that rapidly increases the amount of greenhouse gases in the atmosphere.

saturation vapour pressure The pressure on the saturation line of a substance, i.e. the line in the *phase diagram* on which liquid and gas, or solid and gas, coexist.

secular resonance A gravitational interaction between bodies, averaged over a time much longer than the orbital periods of the bodies involved, that couples the evolution of one or more *orbital elements* of one body with one or more of those of another.

sedimentary rock A *rock* produced from a sediment by chemical action, or pressure.

seismic wave A mechanical wave in a *planetary body* (in smaller bodies too). They also occur in the Sun. Two important types are P waves and S waves.

self-exciting dynamo A mechanism thought to sustain the *magnetic fields* that originate deep in the Sun and in certain planets. It involves the conversion of kinetic to magnetic energy, which also happens in a dynamo.

semimajor axis Half of the long dimension of an *ellipse*.

shield volcano A volcano that in profile resembles a warrior's shield. Shield volcanoes result from a sequence of low-viscosity lava flows, erupted at modest rates.

shock wave A wave with a very steep front, so that the material in front of the wave is undisturbed, whereas the material immediately behind the wave is highly disturbed. One way in which a shock wave is generated is when a projectile encounters a body at a speed greater than that of *seismic waves* in the body.

short-period comet A *comet* with an orbital period of less than 200 years.

sidereal orbital period The time it takes a body to complete one orbit from a viewpoint fixed with respect to the distant stars. For the Earth this is the sidereal year.

sidereal rotation period The rotation period with respect to the distant stars. For the Earth it is called the mean rotation period.

siderophiles Chemical elements that tend to be present with iron, e.g. nickel.

silicate A chemical compound that has a basic unit consisting of atoms of one or more metallic elements and atoms of silicon and oxygen.

solar activity A collective term for those solar phenomena that vary (currently) with a period of about 11 years. The number of *sunspots*, flares, and prominences, and the luminosity of the Sun, are among various aspects of this activity.

solar day The period of rotation of the Earth with respect to the Sun. The mean solar day is the mean length of the solar days averaged over a year. The mean solar day varies very slightly, whereas the *day* is exactly $24 \times 60 \times 60$ seconds in length.

solar eclipse See *eclipse*.

solar nebula The disc of gas and dust that is presumed to have encircled the young Sun. In solar nebular theories, the Solar System forms from such a disc.

solar wind A thin, gusty stream of high-speed particles (mainly protons and electrons) that escapes from the Sun.

solstice A point in an orbit where one of the poles of a body is pointed maximally towards the Sun (the other pole is then pointed maximally away from the Sun).

spectrometry The measurement of the radiation received from a source in numerous, narrow, contiguous wavelength ranges. Compare *photometry*.

spherical symmetry When the only variation in a quantity is with radius from some centre. In a body with a spherically symmetrical mass distribution, the density varies only with radius from the centre of the body.

steady state A system is in a steady state with respect to certain parameters (e.g. temperature), when the values of the parameters are not changing. A steady state is achieved when inputs that tend to increase a parameter are balanced by those that tend to decrease it.

stony-iron meteorite A type of *meteorite* that consists mostly of silicates and iron–nickel alloy.

stony meteorite A type of *meteorite* that consists mostly of silicates, often with small quantities of iron–nickel alloy.

stratosphere A layer peculiar to the Earth's atmosphere, immediately above the *troposphere*, where the temperature increases with increasing altitude. The *mesosphere* lies above the stratosphere, and in it the temperature decreases with increasing altitude. The temperature bulge is the result of the absorption of solar ultraviolet radiation by ozone.

sublimation A phase change in which a substance goes directly from solid to gas.

sunspot A small patch on the Sun's *photosphere* at a lower temperature than the rest of the photosphere. The number visible goes through a cycle with a mean period (currently) of 11.1 years. Sunspots are one aspect of *solar activity*.

synchronous orbit An orbit for which the period equals the rotation period of the body that is being orbited.

synchronous rotation When the rotation period of a body A with respect to a body B equals the orbital period of A around B so that A keeps the same face to B.

synchrotron emission Electromagnetic radiation emitted by electrons travelling at very high speeds through a magnetic field.

synodic period For the Earth and another body, the time interval between similar spatial configurations, e.g. *oppositions* and conjunctions. The term can be applied to vantage points other than the Earth.

synthetic aperture radar A particular radar technique used to determine the topography of a landscape with high three-dimensional resolution.

T Tauri phase The phase in a star's lifetime just before the main sequence phase (see *main sequence star*). It is characterised by instability, with powerful stellar winds (the T Tauri wind) and high ultraviolet flux.

tectonic processes Processes that cause relative motion or distortion of the *lithosphere*.

terrestrial bodies The large bodies that consist largely of *rocky materials*, i.e. the *terrestrial planets* plus the Moon, Io, and Europa.

terrestrial planet One of the four planets Mercury, Venus, the Earth, Mars. They consist largely of *rocky materials*.

thermal conduction The process by which *heat* is transferred through the direct contact of two regions that are at different temperatures.

thermal escape Escape of an atmospheric constituent arising from the *Maxwell distribution* of speeds for speeds in excess of the *escape speed*.

thermal tide A component of atmospheric circulation whereby the atmosphere flows to regions where the pressure has been lowered by cooling. This mass redistribution makes the atmosphere subject to an extra *torque* exerted by *tidal forces*.

thermosphere An upper layer in a planetary atmosphere where temperature increases with altitude, to reach high values.

tidal force A differential gravitational force. Any distortion produced by a tidal force is called a tide. Tidal energy is an important source of *internal energy* in some bodies.

Titius–Bode rule An empirical rule that describes the increasing spacing of the planetary orbits with increasing distance from the Sun.

torque A system of forces that tends to cause twisting or rotation. *Tidal forces* can exert torques. One outcome is *precession of the equinoxes*.

total solar eclipse See *eclipse*.

transit technique   The means of detecting a body when it passes between us and the *photosphere* of a star, thus causing a (slight) reduction in the apparent brightness of the star.

Trojan asteroids   A group of asteroids near the L_4 and L_5 *Lagrangian points* of Jupiter.

troposphere   The lowest layer in a planetary atmosphere where temperature is decreasing with altitude and where convection is usually pervasive.

uncompressed mean density?   The mean density that a body would have were its interior not compressed by the general increase of pressure with depth.

Van Allen radiation belts   The belts of *plasma* that surround the Earth. They lie within the Earth's *magnetosphere.*

vernal equinox   The *equinox* that occurs in March. Also, the position of the Earth in its orbit at this equinox. See also the *first point of Aries.*

volatility   A property of a substance that can be measured by the maximum temperatures at which it is in equilibrium in a condensed phase; the lower these temperatures, the greater the volatility.

volcanism   The processes by which gases, liquids, or solids are expelled from the interior of a body.

X-ray fluorescence spectrometry   The investigation of the chemical composition of a substance using the X-rays emitted when electrons in its atoms return to the states they were in before they were excited into higher energy states.

Yarkovsky effect   The slow change in the *semimajor axis* of the orbit of a rotating body as a result of the greater *photon* flux emitted (in the infrared) by its warmer, afternoon side. It is significant only for bodies less than about 100 m across.

year   Used here to mean the tropical year. The tropical year is the time interval between the Earth's *vernal equinoxes.* Its duration is 365.242 190 *days.* The sidereal year is the Earth's orbital period with respect to the distant stars, and its duration is 365.256 363 days. These years differ because of the *precession of the equinoxes.*

Electronic Media

There is a huge range of electronic media and a high rate of updating and obsolescence. An annotated list of free astronomy software, websites, news, and much else is available at the Sky Publishing Corporation website http://skytonight.com/. It is updated monthly. We strongly recommend it.

CD-ROMs and DVDs – Commercial

The list below is a *small* selection of what is currently available on the Solar System, though many products include much else. In most cases we have specified no supplier website. In such cases you can find a supplier, with a full product description, often with reviews, via http://www.google.com/ by entering the name of the product into the search field.

Welcome to the Planets, CD-ROM for PCs or Macs, NASA/Planetary Data System/Data Distribution Laboratory. 190 images of Solar System bodies, with narration and data. http://pds.jpl.nasa.gov/planets/welcome/cdrom.htm

Dance of the Planets, CD-ROM for PCs (and Macs with Windows), ARC Science Simulation, Colorado, USA. An extensive planetary tour.

Starry Night Deluxe, CD-ROM for Macs or PCs, Sienna Software Inc., Ontario, Canada. Plenty on the Solar System, and lots more besides.

RedShift 3, CD-ROM for PCs or Macs, Piranha Interactive Publishing Inc., Arizona, USA. Covers the Solar System and lots more.

A Few Websites

http://www.google.com/ A very useful website for obtaining information on anything that can be expressed in a word or in a few words.

http://wikipedia.org/ An on-line encyclopaedia, with entries supplied by anyone. Generally OK, but the entries are not refereed.

http://www.skytonight.com/ The website of Sky Publishing Corporation, publishers of *Sky and Telescope*, the leading popular astronomy monthly in the USA and elsewhere. See above also.

http://www.astronomynow.com/ The website of *Astronomy Now*, the leading UK popular astronomy monthly. Much information and news.

http://www.nasa.gov/search/index.html The NASA web search engine.

ftp://ftp.hq.nasa.gov/pub/pao/pressrel/ NASA press releases, from 1990 onwards.

http://www.jpl.nasa.gov/ Lots of information available here, including news of missions.

http://photojournal.jpl.nasa.gov/ A huge selection of NASA images.

http://nssdc.gsfc.nasa.gov/ The National Space Science Data Center – NASA's permanent archive for space science mission data. A mine of information on NASA, including a photo gallery.

http://www.stsci.edu/ The Space Telescope Science Institute website. Many resources on offer.

http://www.esrin.esa.it/ The European Space Agency website. Many resources on offer.

http://www.nineplanets.org/ *The nine planets.* A multimedia tour of the Solar System. Bill Arnett.

http://www.solarviews.com/eng/homepage.htm *Views of the Solar System*, C J Hamilton. Many images, much data, and several animations.

http://www.exoplanet.eu/ Information and listings of exoplanets, with much background information and many links.

http://www.popastro.com/ The website of the UK Society for Popular Astronomy (SPA). This society, with over 3000 members, promotes astronomy as a hobby, with a particular focus on beginners, but catering also up to the highest levels.

http://www.britastro.org/baa/ The website of the British Astronomical Association (BAA). It has about 3000 members and promotes astronomy as a hobby with a focus at rather higher levels than the SPA.

http://www.astrosociety.org/ The Astronomical Society of the Pacific's website. This caters for amateur astronomers at all levels.

Further Reading

This is a selection of the many good books available. A visit to a university bookshop, or a scan of Amazon on the Internet, will reveal further good titles, as will use of the Internet search engine Google http://www.google.com/.

Books at the 'popular astronomy' level are largely excluded from this list, as are books aimed at amateurs who wish to make their own observations. With a few exceptions, picture books are excluded, as are books published before 1999.

In some cases the ISBN is for a hardback edition – check with your supplier whether a paperback edition is available.

Books more advanced than *Discovering the Solar System* are marked with an asterisk.

Dictionaries and Atlases

The Compact NASA Atlas of the Solar System, Ronald Greeley and Raymond Batson, Cambridge University Press 2001, ISBN 0 521 80633X

Oxford Dictionary of Astronomy, Ian Ridpath (editor), Oxford University Press 2004, ISBN 019 860513 7

Collins Internet-linked Dictionary of Astronomy, John Daintith and William Gould, HarperCollins 2006, ISBN 0 00 722092 7

Books Covering the Whole Solar System

The New Solar System (4th edition), J Kelly Beatty, Carolyn C Petersen, Andrew L Chaikin (editors), Cambridge University Press and Sky Publishing Corporation 1999, ISBN 0 933346 86 7

Solar System Evolution: A New Perspective (2nd edition), Stuart R Taylor, Cambridge University Press 2001, ISBN 0 521 64130 6

The Earth in Context: A Guide to the Solar System, David M Harland, Springer-Praxis 2001, ISBN 1 85233 375 8

Planetary Science: The Science of Planets around Stars, George H A Cole and Michael M Woolfson, Institute of Physics Publishing 2002, ISBN 0 7503 0815 X

The Cambridge Guide to the Solar System, Kenneth R Lang, Cambridge University Press 2003, ISBN 0 521 81306 9

The Physics of the Solar System, Bruno Bertotti, Paolo Farinella, David Vokrouhlicky, Kluwer Academic Publishers 2003, ISBN 1 4020 1509 7

The Solar System (3rd edition), Teresa Encrenaz *et al.*, Springer 2003, ISBN 3 540 00241 3

An Introduction to the Solar System, Neil McBride and Iain Gilmour (editors), The Open University (UK) and Cambridge University Press 2004, ISBN 0 521 54620 6

Discovering the Solar System, Second Edition   Barrie W. Jones
© 2007 John Wiley & Sons, Ltd

Books on Particular Topics

Solar System Dynamics, Carl D Murray and Stanley F Dermott, Cambridge University Press 1999, ISBN 0 521 57597 4

Orbital Motion (4th edition), Archie E Roy, Institute of Physics Publishing 2005, ISBN 0 7503 1015 4

Atmospheric Science (2nd edition), John M Wallace and Peter V Hobbs, Elsevier (Academic Press) 2006, ISBN 0 12 732951 X

Planetary geology: an introduction, Claudio Vita-Finzi, Terra 2005, ISBN 1 903544 20 3

Books on Particular Bodies/Groups of Bodies

Guide to the Sun, Kenneth J H Phillips, Cambridge University Press 1995, ISBN 0 521 39788 X

The Sun and Stars, Simon F Green and Mark H Jones (editors), The Open University (UK) and Cambridge University Press 2004, ISBN 0521 546 622

Sun, Earth, and Sky (2nd edition), Kenneth R Lang, Springer 2006, ISBN 978 0387 30456 4

Exploring Mercury: The Iron Planet, Robert G Strom and Ann L Sprague, Springer-Praxis 2003, ISBN 1 85233 733 1

Earth (4th edition), Frank Press and Raymond Siever, W H Freeman 1986, ISBN 0 7167 1743 3

Earth: Evolution of a Habitable World, Jonathan I Lunine, Cambridge University Press 1999, ISBN 0 521 64423 2

Understanding the Earth (5th edition), John Grotzinger, Thomas H Jordan, Frank Press, Raymond Siever, W H Freeman 2007, ISBN 0 7167 6682 5

Mars: A Warmer, Wetter Planet, Jeffrey S Kargel, Springer-Praxis 2004, ISBN 1 85233 568 8

The Martian Climate Revisited: Atmosphere and Environment of a Desert Planet, Peter L Read and Stephen R Lewis, Springer-Praxis 2004, ISBN 3 540 40743 X

Jupiter: The Planet, Satellites and Magnetosphere, Fran Bagenal, Timothy Dowling, William McKinnon (editors), Cambridge University Press 2004, ISBN 0 521 81808 7

Giant Planets of our Solar System: An Introduction, Patrick Irwin, Springer-Praxis 2006, ISBN 354 031 317 6

Europa: The Ocean Moon, Richard Greenberg, Springer-Praxis 2005, ISBN 3 540 22450 5

Asteroids, Comets, and Meteors, Sylvio Ferraz-Meelo and Julio Angel Fernández (editors), Cambridge University Press 2006, ISBN 0521 85200 5

Comets and the Origin and Evolution of Life (2nd edition), Paul J Thomas *et al.* (editors), Springer 2006, ISBN 354 033 0860

Planetary Rings, Larry Esposito, Cambridge University Press 2006, ISBN 0 521 36222 9

Meteorites: a Petrologic, Chemical, and Isotopic Analysis, Robert Hutchison, Cambridge University Press 2004, ISBN 0 521 47010 2

Meteors and Meteorites, Martin Beech, The Crowood Press 2006, ISBN 1 86126 825 4

Astrophysical Techniques (4th edition), Christopher R Kitchin, Institute of Physics Publishing 2003, ISBN 0 7503 0946 6

Astronomy Methods, Hale Bradt, Cambridge University Press 2004, ISBN 0 521 53551 4

Observational Astronomy (2nd edition), D Scott Birney, Guillermo Gonzalez, David Oesper, Cambridge University Press 2006, ISBN 0 521 85370 2

Astrobiology

Life in the Universe, Jeffrey Bennett, Seth Shostak, Bruce Jakosky, Addison-Wesley 2003, ISBN 0 8053 8577 0

An Introduction to Astrobiology, Iain Gilmour and Mark A Sephton (editors), The Open University (UK) and Cambridge University Press 2004, ISBN 0 521 54621 4

Life in the Solar System and Beyond, Barrie W Jones, Springer-Praxis 2004, ISBN 1 85233 101 1

Astrobiology, Jonathan I Lunine, Pearson/Addison-Wesley 2005, ISBN 0 8053 8042 6

The Living Universe: NASA and the Development of Astrobiology, Steven J Dick and James E Strick, Rutgers University Press 2005, ISBN 978 0 8135 3733 7

Looking for Life: Searching the Solar System, Paul Clancy, André Brack, Gerda Horneck, Cambridge University Press 2005, ISBN 0 521 82450 8

Astrobiology: A Brief Introduction, Kevin W Plaxco and Michael Gross, Johns Hopkins University Press 2006, ISBN 0 8018 8367 9

Periodicals

Sky and Telescope (monthly), *Astronomy* (monthly), and *Astronomy Now* (monthly), are all aimed at a wide readership and contain articles on the Solar System, plus other areas of astronomy. *Scientific American* (monthly) and *New Scientist* (weekly), also aimed at a wide readership, contain occasional articles and short items on the Solar System. There are a large number of technical periodicals, available at university libraries, and aimed at the research community. One of more general interest is the *Annual Review of Earth and Planetary Sciences*.

Index

Note that in this index, general features and principles are separated from specific bodies. For example, under "atmospheres" there are no entries for specific atmospheres, such as that of the Earth. To find entries for the Earth's atmosphere you would need to look under "Earth". You should also check under groups of bodies e.g. "terrestrial planets". Many terms are subsumed under umbrella terms e.g. "hydrodynamic escape" will be found under "atmospheres" – it has no separate entry. For Table pages see p. xiii.

magnetosphere (*Continued*)
  magnetopause 191
  magnetosheath 192
  magnetotail 191
  reconnected lines 191
  size 191
  sources of plasma 192
  synchrotron emission 192
Magnya 120
main sequence star 7, 55
mantle 163
Mars
  aeolian/features processes 238–239, 253–254
  ages (surfaces) 241, 244
  albedo features 238–239
    dark areas 238–239
    light areas 238
  altitude (of surface) 240
    zero 200, 239
  Argyre 243–244
  atmosphere
    circulation 339–340
    dust 238–239, 337
    evolution (and climate change) 360–361
    origin, *see* terrestrial planets
    properties Tables 9.1 and 9.2
    reservoirs, gains and losses 338–339
    vertical structure, heating and cooling
        336–338
    winds 238–239, 337
  central pressure, temperature, density
      Table 5.2
  chaotic terrain 246–247
  climate change 252, 340–341
  clouds 336–337
  contrasting hemispheres 239–241
    boundary 245
    northerly hemisphere 239–243
    southerly hemisphere 239–240, 243–245
  core 164
    composition 164
  crust and surface
    composition 164, 171–172, 239, 241–243,
        253–256
    thicknesses (crust) 239
  differentiation 172
  dust storms 239
  evolution of interior and surface 256–257
  fretted terrain 245
  greenhouse effect 338
  Hellas basin 243, 248, 251
  hypsometric distribution 239–240
  impact basins 241, 243
  impact craters 241, 243
  interior model 164, 172–173
  internal temperatures 172

  life? 361–362
  lithosphere 172
  magnetic field 172
    generation 172
  mantle 164
    composition 164, 172–173
  meteorites from Mars 120–121, 172,
        255–257, 361
  observational data on the interior
        Table 4.2
  observations at the surface 171, 253–256
  Olympus Mons 243
  origin of satellites 69
  polar regions/caps 238, 245–247, 338
    composition 245
    growth and retreat 238
    layered sediments 245–247
    temperatures 245
  regional domes 242
  seasonal effects 238
  surface temperatures 238, 338
  Tharsis region 242–243
  Valles Marineris 242
  volatiles (*see also* terrestrial planets) 243,
        245, 247–252
    inventories 345–346, 349
    inventory changes 346–347, 360–361
  volcanic activity 242
  volcanic features 242–244
  water related features 247–252
    duricrust 253
    fretted channels 249–250
    gullies 250–252
    lakes/oceans 248, 250–252
    layered deposits 253, 255–256
    minerals 252–256
    outflow/outflow channels 247–249, 253
    unusual ejecta blankets 247–248
    valley networks 250
  weathering 244, 246
mass measurement 128–129
Mathilde 91–93, 95
Matthew effect 66
maximum eastern elongation 35
maximum western elongation 35
Maxwell distribution 315–316
Maxwell, James Clerk 315
mean motion resonance, *see* orbital resonances
Mercury
  age (surface) 235
  altitude range (surface) 232
  atmosphere 362
  Caloris basin 233–234
  central pressure, temperature, density Table 5.2
  core 164
    composition 164, 170–171, 233